ELECTRONIC COMMERCE

Third Annual Edition

Gary P. Schneider, Ph.D., CPA
University of San Diego

THOMSON
COURSE TECHNOLOGY™

Australia • Canada • Mexico • Singapore • Spain • United Kingdom • United States

THOMSON
COURSE TECHNOLOGY

Electronic Commerce, Third Annual Edition

by Gary P. Schneider, Ph.D, CPA

Senior Vice President, Publisher:
Kristen Duerr

Managing Editor:
Jennifer Locke

Senior Product Manager:
Margarita Leonard

Development Editor:
Amanda Brodkin

Production Editor:
Melissa Panagos

Marketing Manager:
Jason Sakos

Associate Product Manager:
Janet Aras

Editorial Assistant:
Christy Urban

Text Designer:
Anne Small-Wills

Cover Designer:
Betsy Young

Manufacturing Manager:
Denise Sandler

BRIEF CONTENTS

Technologies for Electronic Commerce

Integration

TABLE OF CONTENTS

Chapter 5: Business-to-Business Strategies: From Electronic Data Interchange to Electronic Commerce — 178

Technologies for Electronic Commerce

Integration

Electronic Commerce, Third Annual Edition provides complete coverage of the constantly changing field of electronic commerce. The book assumes that readers have no previous electronic commerce knowledge or experience. This book is designed to help you learn about the key business and technology elements of electronic commerce.

In 1998, having worked in electronic commerce research, consulting, and corporate training, I began developing both an undergraduate business school course and an MBA-level course in electronic commerce. Although I had used a variety of materials in my corporate training work, I was concerned that those materials would not work well in university courses because they were written at widely varying levels and did not have the pedagogic organization and features, such as review questions, that are so important to students.

After searching for a textbook that offered balanced coverage of both the business and technology elements of electronic commerce, I concluded that no such textbook existed. The first edition of *Electronic Commerce* was written to fill that void. In the second edition and now, in this third edition, I have worked to improve the book and keep it current with the rapid changes in this dynamic field. The third edition also includes a new feature, "Learning from Failures," that appears in each chapter. This feature highlights some of the more interesting false starts that have occurred in this new way of doing business and suggests a lesson that can be learned from each.

Electronic Commerce, Third Annual Edition introduces readers to both the theory and practice of conducting business over the Internet and World Wide Web.

In this third edition, the chapters were rearranged and one new chapter was added. The chapters now fall into four sections: an introduction, business strategies, technologies, and a summary and integration. A comparison of the organization of the previous and current edition of *Electronic Commerce* is provided in this table:

Chapter in *Electronic Commerce, Third Annual Edition*		Chapter in *Electronic Commerce, Second Annual Edition*	
Introduction			
1	Introduction to Electronic Commerce	1	Introduction to Electronic Commerce
2	Technology Infrastructure: The Internet and the World Wide Web	2	Infrastructure for Electronic Commerce
Business Strategies for Electronic Commerce			
3	Selling on the Web: Revenue Models and Building a Web Presence	8	Strategies for Marketing, Sales, and Promotion
4	Marketing on the Web	New	
5	Business-to-Business Strategies: From Electronic Data Interchange to Electronic Commerce	9	Strategies for Purchasing and Support Activities: From Electronic Data Interchange to Electronic Commerce
6	Web Auctions, Virtual Communities, and Web Portals	10	Web Auctions, Virtual Communities, and Web Portals
7	The Environment of Electronic Commerce: International, Legal, Ethical, and Tax Issues	11	The Environment of Electronic Commerce: International, Legal, Ethical, and Tax Issues
Technologies for Electronic Commerce			
8	Web Server Hardware and Software	3	Web-Based Tools for Electronic Commerce
9	Electronic Commerce Software	4	Electronic Commerce Software
10	Security Threats to Electronic Commerce	5	Security Threats to Electronic Commerce
11	Implementing Electronic Commerce Security	6	Implementing Electronic Commerce Security
12	Payment Systems for Electronic Commerce	7	Electronic Payment Systems
Integration			
13	Planning for Electronic Business	12	Planning for Electronic Business
Appendix	A Comparison of HTML and XML Features	New	

Introduction

The book's first section includes two chapters. Chapter 1, "Introduction to Electronic Commerce," begins with an explanation and overview of commerce, then defines electronic commerce and describes how companies use it to create new products and services and improve many other standard business activities. Chapter 1 also describes the history of the Internet and the Web, provides an overview of the economic structures in which businesses operate, and describes how electronic commerce fits into those structures. Two themes are introduced in this chapter that recur throughout later chapters: examining a firm's value chain can suggest opportunities for electronic commerce initiatives, and reductions in transaction costs are important elements of many electronic commerce initiatives.

Chapter 2, "Technology Infrastructure: The Internet and the World Wide Web," introduces the technologies used to conduct business online, including topics such as Internet infrastructure, protocols, packet-switched networks, text markup languages, and Internet application and utility programs. Chapter 2 also describes the language of the Web, hypertext transfer protocol, and discusses Internet connection options and tradeoffs.

Business Strategies for Electronic Commerce

The second section of the book includes five chapters that describe the business strategies that companies and other organizations are using to do business online. Chapter 3, "Selling on the Web: Revenue Models and Building a Web Presence," explains that by understanding how the Web differs from other media, businesses can create an effective Web presence that delivers value to visitors. The chapter describes how firms that understand the nature of communication on the Web can identify and reach the largest possible number of qualified customers.

Chapter 4, "Marketing on the Web," provides an introduction to the subject becoming known as Internet marketing. It includes coverage of technology-enabled relationship management, rational branding, viral marketing, and permission marketing. The chapter also explains how some businesses on the Web share and transfer brand benefits through affiliate marketing and cooperative efforts among brand owners.

Chapter 5, "Business-to-Business Strategies: From Electronic Data Interchange to Electronic Commerce," explores the variety of methods that companies are using to improve their purchasing and logistics primary activities with Internet and Web technologies, and are making similar improvements in a wide range of support activities. Chapter 5 also provides an overview of EDI and explores how the Internet now provides an inexpensive EDI communications channel that allows smaller businesses to reap EDI's benefits. Chapter 5 also explains how the Internet and the Web have become an important force driving the adoption of supply chain management techniques in a variety of industries.

Chapter 6, "Web Auctions, Virtual Communities, and Web Portals," outlines how companies now use the Web to do things that they have never done before, such as operating auction sites, creating virtual communities, and serving as Web portals. The chapter describes how firms are using Web auction sites to sell goods to their customers and generate advertising revenue. The chapter explains how new companies formed to take advantage of the Web's ability to bring together people and organizations that share narrow interests but are geographically dispersed. Businesses are creating virtual communities with their customers and suppliers and are using these communities to sell goods and services. The chapter describes how the major Web search engine sites evolved into Web portals and discusses some of the problems these sites now face as advertising revenue declines.

Chapter 7, "The Environment of Electronic Commerce: International, Legal, Ethical, and Tax Issues," discusses the challenges posed to businesses by differing language, culture, laws, and infrastructure when they conduct electronic commerce across international borders. Chapter 7 notes that variations and inadequacies in the infrastructure that supports the Internet worldwide can make it challenging to conduct electronic commerce in certain countries. Chapter 7 explains that although companies conducting electronic commerce are subject to the same laws and taxes as other companies, those that engage in electronic commerce face a large number of laws and taxes sooner than traditional companies. The large number of government units that have jurisdiction and power to tax makes it essential that companies doing business on the Web understand the potential liabilities of doing business with customers in those jurisdictions.

Technologies for Electronic Commerce

Chapter 8, "Web Server Hardware and Software," covers Web-based tools for electronic commerce, including Web server hardware and software options and their various strengths and scalability. Web site hosting options and their implications are discussed. Chapter 8 also presents overviews of several Web performance evaluation and tuning tools.

Chapter 9, "Electronic Commerce Software," describes the basic electronic commerce software functions, building on the Web server software and hardware presented in previous chapters. Also covered in Chapter 9 are fundamental services, including catalog display, transaction processing, and shopping carts. The chapter discusses a wide range of software commerce choices for a range of businesses from small stores to enterprise-class firms.

Chapter 10, "Security Threats to Electronic Commerce," discusses the many internal and external security threats to electronic commerce. Chapter 10 also covers the roles of copyright and intellectual property security and threats to them, including cybersquatting and domain name stealing. The vulnerability of the communication channels carrying information between one location and another on the Web and the vulnerabilities of the Web servers that companies use to conduct electronic commerce are also discussed.

Chapter 11, "Implementing Electronic Commerce Security," describes security threat countermeasures, including antivirus software and encryption, to combat the threats presented in Chapter 10. Also covered are Internet protocols and authentication codes that provide message protection and message deletion protection. Chapter 11 discusses digital certificates and certification authorities, and how each is used to verify user identification. In addition to these protections, the chapter also describes intellectual property threats and methods to protect graphics and audio resources for sale on the Internet.

Chapter 12, "Payment Systems for Electronic Commerce," presents a discussion of electronic payment systems, including electronic cash, electronic wallet technologies, stored-value cards, credit cards, debit cards, and charge cards. In this chapter, both failed electronic payment systems and related organizations and new, promising payment methods are covered. Chapter 12 also describes how these payment systems operate, including approval of transactions and disbursements to merchants.

Integration

Chapter 13, "Planning for Electronic Business," presents an overview of key elements that are typically included in business plans for electronic commerce implementations. These elements include the setting of objectives and estimated costs and benefits of the project. Chapter 13 further describes how companies develop and implement an outsourcing strategy for electronic commerce projects and also covers the use of project management as a formal way to plan and control specific tasks and resources used in electronic commerce projects. This chapter also includes a discussion of staffing strategies and describes the critical staffing areas of business management, application specialists, customer service staff, systems administration, network operations staff, and database administration.

The book closes with an Appendix that describes the similarities and differences between HTML and XML.

FEATURES

Electronic Commerce, Third Annual Edition is unique in its field because it includes the following features:

- **Business Case Approach** A business case introduces each chapter and provides a unifying theme for the chapter. The case provides a backdrop for the material described in the chapter. Each case has been chosen carefully to illustrate the role and use of electronic commerce.
- **Learning From Failures** Not all electronic commerce initiatives have been successful. Each chapter in the book includes a short summary of an electronic commerce failure related to the content of that chapter. We all learn from our mistakes—this feature is designed to help readers understand the missteps of electronic commerce pioneers who learned their lessons the hard way.

- **Summaries** Each chapter concludes with a Summary that concisely recaps the most important concepts in the chapter.

- **Online Companion** The Online Companion is a set of Web pages maintained by the publisher for readers of this book. The Online Companion complements the book and contains links to Web sites referred to in the book and to other online resources that further illustrate the concepts presented. The Web is constantly changing and the Online Companion is continually monitored and updated for those changes so that its links continue to lead to useful Web resources for each chapter. You can find the Online Companion for this book at *http://www.course.com/downloads/sites/ecommerce3/* or by visiting Course Technology's Web site at *www.course.com/* and searching on *Electronic Commerce*.

- **Online Companion References in Text** Throughout each chapter, there are Online Companion References that indicate the name of a link included in the Online Companion. Text set in bold, sans-serif letters ("**Metabot Pro**") indicates a like-named link in the Online Companion. The links in the Online Companion are organized under chapter and subchapter headings that correspond to those in the book. The Online Companion also contains many supplemental links to help students explore beyond the book's content.

- **Review Questions and Exercises** Every chapter concludes with meaningful review materials including both conceptual discussion questions and hands-on exercises. The review questions are ideal for use as the basis for class discussions or as written homework assignments. The exercises give students hands-on experiences that yield a computer output or a written report.

- **For Further Study and Research** Each chapter concludes with a comprehensive list of the resources that were consulted during the writing of the chapter. These references to publications in academic journals, books, and the IT industry and business press provide a sound starting point for readers who want to learn more about the topics contained in the chapter.

TEACHING TOOLS

When this book is used in an academic setting, instructors may obtain the following teaching tools from Course Technology:

- **Instructor's Manual** The Instructor's Manual has been carefully prepared and tested to ensure its accuracy and dependability. The Instructor's Manual is available through the Course Technology Faculty Online Companion on the World Wide Web (call your customer service representative for the exact URL and to obtain your username and password).
- **ExamView** This textbook is accompanied by ExamView, a powerful testing software package that allows instructors to create and administer printed, computer (LAN-based), and Internet exams. ExamView includes hundreds of questions that correspond to the topics covered in this text, enabling students to generate detailed study guides that include page references for further review. The computer-based and Internet testing components allow students to take exams at their computers, and also save the instructor time by grading each exam automatically.
- **Classroom Presentations** Microsoft PowerPoint presentations are available for each chapter of this book to assist instructors in classroom lectures or to make available to students. The Classroom Presentations are included on the Instructor's CD.

ACKNOWLEDGMENTS

I owe a great debt of gratitude to my good friends at Course Technology who made this book possible. Course Technology remains the best publisher with which I have ever worked. Everyone at Course Technology put forth tremendous effort to publish this edition on a very tight schedule. My heartfelt thanks go to Kristen Duerr, Senior Vice President, Publisher; Jennifer Locke, Managing Editor; Margarita Leonard, Senior Product Manager; Melissa Panagos, Production Editor; Toby Shelton, Marketing Manager; and Janet Aras, Associate Product Manager, for their tireless work and dedication to the project. I am deeply indebted to Amanda Brodkin, Development Editor extraordinaire, for her outstanding contributions to three editions of this book. Amanda performed the magic of turning my manuscript drafts into a high-quality textbook. She added the organizational touches that make the book easy for students to read and use. Amanda was always ready with encouragement and fresh ideas when I was running low on them. Many of the best elements of this book resulted from Amanda's ideas and inspirations.

I want to thank the following reviewers for their insightful comments and suggestions on this and previous editions: Tina Ashford, Macon State College; Robert Chi, California State University-Long Beach; Roland Eichelberger, Baylor University; Milena Head, McMaster University; Perry M. Hidalgo, Gwinnett Technical Institute; Cheri L. Kase, Legg Mason Corporate Technology; William Lisenby, Alamo Community College; Diane Lockwood, Albers School of Business and Economics, Seattle University; Michael P. Martel, Culverhouse School of Accountancy, University of Alabama; Martha Myers, Kennesaw State University; Pete Partin, Forethought Financial Services; and William E. McTammany, Florida Community College at Jacksonville. Special thanks go to reviewer A. Lee Gilbert of Nanyang Technological University in Singapore, who provided extremely detailed comments and many useful suggestions for improving Chapter 13. My thanks also go to the many professors who have used the previous editions in their classes and who have sent me suggestions for improving the text. In particular, I want to acknowledge the detailed recommendations made by David Bell of Pacific Union College regarding the coverage of IP addresses in Chapter 2.

I appreciate the role that the University of San Diego had in making this book possible. The University made available research funding that provided time to work on the first edition of this book and gave me fellow faculty members who were always happy to discuss and critically evaluate ideas for the book. Of these faculty members, my thanks go first to Jim Perry, both colleague and friend, for his contributions as co-author on the first two editions of this book. Tom Buckles, now a professor of marketing at Biola University, provided many useful suggestions, pointed out a number of valuable research resources, and was willing to sit and discuss ideas for this book long after everyone else had left the building. Rahul Singh, now teaching at the University of North Carolina-Greensboro, provided suggestions regarding the book's coverage of electronic commerce infrastructure. Carl Rebman made recommendations on a number of networking, telecommunications, and security topics. The University of San Diego School of Business Administration also provided the research assistance of several graduate students. Among those students were Anthony Coury, who applied his considerable legal knowledge to reviewing Chapter 7 and suggesting many improvements. Other students who provided assistance and suggestions include Sebastian Ailioaie, Adrian Boyce, Emilie Johnson Hersh, Chad McManamy, Dan Mulligan, Suzanne Phillips, Susan Soelaiman, and Leila Worthy.

Finally, I want to express my deep appreciation for the support and encouragement of my wife, Cathy Cosby, and our children, Ben, Annie, and Maggie. Without their support and patience, writing this book would not have been possible.

DEDICATION

To Cathy, Ben, Annie, and Maggie

Gary P. Schneider

About the Author

Gary Schneider is an Associate Professor of Accounting and Information Systems at the University of San Diego, where he teaches courses in electronic commerce, database design, and management control systems. He has published more than 25 books and 70 research papers on a variety of accounting, information systems, and management topics. Gary's research has been funded by the Irvine Foundation and the U.S. Office of Naval Research. His work has appeared in the *Journal of Information Systems*, *Interfaces*, and the *Information Systems Audit & Control Journal*. He has served as editor of the *Accounting Systems and Technology Reporter*, as associate editor of the *Journal of Global Information Management*, and on the editorial boards of the *Journal of Information Systems*, the *Journal of Database Management*, and the *Information Systems Audit & Control Journal*. Gary has lectured on electronic commerce topics at universities and businesses in the United States, Europe, South America, and Asia. He has provided consulting and training services to a number of major clients, including the GartnerGroup, Gateway, Honeywell, LexFusion, and Qualcomm. In 1999, he was named a Fellow of the Gartner Institute. Gary is a licensed CPA in Ohio, where he practiced public accounting for 14 years. He holds a Ph.D. in accounting information systems from the University of Tennessee, an M.B.A. in accounting from Xavier University, and a B.A. in economics from the University of Cincinnati.

INTRODUCTION TO ELECTRONIC COMMERCE

INTRODUCTION

Very few people in the United States truly enjoy their hunt for a new or used car. Although many auto dealers have worked to improve their customers' experiences by introducing fixed pricing and "no haggle" policies, a number of auto dealers continue to use aggressive sales approaches that can leave buyers exhausted, confused, and even worried that they might have been cheated in the transaction. In 1995, **Autobytel** launched an online car-buying service that promised purchasers a haggle-free buying experience and offered car dealers a way to increase their new vehicle sales volumes and reduce their selling costs.

Buying a car with the assistance of Autobytel requires that the buyer register with the Autobytel Web site and specify the desired auto in detail, usually after researching the vehicle's options and features on the Internet or by visiting local dealers. Autobytel provides the buyer with a firm price quote for the selected car, then forwards the buyer's contact information to a local participating dealer. Dealers pay Autobytel a subscription fee to receive exclusive rights to referrals from a particular geographic area for the brands of vehicles that they sell. The dealer contacts the buyer, who then completes the purchase transaction at the dealer's location.

The buyer benefits from a speedy and predictable buying process. The dealer benefits by selling more automobiles and by not paying a commission to a salesperson. Autobytel receives a monthly subscription

1

fee from each dealer that it has under contract and sells advertising to insurance and finance companies on its Web site. In six years, Autobytel has processed more than 6 million requests for quotes. It currently processes more than $15 billion in car sales every year through its site to its 8 million registered users, generating revenues of about $60 million per year. Autobytel sales accounted for approximately 4 percent of all U.S. new vehicle sales in 2000.

Autobytel has established a business by replacing the salesperson in consumer car-buying transactions. Although auto manufacturers are moving to build effective online selling opportunities on their Web sites, Autobytel believes it can continue to offer value to both buyers and dealers that the manufacturers' sites cannot. A National Bureau of Economic Research study in 2000 (see the Morton, Zettelmeyer, and Risso citation in the "For Further Study and Research" section at the end of this chapter for a full reference) concluded that the average Internet car buyer pays about 2 percent less for a car than other buyers. The study attributed the savings to a combination of negotiating power and transaction efficiency, which means that the dealer can be making more profit despite giving the buyer a better price and paying a subscription fee to Autobytel or a similar service.

LEARNING OBJECTIVES

In this chapter, you will learn about:

- The basic elements of electronic commerce
- Differences between electronic commerce and traditional commerce
- Advantages and disadvantages of using electronic commerce to conduct business activities
- The international nature of electronic commerce
- Economic forces that have created a business environment that fosters electronic commerce
- The ways in which businesses use value chains to identify electronic commerce opportunities
- The ways in which the growth of the Internet and the World Wide Web has stimulated the emergence of electronic commerce

TRADITIONAL COMMERCE AND ELECTRONIC COMMERCE

In this section, you will learn about traditional commerce and the activities it includes. Then, you will learn how electronic commerce uses various technologies to implement some or all of those activities.

To many people, the term electronic commerce (sometimes shortened to e-commerce) means shopping on the part of the Internet called the World Wide Web (the Web). The **Pew Internet and American Life Project** (Note: This typeface indicates a corresponding link to a related Web page in the book's Online Companion) is funded by the Pew Charitable Trusts, and began conducting several long-term research projects in 2000 and 2001 to study the growth of the Internet and its effects on society. Pew Project studies concluded in early 2001 found that about 60 percent of U.S. households had Internet connections, and about 16 percent of those (about 14 million households), were using those connections to buy goods and services.

Although consumer shopping on the Web was running about $50 billion per year in 2001 and is expected to exceed $350 billion by 2004, electronic commerce is much broader and encompasses many more business activities than just Web shopping. For example, businesses conduct transactions with other businesses, with their employees, and with governmental agencies (for example, when they pay their taxes). In fact, the total volume of all business activities on the Web is expected to exceed $5 trillion by 2004. A few electronic commerce research firms have forecasted even higher volumes by 2004—in one case, $7 trillion.

Some people use the term **electronic business** (or **e-business**) when they are talking about electronic commerce in this broader sense. However, most people use the terms electronic commerce and electronic business interchangeably. In this book, we will use the term **electronic commerce** (or **e-commerce**) in its broadest definition: business activities conducted using electronic data transmission technologies such as those used in the Internet and the World Wide Web. Figure 1-1 shows the three main elements of electronic commerce.

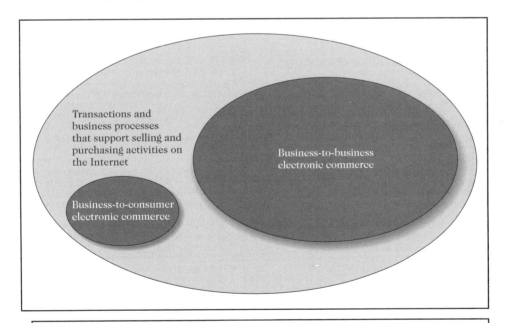

Figure 1-1   *Elements of electronic commerce, broadly defined*

These elements include:

- Consumer shopping on the Web, often called **business-to-consumer** (or **B2C**)
- Transactions conducted between businesses on the Web, often called **business-to-business** (or **B2B**)
- The transactions and business processes that companies, governments, and other organizations undertake on the Internet to support selling and purchasing activities

To understand these elements better, consider a company that manufactures stereo speakers. The company might sell its finished product to consumers on the Web, which would be B2C electronic commerce. It might also purchase the materials it uses to make the speakers from other companies on the Web, which would be B2B electronic commerce. In addition to buying and selling, however, the company must undertake many other activities to convert the purchased materials into speakers. These activities might include, for example, hiring and managing the people who make the speakers, renting or buying the facilities in which the speakers are made and stored, shipping the speakers, maintaining accounting records, purchasing insurance, developing advertising campaigns, and designing new versions of the speakers. An increasing number of these transactions and business processes can be done on the Web.

Some researchers define a fourth category of electronic commerce, called **consumer-to-consumer** (or **C2C**), which includes individuals who buy and sell items among themselves. For example, C2C electronic commerce occurs when a person sells an item through a Web auction site to another person. In this book, we include such sales in the B2C category because the person selling the item acts very much as a business would. In terms of dollar volume and number of transactions, B2B electronic commerce is greater than B2C electronic commerce. However, the dollar volume of the transactions and business processes that support B2C and B2B activities is greater than both of them combined.

Although the Web has made online shopping possible for many businesses and individuals, in a broader sense, electronic commerce has existed for many years. For more than 30 years, banks have been using **electronic funds transfers** (**EFTs**, also called **wire transfers**), which are electronic transmissions of account exchange information over private communications networks.

Businesses also have been engaging in a type of electronic commerce, known as electronic data interchange, for many years. **Electronic data interchange** (**EDI**) occurs when one business transmits computer-readable data in a standard format to another business. In the 1960s, businesses realized that many of the documents they exchanged were related to the shipping of goods—such as invoices, purchase orders, and bills of lading. These documents included the same set of information for almost every transaction. Businesses also realized that they were spending a good deal of time and money entering this data into their computers, printing paper forms, and then reentering the data on the other side of the transaction. Although the purchase order, invoice, and bill of lading for each transaction contained much of the same information—such as item numbers, descriptions, prices, and quantities—each paper form usually had its own unique format for presenting that information. By creating a set of standard formats for transmitting that information electronically, businesses were able to reduce errors, avoid printing and mailing costs, and eliminate the need to reenter the data.

Businesses that engage in EDI with each other are called **trading partners**. The standard formats used in EDI contain the same information that businesses have always included in their standard paper invoices, purchase orders, and shipping documents. Firms such as **General Electric** and **Wal-Mart** have been pioneers in using EDI to improve their purchasing processes and their relationships with suppliers. Other firms, such as Sterling (which is now a part of **Computer Associates**), **Commerce One**, and Harbinger (which is now a part of **Peregrine Systems**) played a key role in facilitating EDI between firms by developing needed software and providing connectivity. The U.S. government, which is one of the largest EDI trading partners in the world, also was instrumental in bringing businesses into EDI. For nine years, ending in 2001, the Defense Logistics Agency operated a number of Electronic Commerce Resource Centers (ECRCs) throughout the country. The ECRCs provided free assistance to many businesses, especially smaller businesses, so they could do EDI with the U.S. Defense Department and other federal agencies.

One serious problem that potential adopters of EDI faced was the high cost of implementation. Until quite recently, doing EDI meant buying expensive computer hardware and software, then either establishing direct network connections (using leased telephone lines) to all trading partners or subscribing to a value added network. A **value added network (VAN)** is an independent firm that offers connection and transaction-forwarding services to buyers and sellers engaged in EDI. Before the Internet came into existence as we know it today, VANs provided the connections between most trading partners and were responsible for ensuring the security of the data transmitted. VANs usually charged a fixed monthly fee plus a per-transaction charge, adding to the already significant expense of implementing EDI. Many smaller firms were unable to afford to participate in EDI and lost important customers, who went elsewhere to buy. **Open Market** was one of the first firms to move EDI traffic to the Internet, but many other EDI software development and consulting firms have joined in this trend. Experts estimate that EDI transaction activity on the Internet will exceed $1 trillion by 2003.

A good way to understand the full range of electronic commerce is to learn about the activities that companies undertake when they do any kind of business (engage in commerce), and then to learn how these firms might undertake these activities electronically. In the next two sections, you will learn about the elements of traditional commerce, and then you will see how these elements are carried out electronically using electronic commerce.

Traditional Commerce

In addition to buying or selling, firms engage in many other activities that keep them in business. For example, the seller of a product must identify demand, promote its product to potential buyers, accept orders, deliver its product, bill and accept payment for its product, and support its customers' use of its product after the sale. In many cases, sellers will customize or create a product to a customer's specifications. Similarly, buyers of a product also engage in additional activities. They must examine their needs, identify products that might meet those needs, and evaluate those products. Next, buyers must order the selected product, arrange for delivery, and pay for the product. In many cases, buyers need to maintain contact with the seller for warranty and other maintenance on the product. Of course, buyers and sellers can engage in these transactions not just for products, but for services too. When you think broadly about commerce, you see that it can involve individuals, business firms, not-for-profit organizations, and governmental entities as both buyers and sellers.

The origins of traditional commerce occurred before recorded history, when our ancestors first decided to specialize their everyday activities. Instead of each family unit having to grow crops, hunt for meat, and make tools, families developed skills in one of these areas and traded for other needs: The tool-making family would exchange tools for grain from the crop-growing family, and so on. Services were bought and sold in these primitive economies, too. For example, the local shaman would cast spells or intercede with the deities in exchange for food and tools.

Eventually, bartering gave way to the use of currency, making transactions easier to settle; however, the basic mechanics of trade were the same. One member of society created something of value that another member of society desired. **Commerce**, or **doing business**, is a negotiated exchange of valuable objects or services between at least two parties and includes all activities that each of the parties undertakes to complete the transaction.

The Buyer

You can examine any commerce transaction from either the buyer's or the seller's viewpoint. The elements of traditional commerce for buyers appear in Figure 1-2.

A buyer begins by identifying a need. This may be a simple need, such as when an individual decides "I'm hungry and would like to find some lunch," or it may be very complex, such as when a city council decides "We must find a way to generate clean power that will meet our city's needs in 25 years." The need-identification process for a hungry individual may require no more than a quick consideration of which fast-food outlets are nearby and open. To identify the specific needs for the power-generation example could require many people working in an organized way for a long period of time. Most need-identification processes fall between these two extremes.

Once buyers have identified their specific needs, they must find products or services that will satisfy those needs. In traditional commerce, buyers use a variety of search techniques. They may consult catalogs, ask friends, read advertisements, or examine directories. The Yellow Pages is a good example of a directory that buyers often use to find products and services. Buyers may consult salespersons to gather information about specific features and capabilities of products they are considering. Companies often have highly structured procedures for finding products and services that satisfy recurring needs of their businesses.

After buyers have selected a product or service that will meet the identified need, they must select a vendor who can supply that product or service. Buyers in traditional commerce contact vendors in a variety of ways, including by telephone, by mail, and by contact at trade shows. After choosing a vendor, the buyer negotiates a purchase transaction. This transaction may have many elements, including delivery date, method of shipment, price, warranty, and payment terms, and will often include detailed specifications to be confirmed by inspection when the product is delivered or the service is performed. When the buyer is a business, the negotiation of a purchase transaction can be very complicated. Imagine, for example, the complex ordering, delivery, and inspection activities that must occur when an airline buys a new airplane from an aircraft manufacturer. Businesses often have entire departments devoted to negotiating purchase transactions with their suppliers. These departments are usually named **supply management** or **procurement**.

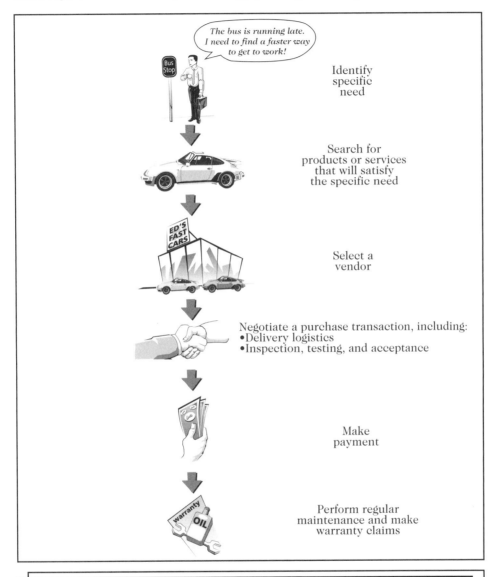

The bus is running late. I need to find a faster way to get to work!

Identify specific need

Search for products or services that will satisfy the specific need

Select a vendor

Negotiate a purchase transaction, including:
• Delivery logistics
• Inspection, testing, and acceptance

Make payment

Perform regular maintenance and make warranty claims

Figure 1-2   *Elements of traditional commerce: the buyer's side*

When the buyer is satisfied that the purchased product or service has met the terms and conditions agreed to by both buyer and seller, the buyer will pay for the purchase. After the sale is complete, the buyer may have further contact with the seller regarding warranty claims, upgrades, and regular maintenance.

The Seller

Each action taken by a buyer engaging in commerce has a corresponding action that is taken by the seller. Figure 1-3 shows the elements of commerce from a seller's viewpoint.

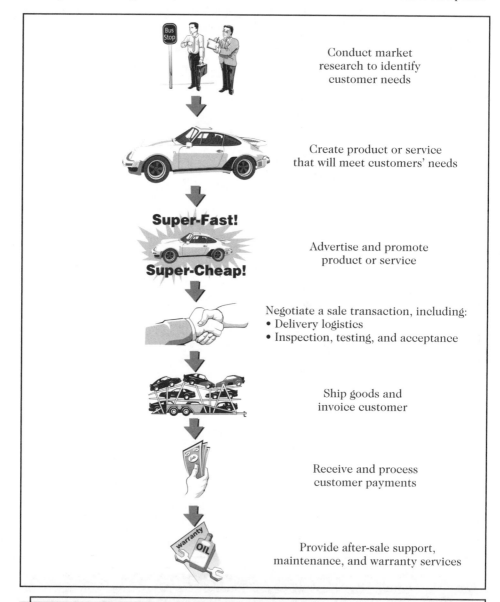

Conduct market research to identify customer needs

Create product or service that will meet customers' needs

Advertise and promote product or service

Negotiate a sale transaction, including:
• Delivery logistics
• Inspection, testing, and acceptance

Ship goods and invoice customer

Receive and process customer payments

Provide after-sale support, maintenance, and warranty services

Figure 1-3   *Elements of traditional commerce: the seller's side*

Sellers often undertake **market research** to identify potential customers' needs. Even businesses that have been selling the same product or service for many years are always looking for ways to improve and expand their offerings. Firms conduct surveys, have salespersons talk with customers, run focus groups, and hire outside consultants to help them in this identification process.

Once customer needs have been identified, sellers create the products and services that they believe will meet those needs. This creation activity includes design, testing, and production activities.

The next step for sellers is to make potential customers aware that the new product or service exists. Sellers engage in many different kinds of advertising and promotional activities to communicate information about their products and services to existing and potential customers.

Once a customer responds to the seller's promotional activities, the two parties must negotiate the details of a purchase transaction. In some cases, this is simple; for example, many retail transactions involve nothing more than a buyer entering a seller's store, selecting and inspecting items to purchase, and paying for them. In other cases, purchase transactions require prolonged negotiations to settle the terms of delivery, inspection, testing, and acceptance.

After the seller and buyer resolve delivery logistics, the seller ships the goods or provides the service and sends an invoice to the buyer. In some businesses, the seller will also provide a monthly billing statement to each customer summarizing the invoicing and payment activity of that customer. In some cases, the seller will require payment before or at the time of shipment. However, most businesses sell to each other on credit, so the seller must keep a record of the sale and wait for the customer to pay. Most businesses maintain sophisticated systems for receiving and processing customer payments. They want to track the amounts they are owed and ensure that payments they receive are credited to the proper customer and invoice.

Following the conclusion of the sale transaction, the seller will often provide continuing after-sale support for the product or service. In many cases, the seller is bound by contract or statute to guarantee or warrant that the product or service sold will perform satisfactorily for a specific period of time. The seller provides support, maintenance, and warranty work to help ensure that the customer is satisfied and will return to buy again.

Activities, Transactions, and Business Processes

Business researchers have been studying the ways people behave in businesses for more than 70 years. This research has helped managers understand better how workers do their jobs. The research results have also helped managers, and increasingly, the workers themselves, improve their job performance. By changing the nature of jobs, managers and workers can, as the saying goes, "work smarter, not harder." An important part of doing these job studies is to learn what activities each worker is performing. In this setting, an **activity** is a task performed by a worker in the course of doing his or her job.

For a much longer time—centuries, in fact—business owners have kept records of how well their businesses are performing. The formal practice of accounting, or recording transactions, dates back to the 1400s. A **transaction** is an exchange of value, such as a purchase, a sale, or the conversion of raw materials into a finished product. By recording transactions, accountants helped business owners keep score and measure how well they were doing. All transactions involve at least one activity, and some transactions involve many activities. Not all activities result in measurable (and therefore recordable) transactions. Thus, a transaction will always have one or more activities associated with it, but an activity might not be related to a transaction.

The group of logical, related, and sequential activities and transactions in which businesses engage are often collectively referred to as **business processes**. Transferring funds, placing orders, sending invoices, and shipping goods to customers are all types of activities or transactions. For example, the business process of shipping goods to customers might include a number of activities (or tasks, or transactions), such as inspecting the goods, packing the goods, negotiating with a freight company to deliver the goods, creating and printing the shipping documents, loading the goods onto the truck, and sending a check to the freight company.

Electronic Commerce

Over the thousands of years that people have engaged in commerce with one another, they have adopted the tools and technologies that became available. For example, the advent of sailing ships in ancient times opened new avenues of trade to buyers and sellers. Later innovations, such as the printing press, the steam engine, and the telephone, have each changed the way in which people conduct commerce activities.

For decades, firms have used various electronic communications tools to conduct different kinds of business transactions. Banks have used EFTs to move customers' money around the world, all kinds of businesses have used EDI to place orders and send invoices, and retailers have used television advertising to generate telephone orders from the general public for various types of merchandise.

Our definition of electronic commerce, stated earlier in this chapter, mentions the use of electronic data transmission to implement or enhance business processes. Some people use the term **Internet commerce** to mean electronic commerce that specifically uses the Internet or the Web as its data transmission medium. Other terms exist for what we call electronic commerce in this book. Since the field of electronic commerce is so new, people and businesses sometimes use terms in different ways. For example, **IBM** has defined electronic business as "the transformation of key business processes through the use of Internet technologies."

This book will show you how to use electronic data transmission technologies, primarily those that are part of the Internet, to improve existing business processes and identify new business opportunities. An important aspect of electronic commerce is that it can be used by firms to adapt to change. The business world is changing more rapidly now than ever before. Although much of this book is devoted to explaining technologies, its focus is on the business of electronic commerce; the technologies only enable the business processes. Figure 1-4 compares a traditional commerce version of a typical business transaction with an electronic commerce version of the same transaction.

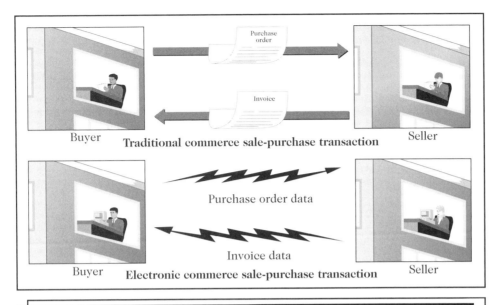

Figure 1-4   *Comparing traditional commerce and electronic commerce*

Business Processes in Commerce

An important function of this book is to help you learn how to identify those business processes that firms can accomplish more effectively by using electronic commerce technologies. In many cases, business processes use traditional commerce activities very effectively, and technology cannot improve upon them. Products that buyers prefer to touch, smell, or examine closely are difficult to sell using electronic commerce. For example, customers might be reluctant to buy high-fashion clothing and perishable food products, such as meat or produce, if they cannot examine the products closely before agreeing to purchase them.

Retail merchants have years of traditional commerce experience in creating store environments that help convince customers to buy. This combination of store design, layout, and product display knowledge is called **merchandising**. In addition, many salespeople have developed skills that allow them to identify customer needs and find products or services that meet those needs. The arts of merchandising and personal selling can be difficult to practice remotely.

Some products, such as books or CDs, are good candidates for electronic commerce because customers do not need to experience the physical characteristics of the particular item before they buy it. Since one copy of a new book is identical to other copies, and since the customer is not concerned about fit, freshness, or other such qualities, customers are usually willing to order a title without examining the specific copy they will receive. The advantages of electronic commerce, including the ability of one site to offer a wider selection of titles than even the largest physical bookstore, can outweigh the advantages of a traditional bookstore—for example, the customer's ability to browse. In later chapters, you will learn how to evaluate the advantages and disadvantages of using electronic commerce for specific business processes.

Figure 1-5 lists 13 examples of business processes. Five of these are well suited to electronic commerce, and four are better suited to traditional commerce. The other four business processes, listed in the middle column, are well suited to a combination of the two approaches.

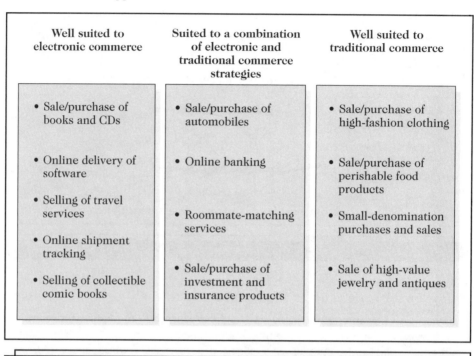

Well suited to electronic commerce	Suited to a combination of electronic and traditional commerce strategies	Well suited to traditional commerce
• Sale/purchase of books and CDs	• Sale/purchase of automobiles	• Sale/purchase of high-fashion clothing
• Online delivery of software	• Online banking	• Sale/purchase of perishable food products
• Selling of travel services	• Roommate-matching services	• Small-denomination purchases and sales
• Online shipment tracking		
• Selling of collectible comic books	• Sale/purchase of investment and insurance products	• Sale of high-value jewelry and antiques

Figure 1-5   *Business process suitability to type of commerce*

Of course, these suitability classifications depend on the current state of available technologies and thus might change as new tools for implementing electronic commerce emerge. For example, low-denomination transactions have not been well suited to electronic commerce because no standard method for transferring small amounts of money on the Web has yet evolved. If a company develops a good method for doing this and it gains general acceptance, low-denomination transactions could move from the traditional commerce column to the electronic commerce column.

One business process that is especially well suited to electronic commerce is the selling of commodity items. A **commodity** item is a product or service that is hard to distinguish from the same products or services provided by other sellers; its features have become standardized and well-known. Gasoline, office supplies, soap, computers, and airline transportation are all examples of commodity products or services, as are the books and CDs sold by Amazon.com and other online bookstores.

Another key factor that can make an item well suited to electronic commerce is the product's shipping profile. A **shipping profile** is the collection of attributes that affect how easily a product can be packaged and delivered. A high value-to-weight ratio can help by making the overall shipping cost a small fraction of the selling price. An airline ticket is an excellent example of an item that has a high value-to-weight ratio. Products that are consistent in size, shape, and weight can make warehousing and shipping much simpler and less costly. The shipping profile is only one factor, however. Expensive jewelry has a high value-to-weight ratio, but few people would buy it sight unseen.

A product that has a strong brand identity—such as a Sony CD player—is easier to sell over the Web than an unbranded item, because the brand's reputation reduces the buyer's concerns about quality when buying that item sight unseen. Other items that are well suited to electronic commerce are items that appeal to small, but geographically dispersed, groups of customers. Collectible comic books are an example of this type of product.

When personal selling skills are a factor, as in commercial real estate sales; or when the condition of the products is difficult to determine without making a personal inspection, as in purchases of high-fashion clothing, antiques, or perishable food products, traditional commerce is a better way to sell the items or services.

A combination of electronic and traditional commerce strategies works best when the business process includes both commodity and personal inspection elements. For example, many people are finding information on the Web about new and used automobiles. Autobytel has had much success handling new car transactions, but few people are willing to buy a used car without driving it and inspecting it personally. In this case, electronic commerce provides a good way for buyers to obtain information about available models, options, reliability, prices, and dealerships; but the variability of individual used cars makes the traditional commerce component of personal inspection a key part of the transaction negotiation. It is still not possible to take a test drive on the Web. The next two sections summarize some advantages and disadvantages of electronic commerce.

Advantages of Electronic Commerce

Firms are interested in electronic commerce because, quite simply, it can help increase profits. All the advantages of electronic commerce for businesses can be summarized in one statement: Electronic commerce can increase sales and decrease costs. Advertising done well on the Web can get even a small firm's promotional message out to potential customers in every country in the world. A firm can use electronic commerce to reach narrow market segments that are geographically scattered. The Web is particularly useful in creating virtual communities that become ideal target markets for specific types of products or services. A virtual community is a gathering of people who share a common interest, but instead of this gathering occurring in the physical world, it takes place on the Internet. You will learn more about virtual communities in Chapter 6.

A business can reduce the costs of handling sales inquiries, providing price quotes, and determining product availability by using electronic commerce in its sales support and order-taking processes. **Cisco Systems** currently sells almost all of its computer equipment through its Web site. Since no customer service representatives are involved in making these sales, Cisco operates very efficiently. In 1998, the first year in which its online sales initiative was fully operational, Cisco made 72 percent of its sales on the Web. Cisco estimated that it avoided handling 500,000 calls per month and saved $500 million in that year alone. Many other businesses have emulated the **Cisco Global Networked Business Model** since 1998.

Just as electronic commerce increases sales opportunities for the seller, it increases purchasing opportunities for the buyer. Businesses can use electronic commerce to identify new suppliers and business partners. Negotiating price and delivery terms is easier in electronic commerce, because the Web can provide competitive bid information very efficiently. Electronic commerce increases the speed and accuracy with which businesses can exchange information, which reduces costs on both sides of transactions.

Electronic commerce provides buyers with a wider range of choices than traditional commerce, because buyers can consider many different products and services from a wider variety of sellers. This wide variety is available for consumers to evaluate 24 hours a day, every day. Some buyers prefer a great deal of information in deciding on a purchase; others prefer less. Electronic commerce provides buyers with an easy way to customize the level of detail in the information they obtain about a prospective purchase. Instead of waiting days for the mail to bring a catalog or product specification sheet, or even minutes for a fax transmission, buyers can have instant access to detailed information on the Web. Some products, such as software, audio clips, or images, can even be delivered through the Internet, which reduces the time buyers must wait to begin enjoying their purchases.

The benefits of electronic commerce extend to the general welfare of society. Electronic payments of tax refunds, public retirement, and welfare support cost less to issue and arrive securely and quickly when transmitted over the Internet. Furthermore, electronic payments can be easier to audit and monitor than payments made by check, providing protection against fraud and theft losses. To the extent that electronic commerce enables people to work from home, we all benefit from the reduction in commuter-caused traffic and pollution. Electronic commerce can also make products and services available in remote areas. For example, distance education is making it possible for people to learn skills and earn degrees no matter where they live or which hours they have available for study.

Disadvantages of Electronic Commerce

Some business processes may never lend themselves to electronic commerce. For example, perishable foods and high-cost, unique items, such as custom-designed jewelry and antiques, may be impossible to inspect adequately from a remote location, regardless of any technologies that might be devised in the future. Most of the disadvantages of electronic commerce today, however, stem from the newness and rapidly developing pace of the underlying technologies. These disadvantages will disappear as electronic commerce matures and becomes more available to and accepted by the general population.

Many products and services require that a critical mass of potential buyers be equipped and willing to buy through the Internet. For example, online grocers such as **Peapod** offer their delivery services only in a few cities. As more of Peapod's potential customers become connected to the Internet and begin to feel comfortable with purchasing online, the business might be able to expand into more geographic areas.

Peapod is a good example of how difficult it can be to build a business in an industry that requires this kind of critical mass. Although it was one of the first online groceries, Peapod has had a difficult time staying in business, and was even offline for a few weeks in mid-2000 before being acquired by a European firm that was willing to invest additional cash to keep it in operation.

Businesses often calculate return-on-investment numbers before committing to any new technology. This has been difficult to do for investments in electronic commerce, because the costs and benefits have been hard to quantify. Costs, which are a function of technology, can change dramatically even during short-lived electronic commerce implementation projects, because the underlying technologies are changing so rapidly. Many firms have had trouble recruiting and retaining employees with the technological, design, and business process skills needed to create an effective electronic commerce presence. Another problem facing firms that want to do business on the Internet is the difficulty of integrating existing databases and transaction-processing software designed for traditional commerce into the software that enables electronic commerce.

In addition to technology and software issues, many businesses face cultural and legal obstacles to conducting electronic commerce. Some consumers are still somewhat fearful of sending their credit card numbers over the Internet or having online merchants—merchants they have never met—know so much about them. You will learn more about electronic commerce security, privacy issues, and payment systems later in this book. Other consumers are simply resistant to change and are uncomfortable viewing merchandise on a computer screen rather than in person. The legal environment in which electronic commerce is conducted is full of unclear and conflicting laws. In many cases, government regulators have not kept up with technologies. As you will learn in Chapter 7, laws that govern commerce were written when signed documents were a reasonable expectation in any business transaction. However, as more businesses and individuals find the benefits of electronic commerce to be compelling, many of these technology and culture-related disadvantages will be resolved or seem less problematic.

LEARNING FROM FAILURES

PETS.COM

In February 1999, Pets.com launched its Web site with the hopes of making substantial sales to the 60 percent of U.S. households that own pets and spend more than $20 billion each year feeding, entertaining, and caring for them. More than 10,000 stores sold pet supplies. These stores included small retail outlets, grocery stores, discount retailers (such as Wal-Mart and Costco), and a new generation of pet superstores. Pets.com had acquired an excellent domain name and intended to exploit the opportunities presented by high levels of investor interest in funding electronic commerce companies. The plan for Pets.com was to spend heavily to develop a brand and a Web presence that would rapidly make the company the premier online source for pet-related products.

After launching the site, Pets.com raised $110 million from private investors in 1999, and another $80 million in a public sale of stock in early 2000. Pets.com spent more than $100 million of the money on advertising during its short life. It also spent significant sums to create a Web store that offered more than 12,000

(continued)

different products. In November 2000—less than two years after launching its Web site—Pets.com went out of business.

Pets.com had created an electronic commerce initiative in a line of business in which online business offered few advantages over traditional commerce. The products had a very low value-to-weight ratio. The shipping costs for pet food, one of the company's best-selling product categories, caused it to lose money on every sale. Pet products come in all shapes, sizes, and weights, and are, therefore, difficult to pack and ship efficiently. Pets.com was also spending money rapidly at a time when investors were beginning to question the long-run viability of all electronic commerce businesses. The lesson here is that Pets.com could not develop any sustainable advantage over traditional pet stores. Without such an advantage, customers stayed away from the site.

International Electronic Commerce

The Internet brings together people from every country in the world. It reduces the distance between people in many ways. About 60 percent of all electronic commerce sites on the Web are in English, although sites in other languages and in multiple languages are appearing with increasing frequency. Researchers have found that people who can read English as a second language still prefer to use Web sites that are in their native language. Thus, the demand for non-English Web sites—including electronic commerce Web sites—will continue to increase as the online population becomes more representative of the entire world. Once the language barrier is overcome, the technology exists for any business to conduct electronic commerce with any other business or consumer, anywhere in the world.

Unfortunately, the political structures of the world have not kept up with Internet technology, so doing business internationally presents a number of challenges. Currency conversions, tariffs, import and export restrictions, local business customs, and the laws of each country in which a trading partner resides can make international electronic commerce difficult.

Many of the international issues that arise relate to legal, tax, and privacy concerns. Each country has the right to pass laws and levy taxes on businesses that operate within their jurisdiction. European countries, for example, have very strict laws that limit the collection and use of personal information that companies gather in the course of doing business with consumers. Even within the United States, individual states and counties have the power to levy sales and use taxes on goods and services. In other countries, national sales taxes and value-added taxes are imposed on an even broader list of business activities. Later in this book, you will learn more about the international, legal, and tax issues that can arise when conducting electronic commerce.

The next section discusses some of the broad economic issues that relate to the emergence and future growth of electronic commerce.

Economics is the study of how people allocate scarce resources. One important way that people allocate resources is through commerce (the other major way people allocate resources is through government actions, such as taxes or subsidies). Many economists are interested in how people organize their commerce activities. One way people do this is to participate in markets. Economists use a formal definition of **market** that includes two conditions: first, that the potential sellers of a good come into contact with potential buyers, and second, that a medium of exchange is available. This medium of exchange can be currency or barter. Most economists agree that markets are strong and effective mechanisms for allocating scarce resources. Thus, one would expect most business transactions to occur within markets. However, much business activity today occurs within large **hierarchical business organizations**, which economists generally refer to as **firms**, or **companies**.

Most hierarchical organizations are headed by a top-level president or chief operating officer. Reporting to the president are a number of vice presidents who, in turn, have a larger number of middle managers who report to them, and so on. An organization can have a relatively flat hierarchy, in which there are only a few levels of management, or it can have many reporting levels. In either case, the bottom level includes the largest number of employees and is usually made up of production workers or service providers. Thus, the hierarchical organization will always have a pyramid-shaped structure.

These large firms often conduct many different business activities entirely within the organizational structure of the firm and participate in markets only for the purchase of raw materials and for the selling of their completed products. If markets are indeed highly effective mechanisms for allocating scarce resources, these large corporations should participate in markets at every stage of their production and value-generation processes. Nobel laureate Ronald Coase wrote an essay in 1937 in which he questioned why individuals who engaged in commerce often created firms to organize their activities. He was particularly interested in the hierarchical structure of these business organizations. Coase concluded that transaction costs were the main motivation for moving economic activity from markets to hierarchically structured firms.

Transaction Costs

Transaction costs are the total of all costs that a buyer and a seller incur as they gather information and negotiate a purchase-sale transaction. Although brokerage fees and sales commissions can be a part of transaction costs, the cost of information search and acquisition is often far larger. Another significant component of transaction costs can be the investment a seller makes in equipment or in the hiring of skilled employees to supply the product or service to the buyer.

To understand better how transaction costs occur in markets, consider the following example: A sweater dealer could obtain sweaters by engaging in market transactions with a number of independent sweater knitters. Transaction costs incurred by the dealer would include the costs of identifying the independent knitters, visiting them to negotiate the purchase price, arranging for delivery of the sweaters, and inspecting the sweaters when they arrived. The knitters would also incur costs, such as the purchase of knitting tools and yarn. Since individual knitters could not know whether any sweater dealer would ever buy sweaters from them, the investments they must make to enter the sweater-knitting business would have an uncertain yield. This risk is a significant transaction cost for the knitters.

After purchasing the sweaters, the dealer takes them to a different market in which sweater dealers meet and do business with the retail shops that sell sweaters to the consumer. The dealers can use these market negotiations to find out which sweater colors and patterns are in demand and can then use that information to negotiate price and other terms in the knitters' market. A diagram of this set of markets appears in Figure 1-6.

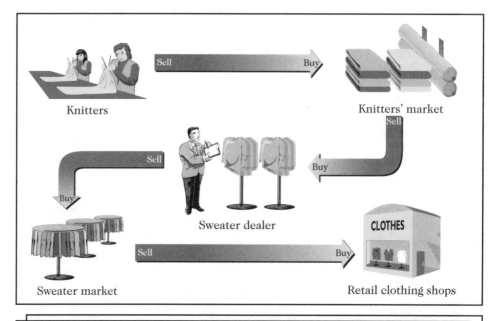

Figure 1-6   *Market forms of economic organization*

Markets and Hierarchies

Coase reasoned that when transaction costs were high, businesspersons would form organizations to replace market-negotiated transactions. These organizations would be hierarchical and would include strong supervision and worker-monitoring elements. Instead of negotiating with individuals to purchase sweaters they had knit, a hierarchical organization would hire knitters, and then supervise and monitor their work activities. This supervision and monitoring system would include flows of monitoring information from the lower levels to the higher levels of the organization. It would also have control of information flowing from the upper levels of the organization to the lower levels. Although the costs of creating and maintaining a supervision and monitoring system are high, they can be lower than transaction costs in many instances. In the sweater example, the sweater dealer would hire knitters, supply them with yarn and knitting tools, and supervise their knitting activities. This supervision could be done mainly by first-line supervisors, who might be drawn from the ranks of the more skilled knitters. The practice of an existing firm replacing one or more of its supplier markets with its own hierarchical structure for creating the supplied product is called **vertical integration**. Figure 1-7 shows how our wool sweater example would look after the knitters were vertically integrated into the hierarchical structure of the sweater dealer's organization.

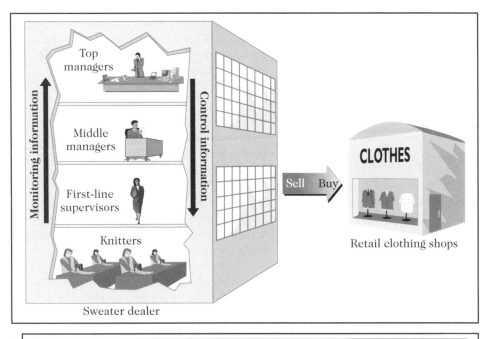

Figure 1-7 *Hierarchically structured form of economic organization*

Oliver Williamson, an economist who extended Coase's analysis, noted that industries with complex manufacturing and assembly operations tended to include many firms that used hierarchical structures and were substantially vertically integrated. Many of the manufacturing and administrative innovations that occurred in businesses during the 20th century increased the efficiency and effectiveness of hierarchical monitoring activities. Assembly lines and other mass-production technologies allowed work to be broken down into small, easily supervised procedures. The advent of computers brought tremendous increases in the ability of upper-level managers to monitor and control the detailed activities of their subordinates. Some of these direct measurement techniques are even more effective than the first-line supervisors on the shop floor.

During the years from the Industrial Revolution through the present, as improvements in monitoring became commonplace, the size and level of vertical integration of firms has increased. In some very large organizations, however, monitoring systems have not kept pace with the organization's increase in size. This has created problems because the economic viability of a firm depends on its ability to track operational activities effectively at the lowest levels of the firm. These firms have instituted decentralization programs that allow business units to function as separate organizations and negotiate transactions with other business units as if they were operating in a market rather than as part of the same firm. These decentralization approaches are simply a return to the highly effective market mechanisms that worked so well before the firm vertically integrated itself.

Exceptions to the general trend toward hierarchies do exist. Many commodities, such as wheat, sugar, and crude oil, are still traded in markets. The commodity nature of the products traded in these markets reduces transaction costs significantly. There are a large number of potential buyers for an agricultural commodity such as wheat, and the farmer does not make any special investment in customizing or modifying the product for a particular customer. Thus, neither buyers nor sellers in commodity markets experience significant transaction costs.

The Role of Electronic Commerce

Businesses and individuals can use electronic commerce to reduce transaction costs by improving the flow of information and increasing the coordination of actions. By reducing the cost of searching for potential buyers and sellers and increasing the number of potential market participants, electronic commerce can change the attractiveness of vertical integration for many firms. It is not clear yet whether widespread adoption of electronic commerce will cause hierarchical organization structures to revert to their former market-based structures, but it certainly is a distinct possibility.

To see how electronic commerce can change the level and nature of transaction costs, consider an employment transaction. The agreement to employ a person has high transaction costs for the seller—the employee who sells his or her services. These transaction costs include a commitment to forego other employment and career development opportunities. Individuals make a high investment in learning and adapting to the culture of their employers. If accepting the job involves a move, the employee can incur very high costs, including actual costs of the move and related costs, such as the loss of a spouse's job. Much of the employee's investment is specific to a particular job and location; the employee could not transfer the investment to a new job. If a sufficient number of employees throughout the world can telecommute—that is, perform their job tasks from any location using electronic commerce technologies—then many of these transaction costs could be reduced or eliminated. Instead of uprooting a spouse and family to move, a worker could accept a new job by simply logging on to a different Internet server!

Some researchers have argued that many companies and strategic business units operate today in an economic structure that is neither a market nor a hierarchy. In this **network economic structure**, companies coordinate their strategies, resources, and skill sets by forming long-term, stable relationships with other companies and individuals based on shared purposes. These relationships are often called **strategic alliances** or **strategic partnerships**, and when they occur between or among companies operating on the Internet, these relationships are also called **virtual companies**.

In some cases, these entities, called **strategic partners**, come together as a team for a specific project or activity. The team dissolves when the project is complete; however, the partners maintain contact with each other through the ensuing period of inactivity. When the need for a similar project or activity arises, the same organizations and individuals build teams from their combined resources. In other cases, the strategic partners form many inter-company teams to undertake a variety of ongoing activities. Later in this book, you will see many examples of strategic partners creating alliances of this sort on the Web. In a hierarchically structured business environment, these types of strategic alliances would not last very long, because the larger strategic partners would buy out the smaller partners and form a larger single company.

Network organizations are particularly well suited to technology industries that are information intensive. In our sweater example, the knitters might organize into networks of smaller organizations that specialize in certain styles or designs. Some of the particularly skilled knitters might leave the sweater dealer to organize their own company to produce custom-knit sweaters. Some of the sweater dealer's marketing employees might form an independent firm that conducts market research on what the retail shops plan to buy in the upcoming months. This firm could sell its research reports to both the sweater dealer and the custom-knitting firm. As market conditions change, these smaller and more nimble organizations could continually reinvent themselves and take advantage of new opportunities that arise in the sweater markets. An illustration of such a network organization appears in Figure 1-8.

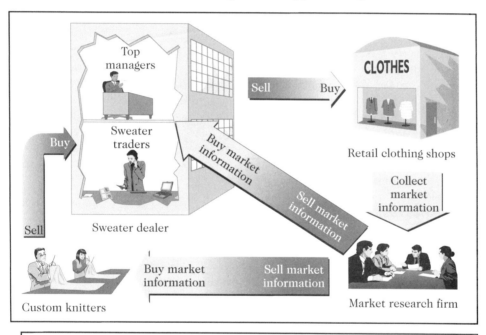

Figure 1-8   *Network form of economic organization*

Electronic commerce can make such networks, which rely extensively on information sharing, much easier to construct and maintain. Some researchers believe that these network forms of organizing commerce will become predominant in the near future. One of these researchers, Manuel Castells, has even predicted that economic networks will become the organizing structure for all social interactions among people. Thomas Petzinger, a columnist for *The Wall Street Journal*, has written extensively about these new patterns of work and commerce in his newspaper columns and in his 1999 book, *The New Pioneers*.

One interesting role for electronic commerce is the improvement of existing markets and the creation of completely new markets. You will learn more about this aspect of electronic commerce in later chapters of this book. Regardless of how businesses in a particular industry organize themselves—as markets, hierarchies, or networks—you will need a way to identify business processes and evaluate whether electronic commerce is suitable for each process. The next section presents one good structure for examining business processes.

VALUE CHAINS IN ELECTRONIC COMMERCE

Electronic commerce includes so many activities and transactions that it can be difficult for managers to decide where and how to use it in their businesses. One way to focus on specific business processes as candidates for electronic commerce is to break the business down into a series of value-adding activities that combine to generate profits and meet other goals of the firm. In this section, you will learn one popular way to analyze business activities as a sequence of activities that create value for the firm.

Commerce is conducted by firms of all sizes. Smaller firms can focus on one product, distribution channel, or type of customer. Larger firms often sell many different products and services through a variety of distribution channels to several types of customers. In these larger firms, managers organize their work around the activities of strategic business units. A **strategic business unit**, or simply **business unit**, is one particular combination of product, distribution channel, and customer type. Multiple business units owned by a common set of shareholders make up a firm, or company, and multiple firms that sell similar products to similar customers make up an **industry**.

Strategic Business Unit Value Chains

In his 1985 book, *Competitive Advantage*, Michael Porter introduced the idea of value chains. A **value chain** is a way of organizing the activities that each strategic business unit undertakes to design, produce, promote, market, deliver, and support the products or services it sells. In addition to these **primary activities**, Porter also includes **supporting activities**, such as human resource management and purchasing, in the value chain model. Figure 1-9 shows a value chain for a strategic business unit engaged in manufacturing a product, including both primary and supporting activities.

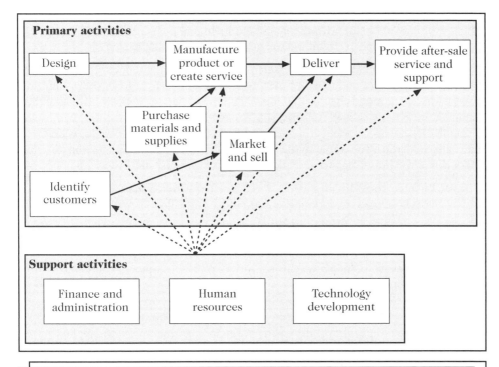

Figure 1-9 *Value chain for a strategic business unit*

The left-to-right flow in Figure 1-9 does not imply a strict time sequence for these processes. For example, a business unit may engage in marketing activities before purchasing materials and supplies. For each business unit, the primary activities are as follows:

- Identify customers: activities that help the firm find new customers and new ways to serve old customers, including market research and customer satisfaction surveys
- Design: activities that take a product from concept to manufacturing, including concept research, engineering, and test marketing
- Purchase materials and supplies: procurement activities, including vendor selection, vendor qualification, negotiating long-term supply contracts, and monitoring quality and timeliness of delivery
- Manufacture product or create service: activities that transform materials and labor into finished products, including fabricating, assembling, finishing, testing, and packaging
- Market and sell: activities that give buyers a way to purchase and that provide inducements for them to do so, including advertising, promoting, managing salespersons, pricing, and identifying and monitoring sales and distribution channels
- Deliver: activities that store, distribute, and ship the final product, including warehousing, handling materials, consolidating freight, selecting shippers, and monitoring timeliness of delivery

- Provide after-sale service and support: activities that promote a continuing relationship with the customer, including installing, testing, maintaining, repairing, fulfilling warranties, and replacing parts

The importance of each primary activity depends on the product or service the business unit provides and to which customers it sells. If a strategic business unit provides a service, its value chain would include a Provide service activity instead of the Manufacture activity shown in Figure 1-9. The other activities in its value chain would be similar to those for a product-manufacturing business unit. Each business unit must also undertake support activities that provide the infrastructure for the unit's primary activities. These support activities appear in Figure 1-9 and are as follows:

- Finance and administration: activities that provide the firm's basic infrastructure, including accounting, paying bills, borrowing funds, reporting to government regulators, and ensuring compliance with relevant laws
- Human resources: activities that coordinate the management of employees, including recruiting, hiring, training, compensation, and providing benefits
- Technology development: activities that help improve the product or service that the firm is selling and that help improve the business processes in every primary activity, including basic research, applied research and development, process improvement studies, and field tests of maintenance procedures

Industry Value Chains

Porter's book also identifies the importance of examining where the strategic business unit fits within its industry. Porter uses the term **value system** to describe the larger stream of activities into which a particular business unit's value chain is embedded. However, many subsequent researchers and business consultants have used the term **industry value chain** when referring to value systems. When a business unit delivers a product to its customer, that customer may, for example, use the product as purchased materials in its value chain. By becoming aware of how other business units in the industry value chain conduct their activities, managers can identify new opportunities for cost reduction, product improvement, or channel reconfiguration. An example of an industry value chain appears in Figure 1-10. This value chain is for a wooden chair and traces the life of the product from trees in a forest to its grave in a landfill or at a sawdust recycler.

Each business unit (logger, sawmill, lumberyard, chair factory, retailer, consumer, and recycler) shown in Figure 1-10 has its own value chain. For example, the sawmill purchases logs from the tree harvester and combines them in its manufacturing process with inputs such as labor and saw blades from other sources. Among the sawmill customers are the chair factory shown in Figure 1-10 and other users of cut lumber. Examining this industry value chain could be useful for the sawmill that is considering entering the tree-harvesting business or the furniture retailer who is thinking about partnering with a trucking line. The industry value chain identifies opportunities up and down the product's life cycle for increasing the efficiency or quality of the product.

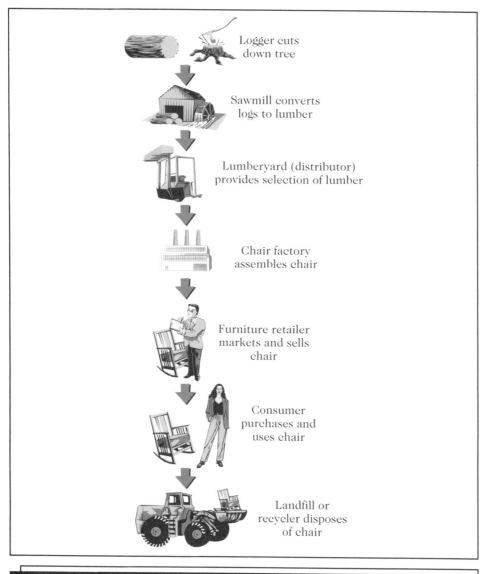

Figure 1-10 *Industry value chain for a wooden chair*

SWOT Analysis: Evaluating Business Unit Opportunities

Now that you have learned how to identify industry value chains and break each value chain down into strategic business units, you can learn one popular technique for analyzing and evaluating business opportunities. Most electronic commerce initiatives add value by either reducing transaction costs, creating some type of network economics effect, or a combination of both. In **SWOT analysis**, you list the strengths and weaknesses of the business unit, and then identify opportunities presented by the markets of the business unit and threats posed by competitors of the business unit. Figure 1-11 shows common issues that arise in a SWOT analysis.

Strengths	Weaknesses
Best resources Strength in market Strategic direction Points of excellence Culture	Worst liabilities Things that could be improved What you do poorly Financial liabilities
Opportunities	**Threats**
New markets Upward trends Customers with potential Technology capabilities	Competitors Obstacles Changing environment Changing technologies New laws and regulations

Figure 1-11   *SWOT analysis issues*

By considering all of the issues that it faces, a business unit can formulate strategies that will make the most of its opportunities by building on its strengths, avoiding the threats, and compensating for its weaknesses. In the mid-1990s, **Dell Computer** used a SWOT analysis to create a strong business strategy that has helped it become a very strong competitor in its industry value chain.

Dell identified its strengths in selling directly to customers and in designing its computers and other products to reduce manufacturing costs. It acknowledged the weakness of having no relationships with local computer dealers. Dell faced threats from competitors such as **Compaq** and IBM, both of which had much stronger brand names and reputations for quality at that time. Dell identified an opportunity by noting that its customers were becoming more knowledgeable about computers and could specify exactly what they wanted without having Dell answer questions or develop configurations for them. It also saw the Internet as a potential marketing tool.

The strategy that Dell followed after doing the SWOT analysis took all four of these factors into consideration. Dell decided to offer customized computers built to order and sold over the phone, and eventually, over the Internet. Dell's strategy capitalized on its strengths and avoided relying on a dealer network. The brand and quality threats posed by Compaq and IBM were lessened by Dell's ability to deliver a higher perceived quality because each computer was custom made for each buyer.

The Role of Electronic Commerce

One opportunity that many businesses are finding as they examine their industry value chains is that electronic commerce can play a role in reducing costs, improving product quality, reaching new customers or suppliers, and creating new ways of selling existing products. For example, a software developer who releases annual updates to programs might consider removing the software retailer from the distribution channel for updates by offering to send the updates through the Internet directly to the consumer. This change would modify the software developer's industry value chain and would provide an opportunity for increasing sales revenue (the software developer would retain the margin a retailer would have added to the price of the update), but it would not appear as part of the software developer business unit value chain. By examining elements of the value chain outside the individual business unit, managers can identify many business opportunities, including opportunities that can be exploited using electronic communications technologies—electronic commerce.

The value chain concept is a useful way to think about business strategy in general. When firms are considering electronic commerce, the value chain can be an excellent way to organize their examination of business processes within their business units and in other parts of their product's life cycle. Using the value chain reinforces the idea that electronic commerce should be a business solution, not a technology implemented for its own sake.

The main technological development that has allowed electronic commerce to grow beyond its beginnings in bank EFTs and business-to-business EDI is the emergence of the Internet and the Web. The next section presents a short overview of this technology and describes how it came to exist and grow to its present state.

THE INTERNET AND WORLD WIDE WEB

Millions of people use the Internet every day, but only a small percentage of them really understand how it works. The **Internet** is a large system of interconnected computer networks that spans the globe. Using the Internet, you can communicate with other people throughout the world by means of electronic mail; read online versions of newspapers, magazines, academic journals, and books; join discussion groups on almost any conceivable topic; participate in games and simulations; and obtain free computer software. In recent years, the Internet has allowed commercial enterprises to connect with one another and with customers. Today, all kinds of businesses provide information about their products and services on the Internet. Many of these businesses use the Internet to market and sell their products and services. The part of the Internet known as the **World Wide Web**, or, more simply, the **Web**, is a subset of the computers on the Internet that are connected to each other in a specific way that makes those computers and their contents easily accessible to each other. The most important thing about the Web is that it includes an easy-to-use standard interface. This interface makes it possible for people who are not computer experts to use the Web to access a variety of Internet resources.

Origins of the Internet

In the early 1960s, the U.S. Department of Defense became very concerned about the possible effects of nuclear attack on its computing facilities. The Defense Department realized that the weapons of the future would require powerful computers for coordination and control. The powerful computers of that time were all large mainframe computers, so the Defense Department began examining ways to connect these computers to each other and also to weapons installations that were distributed all over the world. The Defense Department agency charged with this task hired many of the best communications technology researchers and, for many years, funded research at leading universities and institutes to explore the task of creating a worldwide network that could remain operational, even if parts of the network were destroyed by enemy military action or sabotage. These researchers worked to devise ways to build networks that could operate independently—that is, networks that would not require a central computer to control network operations.

The world's telephone companies were the early models for networked computers, because early networks of computers used leased telephone company lines for their connections. Telephone company systems of that time established a single connection between sender and receiver for each telephone call, and that connection carried all data along a single path. When a company wanted to connect computers it owned at two different locations, the company placed a telephone call to establish the connection, and then connected one computer to each end of that single connection.

The Defense Department was concerned about the inherent risk of this single-channel method for connecting computers, and its researchers developed a different method of sending information through multiple channels. In this method, files and messages are broken into packets that are labeled electronically with codes for their origin, sequence, and destination. The packets travel from computer to computer along the network—each packet following its own path (which is often different from that of the other packets)—until they reach their destination. The destination computer collects the packets and reassembles the original data from the pieces in each packet. Each computer that an individual packet encounters on its trip through the network determines the best way to move the packet forward to its destination. You will learn more about how packet networks operate in Chapter 2.

In 1969, these Defense Department researchers used this network model to connect four computers—one each at the University of California at Los Angeles, SRI International, the University of California at Santa Barbara, and the University of Utah. Over subsequent years, many researchers in the academic community connected to this network and contributed to the technological developments that increased the speed and efficiency with which the network operated. At the same time, researchers at other universities were creating their own networks using similar technologies.

New Uses for the Internet

Although the goals of the Defense Department network were still to control weapons systems and transfer research files, other uses for this vast network began to appear in the early 1970s. In 1972, Ray Tomlinson wrote a program that could send and receive messages over the network. E-mail was born and became widely used very quickly. The number of network users in the military and education research communities continued to grow. Many of these new participants used the networking technology to transfer files and access computers remotely. The network software included two tools for performing these tasks, File Transfer Protocol and Telnet. **File Transfer Protocol (FTP)** enabled users to transfer files between computers, and **Telnet** let users log on to their computer accounts from remote sites. Both FTP and

Telnet are still widely used on the Internet today for file transfers and remote logins, even though more advanced techniques are now available that allow multimedia transmissions such as real-time audio and video clips.

The first e-mail mailing lists also appeared on these networks. A **mailing list** is an e-mail address that forwards any message it receives to any user who has subscribed to the list. In 1979, a group of students and programmers at Duke University and the University of North Carolina started **Usenet**, an abbreviation for **User's News Network**. Usenet allows anyone who connects to the network to read and post articles on a variety of subjects. Usenet survives on the Internet today, with over 1000 different topic areas that are called **newsgroups**. Other researchers even created game-playing software for use on these interconnected networks.

Although the people using these networks were developing many creative applications, use of the networks was limited to those members of the research and academic communities who could access them. Between 1979 and 1989, these new and interesting network applications were improved and tested by an increasing number of users. The Defense Department's networking software became more widely used as academic and research institutes realized the benefits of having a common communications network. The explosion of personal computer use during that time also helped more people become comfortable with computers. In the late 1980s, these independent academic and research networks merged into what we now call the Internet.

Commercial Use of the Internet

As personal computers became more powerful, affordable, and available during the 1980s, companies increasingly used them to construct their own internal networks. Although these networks included e-mail software that employees could use to send messages to each other, businesses wanted their employees to be able to communicate with people outside their corporate networks. The Defense Department network and most of the other academic networks that had teamed up with it were receiving funding from the **National Science Foundation (NSF)**. The NSF prohibited commercial network traffic on its networks, so businesses turned to commercial e-mail service providers to handle their e-mail needs. Larger firms built their own networks that used leased telephone lines to connect field offices to corporate headquarters.

In 1989, the NSF permitted two commercial e-mail services, MCI Mail and CompuServe, to establish limited connections to the Internet for the sole purpose of exchanging e-mail transmissions with users of the Internet. These connections allowed commercial enterprises to send e-mail directly to Internet addresses and allowed members of the research and education communities on the Internet to send e-mail directly to MCI Mail and CompuServe addresses. The NSF justified this limited commercial use of the Internet as a service that would primarily benefit the Internet's noncommercial users. As the 1990s began, people from all walks of life— not just scientists or academic researchers—started thinking of these networks as the global resource that we now know as the Internet. Although this network of networks had grown from four Defense Department computers in 1969 to more than 300,000 computers on many interconnected networks by 1990, the greatest growth of the Internet was yet to come.

Growth of the Internet

In 1991, the NSF further eased its restrictions on commercial Internet activity and began implementing plans to privatize the Internet. The privatization of the Internet was substantially completed in 1995, when the NSF turned over the operation of the main Internet connections to a group of privately owned companies. The new structure of the Internet was based on four **network access points (NAPs)**, each operated by a separate company. The San Francisco NAP is operated by Pacific Bell, the New York NAP is operated by Sprint, the Chicago NAP is operated by Ameritech, and the Washington, D.C. NAP is operated by MFS Corporation. These companies, known as **network access providers**, sell Internet access rights directly to larger customers and indirectly to smaller firms through other companies, called **Internet service providers (ISPs)**.

The Internet was a phenomenon that truly sneaked up on an unsuspecting world. The researchers who had been so involved in the creation and growth of the Internet just accepted it as part of their working environment. However, people outside the research community were largely unaware of the potential offered by a large interconnected set of computer networks. Figure 1-12 shows the number of **Internet hosts**, which are computers directly connected to the Internet, for the years 1991 through 2001. As you can see, the growth has been dramatic.

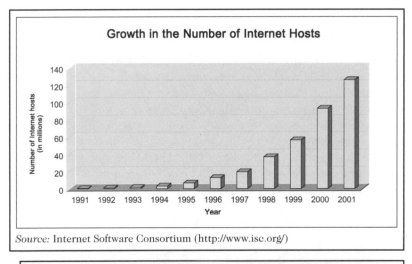

Source: Internet Software Consortium (http://www.isc.org/)

Figure 1-12   *Growth of the Internet, 1991–2001*

In under 30 years, the Internet became one of the most amazing technological and social accomplishments of the last century. Millions of people are using this complex, interconnected network of computers. These computers run thousands of different software packages. The computers are located in almost every country of the world. Every year billions of dollars change hands over the Internet in exchange for all kinds of products and services. All of this activity occurs with no central coordination point or control, which is especially interesting given that the Internet began as a way for the military to maintain control while under attack.

The opening of the Internet to business activity helped increase the Internet's growth dramatically; however, there was another development that worked hand in hand with the commercialization of the Internet to spur its growth. That development was the World Wide Web.

Emergence of the World Wide Web

The Web is more a way of thinking about and organizing information storage and retrieval than it is a technology. As such, its history goes back many years. Two important innovations played key roles in making the Internet easier to use and more accessible to people who were not research scientists. These two innovations were hypertext and graphical user interfaces.

The Development of Hypertext

In 1945, **Vannevar Bush**, who was director of the U.S. Office of Scientific Research and Development, wrote an article in *The Atlantic Monthly* about ways that scientists could apply the skills they learned during World War II to peacetime activities. The article included a number of visionary ideas about future uses of technology to organize and facilitate efficient access to information. Bush speculated that engineers would eventually build a machine that he called the Memex, a memory extension device that would store all a person's books, records, letters, and research results on microfilm. Bush's Memex would include mechanical aids, such as microfilm readers and indexes, that would help users consult their collected knowledge quickly and flexibly. In the 1960s, Ted Nelson described a similar system in which text on one page links to text on other pages. Nelson called his page-linking system **hypertext**. Douglas Engelbart, who also invented the computer mouse, created the first experimental hypertext system on one of the large computers of the 1960s. In 1987, Nelson published *Literary Machines*, a book in which he outlined project Xanadu, a global system for online hypertext publishing and commerce. Nelson used the term *hypertext* to describe a page-linking system that would interconnect related pages of information, regardless of where in the world they were stored.

In 1989, Tim Berners-Lee was trying to improve the laboratory research document-handling procedures for his employer, CERN: European Laboratory for Particle Physics. CERN had been connected to the Internet for two years, but its scientists wanted to find better ways to circulate their scientific papers and data among the high-energy physics research community throughout the world. Berners-Lee proposed a hypertext development project intended to provide this data-sharing functionality.

Over the next two years, Berners-Lee developed the code for a hypertext server program and made it available on the Internet. A **hypertext server** is a computer that stores files written in the hypertext markup language and lets other computers connect to it and read these files. Hypertext servers used on the Web today are usually called **Web servers**. **Hypertext markup language (HTML)**, which Berners-Lee developed from his original hypertext server program, is a language that includes a set of codes (or tags) attached to text. These codes describe the relationships among text elements. For example, HTML includes tags that indicate which text is part of a header element, which text is part of a paragraph element, and which text is part of a numbered list element. One important type of tag is the hypertext link tag. A **hypertext link**, or **hyperlink**, points to another location in the same or another HTML document.

Graphical Interfaces for Hypertext

You can use several different types of software to read HTML documents, but most people use a Web browser such as Netscape Navigator or Microsoft Internet Explorer. A **Web browser** is a software interface that lets users read (or browse) HTML documents and move from one HTML document to another through text formatted with hypertext link tags in each file. If the HTML documents are on computers connected to the Internet, you can use a Web browser to move from an HTML document on one computer to an HTML document on any other computer on the Internet.

An HTML document differs from a word processing document in that it does not specify how a particular text element will appear. For example, you might use word processing software to create a document heading by setting the heading text font to Arial, its font size to 14 points, and its position to centered. The document displays and prints these exact settings whenever you open the document in that word processor. In contrast, an HTML document simply includes a heading tag with the heading text. Many different browser programs can read an HTML document. Each program recognizes the heading tag and displays the text in whatever manner each program normally displays headings. Different programs might display the text differently, but always with the characteristics of a heading.

A Web browser presents an HTML document in an easy-to-read format in the browser's graphical user interface. A **graphical user interface (GUI)** is a way of presenting program control functions and program output to users. It uses pictures, icons, and other graphical elements instead of just displaying text. Almost all personal computers today use a GUI such as Microsoft Windows or the Macintosh user interface.

Berners-Lee called his system of hyperlinked HTML documents the World Wide Web. The Web caught on quickly in the scientific research community, but few people outside that community had software that could read the HTML documents. In 1993, a group of students led by Marc Andreessen at the University of Illinois wrote Mosaic, the first GUI program that could read HTML and use HTML hyperlinks to navigate from page to page on computers anywhere on the Internet. Mosaic was the first Web browser that became widely available for personal computers, and some Web surfers still use it today.

Programmers quickly realized that a functional system of pages connected by hypertext links would provide many new Internet users with an easy way to access information on the Internet. Businesses recognized the profit-making potential offered by a worldwide network of easy-to-use computers. In 1994, Andreessen and other members of the University of Illinois Mosaic team joined with James Clark of **Silicon Graphics** to found Netscape Communications (which is now owned by **AOL Time Warner**). Its first product, the Netscape Navigator Web browser program based on Mosaic, was an instant success. Netscape became one of the fastest growing software companies ever. **Microsoft** created its Internet Explorer Web browser and entered the market soon after Netscape's success became apparent. A number of other Web browsers exist, but these two products dominate the market today.

The number of Web sites has grown even more rapidly than the Internet itself. The number of Web sites is currently estimated to be over 30 million, and individual Web pages number in the billions because each Web site might include hundreds or even thousands of individual Web pages. Therefore, nobody really knows how many Web pages exist. For example, researchers at **BrightPlanet** estimate that the number of Web sites could be more than 500 million. Figure 1-13 shows how the growth rate of the Web has increased dramatically beginning in 1996.

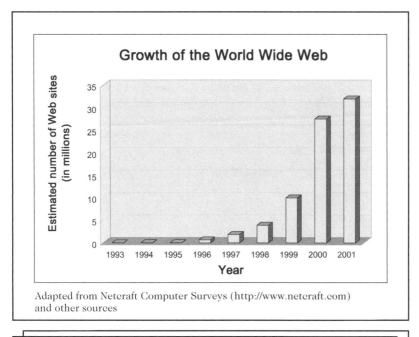

Adapted from Netcraft Computer Surveys (http://www.netcraft.com) and other sources

Figure 1-13   *Growth of the World Wide Web*

As more people gain access to the Web, commercial interest in using the Web to conduct business will increase, and the variety of nonbusiness uses will become even greater. In Chapter 2, you will learn more about how Internet and Web technologies work to enable electronic commerce.

Summary

In this chapter, you have learned that commerce, the negotiated exchange of goods or services, has been practiced in traditional ways for thousands of years. Electronic commerce is the application of new technologies, particularly Internet and Web technologies, to help individuals, businesses, and other organizations better conduct business. Not all activities lend themselves to improvement with these technologies, but many do. Using electronic commerce, some businesses have been able to create new products and services, and others have improved the promotion, marketing, and delivery of existing offerings. Firms have also found many ways to use electronic commerce to improve purchasing and supply activities; identify new customers; and operate their finance, administration, and human resource management activities more efficiently.

We examined an overview of markets, hierarchies, and networks—the economic structures in which businesses operate—and described how electronic commerce fits into those structures. Porter's ideas about value chains at the business unit and industry levels were presented, and you learned how to use value chains and SWOT analysis as ways to organize business processes and analyze their suitability for electronic commerce implementation. You learned about the history of the Internet and the Web, including how these technologies emerged from research projects and grew to be the supporting infrastructure for electronic commerce today.

Key Terms

Activity	Internet
Business processes	Internet commerce
Business unit	Internet host
Business-to-business (B2B)	Internet service providers (ISPs)
Business-to-consumer (B2C)	Mailing list
Commerce	Market
Commodity	Market research
Company	Merchandising
Consumer-to-consumer (C2C)	Network access points (NAPs)
Doing business	Network access providers
Electronic business (e-business)	Network economic structure
Electronic commerce (e-commerce)	Newsgroup
Electronic data interchange (EDI)	Primary activities
Electronic funds transfer (EFT)	Procurement
File Transfer Protocol (FTP)	Shipping profile
Firm	Strategic alliance
Graphical user interface (GUI)	Strategic partner
Hierarchical business organization	Strategic partnership
Hyperlink	Strategic business unit
Hypertext	Supply management
Hypertext link	Supporting activities
Hypertext markup language (HTML)	SWOT analysis
Hypertext server	Telnet
Industry	Trading partners
Industry value chain	Transaction

Transaction costs
Usenet (User's News Network)
Value added network (VAN)
Value chain
Value system
Vertical integration

Virtual company
Web browser
Web server
Wire transfer
World Wide Web (Web)

Review Questions

1. Name three factors that would cause a company to continue doing business in traditional ways and avoid electronic commerce.

2. What two developments contributed to the emergence of the Internet as the technology infrastructure for electronic commerce?

3. What are transaction costs and why are they important?

4. How might electronic commerce help change an industry's economic structure from a hierarchy to a network?

5. How might managers use SWOT analysis to identify new applications in their strategic business units for electronic commerce?

Exercises

1. You have decided to buy a new laser printer for your home office. List specific activities that you must undertake as you gather information about printer capabilities and features. Use the **CompUSA**, **Office Depot**, and **Staples** Web sites to gather information. Write a short summary of the process you undertook so that others who plan to undertake a similar task can use your information.

2. **Germany** and **China** have laws that prohibit the use of certain kinds of political speech and adult images in the conducting of commerce. Each of these countries has recently taken steps to enforce these laws by prosecuting firms and individuals who violate them while conducting electronic commerce. Outline the problems that such enforcement actions can create for firms and their non-German or non-Chinese customers who engage in electronic commerce. Propose a solution for each problem you identify.

3. You are a consultant to a firm that wants to establish an electronic commerce presence selling videos to consumers. You believe that after-sale service and support will be an important way in which this firm can differentiate itself from other firms in this business. You have decided that the online book industry is a good model for the video business, and you want to adapt business processes and practices from the after-sale service and support primary activity of that industry's value chain for your client. Use the Online Companion links for this exercise to examine the Web sites for **Amazon.com**, **Barnes & Noble**, **Books-A-Million**, **eCampus**, and **Internet Bookshop**. Identify three electronic commerce business processes that one or more of these booksellers use and that you feel would work for your client. As you work to identify these business processes, remember that you may be able to observe specific Web site features that result from such processes. You may also be able to infer the existence of business processes by reading statements that these companies make on their Web sites. Explain how you would adapt each process to meet the needs of the video retailing business.

Agrawal, V., L. Arjona, and R. Lemmens. 2001. "E-performance: The Path to Rational Exuberance," *The McKinsey Quarterly*, Number 1, 31–43.

Al-Kibsi, G., K. de Boer, M. Mourshed, and N. Rea. 2001. "Putting Citizens On-Line, Not In-Line," *The McKinsey Quarterly*, Number 2, 65–73.

Bergman, M. 2000. *The Deep Web: Surfacing Hidden Value*. Sioux Falls, SD: BrightPlanet.com. Available online at: (http://www.completeplanet.com/Tutorials/DeepWeb/).

Berry, J. 2000. "E-Business Still an Elusive Goal for Most Companies," *InternetWeek*, April 24, 41–42.

Blackmon, D. 2000. "Where the Money Is," *The Wall Street Journal*, April 17, R30.

Brache, A. and J. Webb. 2000. "The Eight Deadly Assumptions of E-Business," *Journal of Business Strategy*, 21(3), May–June, 13–17.

Bryce, R. 2001. "New Webonomics: What It Means to You," *Interactive Week*, July 17. Available online at: (http://www.zdnet.com/ecommerce/stories/main/0,10475,2789393,00.html).

Castells, M. 1996. *The Rise of the Network Society*. Cambridge, MA: Blackwell.

Champy, J. 2000. "Re-Engineering Redux," *Computerworld*, 34(17), April 24, 47.

Coase, R. 1937. "The Nature of the Firm," *Economica*, 4(4), November, 386–405.

Collett, S. 1999. "SWOT Analysis," *Computerworld*, 33(29), July 19, 58.

Computerworld. 2001. "Autopsy of a Dot Com," January 19. Available online at: (http://www.computerworld.com/cwi/story/0,1199,NAV47_STO56616,00.html).

Corcoran, C. 2000. "Who's on Top, and in Trouble, in E-Tail's Various Categories," *The Wall Street Journal Interactive Edition*, January 4. Available online at: (http://interactive.wsj.com/archive/retrieve.cgi?id=SB946927753582315731.djm).

Dvorak, J. 1999. "E-Biz Busters," *PC Computing*, 12(2), February, 101.

Ernst & Young, LLP. 1999. *The Second Annual Ernst & Young Internet Shopping Study: The Digital Channel Continues to Gather Steam*. New York: Ernst & Young, LLP.

Gantz, J. 2001. "Despite Crash, E-Commerce Will Still Flourish," *Computerworld*, 35(2), January 8, 31.

Garner, R. 1999. "The E-Commerce Connection," *Sales & Marketing Management*, 151(1), January, 40–44.

Glasner, J. 2001. "EToys Epitaph: 'End of an Error.'" *Wired News*, March 8. Available online at: (http://www.wired.com/news/business/0,1367,42078,00.html).

Gosh, S. 1998. "Making Business Sense of the Internet," *Harvard Business Review*, 76(2), March–April, 126–135.

Greenemeier, L. 2000. "Andersen Plans for E-Business Growth," *Informationweek*, April 24, 40.

Halliday, J. 2000. "Shakeout Foreseen for Car-Buying Sites," *Advertising Age*, 71(22), May, 58–60.

Hamel, G. and J. Sampler. 1998. "The E-Corporation," *Fortune*, 138(11), December 7, 80–87.

Hammer, M. and J. Champy. 1993. *Reengineering the Corporation: A Manifesto for Business Revolution*. New York: HarperBusiness.

Hannon, N. 1998. *The Business of the Internet*. Cambridge, MA: Course Technology.

Harrington, H., E. Esseling, and H. van Nimwegen. 1997. *Business Process Improvement Workbook: Documentation, Analysis, Design, and Management of Business Process Improvement*. New York: McGraw-Hill.

Helft, M. 2001. "The E-Commerce Survivors," *The Industry Standard*, July 16. Available online at: (http://www.thestandard.com/article/0,1902,27593,00.html).

Hertzberg, R. 1999. "Four Key Rules of Commerce," *Internet World*, January 11, 8.

Hofman, M. 2000. "Brave New Companies: The Metamorphosis," *Inc*, 22(4), March 14, 52–60.

Kalakota, R. and A. Whinston. 1996. *Readings in Electronic Commerce*. Reading, MA: Addison-Wesley.

Koulopoulos, T. and N. Palmer. 2001. *The X-Economy: Profiting from Instant Commerce*. New York: TEXERE.

LaMonica, M. 2000 "Buying an E-Business Strategy on the Cheap," *InfoWorld*, 22(17), April 24, 5.

Laseter, T. and M. Turner. 2001. "Operations at the Core: What Amazon Offers Category Killers," *strategy+business enews*, May 18. Available online at: (http://www.strategy-business.com/enews/051801/051801.html).

Lewis, M. 2001. "Faking It: The Internet Revolution Has Nothing to Do with the Nasdaq," *The New York Times*, July 15. Available online at: (http://www.nytimes.com/2001/07/15/magazine/15INTERNET.html).

Lico, N. 2001. "Can Cars Sell Via Web?" *Advertising Age*, 72(15), April 9, 20.

Morton, F., F. Zettelmeyer, and J. Risso. 2000. "Internet Car Retailing," *National Bureau of Economic Research Working Paper*, December. Available online at: (http://faculty.haas.berkeley.edu/~florian/).

Nelson, T. H. 1987. *Literary Machines*. Swarthmore, PA: Nelson.

Neuborne, E. 2000. "E-Tail: Gleaming Storefronts with Nothing Inside," *Business Week*, May 1, 94–95.

Oh, J. "Looking Good," *The Industry Standard*, March 5. Available online at: (http://www.thestandard.com/article/0,1902,22337,00.html).

Perdue, L. 2001. "A Bright Future: After the Train Wreck," *Inc. Magazine*, 23(4), March 15, 51–53.

Petzinger, T. 1999. *The New Pioneers: The Men and Women Who Are Transforming the Workplace and Marketplace*. New York: Simon & Schuster.

Porter, M. 1985. *Competitive Advantage*. New York: Free Press.

Porter, M. 1998. "Clusters and the New Economics of Competition," *Harvard Business Review*, 76(6), November–December, 77–90.

Porter, M. 2001. "Strategy and the Internet," *Harvard Business Review*, 79(3), March, 63–78.

Powell, W. 1990. "Neither Market nor Hierarchy: Network Forms of Organization," *Research in Organizational Behavior*, 12(3), 295–336.

Rainie, L. and D. Packel. 2001. *More Online, Doing More*. Washington: The Pew Internet and American Life Project.

Rayport, J. and B. Jaworski. 2001. *e-Commerce*. New York: McGraw-Hill/Irwin.

Ring, R. and A. Van de Ven. 1992. "Structuring Cooperative Relationships Between Organizations," *Strategic Management Journal*, 13(4), 483–498.

Shapiro, A. 1999. *The Control Revolution: How the Internet Is Putting Individuals in Charge and Changing the World We Know*. New York: The Century Foundation.

Shapiro, C. and H. Varian. 1999. *Information Rules: A Strategic Guide to the Network Economy*. Boston: Harvard Business School Press.

Schonfeld, E. 1999. "The Exchange Economy," *Fortune*, 139(3), February 15, 67–68.

Schulz, J. 2000. "The World's Big Four," *Journal of Commerce*, May 8, 29.

Schwartz, N. 2000. "Playing the Internet's Next Gold Rush," *Fortune*, 141(10), May 15, 178–183.

Taylor, D. and A. Terhune. 2001. Doing E-Business: Strategies for Thriving in an Electronic Marketplace. New York: John Wiley & Sons.

The Wall Street Journal. 2001. "Autobytel Expects Profits in Second Half, Says CEO Lorimer," February 5, B15.

Weernink, W. 2001. "Carmakers Target Online Middlemen," *Automotive News Europe*, April 9, 1–2.

Weintraub, A. 2001. "Make or Break for Autobytel," *Business Week*, July 9, EB30–EB32.

Williamson, O. 1985. *The Economic Institutions of Capitalism*. New York: Free Press.

Williamson, O. 1975. *Markets and Hierarchies: Analysis and Antitrust Implications*. New York: Free Press.

Willis, C. and S. Donahue. 1998. "Does Amazon.com Really Matter?" *Forbes*, 161(7), April 6, 55–58.

TECHNOLOGY INFRASTRUCTURE: THE INTERNET AND THE WORLD WIDE WEB

INTRODUCTION

Dell Computer Corporation is one of the most successful personal computer retailers in the history of PC sales. One of the world's largest computer manufacturers, Dell recently exceeded $31 billion in annual sales. Dell's customers include large and small corporations, government agencies, educational institutes, and individuals.

Through the early 1990s, Dell sold directly to the consumer through its toll-free telephone line. A few years ago, Dell expanded its sales to the Internet and has increasingly logged a significant percentage of its overall sales from the Internet. Thousands of customers now visit and place orders through the **Dell Computer Corporation Electronic Commerce Site**. Dell records millions of dollars in online sales each day. In addition to increasing business, selling through the Web has lowered overhead for the company. Sales made directly through the Dell Web site mean that fewer humans are involved in the transactions. Technical support, including answers to the most frequently asked questions, is available online as well. The Web site is a significant part of Dell's strategy for continued growth in the 21st century. Company officials predict that within the next few years, more than half of Dell's sales will be from the Web.

Supporting Dell's burgeoning online sales enterprise is a robust infrastructure of communication devices and networks, Dell server computers, and electronic commerce software from Microsoft. Both the hardware infrastructure and the software have been chosen for their capabilities and their ability to scale up as the customer base increases. Dell's state-of-the-art Web site will help it maintain its status as an extremely successful PC retailer.

LEARNING OBJECTIVES

In this chapter, you will learn about:

- The general technical structure of the Internet
- Internet utility programs that can help you trace, locate, and verify the status of Internet host sites
- Common Internet applications, including electronic mail, Telnet, and File Transfer Protocol (FTP)
- The history and use of markup languages, including SGML, HTML, and XML
- How HTML tags and links work on the Web
- The client-server architecture of the Web
- The differences among internets, intranets, and extranets
- Options for connecting to the Internet, including cost and bandwidth trade-offs

TECHNOLOGY OVERVIEW

Computer networks and the Internet, which interconnects networks around the world, form the basic technology structure for electronic commerce. The computers in these networks run a variety of software packages such as operating systems, database managers, encryption software, multimedia creation and viewing software, and the graphical user interface provided by World Wide Web browser and server software. In addition to the computers themselves, the Internet includes the networking hardware that connects the computers to each other and the internetworking hardware that connects the resulting networks to each other.

All of these hardware and software technologies are in a constant state of change—often rapid change. Businesses that engage in electronic commerce must implement new Internet technologies as they become available. The rapid pace of change in these technologies requires that businesses be nimble and willing to accept changes in the way they conduct business on the Web. Firms that are not flexible can lose their Web-based commerce to other companies that are flexible. Online consumers say that poor Web site performance, such as slow response time, often causes them to leave cumbersome electronic commerce sites and find others that better meet their needs.

This chapter introduces you to the hardware and software technologies that make electronic commerce possible. First, you will learn how information moves through the Internet. Then, you will learn about the other technologies that support the Internet, the Web, and electronic commerce. In this chapter, you will learn in a general way how many complex networking technologies work. If you are interested in learning more about exactly how computer networks operate, you can consult one of the computer networking books cited in the "For Further Study and Research" section at the end of this chapter or take courses in data communications and networking.

PACKET-SWITCHED NETWORKS

A network of computers that are located close together—for example, in the same building—is called a **local area network**, or a **LAN**. Networks of computers that are connected over greater distances are called **wide area networks**, or **WANs**.

The early models (dating back to the 1950s) for WANs were the circuits of the local and long distance telephone companies of the time, because the first early WANs used leased telephone company lines for their connections. A telephone call establishes a single connection path between the caller and the receiver. Once that connection is established, data travels along that single path. Telephone company equipment (originally mechanical, now electronic) selects specific telephone lines to connect to each other by closing switches. These switches work very much like the switches you use to turn lights on and off in your home, except that they open and close much faster and are controlled by mechanical or electronic devices instead of human hands. The combination of telephone lines and the closed switches that connect them to each other is called a **circuit**. This circuit forms a single electrical path between caller and receiver. This single path of connected circuits that have been switched into each other is maintained for the entire length of the call. This type of centrally controlled, single-connection model is known as **circuit switching**.

Although circuit switching works well for telephone calls, it does not work as well for sending data across a large WAN or an interconnected network like the Internet. The Internet was designed to be resistant to failure. In a circuit-switched network, a failure in any one of the connected circuits causes the connection to be interrupted and data to be lost. The Internet uses packet switching to move data between two points. In a **packet-switched** network, files and e-mail messages are broken down into small pieces, called **packets**, that are labeled electronically with their origin, sequence, and destination addresses. Packets travel from computer to computer along the interconnected networks until they reach their destination. Each packet can take a different path through the interconnected networks, and the packets may arrive out of order. The destination computer collects the packets and reassembles the original file or e-mail message from the pieces in each packet.

Routing Packets

As an individual packet travels from one network to another, the computers through which the packet travels determine the best route for getting the packet to its destination. The computers that decide how best to forward each packet are called **routing computers**, **router computers**, **routers**, or **gateway computers** (because they act as the gateway from a LAN or WAN to the Internet). The programs on router computers that determine the best path on which to send each packet contain rules called **routing algorithms**. The programs apply their routing algorithms to information they have stored in **routing tables** or **configuration tables**. This information includes lists of connections that lead to particular groups of other routers, rules that specify which connections to use first, and rules for handling instances of heavy packet traffic and network congestion.

Individual LANs and WANs can use a variety of different rules and standards for creating packets within their networks. The pieces of hardware that move packets from one part of a network to another, or to a different network, are called switches, bridges, and routers. You can take a data communications and networking class to learn more about what functions network elements perform and how they work. When packets leave a network to travel on the Internet, they must be translated into a standard format. Routers usually perform this translation function. As you can see, routers are a very important part of the infrastructure of the Internet. When a company or organization becomes a part of the Internet, it must connect at least one router to the other routers (owned by other companies or organizations) that make up the Internet. Figure 2-1 is a diagram of a small portion of the Internet that shows its router-based architecture.

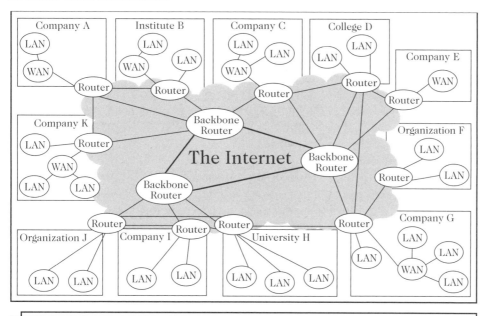

Figure 2-1   Router-based architecture of the Internet

The Internet also has routers that handle packet traffic along the Internet's main connecting points. These routers and the telecommunications lines connecting them are collectively referred to as the **Internet backbone**. These routers are very large computers that can each handle more than 50 million packets per second! These routers are sometimes called **backbone routers**. You can see in Figure 2-1 that a router will always have more than one path to which it can direct a packet. By building in multiple packet paths, the designers of the Internet created a degree of redundancy in the system that allows it to keep moving packets even if one or more of the routers or connecting lines fails.

DATA TRANSMISSION PROTOCOLS

The earliest packet-switched network, called ARPANET, connected only a few universities and research centers. This experimental network grew over the next few years and used the **Network Control Protocol (NCP)**. A **protocol** is a collection of rules for formatting, ordering, and error-checking data sent across a network. For example, protocols determine how the sending device indicates that it has finished sending a message, and how the receiving device indicates that it has received (or not received) the message. The **open architecture** philosophy developed for the evolving ARPANET, which later became the core of the Internet, included four key rules that have contributed to the success of the Internet:

- Independent networks should not require any internal changes to be connected to the network.
- Packets that do not arrive at their destinations must be retransmitted from their source network.
- Router computers act as receive-and-forward devices; they do not retain information about the packets that they handle.
- No global control exists over the network.

Thus, the ARPANET and its successor, the Internet, use routers to isolate each LAN or WAN from the other networks to which they are connected. Each LAN or WAN can use its own set of protocols for packet traffic within the LAN or WAN, but can use the router to move packets onto the Internet in its standard format. Following these simple rules makes the connections between the interconnected networks operate effectively.

Internet Protocols

The Internet uses two main protocols: the **Transmission Control Protocol (TCP)** and the **Internet Protocol (IP)**. Developed by Internet pioneers Vincent Cerf and Robert Kahn, these protocols are the rules that govern how data moves through the Internet and how network connections are established and terminated. The acronym **TCP/IP** is commonly used to refer to the two protocols.

The TCP protocol controls the disassembly of a message or a file into packets before it is transmitted over the Internet, and it controls the reassembly of those packets into their original formats when they reach their destination. The IP protocol specifies the addressing details for each packet, labeling each with the packet's origination and destination addresses. Soon after the new TCP/IP protocol set was developed, it replaced the NCP that ARPANET originally used.

In addition to its Internet function, TCP/IP is used today in many LANs. The TCP/IP protocol is provided in most personal computer operating systems commonly used today, including Macintosh systems, Linux systems, and the Windows 95, Windows 98, Windows NT, Windows 2000, Windows Me, and Windows XP Microsoft operating systems.

IP Addresses

The version of IP that has been in use for the past 20 years on the Internet is **IP version 4**, abbreviated **IPv4**. It uses a 32-bit number to identify the computers connected to the Internet. This address is called an **IP address**. Computers do all of their internal calculations using a **base 2** (or **binary**) number system—each digit is either a 0 or a 1, corresponding to a condition of either "off" or "on." IPv4 uses a 32-bit binary number that allows over 4 billion different addresses ($2^{32} =$ 4,294,967,296).

When a router breaks a message into packets before sending it on to the Internet, the router marks each packet with both the source IP address and the destination IP address of the message. To make them easier to read, IP numbers (addresses) appear as four numbers separated by periods. This notation system is called **dotted decimal** notation. An IPv4 address is a 32-bit number, so each of the four numbers is an 8-bit number ($4 \times 8 = 32$). In most computer applications, an 8-bit number is called a **byte**; however, in networking applications, an 8-bit number is often called an **octet**. In binary, an octet can have values from 00000000 to 11111111; the decimal equivalents of these binary numbers are 0 and 255, respectively.

Since each of the four parts of a dotted decimal number can range from 0 to 255, IP addresses range from 0.0.0.0 (which would be written in binary as 16 zeros) to 255.255.255.255 (which would be written in binary as 16 ones). Although many people find dotted decimal notation to be somewhat confusing at first, most do agree that writing, reading, and remembering a computer's address as 216.115.108.245 is easier than 10101000111011000110100101 1000 or its full decimal equivalent, which is 674,962,008.

Today, IP addresses are assigned by three not-for-profit organizations: the **American Registry for Internet Numbers (ARIN)**, the **Reséaux IP Européens (RIPE)**, and the **Asia-Pacific Network Information Center (APNIC)**. These registries assign and manage IP addresses for various parts of the world: ARIN for North America, South America, the Caribbean, and sub-Saharan Africa; RIPE for Europe, the Middle East, and the rest of Africa; and APNIC for countries in the Asia-Pacific area. These organizations took over IP address management tasks from the **Internet Assigned Numbers Authority (IANA)**, which had performed them under contract with the U.S. government when the Internet was an experimental research project.

In the early days of the Internet, the 4 billion addresses provided by the IPv4 rules certainly seemed to be more addresses than an experimental research network would ever need. However, about 2 billion of those addresses today are either in use or unavailable for use because of the way blocks of addresses were assigned to organizations. The new kinds of devices on the Internet's many networks, such as wireless personal digital assistants and cell phones that can access the Web, promise to keep the demand for IP addresses high.

Network engineers have devised a number of stop-gap techniques such as **subnetting**, which is the use of reserved private IP addresses within LANs and WANs to provide additional address space. **Private IP addresses** are a series of IP numbers that have been set aside for subnet use and are not permitted on packets that travel on the Internet. In subnetting, a computer called a **network address translation (NAT) device** converts those private IP addresses into normal IP addresses when it forwards packets from those computers to the Internet.

The **Internet Engineering Task Force (IETF)** worked on several new protocols that could solve the limited addressing capacity of IPv4, and in 1997, approved Internet Protocol version 6 (**IPv6**) as the protocol that would replace IPv4. The new

IP will be implemented gradually because the two protocols are not directly compatible. However, network engineers have devised ways to run both protocols together on interconnected networks. The major advantage of IPv6 is that it uses a 128-bit hexadecimal (base 16) number for addresses instead of the 32-bit binary (base 2) number used in IPv4. A **hexadecimal** numbering system uses 16 digits (0, 1, 2, 3, 4, 5, 6, 7, 8, 9, a, b, c, d, e, and f). For example, the dotted decimal IP address 216.115.108.245 would be written as 283b1a58 in hexadecimal. The number of available addresses in IPv6 (16^{128}) is 134 followed by 152 zeros—many billions of times larger than the address space of IPv4. The new IP also changes the format of the packet itself. Improvements in networking technologies over the past 20 years have made many of the fields in the IPv4 packet unnecessary. IPv6 eliminates those fields and adds fields for security and other optional information.

Domain Names

The founders of the Internet were concerned that users might find the dotted decimal notation difficult to remember. To make the numbering system easier to use, they created an alternative addressing method that uses words.

An address such as www.course.com is called a **domain name**. Domain names contain two or more word groups separated by periods. The leftmost part of a domain name (excluding www) is the most specific portion of the name. Each part of a domain name becomes more general as you move to the right, and the rightmost portion of a domain name is the most general. For example, the domain name www.breezy.compsci.nebraska.edu contains five parts separated by periods. Beginning at the left, www indicates that the domain is a part of the World Wide Web (Note: Most, but not all, Web addresses follow that naming convention). Next, breezy is the name of a computer owned by the computer science department, which the next name, compsci, indicates. The institution University of Nebraska is identified by nebraska. Finally, the name edu indicates that the computer belongs to a four-year educational institution.

The last part of a domain name, which is the most general identifier in the name, is called a **top-level domain** (or **TLD**). For many years, these domains have included a group of general domains, such as .edu, .com, and .org; and a set of country domains. Since 1998, the **Internet Corporation for Assigned Names and Numbers (ICANN)** has had responsibility for managing domain names and coordinating them with the IP address registrars. ICANN is also responsible for setting standards for the router computers that make up the Internet. In 2000, ICANN added seven new TLDs to the general domain category. Although these new domain names were chosen after much deliberation and consideration of more than 100 possible new names, a number of people were highly critical of the selections (see, for example, the **ICANNWatch** Web site). Figure 2-2 presents a list of the general TLDs, including the seven new additions, and some of the more popular country TLDs.

Original General TLDs		Country TLDs		New (2000) General TLDs	
TLD	**Use**	**TLD**	**Country**	**TLD**	**Use**
.com	Commercial	.au	Australia	.aero	Air-transport industry
.edu	Four-year educational institution	.ca	Canada	.biz	Businesses
		.de	Germany	.coop	Cooperatives
.gov	U.S. Federal government	.fi	Finland	.info	General use
.mil	U.S. Military	.fr	France	.museum	Museums
.net	General use	.jp	Japan	.name	Individual persons
.org	Not-for-profit organization	.se	Sweden	.pro	Professionals (accountants, lawyers, physicians)
		.uk	United Kingdom		

Figure 2-2 *Top-level domain names*

INTERNET SERVICES PROTOCOLS

The Internet provides a variety of services to users. Sometimes known as **application services**, these include Web page delivery, network management tools, remote login, file transmission, electronic mail, and directory services. Some application services are used very frequently and others are used only occasionally. Several of the more popular services are outlined in the next sections.

Web Page Delivery

The set of rules for delivering Web pages over the Internet are collected in a protocol called the **Hypertext Transfer Protocol (HTTP)**. The original HTTP that Tim Berners-Lee developed in 1991 was quite simple, but it has been continually evolving. Like other Internet protocols, HTTP employs the client/server model in which a user's (the client's) Web browser opens an HTTP session and sends a request for a Web page to a remote server. In response, the server creates an HTTP response message that is sent back to the client's (requester's) Web browser. A **client/server model** is a network in which each computer on the network is either a client or a server. A **client** is a PC or workstation on which users run applications. A **server** is a powerful computer dedicated to managing disk drives, printers, or network traffic. When you type a domain name (for example, "www.yahoo.com") into your Web browser, the browser has your computer send an HTTP-formatted message to the computer at Yahoo! that stores Web page files. The computer at Yahoo! then replies by sending a set of files (one for the Web page, and others for each graphic object, sound, or video clip included on the page) back to your computer. These files are also in HTTP format.

When you enter a domain name in your Web browser, you type the name of the protocol, followed by the characters "//:" before the domain name. Thus, you would type http://www.yahoo.com to go to the Yahoo! Web site, although most Web browsers today will automatically insert the http:// if you do not include it. The combination of the protocol name and the domain name is called a **Uniform Resource Locator (URL)**.

SMTP, POP, MIME, and IMAP

Electronic mail, or **e-mail**, that is sent across the Internet must also be formatted according to a common set of rules. People use a variety of e-mail software programs to read and send e-mail, including Microsoft Outlook, Netscape Messenger, Eudora, PINE, and many others. An increasing number of people use e-mail software that is offered as part of a Web site, such as Yahoo! Mail or Hotmail. If e-mail messages did not follow standard rules, an e-mail created by a person in one company (or using one Web site) could not be read by a person at another company (or Web site) using different e-mail software.

SMTP and POP are two common protocols used for sending and retrieving e-mail. **Simple Mail Transfer Protocol (SMTP)** specifies the exact format of a mail message and describes how mail is to be administered at the Internet and network level. An e-mail program running on a user's computer can request mail from the company's (or the school's) main e-mail computer using the **Post Office Protocol (POP)**. A POP message can tell the e-mail computer to send mail to the user's computer and delete it from the e-mail computer; send mail to the user's computer and not delete it; or simply ask whether new mail has arrived. The POP protocol provides support for **Multipurpose Internet Mail Extensions (MIME)**, which allows the user to attach binary files, such as formatted word processing documents or spreadsheets, to e-mail.

IMAP, the **Interactive Mail Access Protocol**, is a newer protocol that will likely one day replace POP. IMAP performs the same basic functions as POP, but includes additional features. For example, IMAP can instruct the e-mail computer to download only selected e-mail messages instead of all messages. IMAP also allows the user to view only the headers and the e-mail sender's name before deciding to download the entire message. POP allows users to search for and manipulate only those e-mail messages that they have downloaded to their computers. IMAP lets users create and manipulate mail folders (also called mailboxes on some systems), delete messages, and search for certain parts of a message while the e-mail is still on the e-mail computer; that is, the user does not need to download the e-mail before working with it in many ways. An increasing number of people access their e-mail from different computers at different times. IMAP lets users manipulate and store their e-mail on one computer and access it from any number of other computers. If these people used POP, they would have access to new e-mail messages from only one PC after they had downloaded the old messages to another PC. You can learn more about IMAP at the University of Washington's **IMAP Connection** Web site. That site also offers software that lets a computer use POP and still relay commands to a separate computer that uses IMAP. That way, users can have the benefits of both systems.

INTERNET UTILITY PROGRAMS

The TCP/IP protocol set supports a wide variety of utility programs, or tools. Many of these utility programs allow people to use the Internet more efficiently. Some of the more popular utility programs include Finger, Ping, Talk, Tracert, and VisualRoute. In this section, you will learn about several of these programs and see examples of how they work. Most of these utility programs are available from many different software producers. You can search the Internet and download several different developers' offerings for programs that interest you. Sites such as **TUCOWS** or **Download.com** provide convenient access to thousands of programs from one Web site. The Online Companion includes a list of links to several of the programs described in this section.

Finger

Finger is an Internet utility program that runs on UNIX computers and allows a user to obtain some limited information about other network users. You can issue a Finger command to determine which users are logged on to a given network or ascertain the last time a user logged on to the network. Many organizations have disabled the Finger command on their systems for privacy and security reasons. For example, if you issue a Finger command for Microsoft Corporation (finger www.Microsoft.com), you will receive no response. Some e-mail programs have the Finger program built into them, so you can send the command while reading your e-mail. Figure 2-3 shows an annotated example of the Finger command and its output. Finger typically produces five columns. The first column (Login) contains the user's login name. The second column (Name) contains a description of the user's department or supervisory unit. Column three (TTY) contains an internal machine address of each user's computer. The fourth column (Idle) indicates how long the user has been logged on to the computer. The fifth column (When) indicates the day and time when the user first logged on. (Some people log on and remain connected for days, weeks, and even months at a time.)

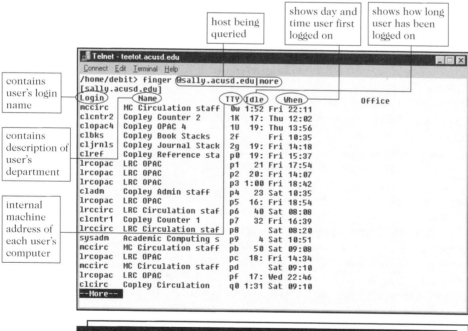

Figure 2-3 *Finger program output*

Ping

A program called **Ping**, short for **Packet Internet Groper**, tests the connectivity between two computers connected to the Internet. Ping provides performance data about the connection between Internet computers, such as the number of computers (**hops**) between them. It sends two packets to the specified address and waits for a reply. Network technicians often use Ping to troubleshoot Internet connections. Many freeware and shareware Ping programs are available on the Internet. It is a fast and helpful utility that can be used, for example, to determine whether an electronic commerce site is online. If you suspect that the site www.census.gov is offline, you can send out a Ping by opening an MS-DOS window and typing

```
ping 148.129.129.31
```

and waiting for the reply. If you do not receive a response within a second or two, then this site (the U.S. Census Bureau) is probably offline.

You can experiment with ping by trying one of the Ping client programs available from Internet download sites such as **TUCOWS** or **Download.com**.

Tracert and Other Route-Tracing Programs

Tracert (TRACE RouTe) sends data packets to every computer on the path (Internet) between your computer and another computer and clocks the packets' round-trip times. This provides an indication of the time it takes a message to travel from your computer to a remote one and back, ensures that the remote computer is online, and pinpoints any data traffic jams. Route-tracing programs also calculate and display the number of hops between computers and the time it takes to traverse the entire one-way path between machines. Route-tracing programs such as Tracert work by sending a series of packets to a particular destination. Each computer (called a router) along the Internet path between your computer and the destination computer sends back to your machine the IP address of each router and the round-trip time it took to reach it. After the program completes its execution, it displays the number of hops and how much time it took to reach each node and to travel the entire path.

Graphical user interface route-tracing programs provide a plot of the packets' route on a map. You can use route-tracing programs to determine where the greatest delays are. Companies that provide Internet connections to customers often run route-tracing programs to monitor and improve their services. However, even if you are not interested in locating delay occurrences, you can use a route-tracing program to see how the Internet works in some detail. Visualware offers its **VisualRoute** route-tracing program for download, trial, and purchase. The site also offers a demonstration of VisualRoute that you can run on the Web site without downloading any software. Figure 2-4 shows an example of a route traced between a West coast (San Diego) computer facility and an East coast computer facility (Harvard University) using the VisualRoute program.

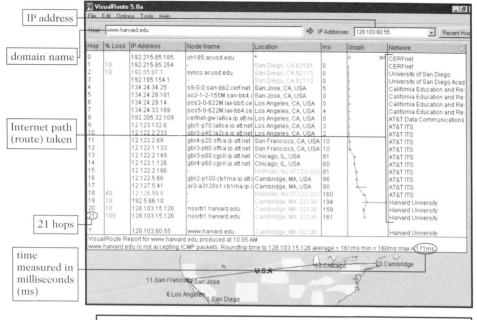

Labels on the figure (left side, top to bottom):
IP address
domain name
Internet path (route) taken
21 hops
time measured in milliseconds (ms)

Figure 2-4   *Tracing a path between two computers on the Internet*

INTERNET APPLICATIONS

Three important Internet applications were developed in the very early days of the ARPANET: electronic mail, Telnet, and FTP. You can use these applications to access the Internet and check business information using any Internet-connected computer. Today, very few populated areas on earth are out of range of the Internet.

Electronic Mail

Electronic mail originated in the 1970s on the ARPANET. Although the goals of the ARPANET were to control weapons systems and transfer research files, general communications uses emerged on the network. In 1972, Ray Tomlinson, an ARPANET researcher, wrote a program that could send and receive messages over the network. Today, e-mail is the most popular form of business communication—surpassing the telephone, conventional mail, and fax in volume.

Not only was e-mail one of the first Internet applications, it was also one reason that many people were originally attracted to the Internet. E-mail conveys messages from one destination to another in a few seconds. Messages can contain simple ASCII text, or they can contain character formatting similar to word processing programs.

One useful feature of e-mail is that you can send documents, pictures, movies, worksheets, or other information along with the message itself. These **attachments** are frequently the most important part of the message. A business-to-business e-mail message attachment might contain an invoice; the latest 200-page wholesale catalog; or a complete, compressed set of Web pages describing a company's products being sold online.

Figure 2-5 shows a typical e-mail client program (Microsoft's Outlook Express) with an attached file ready to be sent over the Internet.

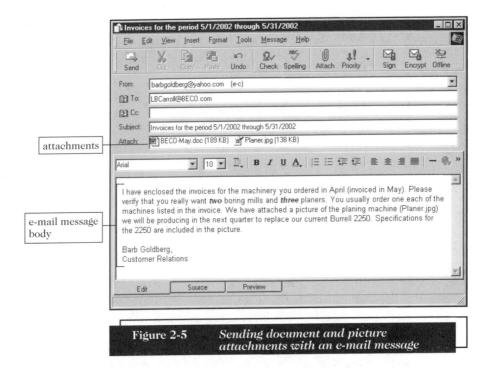

attachments

e-mail message body

Figure 2-5 *Sending document and picture attachments with an e-mail message*

Many electronic commerce sites use e-mail to confirm the receipt of customer orders and to confirm shipment of items ordered. Software vendors can also use e-mail to send information about a purchase to the buyer.

Despite its many benefits, e-mail does have some drawbacks. One significant annoyance is spam. **Spam**, also known as **bulk mail**, is electronic junk mail and can include solicitations, advertisements, or e-mail chain letters. Besides wasting your time and disk space, spam can consume large amounts of Internet capacity. If one person sends a useless e-mail to 100,000 people, that unsolicited mail consumes Internet resources for a few moments that would otherwise be available to other users.

Many grass-roots and corporate organizations have decided to fight spam aggressively. America Online, for example, has taken an active role in limiting spam through legal channels (see **AOL Policy on Unsolicited Bulk E-mail** or **Junkbusters** for more information). The origin of the term *spam* is generally believed to have come from a Monty Python song about Hormel's canned meat product, SPAM®. In the song, an increasing number of people join in the chorus: "Spam spam spam spam, spam spam spam spam, lovely spam, wonderful spam..." Just as in the song, e-mail spam is a tiresome repetition of meaningless text.

A second major irritation brought by e-mail is the **computer virus**, more simply known as a **virus**, which is a program that attaches itself to another program and can cause damage when the host program is activated. Recall that e-mail messages can carry attachments. Although most of these attachments are useful files and programs, they can just as easily be a virus program. Using virus protection software and dealing with the occasional e-mailed virus is a cost we all must bear for the convenience of using e-mail. You will learn more about computer viruses and how to control them later in this book.

A third annoyance is the amount of time that businesspersons spend answering e-mail today. Researchers have found that most managers can deal with e-mail messages at an average rate of about five minutes per message. Some messages can be deleted within a few seconds, but those are balanced by the e-mails that require the manager to spend much more time finding facts, checking files, making phone calls, and doing other tasks as part of answering e-mail. Researchers have found that most people (not including those people who answer e-mails as a full-time job) begin to resent the time that e-mail consumes when they start getting more than 20 or 30 messages a day. At that point, the average manager is spending about two hours a day answering e-mail.

Telnet

Telnet is an application that allows you to log on to a remote computer that is attached to the Internet. If you want to run software that does not have a Web interface (which includes many older programs) on a remote computer, you must use Telnet or a similar application. Several Telnet clients are available as free or low-priced downloads on the Internet, but you may be satisfied with the one supplied with Windows systems: Telnet.exe, which is limited but adequate. Telnet lets your computer give commands to programs running on the remote host. For example, you can log on to the United States Library of Congress network via Telnet using any Telnet client and entering the domain name locis.loc.gov. You can also use your Web browser as a Telnet client by entering the URL telnet://locis.loc.gov and pressing Enter.

Figure 2-6 shows an example of a Telnet session using the Windows-supplied Telnet program.

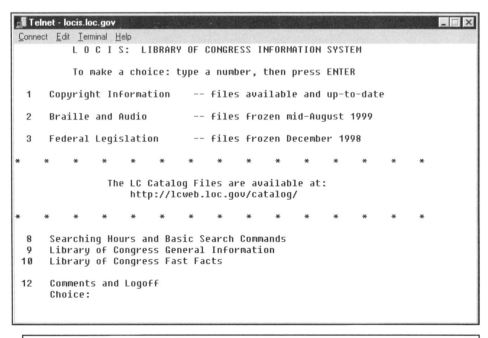

Figure 2-6 *Example of a Telnet session*

The first screen of the Library of Congress site is displayed in the figure. Logging on to a remote computer is easy on most Telnet clients. You simply enter the URL of the remote computer and click a connect button or press the Enter key. To end the Telnet session, you can type quit or follow the instructions from the host computer that appear on the Telnet client's screen. For UNIX-based systems, the command you enter to quit is usually either logout or logoff. Salespeople traveling in the United States or in many foreign countries can use the Internet and a Telnet client to log on to their company's computer to check order status and inventory availability, or to access other business information. As more companies place information on Web pages, which are accessible through any Web browser, the use of Telnet will continue to decrease.

FTP

The **File Transfer Protocol (FTP)** is the part of TCP/IP that defines the formats used to transfer files between TCP/IP-connected computers. FTP transfers both binary data and ASCII text. All FTP programs allow you to select either of the two transfer modes; many FTP programs automatically detect the format of the file and make the selection automatically. **Binary data** includes files containing word-processed documents, worksheets, graphics, and other data. **ASCII text files** contain only characters available through the keyboard but no formatting—such as a file created by the Windows program Notepad.

FTP can transfer files one at a time, or it can transfer many files at once. FTP also provides other useful services, such as displaying remote and local computers' directories, changing the current client's or server's active directory, and creating and removing local and remote directories. FTP uses the TCP protocol and its built-in error controls to copy files reliably and accurately from one computer to another.

Accessing a remote computer with FTP requires that you log on to the remote computer. You can use one of the free-standing FTP programs that exist, or you can use your Web browser by typing the protocol name, ftp://, before the domain name of the computer to which you'd like to connect. For example, if the files you want are on a computer named simtel.net, you would type ftp://simtel.net in your Web browser's address window. If you have an account on the remote computer, you can supply the FTP program with your username and password or type them into the dialog box that your Web browser will display. FTP then establishes contact with the remote computer and logs you on to your account on that computer. This **full privilege FTP** access allows you to send files to the remote computer and download files from it. Another way to access a remote computer is called anonymous FTP. **Anonymous FTP** allows you to log on as a guest. By entering the username *anonymous* and your e-mail address as a password, you can access files that are on the remote computer.

Perhaps the most popular use of FTP is for the sale and delivery of software packages and updates. Microsoft, for instance, offers a large number of free software updates online. Figure 2-7 illustrates an FTP session in which the latest update for antivirus software files is being downloaded from McAfee.

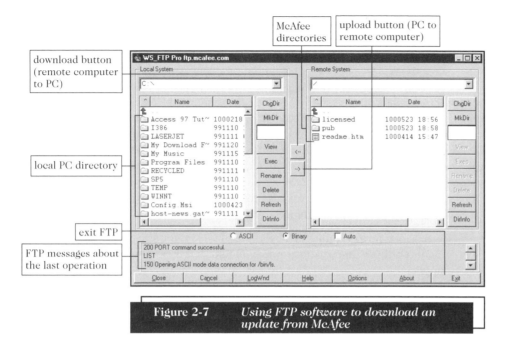

Figure 2-7 *Using FTP software to download an update from McAfee*

In the left panel are the folders and files found on the local computer's disk. The right panel displays folders and files on the remote computer's disk (McAfee's computer, in this example). Buttons in both panels allow you to switch directories, create new directories, and perform other directory maintenance tasks on both the local and remote computers.

MARKUP LANGUAGES AND THE WEB

One of the most popular uses of the Internet is to request and deliver World Wide Web pages, which number in the billions. This section discusses how Web pages are constructed.

The term *markup* describes the annotations and handwritten notes found on paper manuscript pages that tell a compositor or typist how to lay out or typeset a particular page using a set of universal copyediting symbols. Similarly, electronic Web pages are marked with tags to indicate the display and formatting of page elements. Three markup languages are presented in this chapter: SGML, the grandfather of markup languages; HTML, a derivative of SGML; and XML, a more recent derivative of SGML. In addition to its role as a markup language, SGML is a **meta language**, which is a language that can be used to define other languages.

Standard Generalized Markup Language

Since the 1960s, publishers have used markup languages to create documents that can be formatted once, stored electronically, and then printed many times in different layouts that each interpret the formatting differently. U.S. Department of Defense contractors also used early markup languages to create manuals and parts lists for weapons systems. These documents contained many information elements that were often reprinted in different versions and formats. Using electronic document storage and programs that could interpret the formats to produce different layouts saved a tremendous amount of retyping time and cost.

A Generalized Markup Language (GML) emerged from these early efforts to create standard formatting styles for electronic documents. In 1986, after many elements of the standard had been in use for years, the **International Organization for Standardization (ISO)** adopted a version of GML called **Standard Generalized Markup Language (SGML)**. SGML offers a system of marking up documents that is independent of any software application. Many organizations—especially those with complex document management requirements, such as the Association of American Publishers, Hewlett-Packard, and Kodak—use SGML.

SGML has the following advantages:

- It has long-term viability as a regulated ISO standard.
- Nonproprietary and platform-independent, it will outlive most current applications.
- User-defined tags and architectures are supported.

Although it is a complete technical specification, SGML is not well suited to certain tasks, such as the rapid development of Web pages. SGML's disadvantages are as follows:

- Set up is costly and complicated, requiring dedicated technical expertise.
- Tools for creating and editing SGML are relatively expensive.
- Creating document-type definitions in SGML can be expensive and time consuming.
- Learning time can be extensive.

Hypertext Markup Language

Recall from Chapter 1 that Tim Berners-Lee created Hypertext Markup Language (HTML), a simplified subset of SGML. HTML includes tags that define the format and style of text elements in an electronic document. HTML also has tags that can create relationships among text elements within one document or among several documents. The text elements that are related to each other are called **hypertext elements**.

HTML is an instance of one particular SGML document type—Document Type Definition (DTD)—that is both easier to learn and use than SGML. The HTML DTD is used by all the documents on the Web. The first versions of HTML provided the tools needed to create text-based electronic documents with headings, title bar titles, bullets, lines, and ordered lists. As the use of HTML and the Web itself grew, Berners-Lee turned over the job of maintaining versions of HTML to the **World Wide Web Consortium (W3C)**, a not-for-profit group that maintains standards for the Web. Later versions of HTML included tags for tables, frames, and other features that helped Web designers create more complex page layouts. The W3C maintains detailed information about HTML versions and related topics on its **W3C HTML Page**.

Because the process for approval of new HTML features did take considerable time, Web browser developers created some features, called **HTML extensions**, that would only work in their browsers. Microsoft added features that work only in its Internet Explorer browser, and Netscape added features that work only in Netscape

Navigator. Eventually, some of these HTML extensions have become part of later W3C HTML versions, but some of them still only work in one browser or another. You may have seen statements on Web pages indicating that "these pages are best viewed with Internet Explorer (or Netscape Navigator)." These statements serve as warnings that the Web pages include one or more of these proprietary HTML extensions.

Extensible Markup Language

HTML has been extremely useful for marking up text in Web pages. HTML tags provide a Web browser with information about how to display the text contained in a Web page. However, when people began using the Web for business and commerce, the meaning of the text in Web pages started to become as important (or more important) than how it appeared in the Web browser display. **EXtensible Markup Language (XML)** uses markup tags to describe the meaning, or **semantics**, of the text rather than its display characteristics. The W3C presented its first draft form of XML in 1996; the W3C issued its first formal version recommendation in 1998. Thus, it is a much newer markup language than HTML. XML, like HTML, is derived from SGML. Refer to the Appendix at the end of this book for a comparison of XML and HTML features. In 2000, the W3C released the first version of a recommendation for a new markup language called **extensible hypertext markup language (XHTML)**, which is a reformulation of HTML version 4.0 as an XML application. Figure 2-8 shows the relationships between SGML and its descendant markup languages, HTML, XML, and XHTML.

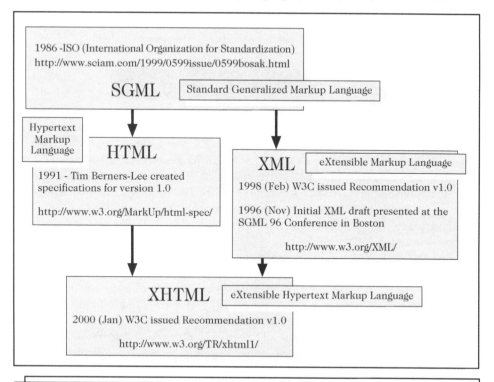

Figure 2-8 *Markup languages*

XML uses paired start and stop tags in much the same way as database software defines a record structure. For example, suppose your company sells products on the Web. Your Web pages contain descriptions and photos of the various products you sell. The Web pages are written in HTML. Unlike the product descriptions, the product information elements themselves—prices, identification numbers, quantities on hand—are formatted with XML tags. The XML document is embedded within the HTML document. Figure 2-9 shows an example of two records in a coffee inventory database in which each product has four attributes (characteristics).

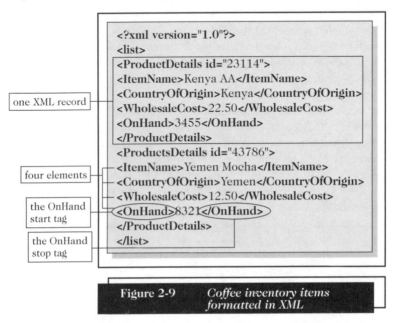

| Figure 2-9 | *Coffee inventory items formatted in XML* |

The first record contains details about the coffee named Kenya AA, and the second record contains details about Yemen Mocha. Each of these coffee records contains four elements: ItemName, CountryOfOrigin, WholesaleCost, and OnHand.

Although these XML tags indicate the meaning of each text element, the tags provide no information about how the data is to be displayed or formatted (size, position, color). Formatting information, if required, would be provided by the HTML in a Web page, a software application receiving the XML file, or another electronic document formatted in SGML.

The XML tags shown in Figure 2-9, such as ItemName, were created for this coffee database. The strength of XML is that it is **extensible**, which means that users can extend the language by creating their own tags. A different company might define the first element in its coffee database as "Name" or "ProductName" rather than "ItemName," as in this XML implementation. When companies want to use XML-formatted data to manage data across organizations, the extensibility strength of XML becomes a weakness, because problems can occur when companies develop XML versions that are incompatible with each other. A number of industry groups have formed to create standard XML tag definitions that can be used by all companies in that industry. The **Open Buying on the Internet (OBI) Consortium** and

RosettaNet are two examples of such industry groups. In 2001, the W3C released a set of rules for XML document interoperability that many researchers believe will help resolve this incompatibility problem. You can learn more about XML by reading the **W3C XML Pages**.

The next section provides an overview of HTML and illustrates how Web pages can be formatted. Formatting differences between the two major Web browsers are also discussed.

More about HTML

The World Wide Web organizes interlinked pages of information residing on sites around the world. Hypertext Markup Language defines the display characteristics of those pages, and the HTTP protocol delivers pages between servers and applications. Hyperlinks within document pages form a "web" of document pages. You can traverse the interwoven pages by clicking hyperlinked text in one page to move to another page in the web of pages. You can choose to read the pages of a document in serial order, or you can read them in whatever order you prefer by following hyperlinks. Figure 2-10 illustrates the differences between reading a paper catalog in a linear way and reading hypertext in a nonlinear way.

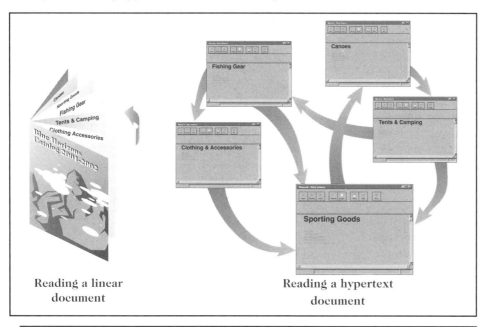

Reading a linear document

Reading a hypertext document

Figure 2-10   *Linear vs. nonlinear paths through documents*

HTML Tags

An HTML document contains document text and elements. The **elements** are the HTML codes that specify how a document or a part of a document should be formatted and arranged onscreen. HTML uses one or two **tags**, which are codes that indicate where an HTML element starts and (if necessary) where it ends, to define elements. Most HTML tags are used in pairs and have a simple structure. The general form of an HTML element is

`<tagname properties>` Displayed information affected by tag `</tagname>`

A specific example of a tag pair is a boldface character-formatting tag that causes the word *best* to appear in boldface type:

The `<B>`**best**`</B>` way to view this document is with a Web browser.

The Web page would display this:

The **best** way to view this document is with a Web browser.

Tags are easily distinguished in an HTML document, because each tag is enclosed in brackets (`<>`). You can write tags in either lowercase or uppercase letters; the tag `<b>` has the exact same meaning as the tag `<B>`. Though most tags are **two-sided**—requiring both an opening and a closing tag—some are not. Tags that only require opening tags are known as **one-sided tags**. The tag that creates a line break (`</br>`) is a common one-sided tag. Some tags, such as the paragraph tag (`<p>` … `</p>`) are two-sided tags for which the closing tag is optional. Designers often omit the optional closing tags, although some HTML purists argue that this practice is poor markup style.

In a two-sided tag set, the **opening tag** comes first, followed by text that the tag affects. The **closing tag** is identified by the slash (/) that precedes the tag's name. For example, if you were to omit the closing bold tab in the preceding example, the entire sentence, and more, would be bolded. Sometimes an opening tag contains one or more property modifiers that further refine how the tag operates. A tag's property may modify a text display, or it may designate where to find a graphic element. For example, the paragraph tag, which is a one-sided tag, defines the beginning of a text paragraph. One of its optional properties indicates a paragraph's alignment. For instance, the following HTML segment displays and right-aligns a paragraph in the browser window.

```
<P align="right"> This will right-align the paragraph,
based on the width of the user's browser and computer screen,
so that the end of each line lines up with the right border
of the screen. The left ends of each line will not be
aligned. This is known as ragged left alignment.
```

Figure 2-11 illustrates the appearance of the previous paragraph when Microsoft's Internet Explorer Web browser displays it. Because the window is not maximized, each line of text automatically wraps to a new line to accommodate the browser's window size. The paragraph automatically readjusts its margins if you change the browser's window size.

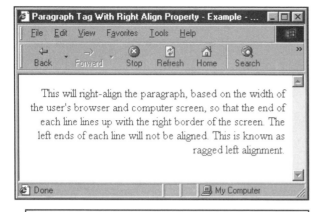

Figure 2-11   *Effects of a paragraph tag with its align property set to "right"*

A number of good Web sources (such as the **W3C Getting Started with HTML** page) and textbooks are available describing HTML tags and their uses, and you may wish to consult them for an in-depth look at HTML. Here, you will examine a few of the tags to gain a broad understanding of HTML document structures without becoming overly concerned with HTML code details.

HTML code defines the structure and formatting of a Web page, but a page may look different in two different browsers. Examine the underlying HTML code in Figure 2-12 and then compare it to the displayed page shown in Figure 2-13.

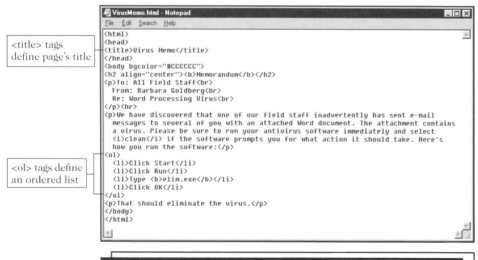

Figure 2-12   *HTML to format a memo*

When you examine Figure 2-12 and then review the Web browser result in Figure 2-13, you can see how each tag pair formats the page. The <title> tag, for example, creates the title that appears in the browser's title bar. (The text of a Web page's title is one of the key elements that is indexed by search engines when

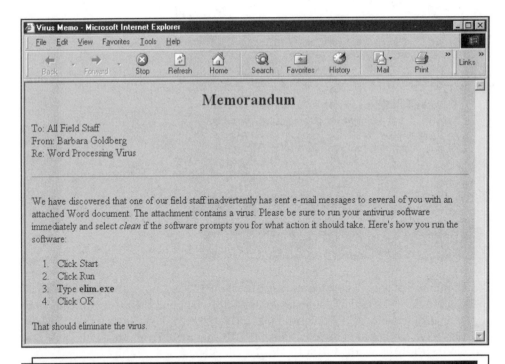

Figure 2-13 *Internet Explorer display of the HTML-formatted memo*

they scan Web pages. Notice that the opening `<body>` tag contains the modifier *bgcolor*. You probably guessed that it stands for background color. The value of the color code is in hexadecimal, the base 16 numbering scheme. It indicates how much red (first), green (second), and blue (third) to mix together to get the desired color. *CC* is 204 in decimal notation. Near the bottom is a numbered list delimited by the `<ol> </ol>` opening and closing tags. Between those tags are three list items (`<li>` tags) that are automatically ordered sequentially, beginning at 1. You may find it interesting to examine the underlying HTML code of other pages. You can examine the HTML in any Web page you are viewing in a browser by executing View, Source (Internet Explorer) or View, Document Source (Netscape Navigator) on the menu bar.

HTML Links

What makes the Web something more than an electronic document storage facility for a collection of independent documents is the HTML hyperlink feature. Hyperlinks connect the current document to another location in the same document, to another document on the same host machine, or to another document somewhere else on the Internet. Linking documents to one another through hyperlinks creates the web of documents called the World Wide Web. Hyperlinks are created using the HTML **anchor tag**. Whether you are linking to text within the same document or to a document on a distant computer, the anchor tag has the same basic form:

```
<A HREF="address">Visible link text</A>
```

Like other HTML tags, anchor tags have opening and closing tags. The opening tag also specifies the hypertext reference (HREF) property, which specifies the remote or local document's address. Clicking the text following the opening link transfers control to the HREF address—wherever that happens to be. Suppose you are creating an electronic, Web-based résumé with your university's name and address under the Education heading. Instead of presenting the university name as plain text in the résumé, you can create a hyperlink and connect it to the university name. Anyone viewing your résumé can click the link, which will lead the reader to your university's home page. The following example shows the HTML code to create a hyperlink to another Web server:

```
<A HREF="http://www.sandiego.edu">University of San Diego</A>
```

Similarly, you could create a local link to another part of the same document—perhaps page 3 of your résumé—with the following link and HTML code:

```
<A HREF="#references">References are found here</A>
```

In both preceding examples, the text *between* the anchors appears on the Web page as a hyperlink. Most browsers display the link in blue and underline it. But no matter how the link appears, whenever you move your mouse over a hyperlink, the mouse pointer changes from an arrow to a pointing hand.

An electronic commerce application, like any other Web application, uses links to direct customers to pages on the company's server and to other secure servers. The way links lead customers through pages can affect the user-friendliness rating of the site, and that can impact the customer's impression of the company. You can use different link structures on your Web site. Experience and customer response will help you decide which is best for your organization.

Two popular link structures are linear and hierarchical. A **linear hyperlink structure** resembles conventional paper documents in that the reader begins on the first page and clicks a "next" button to move to the next page in a serial fashion. Few other paths are provided. This structure works well when customers fill out forms prior to a purchase or other agreement. In this case, the customer reads and responds to page one, and then moves on to the next page. This process continues until the entire form is completed. The only Web page navigation choices the user typically has are "back" and "continue."

Another very popular link arrangement is called a hierarchical structure. In a **hierarchical hyperlink structure**, the Web user opens an introductory or home page. That page contains one or more links to other pages, and those pages, in turn, link to other pages. This hierarchical arrangement resembles an inverted tree in which the root is at the top and the branches are below it. Hierarchical structures are good for leading customers from general topics or products to specific product models and quantities. A company's home page might contain links to help, company history, company officers, order processing, frequently asked questions, and product catalogs. Figure 2-14 illustrates the linear and hierarchical structures. Of course, pages combining linear page with hierarchical structures are also possible.

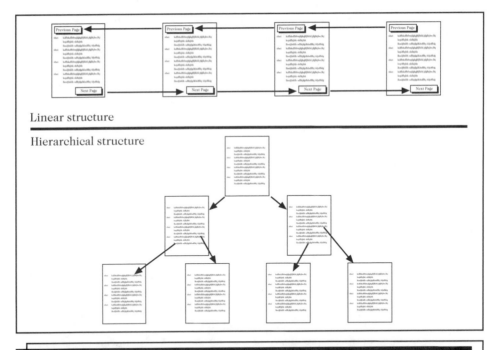

Linear structure

Hierarchical structure

| Figure 2-14 | *Two alternative hyperlink structures* |

Scripting Language and Style Sheet Capabilities

The most recent full version of HTML released by the W3C (Version 4.0, in 1997) includes support for an HTML tag called the OBJECT tag and for Cascading Style Sheets. Web designers can use the OBJECT tag to embed scripting language code in HTML pages. The most common scripting languages used in Web pages are JavaScript, Jscript, Perl, and VBScript. Scripts written in these languages and embedded in Web pages can execute programs on computers that display those pages. You can learn more about embedding script languages in HTML documents (also called **client-side scripting**) by taking a course in Web programming or reading a book such as Kathleen Kalata's *Internet Programming* (a full reference for this book appears in the "For Further Study and Research" section at the end of this chapter).

Cascading Style Sheets (CSS) give Web developers more control over the format of displayed pages. Similar to predefined document styles in word processing programs, CSSs let designers define formatting styles that can be reapplied to multiple Web pages. The term *cascading* means that designers can apply many style sheets to the same Web page.

HTML Editors

You can create HTML documents in any general-purpose text editor or word processor. However, a number of special-purpose HTML editors are available that can help you create Web pages much more easily than you can using a general-purpose editor or word processor. There are dozens of freeware, shareware, and commercial HTML editors available for download on the Internet including CoffeeCup, HotDog Professional, HomeSite, CuteHTML, and HoTMetaL Pro.

At a higher level of sophistication, you can buy HTML editors as part of Web site builder programs. With these, you can create full-scale commercial-grade Web sites, complete with database access, graphics, and fill-in forms. These tools provide a rich editing environment that displays the Web page along with the HTML code. You can drag and drop objects such as graphics, buttons, or lines onto the page. The software includes tools that generate the HTML code for the pages. When you have created a set of pages for your site, the Web site builder software can upload the pages to your Internet Web server from the PC on which you developed them. Examples of Web site builders include **Microsoft FrontPage** and **Macromedia Dreamweaver**. Figure 2-15 shows an example of a Dreamweaver Web page.

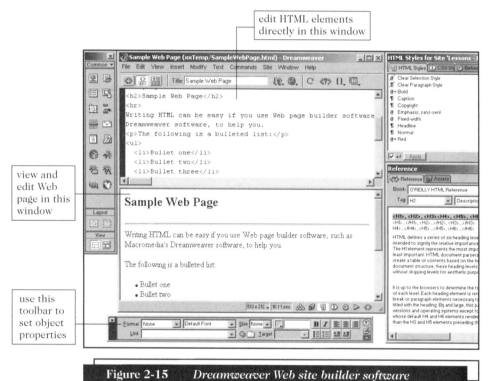

Figure 2-15 *Dreamweaver Web site builder software*

WEB CLIENTS AND WEB SERVERS

When you use your Internet connection to become part of the Web, your computer becomes a **Web client** in a worldwide client-server network. Your Web browser software—Internet Explorer or Netscape Navigator, for example—is the software that makes your computer work as a Web client. The Internet connects many different types of computers running different types of operating system software. Because Web software is platform-neutral, it lets your computer communicate with all of these different types of computers easily and effectively. This platform neutrality is a critical ingredient in the rapid spread and widespread acceptance of the Web.

Interlinked Documents

Computers that are connected to the Internet and that contain documents that their owners have made publicly available through their Internet connections are called Web servers. Unfortunately, the term "server" is used in many different ways by information systems professionals. These multiple uses of the term can be confusing to people who do not have a strong background in computer technology. You are likely to encounter four uses of "server," as described in the next paragraph.

A **server** is any computer used to provide (or "serve") files or programs to other computers connected to it through a network (such as a LAN or a WAN). The software that the server computer uses to make these files and programs available to the other computers is called **server software**. Some servers are connected through a router to the Internet. These servers can run software, called **Web server software**, that makes files available to other computers on the Internet. When a server computer is connected to the Internet and is running Web server software (usually in addition to the server software it runs to serve files to client computers on its own network), it is called a **Web server**. Thus, the word "server" is used to describe several types of computer hardware and software. If you hear a computer technician say "The server is down today," the problem might be in the hardware, in the software, or in a combination of the two!

Figure 2-16 shows how the Web's client-server architecture provides multiple interconnections among a wide variety of client and server computers.

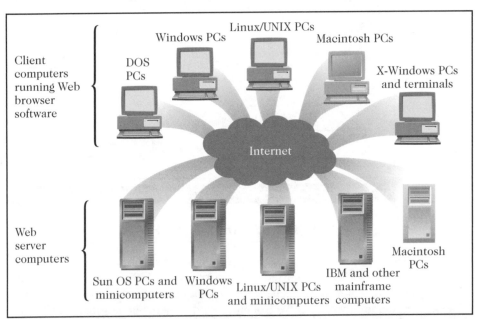

Figure 2-16   *The client-server architecture of the Web*

Documents publicly available on the Web—Web pages—consist of text and HTML code. HTML hyperlinks between documents located on the same computer and on widely dispersed computers form the web of interconnected documents. A document stored on a computer in Kansas City, Missouri, can contain a hyperlink to a document stored on a computer in Athens, Greece. The document in Athens, in turn, can have links to documents at other locations in the world. But Web pages would be static (without links to other pages, customized client data, or features such as animation) if it were not for the client-server architecture of the Internet on which they travel. When someone using a Web browser clicks a hyperlink, another page from the target URL quickly appears on the user's screen. How does that happen? What interaction is there between the browser and the target computer? The next section describes the nature of the interaction between a Web client and a server.

Web Client-Server Architecture

Client-server architecture is used in LANs, WANs, and the Web. In each case, the client computers typically request services such as printing, information retrieval, and database access. The partner in these activities is the server, which processes the clients' requests. The computers that perform the server function usually have more memory and larger, faster disk drives than the client computers they serve.

Web Client-Server Communication

Web clients request files from a distant Web server. Using the Internet as the transportation medium, the request is formatted using the HTTP protocol and sent to the server computer. A moment later, when the server receives the request, it retrieves the file containing the Web page or other information that the client requested, formats it using the HTTP protocol, and sends it back to the client over the Internet.

When the requested information—a file containing the text and markup tags of a Web page, in this instance— arrives at the client computer, the Web browser software determines that the information is an HTML page. It displays the page on the client machine according to the directions defined in the page's HTML code. This same general scenario is carried out repeatedly as the client requests, the server responds, and the client displays the result. Sometimes, a single client request results in dozens or even hundreds of separate server responses to locate and deliver information.

A Web page containing many graphics and other objects can be slow to appear on the client's Web browser window because each page element (such as graphics or multimedia files that play as sound or video in the browser) requires a separate request and response.

Two-Tier Client-Server Architecture

The basic client-server model is a two-tier model because it has only one client and one server. All communication takes place on the Internet between the client and the server. Of course, other computers are involved in transporting packets of information across the Internet, but the messages are created and read by the client and the server computers only in a **two-tier client-server architecture**. Figure 2-17 presents a high-level view of the communication that occurs between a Web client and a Web server in a two-tier client-server architecture.

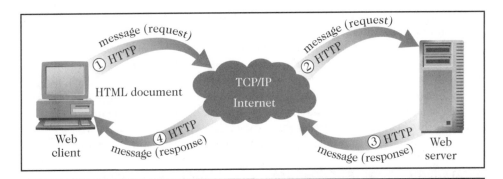

Figure 2-17 *Message flow in a two-tier client-server architecture*

The message that a Web client sends to request a file or files from a Web server is called a **request message**. A typical request message from a client to a server consists of three major parts:

- Request line
- Optional request headers
- Optional entity body

The **request line** contains a command, the name of the target resource (a file-name and a description of the path to that file on the server), and the protocol name and version number. Optional **request headers** can contain information about the types of files that the client will accept in a response to this request. Finally, an optional **entity body** is sometimes used to pass bulk information to the server. A blank line separates the entity body from the request header information. This is an example of a request message containing a request line and two request headers:

```
GET /whatsnew/rfc/rfc1939.html HTTP/1.1   request line
Accept: text/html                         request header 1
Accept: audio/x                           request header 2
```

The preceding example of a client request begins with the GET command on the request line. The GET command asks the server to retrieve and send a file to the requesting client computer. Following the GET command is the path and filename of the needed file. The last item in the request line is the protocol and version information, HTTP/1.1, which indicates that the client is using version 1.1 of the hypertext transfer protocol. The second and third message lines indicate that the client will accept text in HTML format and will accept a specific audio format. TCP/IP is responsible for transporting the message safely and intact to the target server.

When the server receives the request, it executes the command (in this case, it sends a particular Web page file back to the client). The server does this by retrieving the Web page file from its disk (or another disk on a network to which it is connected) and then creating a properly formatted **response message** to send back to the client. A server's response consists of three parts that are identical in structure to a request message: a response header line, one or more response header fields, and an optional entity body. In the response, however, each part has a slightly different function than it does in the request. The **response header line** indicates the

HTTP version used by the server, the status of the response (whether the server found the file that the client wanted), and an explanation of the status information. Response header fields follow the response header line. A **response header field** returns information describing the server's attributes. The entity body returns the HTML page requested by the client machine. Though the entity body is optional, it is almost always present because it is the HTML file (that is, the Web page) that the client requested. Figure 2-18 shows an example of a Web server response message.

```
HTTP/1.1 200 OK
Date: Thu, 25 Jul 2002 22:51:05 GMT
Server: Apache/1.3.4 (Unix)
Set-Cookie: visit=1102002283923049566;
path=/;
expires=Fri, 26 Jul 2002 22:51:05 GMT
Last-Modified: Tue, 23 Jul 2002 14:17:23 GMT
Accept-Range: bytes
Content-Length: 2000
Content-Type text/html
<HTML>
<HEAD>
<TITLE>All About Servers </TITLE>
</HEAD>
<BODY>

...

<HTML>
```

Figure 2-18   *Web server response message*

Three-Tier and N-Tier Client Server Architectures

Although the two-tier client-server architecture works well for the delivery of Web pages, a Web site that supports electronic commerce must do more than deliver Web pages. A **three-tier architecture** extends the two-tier architecture to allow additional processing to occur before the Web server responds to the Web client's request. Higher-order architectures, that is, those that have more than three tiers, are usually called **n-tier architectures**. The third tier usually includes software applications that supply information to the Web server. The Web server can then use the output of these software applications when responding to client requests instead of just looking up a Web page or other type of file on its disk drive. Architectures that have four, five, or even more tiers include software applications (just as the three-tier systems), but they also include the databases and database management programs that work with the software applications to generate information that the Web server can turn into Web pages, which it then sends to the requesting client.

A good example of services supported by a database in an n-tier architecture is a catalog-style Web site with search, update, and display functions. Assume that a user requests a display of your company's exotic fruit selections. The client request is formulated into an HTTP message by the Web browser, sent over the Internet to the Web server, and examined by the Web server. The Web server analyzes the request and determines that responding to the request will require the help of the server's database. The server sends a request to the database management software to search for, retrieve, and return all information about exotic fruit in the catalog database. The database information flows back through the database management software system to the server, which formats the response into an HTML document and sends that document, formatted with the HTTP protocol, back to the client over the Internet.

Third tier and n-tier systems can track customer purchases stored in shopping carts, look up sales tax rates, keep track of customer preferences, query inventory databases, and keep the company catalog current. Figure 2-19 shows an overview of information flows in a three-tier architecture. Numbers on the flow arrows indicate the order in which the messages flow over the indicated paths.

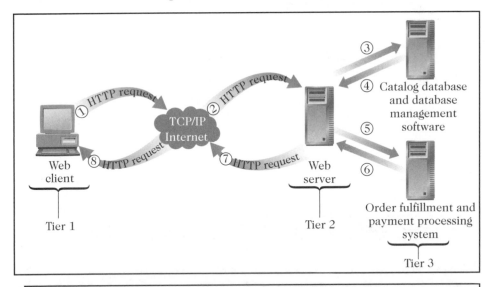

Figure 2-19 *Message flow in a three-tier client-server architecture*

INTRANETS AND EXTRANETS

Not all TCP/IP networks connect to the Internet. Many companies build internets (small "i"), or interconnected networks, that do not extend beyond their organizational boundaries. An **intranet** is an interconnected network, (or internet), usually one that uses the TCP/IP protocol set, that does not extend beyond the organization that created it. An **extranet** is an intranet that has been extended to include specific entities outside the boundaries of the organization, such as business partners, customers, or suppliers. Although fax, telephone, e-mail, and overnight express carriers have been the main communication tools for business for many years, extranets can replace many of them at a lower cost.

Intranets

Intranets are an extremely popular and low-cost way to distribute corporate information. Based on the client-server model, intranet requests for files, documents, or schematic drawings proceed the same way as on the Internet. An intranet uses Web browsers and Internet-based protocols, including TCP/IP, FTP, Telnet, HTML, and HTTP. Because intranets are compatible with the Internet, information from intranets can be shared among departments that use different technologies as well as among external consumers.

Intranets are not expensive to implement because their infrastructure requirements are already in place if a company's PCs connect to the Internet. In addition to standard Internet hardware and software, the intranet infrastructure often includes a **firewall**, which can be either hardware or software that provides security between the outside world and the private corporate intranet. You will learn more about firewalls later in this book.

Intranets are often the most efficient way to distribute internal corporate information, because producing and distributing paper is usually slower and more expensive than using Web-based communications.

Companies can also use intranets to reduce software maintenance and update costs for their employees' computer workstations. Computing staff can place software updates and patches on the intranet and then provide a script to update employee workstations automatically the next time they log on.

Extranets

Secure networks that connect companies with suppliers, business partners, or other authorized users are called extranets; they can be either public networks, secure (private) networks, or virtual private networks (VPNs). Each has the same technical capabilities of sharing information among companies. An extranet may be set up through the Internet, or it can use a separate network. Extranets can use the Internet for communicating among themselves using traditional Internet protocols, including TCP/IP. Even private networks—separate from the Internet—use Internet protocols and technology.

Some extranets start out as intranets that eventually provide access of intranet data to Internet users. For example, FedEx for many years used a system by which customers could track their packages by calling a FedEx toll-free number and then giving the operator a tracking number. A few years ago, FedEx gave package-tracking software to anyone who wanted it. Once it was installed on the customer's computer, the software dialed the FedEx computer using a modem, queried the status of the customer's package, and displayed the results on the customer's computer with no operator required. FedEx has now eliminated client-machine software and has made package tracking available on its Web site. Instead of having thousands of programs running on customers' computers, FedEx has its customers use their Web browsers to access its Web site. This Web-based system is called **FedEx Ship Manager** and it gives customers Web access (from any browser on any computer on the Web) to package tracking, airbill creation, shipment logging, and FedEx supply shipments. Critical information, such as a package's location, is stored and made available to customers through the FedEx Ship Manager section of the FedEx extranet.

Public Network

A **public network** is any computer or telecommunications network that is available to the public. The Internet is one example of a public network. Although a company can operate its extranet using a public network, very few do because of the high level of security risks. The Internet, as you will learn in later chapters, does not provide a high degree of security in its basic structure.

Private Network

A **private network** is a private, leased-line connection between two companies that physically connects their intranets to one another. A **leased line** is a permanent telephone connection between two points. Unlike the normal telephone connection you create when you dial a telephone number, a leased line is always active. The advantage of a leased line is security. Only the two parties that have leased the line to create the private network have access to the connection. The largest drawback to a private network is cost. Leased lines are expensive. Every pair of companies wanting a private network between them requires a separate line connecting them. For instance, if a company wants to set up an extranet connection over a private network with seven other companies, the company must pay the cost of seven leased lines, one for each company. If the extranet expands to 20 other companies, the extranet-sponsoring company must rent another 13 leased lines. As each new company is added, costs increase by the same amount and soon become prohibitive. Vendors refer to this as a **scaling problem**—increasing the number of leased lines in private networks is difficult, costly, and time-consuming. As the number of companies that need to join the extranet increases, other networking options become appealing.

Virtual Private Network (VPN)

A **virtual private network (VPN)** extranet is a network that uses public networks and their protocols to send sensitive data to partners, customers, suppliers, and employees using a system called IP tunneling or encapsulation. **IP tunneling** creates a private passageway through the public Internet that provides secure transmission from one extranet partner to another. The passageway is created by VPN software that encrypts the packet content, and then places the encrypted packets inside an IP wrapper in another packet; this process is called **encapsulation**. The Web server sends the encapsulated packets to their destination over the Internet. The computer that receives the packet unwraps it and decrypts the message using VPN software that is compatible with the VPN software used to encrypt and encapsulate the packet at the sending end.

A VPN provides security shells, with the most sensitive data under the tightest control. The VPN is like a separate, covered commuter lane on a highway (the Internet) in which the passengers are protected from being seen by the vehicles traveling in the other lanes. Company employees in remote locations can send sensitive information to company computers using the VPN private tunnels established on the Internet. This protected tube arrangement between cooperating extranet partners scales very well and is quite inexpensive.

When a company wishes to establish a closer relationship with a supplier or trading partner, a VPN serves to connect them. Establishing VPNs does not require a leased line. The only infrastructure required outside each company's intranets is the Internet. Companies such as **Aventail**—a company providing extranet services—are making VPNs simpler to install and maintain.

Extranets are often confused with VPNs. Although a VPN is an extranet, not every extranet is a VPN. Virtual private networks are designed to save money, although their main purpose is to create a competitive advantage with the alliance formed between cooperating companies. Unlike private networks using leased lines, VPNs establish short-term logical connections in real time that are broken once the communication session ends. The "virtual" part of VPN means that the connection seems to be a permanent, internal network connection, but the connection is actually temporary. Each transaction between two intranets using a VPN is created, carries out its work over the Internet, and is then terminated. Figure 2-20 shows a diagram of a VPN.

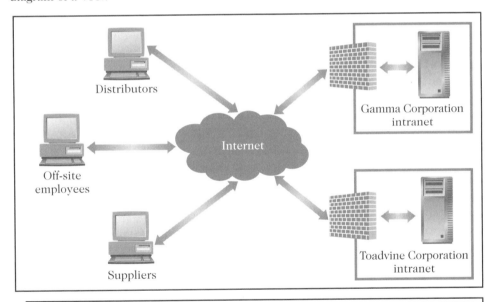

Figure 2-20 *Example of VPN architecture*

INTERNET CONNECTION OPTIONS AND TRADE-OFFS

The Internet is a set of interconnected networks. A corporation or individual cannot become part of the Internet without a telephone connection or a connection to a LAN or an intranet. Larger firms that provide Internet access to other businesses, called **Internet access providers (IAPs)** or **Internet service providers (ISPs)**, usually offer several connection options. This section briefly covers several connection choices and presents their advantages and drawbacks.

Connectivity Overview

ISPs offer several ways to connect to the Internet. The most common connection options are telephone, broadband, leased-line, and wireless. One of the major distinguishing factors between various ISPs and their connection options is the bandwidth they offer. **Bandwidth** is the amount of data that can travel through a communication line per unit of time. The higher the bandwidth, the faster data files travel and the faster Web pages appear on your screen. Each connection option offers different bandwidth, and each ISP offers varying bandwidth for each connection option. Traffic on the Internet and at your local service provider greatly affects **net bandwidth**—the actual speed that information travels. When few people are competing for service from an ISP, net bandwidth will approach the carrier's upper limit. On the other hand, you will experience slowdowns during high-traffic periods.

Depending on the type of connection you have, bandwidth can differ for data traveling to or from the ISP. **Symmetric connections** provide the same bandwidth in both directions. **Asymmetric connections** provide different bandwidths for each direction. **Upstream bandwidth**, sometimes called **upload bandwidth**, occurs when you send information from your connection to your ISP. **Downstream bandwidth**, also called **download** or **downlink bandwidth**, occurs when information travels to your computer from your ISP (for example, when a Web page is sent to your machine). Upstream bandwidth differs from downstream bandwidth for satellite, cable, and some other broadband connections.

Voice-Grade Telephone Connections

The most common way to connect to an ISP is through a modem connected to your local telephone service provider. **POTS**, or **plain old telephone service**, uses existing telephone lines and an analog modem to provide a bandwidth of between 28 and 56 Kbps. Some telephone companies offer a higher grade of service called **Digital Subscriber Line** or **Digital Subscriber Loop (DSL)** protocol. DSL connection methods do not use a modem. They use a piece of networking equipment that is similar to a network switch, but almost everyone calls this piece of equipment (incorrectly) a "DSL modem." **Integrated Services Digital Network (ISDN)** was the first technology developed to use the DSL protocol suite and has been available in the United States since 1984. ISDN is more expensive than regular telephone service and offers bandwidths of between 128 Kbps and 256 Kbps.

Broadband Connections

Connections that operate at speeds of greater than about 200 Kbps are called **broadband** services. One of the newest technologies that uses the DSL protocol to provide service in the broadband range is **Asymmetric Digital Subscriber Line (ADSL**, also abbreviated **DSL**). It provides transmission bandwidths from 100 to 640 Kbps upstream and from 1.5 to 9 Mbps (million bits per second) downstream. For businesses, a **high-speed DSL (HDSL)** connection service is now available that provides 768 Kbps of symmetric bandwidth.

Cable modems—connected to the same broadband coaxial cable that serves a television—typically provide transmission speeds between 300 Kbps and 1 Mbps from the client to the server. The downstream transmission rate can be as high as 10 Mbps. (See the Online Companion for links to cable Web pages.) In the United States alone, more than 100 million homes and organizations have broadband cable service available, and over 70 million homes subscribe to cable television. The latest estimates indicate that there are more than 9 million cable modem subscribers in the United States and another 3 million households that have broadband DSL or satellite connections.

ADSL is a private line with no competing traffic. Unlike ADSL, cable modem connection bandwidths vary with the number of other subscribers competing for the shared resource. Transmission speeds can decrease dramatically in heavily subscribed neighborhoods at prime times—in neighborhoods where many people are using cable modems simultaneously.

Connection options based on cable or telephone line connections are wonderful for urban and suburban Web users, but those living in rural areas often have very limited telephone service and no cable access at all. The telephone lines used to cover the vast distances between rural customers are usually **voice-grade lines,** which cost less than telephone lines designed to carry data, are made of lower-grade copper, and were never intended to carry data. These lines can carry only limited bandwidth—usually less than 14 Kbps. Telephone companies have wired most urban and suburban areas with **data-grade lines** (made more carefully of higher grade copper than voice-grade lines) because the short length of the lines in these areas makes it less expensive than it would be in rural areas where the connection distances are much longer. It is also likely that urban and suburban lines will someday be leased to companies willing to pay the higher fees charged for data-grade lines.

For many people in rural areas, satellite microwave transmissions have made connections to the Internet possible for the first time. In the first satellite technologies, the customer would place a satellite receiving dish antenna on the roof or in the yard and point it at the satellite. The satellite would send microwave transmissions to handle Internet downloads at speeds of around 500 Kbps. Uploads were handled by a POTS modem connection. For Web browsing, this was not too bad, since most of the uploaded messages were small text messages (e-mails and Web page requests). People who wanted to send large e-mail attachments or transfer files over the Internet found this asynchronous solution unsatisfactory.

Recently, two companies have begun offering products, **DirecPC** and **StarBand**, that include a microwave transmitter for Internet uploads. The installation charges are much higher than for other residential Internet connection services because a professional installer must carefully aim the transmitter's dish antenna at the satellite; however, once installed, the service provides upload speeds as high as 150 Kbps.

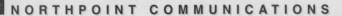

NORTHPOINT COMMUNICATIONS

In 1997, Michael Malaga was a successful telecommunications executive with an idea. He wanted to sell broadband Internet access to small businesses in urban areas. DSL technology was just gaining acceptance, and leased telephone lines were available from telephone companies. He wanted to avoid residential customers because they would soon have inexpensive cable modem access to meet their broadband needs. He also wanted to avoid suburban and rural businesses to keep the telephone line leasing costs low (lease charges are higher for longer distances). He and five friends started NorthPoint Communications with $500,000 of their combined savings and raised another $11 million within a few months. After six months, the company had raised more money from investors and had acquired 1500 customers, but it was posting a net loss of $30 million. On the strength of its number of customers, the company began the task of raising the $100 million that Malaga estimated it would need to create the network infrastructure.

In the DSL business, independent providers such as NorthPoint were pressed by customers to install service rapidly, but had to rely on local telephone companies to make sure the lines were capable of supporting the technology. In many cases, the telephone companies had to install switches and other equipment to make DSL work on a particular line. The telephone companies often were in no rush to do this, since they also sold DSL service, and speedy service would be helping a competitor. The delays led to unpredictable installation holdups and many unhappy NorthPoint customers. Customers with problems after the service was installed often were bounced from the telephone company to NorthPoint without obtaining a satisfactory or timely resolution of their problems.

While struggling to make each customer profitable, Malaga and his team raised money with a vengeance. The capital markets were hot. By the time the company sold its first stock to the public, it had raised $162 million. The stock offering in 1999 brought in an additional $387 million. The company had 13,000 customers in 17 cities. By the end of 1999, NorthPoint had spent $300 million of the cash it had raised to build out its network infrastructure and reported losses of $184 million. It had customers in 28 cities.

During the next year, the company continued to grow rapidly, raise additional funds, and lose money on each customer. In August 2000, the telephone company Verizon agreed to purchase 55 percent of the company for $800 million paid in installments. The total funding that NorthPoint had obtained by the end of 2000, including the partial payments received from Verizon, totaled $1.2 billion. By the end of the year, NorthPoint was in 109 cities and needed to spend $66 million in cash per month just to stay in business. Verizon withdrew from the purchase agreement, the stock plunged, and the layoffs began.

NorthPoint filed for Chapter 11 bankruptcy protection in January 2001 and sold its networking hardware to AT&T in March for $135 million. AT&T was not interested in continuing the DSL business (it just wanted the hardware), so

NorthPoint's 87,000 small business customers lost their Internet service overnight. In many of the cities that NorthPoint had served, there were no competitors to pick up the service.

If NorthPoint had been able to grow even faster and raise even more money, it might have had a chance to obtain enough customers to support the high fixed costs of creating and maintaining its network infrastructure. Alternatively, if the capital markets had not been so anxious to pour money into new ventures, NorthPoint may have had a chance to figure out how to make each customer profitable and become successful on a much smaller scale.

Leased-Line Connections

Large firms with large amounts of Internet traffic can connect to an ISP using higher-bandwidth connections that they can lease from telecommunications carriers. These connections use a variety of technologies and are usually classified by the equivalent number of telephone lines they include (the connection technologies they use were originally developed to carry large numbers of telephone calls). A telephone line designed to carry one digital signal is called DS0 (digital signal zero, the name of the signaling format used on those lines) and has a bandwidth of 56 Kbps. A **T1** line (also called a DS1) carries 24 DS0 lines and operates at 1.544 Mbps. Some telecommunications companies offer **fractional T1**, which provides service speeds of 128 Kbps and upward in 128-Kbps increments. **T3** service (also called DS3) offers 44.736 Mbps (the equivalent of 30 T1 lines or 760 DS0 lines). These connections are much more expensive than POTS, ISDN, or DSL connections.

Large organizations that need to connect hundreds or thousands of individual users to the Internet require very high bandwidth. **Network Access Providers (NAPs)**, organizations that sell Internet access to Internet service providers and large commercial customers, use T1 and T3 lines. NAPs and the computers that perform routing functions on the Internet backbone also use technologies such as **Frame Relay** and **Asynchronous Transfer Mode (ATM)** connections and **optical fiber** (instead of copper wire) connections with bandwidths determined by the class of the fiber-optic cable used. An OC3 (optical carrier 3) connection provides 156 Mbps, an OC12 provides 622 Mbps, an OC48 provides 2.5 Gbps (gigabits, or one billion bits per second), and an OC192 provides 10 Gbps.

Wireless Connections

No discussion of Internet connections would be complete without mention of the wireless devices that can be connected to the Internet. People today use mobile phones, personal digital assistants (PDAs), tablet computers, and even laptops with wireless network cards to connect to networks that, in turn, are connected to the Internet. There are several wireless standards, including **Bluetooth**, which is used for short distances and lower bandwidth connections; and **wireless Ethernet**, also called **802.11b** (the number of its **network specification**, which is the set of rules that equipment connected to the network must follow), which can connect devices over several hundred feet through many kinds of walls and other barriers.

Many researchers and business managers see great potential for these wireless networks and the devices connected to them. They use the term **mobile commerce** or **m-commerce** to describe the kinds of resources people might want to access (and pay for) using devices that have wireless connections. Already, people are obtaining stock quotes, directions, weather forecasts, airline flight schedules, and other information on their mobile phones and PDAs. In Japan, NTT's **DoCoMo I-Mode** is a highly successful Internet cellular phone that people avidly use to send instant messages and play games (in addition to the other uses already mentioned). Many of these services are not free, and NTT, which is a telephone company, charges for connect time to the Internet by the minute.

Internet 2

At the other end of the spectrum, a group of network research scientists from nearly 200 universities and a number of major corporations joined together in 1996 to recapture the original enthusiasm of the ARPANET with an advanced research network called Internet 2. When the National Science Foundation turned over the Internet backbone to commercial interests in 1995, many scientists felt that they had lost a large, living laboratory. **Internet 2** is the replacement for that laboratory. An experimental test bed for new networking technologies that is separate from the original Internet, Internet 2 has achieved bandwidths of 10 Gbps and more in parts of its network.

Internet 2 is also used by universities to conduct large collaborative research projects that require several supercomputers connected at very fast speeds or multiple video feeds—things that would never have been possible on the Internet itself with its lower bandwidth limits. Internet 2 promises to be the proving ground for new technologies and applications of those technologies that will eventually find their way to the Internet. As you can see, the industry that has grown up around the Internet to provide connectivity has produced a wide range of choices from which individuals, businesses, schools, and other organizations can choose to best meet their specific needs. Figure 2-21 summarizes speed and cost information about the different options for connecting your home or business to the Internet.

Service	Upstream speed (Kbps)	Downstream speed (Kbps)	Capacity (number of simultaneous users)	One-time startup costs	Continuing Monthly Costs
Residential and small business services					
Modem (POTS)	28 - 56	28 - 56	1	$0 - $20	$12 - $20
ISDN	128 - 256	128 - 256	1 - 3	$60 - $300	$50 - $90
ADSL	100 - 640	4,500 - 9,000	1 - 4	$50 - $100	$40 - $90
Cable modem	300 - 1,000	1,000 - 10,000	1 - 4	$0 - $100	$40 - $70
Satellite	125 - 150	400 - 500	1 - 3	$600 - $1,200	$60 - $70
Business Services					
Leased digital line (DS0)	64	64	1 - 10	$50 - $200	$40 - $150
Fractional T1 leased line	128 - 1,544	128 - 1,544	5 - 180	$50 - $800	$100 - $1,000
HDSL	768	768	50 - 100	$300 - $1,500	$400 - $900
T1 leased line	1,544	1,544	100 - 200	$100 - $2,000	$900 - $1,600
T3 leased line	44,700	44,700	1,000 - 10,000	$1,000 - $9,000	$5,000 - $12,000
Large Business, ISP, NAP, and Internet 2 Services					
OC3 leased line	156,000	156,000	1,000 - 50,000	$3,000 - $12,000	$9,000 - $22,000
OC12 leased line	622,000	622,000	Backbone	Negotiated	$25,000 - $100,000
OC48 leased line	2,500,000	2,500,000	Backbone	Negotiated	Negotiated
OC192 leased line	10,000,000	10,000,000	Backbone	Negotiated	Negotiated

Figure 2-21   *Internet connection options*

Summary

In this chapter, you learned about the protocols, programs, languages, and architectures that support the Internet and the World Wide Web. TCP/IP is the protocol set used to create and transport information packets across the Internet. IP addresses identify computers on the Internet. Domain names such as www.amazon.com also identify computers on the Internet, but those names are translated into IP addresses by the routing computers on the Internet. HTTP, or Hypertext Transfer Protocol, is the set of rules for transferring Web pages and requests for these Web pages on the Internet. POP SMTP, and IMAP are Internet protocols that help manage e-mail. The file transfer protocol, FTP, is used to format files when they are transferred from one computer to another. The Telnet protocol and related software provides remote computer access through the Internet. People use utility programs such as Finger and Ping to determine if another computer is connected to the Internet and to identify the path to it from the computer on which the utility is installed. E-mail is a popular Internet application that formats, sends to, and receives mail from an e-mail server.

Hypertext Markup Language, or HTML, was derived from the more generic meta language SGML. HTML defines the structure and content of Web pages using markup symbols called tags. Over time, HTML has evolved to include a large number of tags that accommodate a rich variety of elements, including graphics, cascading style sheets, and frames. Hyperlinks are HTML tags that contain a URL. The URL can be a local or remote computer. The better HTML editors facilitate Web page construction with helpful tools and drag-and-drop capabilities.

Extensible Markup Language, or XML, is also derived from SGML. However, unlike HTML, XML uses markup tags to describe the meaning, or semantics, of the text rather than its display characteristics. XML offers businesses hope for a common language that they will be able to use to describe products, services, and even business processes to each other in common, shared databases. A common language for exchanging data between companies could help companies dramatically reduce the costs of handling intercompany information flows.

The Web uses a client-server architecture in which the client computer requests a Web page and a server computer that is hosting the requested page locates and sends a page back to the client. For simple HTTP requests, a two-tier architecture works well. Tier 1 is the client computer and tier 2 is the server. For more complicated Web interactions, such as electronic commerce, requiring services of a database program or other applications, a three-tier architecture is required. Tier 3 adds a backend processor of application programs that compute and deliver information to the Web server in a form that the Web server understands. A backend application could search an inventory database and return catalog entries to a server. The server formats the page and sends it back to the requesting client computer.

Intranets are private internal networks that use the same protocols as the Internet. Employees can access the intranet and view or print information they need. When companies want to gain a competitive advantage and collaborate with suppliers, partners, or customers, they can connect their intranets to each other and form an extranet. There are three types of extranets: public network, private network, and virtual private network. VPNs, or virtual private networks, provide security at a low cost, whereas public network extranets have no security at all. Private network extranets are dedicated, pairwise connections and provide security because the network is not shared; however, they are expensive and not easily scalable.

Internet service providers offer many different connections to the Internet. Basic telephone connections are the most economical and easiest to install, but they are the slowest. Broadband cable, satellite microwave transmission, and DSL services provide Internet access at relatively high speeds. Other, more expensive options provide the bandwidth that larger businesses need. Internet 2 is an experimental network built by a consortium of research universities and businesses that provides a test bed for creating and perfecting the networking technologies of tomorrow.

Key Terms

Anchor tag
Anonymous FTP
Application services
ASCII text file
Asymmetric connection
Asymmetric Digital Subscriber Line (ADSL or DSL)
Asynchronous transfer mode (ATM)
Attachment
Backbone router
Bandwidth
Base 2
Binary
Binary data
Bluetooth
Broadband
Browser
Bulk mail
Byte
Cascading Style Sheets (CSS)
Circuit
Circuit switching
Client
Client-server model
Client-side scripting
Closing tag
Computer virus (virus)
Configuration table
Data-grade lines
Digital Subscriber Line
Digital subscriber loop (DSL)
Domain name
Dotted decimal
Download
Downstream
Downstream bandwidth (downlink bandwidth)
Electronic mail (e-mail)
Element
Encapsulation
Entity body
Extensible
Extensible hypertext markup language (XHTML)
Extensible markup language (XML)
Extranet
File Transfer Protocol (FTP)
Finger

Firewall
Frame relay
Full privilege FTP
Gateway computer
Hexadecimal
Hierarchical hyperlink structure
High-speed DSL (HDSL)
Hops
HTML extensions
Hypertext element
Hypertext Transfer Protocol (HTTP)
Integrated Services Digital Network (ISDN)
Interactive Mail Access Protocol (IMAP)
Internet 2
Internet access provider (IAP)
Internet backbone
Internet Protocol (IP)
Internet service provider (ISP)
Intranet
IP address
IP tunneling
IP version 4 (IPv4)
IP version 6 (IPv6)
Leased line
Linear hyperlink structure
Local area network (LAN)
Meta language
Mobile commerce (M-Commerce)
Multipurpose Internet Mail Extensions (MIME)
N-tier architecture
Net bandwidth
Network access provider (NAP)
Network address translation (NAT) device
Network Control Protocol (NCP)
Network specification
Octet
One-sided tag
Open architecture
Opening tag
Optical fiber
Packet
Packet-switched
Ping (Packet Internet Groper)
Plain old telephone service (POTS)
Post Office Protocol (POP)

Private IP address
Private network
Protocol
Public network
Request header
Request line
Request message
Response header field
Response header line
Response message
Router
Routing algorithm
Router computer (routing computer)
Routing table
Scaling problem
Semantics
Server
Server software
Simple Mail Transfer Protocol (SMTP)
Spam
Subnetting
Symmetric connection

T1
T3
Tags
TCP/IP
Telnet
Three-tier architecture
Top-level domain (TLD)
Transmission Control Protocol (TCP)
Two-sided tag
Two-tier client-server architecture
Uniform resource locator (URL)
Upload bandwidth
Upstream bandwidth
Virtual private network (VPN)
Voice-grade line
Web client
Web server
Web server software
Wide area network (WAN)
Wireless ethernet (802.11b)
World Wide Web Consortium (WC3)

Review Questions

1. Describe in one or two paragraphs the origins of HTML. Be sure to define elements and tags, and describe the role of at least one person involved with HTML's development.

2. Name two protocols that an electronic mail program might use to send and receive mail. Describe the advantages of each.

3. What are the major roles of the World Wide Web Consortium in the operation of the Internet and the Web today? Use your Web browser to locate the W3C site and conduct your research.

4. Compare the two- and three-tier Web client-server architectures, and indicate the role of each tier in each architecture. Which architecture is the most likely candidate for an electronic commerce site?

5. Use your Web browser and the Online Companion to search for more information about satellite connections, DSL connections, and cable connections. Prepare a three-column table (one column for each technology) in which you list the advantages and disadvantages of each connection method.

Exercises

1. Davidson Wines is a small spirits importer in New York that specializes in Portuguese wines. The company sells to wine retailers throughout the United States. Eddie Davidson has asked you to help the company with its first Web site design project. It would like to show current inventories, which would only be accessible to its established customers on its Web site. Eddie is hoping that this will reduce the number of disappointed Davidson customers who order wines that are not in the New York warehouse and thus, which cannot be delivered within a reasonable time period. Davidson has a small information systems staff that maintains its LAN and the computers that run its accounting and inventory databases, but none of the people in that department have worked with HTML. Eddie would like you to research available HTML editors and make a recommendation. He expects that Davidson will need to buy eight licenses for whichever product you choose. Prepare a report for Eddie in which you briefly review at least six HTML editors, using no more than two paragraphs for each editor. Then choose three of the best HTML editors and write a one-page evaluation of strengths and weaknesses for each of them. Use the Online Companion links to help you search the Web for information.

2. Bridgewater Engineering Company (BECO), a privately held machine shop, makes industrial-quality, heavy-duty machinery for assembly lines in other factories. It sells its horn presses, grinders, and milling equipment using a few inside salespersons and telephones. This traditional approach worked well in the startup years, but BECO is getting a lot of competition from abroad. Because you worked for the company during the summers of your college years, BECO's president, Leonard Carroll, knows you and realizes that you are Web savvy. He wants to form a close relationship with his suppliers—steel companies and small parts manufacturers—so that he can tap into their ordering systems and request supplies when he needs them. Leonard wants you to describe to him how he can use the Internet to set up such an electronic relationship. What are the alternatives? Are there any companies that could help you and him acquire the software and hardware needed to set up some sort of network? Use the Web and the links in the Online Companion to locate information about extranets and VPN networks. Write a short report comparing various network options and indicate, in writing, the names of at least two companies (from the Web) that could help develop the system for Leonard. Limit your report to 500 words or less.

3. Frieda Bannister is the IT manager for the State of Iowa's Department of Transportation (DOT). She is interested in finding ways to reduce the costs of operating the DOT's vehicle repair facilities. These facilities purchase replacement parts and repair supplies for all of the state's cars, trucks, construction machinery, and agricultural equipment (such as weed trimming machinery for the state highways). Frieda has read about XML and thinks that it might give the DOT a good way to send orders to its many suppliers throughout the country. Use the Web and your library to conduct research on the use of XML in state, local, and federal government operations. Provide Frieda with a full report that includes sections that discuss what XML is, why XML shows promise for the ordering application Frieda envisions, what other applications in the DOT might benefit from XML, and the major disadvantages of using XML today for integrating business transactions. Your report should be approximately 1500 words in length. You can use the links in the Online Companion, the "For Further Study and Research" section at the end of this chapter, and resources in your library to help you do this research.

For Further Study and Research

Alschuler, L. 2001. "Getting the Tags In: Vendors Grapple with XML-Authoring, Editing and Cleanup," *Seybold Report on Internet Publishing*, 5(6), February, 5–10.

Babcock, C. 2001. "XML Databases Offer Greater Search Capabilities," *Interactive Week*, 8(18), May 7, 11–13.

Bosak, J. and T. Bray. 1999. "How XML Will Fix the Web: Tags Categorizing Facts, Not Formats, Speed Up Transactions," *Scientific American*, 280(5), May, 89.

Campbell, T. 1998. "The First E- Mail," *Pretext Magazine*, March. Available online at: (http://www.pretext.com/mar98/features/story2.htm).

Chain Store Age. 2001. "XML, ERPs and Point Solutions," 77(2), February, 62–63.

Chen, A. 2001. "How to Grow XML," *eWeek*, 18(8), February 26, 47–49.

Chidi, G. 2001. "DSL Options Starting to Trim Down," *InfoWorld*, 23(18), April 30, 61B–62B.

Clark, D. 2001. "E-Business: Soma Is Close to Rolling Out Technology," *The Wall Street Journal*, June 4, B7.

Cope, J. 2001. "IPv6: Is It Inevitable?" *Computerworld*, 35(22), May 28, 58–59.

Costa, D. "Cable: This Technology Is the Simplest and Most Popular Option," *PC Magazine*, 20(3), February 6, 149–151.

Davis, K. and E. Burt. 2001. "Mad as Hell about DSL," *Kiplinger's Personal Finance Magazine*, 55(7), July 2001, 84–85.

DeJesus, E. 2001. "EDI? XML? Or Both?" *Computerworld*, 35(2), January, 54–56.

Dreazen, Y. 2001. "Battle Over Bells and Broadband Service Heats Up," *The Wall Street Journal*, May 15, A28.

Dyck, T. 2001. "Translating XML Schema," *eWeek*, 18(21), May 28, 1–2.

Garcia, J. and J. Wilkins. 2001. "Cable Is Too Much Better to Lose," *The McKinsey Quarterly*, Number 1, 185–188.

Goldfarb, Charles F. 1981. "A Generalized Approach to Document Markup," *ACM Sigplan Notices*, (16)6, June, 68–73.

Gonsalves, C. 2001. "10 Gigabit Built for Speed," *eWeek*, 18(19), May 14, 20.

Goodin, D. 2001. "E-Business: New Internet Protocol Still Has Many Hurdles to Clear," *The Wall Street Journal*, May 14, B5.

Hane, P. 2001. "W3C Issues XML Schema as a W3C Recommendation," *Information Today*, 18(6), June, 43.

Hawn, C. 2001. "Management By Stock Market: NorthPoint Rode the Web Wave," *Forbes*, 167(10), April 30, 52–53.

Huston, J. 2001. "Scaling the Internet," *Satellite Broadband: The Cutting Edge of Satellite Communications*, 2(3), March, 18–22.

Kalata, K. 2001. *Internet Programming with VBScript and JavaScript*. Boston: Course Technology.

Kandra, A. 2001. "Avoid the DSL Runaround," *PC World*, 19(4), April, 29–31.

Keen, G. 2001. "Embracing the PDA," *Computerworld*, 35(16), April 16, 36.

Kemp, T. 2001. "E-Businesses Get Behind Broadband," *InternetWeek*, May 7, 1–3.

LaBarba, L. 2001. "DSL Pains Reach End Users," *Telephony*, 240(15), April 9, 14–15.

Liebman, L. 2001. "XML's Tower Of Babel," *InternetWeek*, April 30, 25–26.

Mannion, P. 2001. "Mobile Devices Set to Drive IPv6," *Electronic Engineering Times*, 1164, April 30, 26–27.

McCullagh, D. 2000. "ICANN Could Use More Web Suffixes," *The Wall Street Journal*, November 20, A26.

O'Connor, R. 2000. "Under Construction: Two Research Groups Work to Build a Better Internet," *Interactive Week*, 7(46), November 13, 44–48.

Olivia, R. 2001. "The Promise of XML," *Marketing Management*, 10(1), Spring, 46–49.

Patrizio, A. 2001. "XML Passes From Development to Implementation," *Information Week*, March 26. 116–118.

PC Magazine. 2001. "Turn XML into HTML," 20(11), IP01–04.

Pimm, F. 2001. "Boeing Shows How XML Can Help Business," *Computerworld*, 35(11), March 12, 28–29.

Ramteke, T. 2001. *Networks*, Second Edition. Upper Saddle River, NJ: Prentice-Hall.

Robertson, D. 2001. "Tweaking Protocols," *Satellite Broadband: The Cutting Edge of Satellite Communications*, 2(2), February, 26–28.

Rupley, S. 2001. "Don't Write Off DSL," *PC Magazine*, 20(9), May 8, 74.

Rysavy, P. 2001. "E-commerce Unleashed," *Network Computing*," 12(2), January, 56–62.

Smith, R. 2001. "Trends in E-business Technologies," *IBM Systems Journal*, 40(1), 4–7.Spring, T. 2001. "Broadband Access from Above: Satellite Services Beam High-Speed Access Anywhere," 19(2), February, 64.

Tillet, S. 2001. "XML Databases Gain Momentum," *InternetWeek*, May 7, 10–11.

Warwick, M. 2001. "The DSL Bottleneck," *Euronet*, March, 38–41.

Weber, T. 2001. "Speedy Internet Connection Is Still Out of Reach for Some," *The Wall Street Journal*, June 18. Available online at: (http://interactive.wsj.com/articles/SB99282035 58501691305.htm).

White, C. 2001. *Data Communications and Computer Networks: A Business User's Approach*. Cambridge, MA: Course Technology.

Wigfield, M. 2001. "Rural Virginia Town Fights for Broadband Access," *The Wall Street Journal*, June 6, B6.

Woods, D. 2001. "DSL on the Rise," *Network Computing*, 12(6), March, 50–55.

SELLING ON THE WEB: REVENUE MODELS AND BUILDING A WEB PRESENCE

INTRODUCTION

The **Vanguard Group** manages a variety of pooled investment accounts for individuals and institutions. These pooled accounts are called mutual funds. To sell its investment products effectively, any investment management firm must maintain good communications with both current and potential customers. In the past, Vanguard used the telephone and mail to develop and maintain contact with its customers and prospects. Early in the rise of the Internet, Vanguard realized that the Web would give it another good way to stay in touch with customers and prospects. To that end, Vanguard has spent more than $100 million to develop and refine its Web site. In its current version, the site allows customers to obtain account information, manage their current investments, and make further investments in Vanguard mutual funds. Unlike many mutual fund management companies, however, Vanguard does not use the site to tout its products—instead, Vanguard's strategy has been to use its Web site to build customer loyalty.

The stated purpose of Vanguard's Web site is to educate its customers and provide them with a high level of service. Many times, information on the site will discourage customers from buying particular mutual fund shares if those shares are inappropriate for their investment goals. Vanguard's corporate policy is to help its customers make good investment decisions rather than to pick up a quick profit by selling customers on particular investments. The company's Web site reflects that policy.

In this chapter, you will learn about:

- Revenue models for selling on the Web
- Establishing an effective business presence on the Web
- Meeting the needs of Web site visitors
- Creating trust and building loyalty in Web site visitors
- Testing usability in Web site design
- Communicating effectively with customers on the Web

REVENUE MODELS FOR SELLING ON THE WEB

One useful way to think about electronic commerce implementations is to consider how they can generate revenue. Not all electronic commerce initiatives have the goal of providing revenue; some are undertaken to reduce costs or improve customer service. In this section, you will learn about the various models for generating revenue used by Web businesses today, including Web catalog, advertising-supported, advertising-subscription mixed, and fee-based models. These approaches can work for both business-to-consumer (B2C) and business-to-business (B2B) electronic commerce. As you will see, many companies create one Web site to handle both B2C and B2B sales. Even when companies create separate sites (or separate pages within one site), they frequently use the same revenue model for both types of sales. A **revenue model** is a general term for the combination of strategies and techniques that a company uses to generate cash flow into the business from customers.

The Web Catalog Model

The revenue model of selling goods and services on the Web is based on the mail order catalog revenue model that predates the Web; in fact, it is over 100 years old. In 1872, a traveling salesman named Aaron Montgomery Ward started selling dry goods to farmers through a one-page list that was to become the well-known Montgomery Ward & Company catalog. Rural customers were often at the mercy of their one local general store, which had no competitors and often charged much higher prices than were available to urban customers at the time. Richard Sears and Alvah Roebuck began mailing catalogs to and accepting mail orders from farmers and small town residents in 1895. Both Montgomery Ward (which closed in 2001) and Sears, Roebuck & Company grew to become dominant retailers in the United States by the mid-20th century, eventually adding retail stores to serve urban markets in addition to the catalog business that served their rural and small-town markets.

In the Web catalog model, the seller establishes a brand image that conveys quality and low cost, and then uses the strength of that image to sell through printed catalogs mailed to prospective buyers. Buyers place orders by mail or by calling the seller's toll-free telephone number. This revenue model, which is often called the **mail order** or **catalog model**, has proven successful for a wide variety of consumer goods items, including apparel, computers, electronics, housewares, and gifts.

Taking the catalog model to the Web means that the firm replaces or supplements print catalog distribution with information on its Web site. When the catalog model is expanded this way, it is often called the **Web catalog model**. Customers can place orders through the Web site or by telephone. This flexibility has been important because many consumers are still reluctant to buy on the Web. In the first few years of consumer electronic commerce, most shoppers used the Web to obtain information about products and to compare prices and features, but then made the purchase by telephone. These shoppers found early Web sites hard to use and were often afraid to send their credit card numbers over the Internet.

Computer Manufacturers

Many of the most successful Web catalog sales businesses are firms that were in the mail order business and have simply expanded their operations to the Web. Leading personal computer manufacturers such as **Dell** and **Gateway** have had great success selling on the Web. Dell has been a leader in allowing customers to specify exactly the configuration of computers they order on the Web. Dell has created value by designing its entire business around offering this high degree of configuration flexibility to its customers. Other personal computer manufacturers that sell directly to customers on the Web have followed Dell's lead by offering visitors different ways to access product information. These sites usually offer links to specific products and to pages designed for specific categories of customers, such as home, small business, education, or government users.

Luxury Goods

For many types of products, people are still unwilling to buy through a Web site. This is particularly true for luxury goods and high-fashion clothing items. The Web sites of couturiers **Vera Wang** and **Versace**, for example, were not constructed to generate revenue themselves, but to provide information to shoppers who would visit the physical stores to examine items they had seen on the sites. Such sites tend to make heavy use of graphics and animation. **Evian**, the purveyor of premium-priced bottled water, went so far as to create a Web site that works well only on computers that are connected to the Internet by a broadband connection. Evian has intentionally designed its site, shown in Figure 3-1, for a select, affluent group of customers. The Flash animation takes a very long time to download to computers that do not have such a connection.

Figure 3-1   *Evian Web site*

Apparel Retailers

A number of apparel sellers have adapted their catalog sales model to the Web, including **Eddie Bauer**, **Lands' End**, **L. L. Bean**, and **Talbots**. Unlike sellers in the high-fashion clothing category discussed above, these Web stores display photos of casual and business clothing with prices, sizes, colors, and tailoring details. Their intent is to have customers examine the clothing and place orders through the Web site. Lands' End pioneered the idea of online Web shopping assistance with its "Lands' End Live" feature in 1999. A Web customer with a question can initiate a text chat with a customer service representative or can click a button on the Web page to have the representative call. In addition to answering questions, the representative can offer suggestions by pushing Web pages to the customer's browser.

Many of Lands' End's competitors (including Eddie Bauer, L.L. Bean, and Talbots) have added similar text chat and call-back features to their sites. More recently, Lands' End has added personal shopper and virtual model features to its site. The personal shopper is an intelligent agent program that learns the customer's preferences and makes suggestions. The virtual model is a graphic image built from customer measurements on which customers can try clothes. Figure 3-2 shows the Lands' End virtual model in action.

In the fast-changing clothing business, retailers have always had to deal with the problem of overstocks—products that did not sell as well as hoped. Many retailers use outlet stores to sell their overstocks. Lands' End has found that its overstocks Web page has worked so well that it is closing many of its brick-and-mortar outlet stores. An online overstocks store works well because it reaches more people than a physical store and it can be updated more frequently than a printed overstocks catalog.

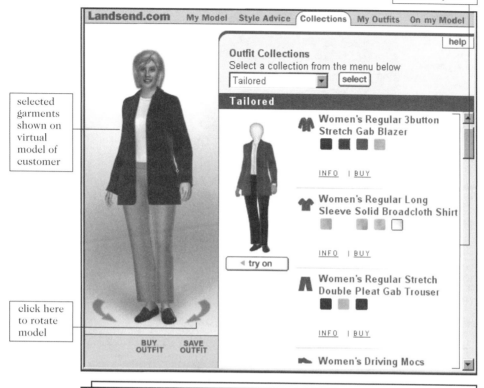

Figure 3-2   *Lands' End virtual model*

One problem that the Web presents for clothing retailers of all types is that the color settings on computer monitors vary widely. It is difficult for customers to get a completely correct idea of what the product's color will look like when it arrives. Until technology solves this problem, most online clothing stores will send a fabric swatch on request. The swatch also gives the customer a sense of the fabric's texture—an added benefit not provided by catalogs. Most Web catalog retailers also have generous return policies that allow customers to return unused merchandise for any reason.

Flowers and Gifts

Gift retailers have also successfully moved or expanded their revenue models to the Web. Florist **1-800-Flowers** has created an online extension to its highly successful telephone order business. **Harry and David**, famous for its trademarked "Fruit-of-the-Month" club, opened an informational Web site to promote its existing catalog business. The company was surprised by the volume of sales leads that the site generated, and quickly added online ordering features to the site, which appears in Figure 3-3.

Figure 3-3   *Harry and David Web page*

General Discounters

A number of new companies have started retail operations on the Web. Some of these completely new businesses operate as Web-based deep discounters. Borrowing a concept from the physical world's Wal-Marts and discount club stores, these discounters sell merchandise such as computer equipment, software, consumer electronics, books, music CDs, and sports equipment at extremely low prices. This category includes firms such as **Beyond.com**, **Buy.com**, and **Cyberian Outpost**. Some of these Web discount retailers originally sold advertising on their sites to subsidize their low product prices. Most of these firms have since abandoned this plan because advertising revenues were not sufficient subsidies. They now rely on the same volume-purchasing strategy as physical world retailers to keep prices low. As in the physical world, the online discount retail business is fiercely competitive and many of these companies operate on thin margins—and consequently, earn little profit. Cyberian Outpost, for example, began business in 1995 as one of the first retailers on the Web. In 2001, after six years of winning awards for customer satisfaction, it merged with **PC Connection**, which provided the cash it needed to continue in business.

WALMART.COM

Wal-Mart is the world's largest retailer, with 4500 stores and annual sales of nearly $200 billion. Founded in 1962 by retailing legend Sam Walton, the company has won numerous awards for business innovation. However, Wal-Mart's move into online retailing has been troubled, to say the least.

Wal-Mart launched its first Web site in July 1996. Like most company sites of that time, it contained some information about the company, but did not offer any products for sale. Wal-Mart did little to develop the Web site over the next three years, but it did add a Web store—just in time to participate in the disastrous 1999 holiday shopping season.

Wal-Mart was not the only Web retailer to have trouble in 1999. A number of companies found that they were ill-prepared for the large number of customers who decided to try electronic commerce in that year's holiday season. Lost orders, unfilled orders, and shipments that failed to arrive until January 2000 were common for many Web retailers that year. Wal-Mart was noted as an industry leader in shipping and logistics management; however, the announcement on its Web site that it could not promise Christmas delivery for items ordered after December 14 was particularly embarrassing.

To make matters worse, Wal-Mart was in the middle of developing a new Web site that it had hoped to launch before the holiday season. The project, which industry analysts estimate cost over $100 million, ran months late and was not operating until January 2000.

After eight months of operating the new Web site, Wal-Mart found itself with low levels of customer traffic (well below those of its major rivals J.C. Penney, Sears, Kmart, and Target) and high levels of criticism from Web site design experts who found the site slow, difficult to use, and lacking customer service features.

In October 2000, Wal-Mart closed the site completely for four weeks. Earlier in the year, it had created Walmart.com, a joint venture with Accel Partners to develop a new Web site, but the new site was not ready to launch until November. Industry analysts widely criticized Wal-Mart's decision to completely shut down its Web operations for such a long time period at the beginning of the holiday shopping season.

The new Web site is a vast improvement over the old site. It is much better organized and offers improved browsing and search functions. Although the number of items offered for sale on the site (about 500,000—several times more than what the physical stores carry) is the same, the company has increased its offerings of consumer electronics, toys, and sporting goods, and has decreased the number of consumable products listed. Behind the scenes is a new distribution center that serves Walmart.com exclusively.

Walmart.com's experience is a testament to how difficult it can be to get Web retailing right. Success eluded the largest and most respected retailer in the world for years. Wal-Mart is estimated to have spent more than $150 million on its Web implementation and is now just at the starting gate.

89

Channel Conflict and Cannibalization

Companies that have existing sales outlets and distribution networks often worry that their Web sites will take away sales from those outlets and networks. For example, **Levi Strauss & Company** sells its Levi's jeans and other clothing products through department stores and other retail outlets. After receiving many complaints from these stores, which had been selling Levi's products for many years, Levi Strauss decided to stop selling products on its own Web site. Such a **channel conflict** can occur whenever sales activities on a company's Web site interfere with its existing sales outlets. The problem is also called **cannibalization**, because the Web site's sales consume the sales that would be made in the company's other sales channels. The **Levi's** Web site now provides product information, but directs customers who want to buy its products to online stores that carry those products. Figure 3-4 shows a recent version of the Levi's Web page with links to online retailers **JCPenney** and **Macy's**.

hyperlinks to online retailers that sell Levi's products

| Figure 3-4 | *Levi's Web page* |

These links take the Web site visitor to the full J.C. Penney and Macy's online department stores. The Levi's site also includes links on its product pages to these two retailers that allow a customer to search for information about a particular product and then click a link on the product page that leads to the retailer's site. Product pages also include links that lead to a store finder page, so that customers who want to shop for Levi's products in a physical store can find stores near them.

Maytag, the manufacturer of home appliances, found itself in the same position as Levi Strauss. It had created a Web site that allowed customers to order directly from Maytag. After less than two years of operating its direct sales outlet and receiving many complaints from its authorized distributors and resellers, Maytag decided to incorporate online partners into its Web site store design. Now, after searching and gathering information about specific products from the Web site, a customer can click a link to one of several online partners that handle the sale transaction. These online partners are authorized Maytag distributors.

Strategic Alliances

When two or more companies join forces to undertake an activity over a long period of time, they are said to create a **strategic alliance**. When companies form a strategic alliance, they are operating in the network form of organization that you learned about in Chapter 1. Companies form strategic alliances for many purposes. An increasing number of businesses are forming strategic alliances to sell on the Web. For example, Levi's has created strategic alliances with J.C. Penney and Macy's by giving them space on the Levi's Web site to sell Levi's products. **Amazon.com**, the largest bookseller on the Web, has forged strategic alliances with **ToysRUs** to sell toys and with **drugstore.com** to sell health and beauty products on the Amazon.com Web site. These two stores appear in Figures 3-5 and 3-6.

drugstore.com loads as a tabbed section within the Amazon.com Web page

Figure 3-5   *Drugstore.com page within Amazon.com*

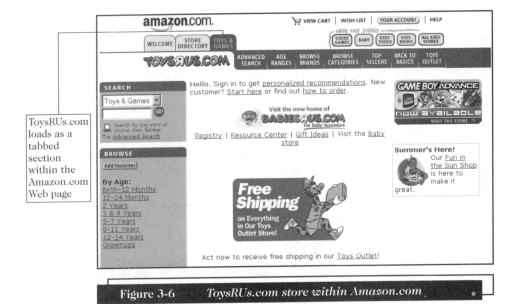

ToysRUs.com loads as a tabbed section within the Amazon.com Web page

Figure 3-6   *ToysRUs.com store within Amazon.com*

Selling Information or Other Digital Content

Firms that own intellectual property or rights to that property have embraced the Web as a new and highly efficient distribution mechanism. **LexisNexis** began as a legal research tool that has been available as an online product for years. Today, LexisNexis offers a variety of information services, including legal information, corporate information, government information, news, and resources for academic libraries. The original legal information product exists on the Web today as **Lexis.com** and provides full-text search of court cases, laws, patent databases, and tax regulations. In the past, law firms had to subscribe and install expensive dedicated computer systems to obtain access to this information. The Web has given LexisNexis customers much more flexibility in how they purchase information. Through the Lexis.com Web site, law firms can subscribe to several versions of the service that are customized for different firm sizes and usage patterns. The Web site even offers a credit card charge option for infrequent users who do not want a subscription. LexisNexis has used the Web to improve the delivery and variety of its existing product line and has been able to devise new products that take advantage of the Web's features. The LexisNexis home page in Figure 3-7 shows the wide array of information services that the company offers on the Web today.

links to products by customer category

links to specific information products

link to per-use charge (non-subscription) service

Figure 3-7   *LexisNexis information services*

ProQuest, a Web site that sells digital copies of published documents, has its roots in two businesses: the former Bell and Howell learning materials business and University Microfilms International (UMI). These firms had acquired reproduction rights to a variety of published and unpublished materials. For example, UMI had contracts with most North American universities to publish all doctoral dissertations and masters theses on demand. ProQuest offers digital versions of these documents for sale, along with a number of newspapers, journals, and other specialized academic publications. Many schools and libraries have subscriptions to ProQuest. **EBSCO Information Services** is a similar type of company that offers subscriptions to digital versions of journals, along with databases and access to electronic journals for individuals, schools, and libraries. **Dow Jones Interactive** is a more business-focused seller of subscriptions to digitized newspaper, magazine, and journal content. The Dow Jones site offers a customized digital clipping service that provides subscribers with a daily e-mail message of news on topics of interest to them.

One of the first academic organizations to make the transition to electronic distribution on the Web was (not surprisingly) the Association for Computer Machinery (ACM). The **ACM Digital Library** offers subscriptions to electronic versions of its journals to its members and to library and institutional subscribers. Academic publishing has always been a difficult business in which to make a profit, because the base of potential subscribers is so small. Even the most highly regarded academic journals often have fewer than 2000 subscribers. To break even, academic journals often must charge each subscriber hundreds or even thousands of dollars per year. Electronic publishing eliminates the high costs of paper, printing, and delivery, and makes dissemination of research results less expensive and more timely.

As was the case for other technologies, such as VCRs and subscription cable television, many of the early commercial users of Web technology were dealers in adult-themed entertainment material. Many of the first profitable sites on the Web were sellers of adult digital content. These sites pioneered the processing of credit card payment transactions (about which you will learn in Chapter 11) and many different digital video technologies that are now used by all types of businesses on the Web.

Encyclopædia Britannica is an excellent example of a company that has transferred an existing brand to the Web. The Encyclopædia Britannica has developed one of the most respected brand names in research and education over its many years in print publishing. It is particularly interesting that the Encyclopædia Britannica began in 1768 as a sort of pre-computer-age frequently asked questions (FAQ) list. A group of academics collected notes they had made while conducting research and decided to publish them as a series of articles.

Encyclopædia Britannica began its online expansion with two Web-based offerings. The Britannica Internet Guide was a free Web navigation aid that classified and rated information-laden Web sites. It featured reviews written by Britannica editors who also selected and indexed the sites. The company's other Web site, Encyclopædia Britannica Online, was available for a subscription fee or as part of its Encyclopædia Britannica CD package. Britannica used the free site to attract users to the paid subscription site.

In 1999, disappointed by low subscription sales, Britannica converted to a free, advertiser-supported site. The first day the new site, **Britannica.com**, became available at no cost to the public, it had over 15 million visitors, forcing Britannica to shut down for two weeks to upgrade its servers.

The Britannica.com site then offered the full content of the print edition in searchable form, plus access to the *Merriam-Webster's Collegiate Dictionary* and the *Britannica Book of the Year*. One of the most successful aspects of the site was the way it integrated the Britannica Internet Guide Web-rating service with its print content. The Britannica Store sold the CD version of the encylopedia along with other educational and scientific products to help generate revenue.

After two years of trying to generate a profit using this advertising-supported model, Britannica faced declining advertising revenues. In 2001, Britannica returned to a mixed model in which it offered free summaries of encyclopedia articles and free access to the *Merriam-Webster's Collegiate Dictionary* on the Web with the full text of the encyclopedia available for a subscription fee of $50 per year or $5 per month.

Britannica has gone from being a print publisher to a seller of information on the Web to an advertising-supported Web site to a mixed advertising subscription model—four major revenue model transitions—in just a few short years. The main value that Britannica has to sell is its reputation and the expertise of its editors, contributors, and advisors. For now, Britannica has decided that the best way to capitalize on that reputation and expertise is through a combined format of subscriptions and advertising support. As you will see in the next two sections, other companies are also struggling to find the right balance between advertising and subscriptions.

Advertising-Supported Model

The **advertising-supported revenue model** is the one used by network television in the United States. Broadcasters provide free programming to an audience along with advertising messages. The advertising revenue is sufficient to support the operations of the network and the creation or purchase of the programs.

Many observers of the Web in its early growth period believed that the potential for Internet advertising was tremendous. However, after a few years of experience in trying to develop profitable advertising-supported revenue models, many of those observers are less optimistic. The success of Web advertising has been hampered by two major problems. First, no consensus has emerged on how to measure and charge for site visitor views. Since the Web allows multiple measurements, such as number of visitors, number of unique visitors, number of click-throughs, and other attributes of visitor behavior, it has been difficult for Web advertisers to develop a standard for advertising charges. In addition to the number of visitors or page views, stickiness is a critical element to creating a presence that will attract advertisers. The **stickiness** of a Web site is its ability to keep visitors at the site and to attract repeat visitors. People spend more time at a **sticky** Web site and are thus exposed to more advertising.

The second problem is that very few Web sites have sufficient numbers of visitors to interest large advertisers. Most successful advertising on the Web is targeted to very specific groups. The characteristics that marketers use to group visitors is called **demographic information**, and includes such things as address, age, gender, income level, type of job held, hobbies, and religion. It can be difficult to determine whether a given Web site is attracting a specific market segment unless that site collects demographic information from its visitors—information that visitors are increasingly reluctant to provide because of privacy concerns.

Web Portals

Only a few general-interest sites have generated sufficient traffic to be profitable based on advertising revenue alone, and with the drop in advertising rates and spending that began in late 2000, even the largest advertising-supported sites are experiencing difficulties. One of the leading general interest sites is **Yahoo!**, which was one of the first Web directories. A **Web directory** is a listing of hyperlinks to Web pages. Because so many people use Yahoo! as a starting point for searching the Web, it has always attracted a large number of visitors. This large number of visitors made it possible for Yahoo! to expand its Web directory into one of the first portal sites. A **portal** or **Web portal** is a site that people use as a launching point to enter the Web (the word "portal" means "doorway"). A portal almost always includes a Web directory and search engine, but it also includes other features that help visitors find what they are looking for on the Web and that make the Web a more useful experience. Most portals include features such as shopping directories, white pages and yellow pages lookup databases, free e-mail, chat rooms, file storage services, games, and personal and group calendar tools.

Because the Yahoo! portal's search engine presents visitors' search results on separate pages, it can include advertising on each results page that is triggered by the terms in the search. For example, when the Yahoo! search engine detects that a visitor has searched on the term "new car deals," it can place a Ford ad at the top of the search results page. Ford is willing to pay more for this ad because it is directed only at visitors who have expressed interest in new cars. This example demonstrates one attractive option for identifying a target market audience without collecting demographic information from site visitors. Unfortunately, only a few high-traffic sites are able to generate significant advertising revenues this way.

Besides Yahoo!, the main portal sites using the advertising-supported revenue model today are **AOL**, **AltaVista**, **Excite**, **Lycos**, **Netscape NetCenter**, and **MSN**. Smaller general-interest sites, such as the Web directory **refdesk.com**, have had much more difficulty attracting advertisers than the larger search engine sites. This may change in the future as more people use the Web. Another type of portal that may be able to earn a profit with smaller numbers of visitors is the portal that offers items of interest to a specialized interest group. The technology portal **C-NET** is one example of this type of portal. You will learn more about portal strategies in Chapter 6 of this book.

Newspaper Publishers

Newspaper publishers have experimented with various ways of establishing a profitable presence on the Web. It is unclear whether a newspaper's presence on the Web helps or hurts the newspaper's business as a whole. Although it provides greater exposure for the newspaper's name and provides a larger audience for advertising that the paper carries, it also can take away sales from the print edition. Like retailers or distributors whose online sales lead to cannibalization of their brick-and-mortar sales, publishers also experience sales losses (cannibalization) as a result of online distribution. Newspapers and other publishers worry about cannibalization, because it is very difficult to measure. Some publishers have conducted surveys in which they ask people whether they have stopped buying the newspaper because the contents they want to see are available online, but the results of such surveys are not very reliable.

Although attempts to create general-interest Web sites that generate sufficient advertising revenue to be profitable have met with mixed results, sites that target niche markets have been more successful. For newspapers, classified advertising is very profitable; thus, Web sites that specialize in providing only classified advertising do have profit potential. This is especially true if they can reach a narrow target market and charge higher rates because the advertising reaches the right audience. The **Internet Public Library Online Newspapers** page includes links to hundreds of newspaper sites around the world.

Employment Sites

One implementation of the advertising-supported revenue model that does appear to be successful is Web employment advertising. **Interbiznet.com**, an Internet recruiting industry analysis firm, estimates that e-recruiting revenues will exceed $40 billion in 2002. Companies such as **CareerSite** offer international distribution of employment ads. As the number of people using the Web increases, these businesses will be able to move out of their current focus on technology and higher-level jobs and include advertising for all kinds of positions. These sites can use the same approach that search engine sites use to offer advertisers target markets. When a visitor specifies an interest in, for example, engineering jobs in Dallas, the results page can include a targeted banner ad for which an advertiser will pay more, because it is directed at a specific segment of the audience.

Employment ad sites can also target specific categories of job seekers by including short articles on topics of interest. This will also keep qualified people who are not necessarily looking for a job coming back to the site—such people are the candidates most highly sought by employers. The **Monster.com** page directed at mid-career managers appears in Figure 3-8. This page offers links to articles that might interest a mid-career executive and a poll customized for that audience.

Figure 3-8 *Monster.com page for mid-career executives*

Advertising-Subscription Mixed Model

In an **advertising-subscription mixed revenue model**, which has been used for many years by newspapers and magazines, subscribers pay a fee and accept some level of advertising. On Web sites that use the advertising-subscription revenue model, subscribers are typically subjected to much less advertising than they are on advertising-supported sites. Firms have had varying levels of success in applying this model. For example, Microsoft's **Slate e-zine** (electronic magazine) returned to using an advertising-only model after failing to attract a sufficient number of subscribers. Other online publishers have had more success with the mixed model. One Web site, **Salon.com**, has even gone in the opposite direction and now offers an optional subscription version of its site. Subscribers pay $30 per year to view a version of the e-zine without the ads.

Two of the world's most distinguished newspapers, **The New York Times** and **The Wall Street Journal**, use a mixed advertising-subscription model. *The New York Times* version is mostly advertising-supported with a small subscription fee for visitors who want online access to the newspaper's crossword puzzles. The paid subscription service originally included the bridge and chess columns, but those are now freely accessible to any registered site visitor (registration is free). *The New York Times* also provides a searchable archive of articles dating back to 1996 and charges a small fee for viewing any article older than one week. *The Wall Street Journal's* mixed model is weighted more heavily to subscription revenue. The site allows nonsubscriber visitors to view the classified ads and certain stories from the newspaper, but most of the content is reserved for subscribers. Visitors who already subscribe to the print edition are offered a reduced rate on subscriptions to the online edition.

Other newspapers, including **The Washington Post** and the **Los Angeles Times**, use another variation of the mixed revenue model. These newspapers do not charge a subscription fee. Instead, they offer current stories (usually within the most recent 30 days) free of charge on their Web sites but require visitors to pay for articles retrieved from their archives.

Business Week offers yet a third variation on the mixed model theme. It offers some free content at its **Business Week online** site, but requires a subscription to access the entire site. Subscribers who want to read archived articles that are more than five years old are levied an additional charge per article. *Business Week* does place content in the subscriber section of its Web site before the magazine appears on the newsstands or is delivered to subscribers.

The **Reuters** wire service also uses a mixed model in its Web offerings. A wire service collects news reports from around the globe; consolidates them; and then sells them to newspapers, radio and television stations, governments, and large companies. Reuters provides some news headlines and stock market quotes on its Web site, but its main Web-related business is providing news feeds and financial market information to large, paying subscribers such as Yahoo!, Lycos, and C-NET. The value added by a wire service is consolidation and filtering of content. Reuters' paying customers expect the news and information to be edited and categorized by Reuters staff. Some companies pay Reuters to provide them with a copy of every story that Reuters collects on the company and its competitors. Reuters offers such services, often called **clipping services**, to corporate clients for a fee.

Sports fans visit the **ESPN** site for all types of sports-related information. Leveraging its brand name from its cable television businesses, ESPN is one of the most visited sites on the Web. It sells advertising and offers a vast amount of free information, but die-hard fans can subscribe to its Insider service to obtain access to even more sports information.

Northern Light is a search engine with a twist. In addition to searching the Web, it searches its own database of journal articles and other publications to which it has acquired reproduction rights. Northern Light's revenue model extends the usual advertising-supported Web search engine to include fee-based services such as the subscription services offered by ProQuest or EBSCO. A Northern Light search results page appears in Figure 3-9.

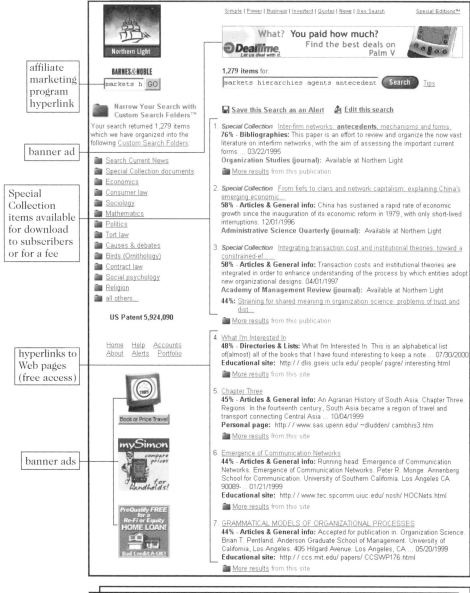

Figure 3-9   *Northern Light search results page*

The search results marked "Special Collection" are included in the Northern Light proprietary database. Site visitors can read the bibliographic information for these entries, which sometimes includes an abstract, at no charge. Download of the full text of any special collection item is available for a small fee, which varies depending on where the article appeared. Alternatively, Northern Light customers can subscribe to one of several different monthly fee plans that offer free or reduced-price downloads.

Fee-for-Transaction Models

In the **fee-for-transaction revenue model**, businesses offer services for which they charge a fee that is based on the number or size of the transactions they process. Some of these services lend themselves well to operating on the Web. To the extent that companies can offer Web site visitors the information they need about the transaction, companies can offer much of the personal service formerly provided by human agents. If customers are willing to enter transaction information into Web site forms, these sites can provide options and execute transactions much less expensively than traditional transaction service providers.

Travel Agents

Travel agents earn commissions on each airplane ticket, hotel reservation, auto rental, or vacation that they book. These commissions are paid to the travel agent by the transportation or lodging provider. The travel agency revenue model involves receiving a fee for facilitating a transaction. The value added by a travel agent is that of information consolidation and filtering. A good travel agent knows many things about the traveler's destination and knows enough about the traveler to select the information elements that will be useful and valuable to the traveler. Computers, particularly computers networked to large databases, are very good at information consolidation and filtering. In fact, travel agents have used networked computers, such as the **Sabre** system, for many years to make reservations for their customers.

When the Internet emerged as a new way to network computers and then became available to commercial users, a number of online travel agencies began doing business on the Web. Existing travel agencies did not, in general, rush to the new medium. They believed that the key value they added, personal customer service, could not be replaced with a Web site. Therefore, the first Web-based travel agencies were new entrants. One of these sites, **Travelocity**, is based on the same Sabre system that traditional travel agents use (Travelocity is also largely owned by Sabre). Microsoft has also established a position in the online travel agency business with its **Expedia** subsidiary.

Travelocity, Expedia, and the **Hotel Reservations Network** are regularly listed among the top electronic commerce sites in surveys and industry analyst rankings. All three are profitable. In 2001, a consortium of five major U.S. airlines (American, Continental, Delta, Northwest, and United) launched a new Web travel site, **Orbitz**. A number of consumer groups and the attorneys general of 20 states expressed concern over possible antitrust issues that could arise with a site such as Orbitz, which is sponsored by competing airlines. For example, Orbitz offers a monetary incentive to airlines that agree to offer Orbitz customers their lowest fares at all times. That means that the airline could not offer a special low fare on its own Web site (or on another site such as Travelocity or Expedia) unless it also made that fare available through Orbitz. The site launch was met with mixed reviews in surveys and criticism from industry analysts. The site encountered some technical difficulties; however, the site did generate significant amounts of visitor traffic immediately.

In addition to earning commissions from the transportation and lodging providers, these sites generate advertising revenue from ads placed on travel information pages. These ads are similar to those on search engine results pages, because advertisers can target them without obtaining demographic details about the site visitor. For example, if you are booking a flight to Chicago, the page that lists airline ticket options may also carry a banner ad for a hotel in Chicago or a car rental company that is running a promotion in the Chicago area.

Automobile Sales

Auto dealers buy cars from the manufacturer and sell them to consumers. They provide showrooms and salespeople to help customers learn about product features, arrange financing, and make a purchase decision. Most auto dealers negotiate the prices at which they sell their cars; thus, the salesperson's job also includes extracting the highest possible price from the consumer. Many people do not like negotiating car prices, especially if they have taken the time to learn about car features, arrange financing, and become ready to purchase a car without further assistance from a salesperson.

As you learned in Chapter 1, Autobytel and other firms offer knowledgeable consumers an option that removes the salesperson from the process. Autobytel and similar firms such as **MSN Carpoint**, **CarsDirect.com**, and **Autoweb.com** provide an information service to car buyers. Each of these firms implements the fee-for-transaction revenue model in a slightly different way. For example, CarsDirect.com offers customers the ability to select a specific car (model, color, options) at a price it determines. CarsDirect.com then finds a local dealer that has such a car and is willing to sell it for the CarsDirect.com price. Alternatively, Autoweb.com and Autobytel locate dealers in the buyer's area that are willing to sell the car specified by the buyer (including make, model, options, and color) for a small premium over the dealer's nominal cost. The buyer can purchase the car from the dealer without negotiating with a salesperson. Autobytel and Autoweb.com charge participating dealers a fee for this service. In effect, firms such as Autobytel, Autoweb.com, and CarsDirect.com are taking the salesperson out of the value chain. To the extent that the salesperson provides little or no value to the consumer, these firms are reducing the transaction costs in the process. The removal of an intermediary from a value chain, such as the salesperson in this case, is called **disintermediation**. The introduction of a new intermediary into a value chain, such as the Web buying service in this case, is called **reintermediation**.

Stockbrokers

Stock brokerage firms also use a fee-for-transaction model. They charge their customers a commission for each trade executed. In the past, stockbrokers offered investment advice and made specific buy and sell recommendations to customers. They did not charge for this advice, but they did charge relatively high commissions on the trades they handled for their customers. After the U.S. government deregulated the securities trading business in the early 1970s, a number of discount brokers opened. These discount brokers distinguished themselves from the established "full line" brokerage houses by not offering any investment advice and charging very low commissions. Because the full line brokers had failed to provide value to some of their customers, those customers were very happy to move their business to the discount brokers.

The Web made it possible for firms such as **E*Trade** to offer investment advice (posted on Web pages) similar to that offered by a full line broker without incurring many of the costs of distributing the advice (such as stockbroker salaries, overhead, and the costs of printing and mailing newsletters). Web-based brokerage firms could also offer fast execution of trades that customers entered into Web page forms. Thus, in the 1990s, discount brokers who had taken business away from full line brokers for 15 years faced new competition from online firms. Of course, the full line brokers found that they were losing business to both the discount brokers and the online brokers. In response, both the discount brokers (such as **Charles Schwab** and **Ameritrade**) and the full line brokers (such as **Merrill Lynch** and **Salomon Smith Barney**) opened new stock trading and information Web sites.

The online brokers are offering customers the same kind of transaction cost reductions as the online auto buying sites. Stockbrokers are finding themselves disintermediated in the same way as car salespersons. Online brokers are offering an alternative service that has greater perceived value for many investors today.

Insurance Brokers

Other sales agency businesses are moving to the Web. Although insurance companies themselves have been slow to offer policies and investments for sale on the Web, a number of intermediaries that sell insurance policies have emerged. **Quotesmith.com**, which began business in 1984 as a policy-quoting service for independent insurance brokers, began in 1996 to offer its policy price quotes directly to the public over the Internet. By quoting policies and accepting applications directly, Quotesmith.com is disintermediating the independent insurance agents with which it formerly worked. The Quotesmith.com home page, which shows the variety of insurance products it offers, appears in Figure 3-10.

Other Web sites that offer insurance policy information, comparisons, and sales include **InsWeb**, **Answer Financial**, **Insurance.com**, and **YouDecide.com**, which was created by the human resources software development company **ProAct Technologies**.

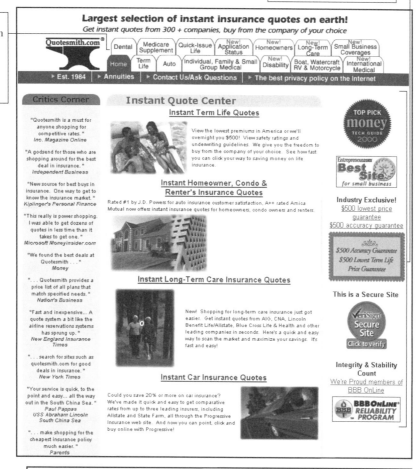

links to information about awards, guarantees, and third-party assurances

links to information about specific types of insurance policies

Figure 3-10   *Quotesmith.com home page*

Event Tickets

Obtaining tickets for concerts, shows, and sporting events has always been challenging. Some venues only offered tickets for sale at their own box offices, and others sold tickets through ticket agencies that could be difficult for patrons to find. The Web offers event-promoters an ability to sell tickets from one virtual location to customers practically anywhere in the world. Traditional ticket agencies such as **Ticketmaster** have opened shop online. Other new entrants, such as **Tickets.com** and **TicketWeb**, also offer a wide variety of tickets for events in many different locations. These electronic commerce initiatives reduce transaction costs for both buyers and sellers of tickets.

Real Estate and Mortgage Loan Brokers

Other fee-for-transaction businesses are also starting to open electronic commerce Web sites, including real estate brokers and mortgage loan brokers. Online real estate brokers provide all of the services that a traditional broker might provide—except that online brokers provide these services through their Web sites. Leading real estate sites include Web pioneers such as **eRealty** and **zipRealty.com**—and industry observers agree that these new online brokers do a much better job selling on the Web than traditional real estate brokers that have opened Web sites, such as **Coldwell Banker** and **Prudential**.

IndyMac Bank Home Lending offers mortgage loan seekers an online credit review and decision in minutes. Approved customers can then print an approval letter from their own computers and take it with them the same day to shop for a new house. This rapid decision-making ability and other customer service features have helped IndyMac become one of the leading mortgage loan sites on the Web, funding more than $12 billion in home loans each year. Other successful mortgage brokers on the Web include **mortgagebot.com** and **E-LOAN**. These sites often provide helpful features such as the IndyMac Bank QuickPricer, a loan payment calculator that appears in Figure 3-11.

Figure 3-11 *IndyMac mortgage loan Quick Pricer*

Online Banking and Financial Services

Since financial services do not involve a physical product, they are easy to offer on the Web. The greatest concerns that most people have when they consider moving their financial transactions to the Web are security and the reliability of the financial institution—the same concerns that exist in the physical world. However, on the Web, it is much more difficult for a firm to establish its reputation for security and trust than it is in the physical world where massive buildings and clearly visible room-sized safes can help create the necessary image.

Some existing banks have opened online "branches" that carry the identification and reputation of the physical world bank's brand (such as Citibank's **Citibank Online**). Other firms have started online banks that are not affiliated with an existing bank (such as the **First Internet Bank of Indiana**).

One interesting experiment was undertaken by Bank One, which opened an online bank under the name Wingspan in 1999. Bank One's expressed intent was to present Wingspan as a new and separate entity, in the spirit of the dot-com gold rush that was underway at that time. After operating Wingspan separately for about two years, Bank One decided to close Wingspan and merge it with the **BankOne.com** site.

Online banks handle only a tiny portion of the world's financial transactions today, but as the reputation and reliability of online banks grows, more customers will accept them as a good way to conduct their banking business. Industry analysts agree that two barriers are preventing a more rapid rate of growth in the online banking business—a lack of bill presentment features and a lack of account aggregation tools.

Today's online banks give customers a way to pay their bills electronically, but the customers still receive most of the bills in the mail. The bills that they can access on the Web are at sites with **bill presentment** features. Unfortunately, most people must visit a different Web site to view each online bill. As online banks add bill presentment features that allow their customers to view all of their bills on the bank's Web site (and pay each of them with a single click), those banks will find more customers willing to do their banking on the Web.

Another important feature that few online banks currently offer is **account aggregation**, which is the ability to obtain bank, investment, loan, and other financial account information from multiple Web sites and display it all in one location at the bank's Web site. Many of a bank's best customers have credit card, loan, investment, and brokerage accounts with multiple financial institutions. Having all of this information in one place would be very useful for these customers. Some non-bank sites, including **Yodlee.com**, **MSN MoneyCentral**, and **OnMoney.com**, are offering account aggregation tools. These companies offer account aggregation services and, in some cases, license their technology to other Web sites. For example, both the **Quicken.com My Finances** service and the **Yahoo! Finance** page are powered by Yodlee.com technology.

Fee-for-Services Revenue Models

Companies are offering an increasing variety of services on the Web for which they charge a fee. These are neither broker services nor services for which the charge is based on the number or size of transactions processed. The fee is based on the value of the service provided. These **fee-for-service revenue models** range from games and entertainment to financial advice and the professional services of accountants, lawyers, and physicians.

Online Games

Although many sites that offer games online use an advertising-supported revenue model, a growing number include premium games in their offerings. Site visitors must pay to play these premium games, either by buying and downloading software to install on their computers, or by paying a subscription fee to enter the premium games area on the site. Microsoft's **Zone.com**, Sony's **Station.com**, Electronic Arts' **EA.com**, and RealNetworks' **RealArcade Central** are among the leading gaming sites that include subscription game services. For example, Sony's EverQuest adventure game has drawn more than 300,000 players who have purchased a $40 software package and pay up to $10 per month to continue playing the game. Most of the gaming sites charge a monthly subscription of between $5 and $10 for access to all their fee-based games offerings. The **Interactive Digital Software Association** estimates that about 150 million people in the United States alone are regular game players, a number that is growing about 15 percent each year.

Concerts and Films

As more households obtain broadband access to the Internet, an increasing number of companies will provide streaming video of concerts and films to paying subscribers. With a revenue model patterned after cable television companies, firms such as **Intertainer** are selling subscriptions for delivery of video content to computers and other devices through cable modem and DSL connections. RealNetwork's **RealPlayer GoldPass** subscription includes sporting events, music videos, comedy, and other entertainment offerings for a one-time charge of $30, plus $10 per month.

The main technological limitation these companies face is that each additional customer who downloads a video stream requires that the provider purchase additional bandwidth from its ISP. Television broadcasters, on the other hand, need only pay the fixed cost of a transmitter—the airwaves are free and carry the transmission to an unlimited number of viewers at no additional cost. In contrast, as the number of an Internet-based provider's subscribers increases, the cost of the provider's Internet connection increases. However, if these Web entertainment companies can charge a high enough monthly fee, they should be able to cover the additional costs of technology upgrades and still make a profit.

Professional Services

State laws have been one of the main forces preventing U.S. professionals from extending their practices to the Web. Since most professionals—including physicians, lawyers, accountants, and engineers—are licensed by individual states, state laws can effectively prevent them from practicing their professions on the Web because their patients or clients could be located in other states. Professionals worry about being charged with unlicensed practice in those other states. State laws that address the location of services are vague—it is difficult to determine where a service provided over the Internet actually occurred. This uncertainty arises because most state professional practice laws were written long before the Internet existed.

The **Law on the Web** site offers legal consultations on a variety of matters for residents of the United Kingdom. Accounting professionals in the United States can be located through the **CPA Directory**, and a number of legal referral sites can refer site visitors to local attorneys. Although a number of Web sites such as **RealAge**, **pain.com**, and **WebMD** (shown in Figure 3-12) offer general health information, physicians and other health care professionals have been reluctant to sell specific advice to specific patients over the Internet. The difficulty of diagnosing medical problems without a physical examination of the patient is a significant barrier to providing health care on the Web.

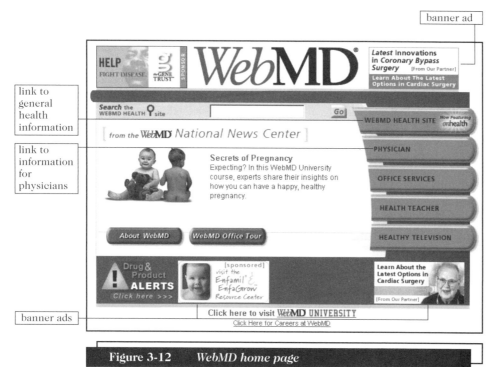

Figure 3-12 *WebMD home page*

CREATING AN EFFECTIVE WEB PRESENCE

Businesses have always created a presence in the physical world by building stores, factories, warehouses, and office buildings. An organization's **presence** is the public image it conveys to its stakeholders. The **stakeholders** of a firm include its customers, suppliers, employees, stockholders, neighbors, and the general public. Most companies tend not to worry much about the image they project until they grow to a significant size—until then, they are too focused on just surviving to spare the effort. On the Web, presence can be much more important. Many customers and other stakeholders of a Web business will know the company only through its Web presence. Creating an effective Web presence can be critical even for the smallest and newest firms operating on the Web.

Identifying Web Presence Goals

When a business creates a physical space in which to conduct its activities, its managers focus on very specific objectives. Few of these objectives are image-driven. The new company must find a location that will be convenient for its customers, with sufficient floor space and features to allow the selling activity to occur. A new business must balance its needs for inventory storage space and employee work space with the costs of obtaining that space. The presence of a physical business location results from satisfying these many other objectives and is rarely a main goal of designing the space.

On the Web, businesses and other organizations have the luxury of building their Web sites intentionally to create distinctive presences. A firm's physical location must satisfy so many other business needs that it often fails to convey a good presence. A good Web site design can provide many image-creation and image-enhancing features very effectively—it can serve as a sales brochure, a product showroom, a financial report, an employment ad, and a customer contact point. Each entity that establishes a Web presence should decide which features the Web site can provide and which of those features are the most important to include.

Making Web Presence Consistent with Brand Image

Different firms, even those in the same industry, might establish different Web presence goals. For example, **Coca Cola** and **Pepsi** are two companies that have established very strong brand images and are in the same business, but have developed very different Web presences. These two companies change their Web pages frequently, but the Coca Cola page usually includes a trusted corporate image such the Coke bottle. Alternatively, the Pepsi page is usually filled with hyperlinks to a variety of activities and product-related promotions.

These Web presences convey the images each company wishes to project. Each presence is consistent with other elements of the marketing efforts of these companies; Coca Cola's traditional position as a trusted classic, and Pepsi's position as the upstart product favored by a younger generation.

Achieving Web Presence Goals

An effective site is one that creates an attractive presence that meets the objectives of the business or other organization. These objectives include:

- Attracting visitors to the Web site
- Making the site interesting enough that visitors stay and explore
- Convincing visitors to follow the site's links to obtain information
- Creating an impression consistent with the organization's desired image
- Building a trusting relationship with visitors
- Reinforcing positive images that the visitor might already have about the organization
- Encouraging visitors to return to the site

Profit-Driven Organizations

The **Toyota** site that appears in Figure 3-13 is a good example of an effective Web presence.

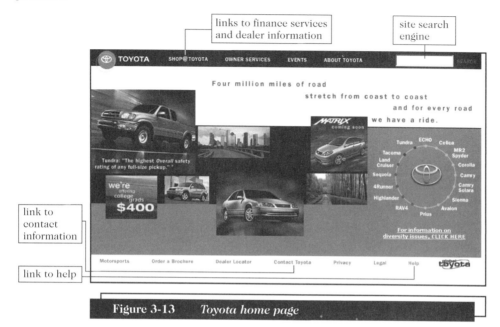

Figure 3-13 *Toyota home page*

The site provides a product showroom feature, links to detailed information about each vehicle model, links to dealers, and links to information about the company and the financing services it offers. The page offers a help link and information about how to contact the company; it also has a site search feature. The presence that Toyota has created with its home page is consistent with its corporate goal of providing cars and trucks that meet the needs of a wide variety of customers. The page presents several different car models, displays all of the model names in a circle at the right side of the page (each name is a link to information about that model), and states that "for every road we have a ride." To the extent that the Web site fulfills that promise, it is an effective extension of the corporate presence that Toyota wants to convey through the Internet to customers and potential customers.

In contrast, **Quaker Oats** has created Web sites that do not offer any strong sense of corporate presence, although they do provide a good selection of information about the firm. Figure 3-14 shows the Quaker Oats Web site as it appeared until 1999.

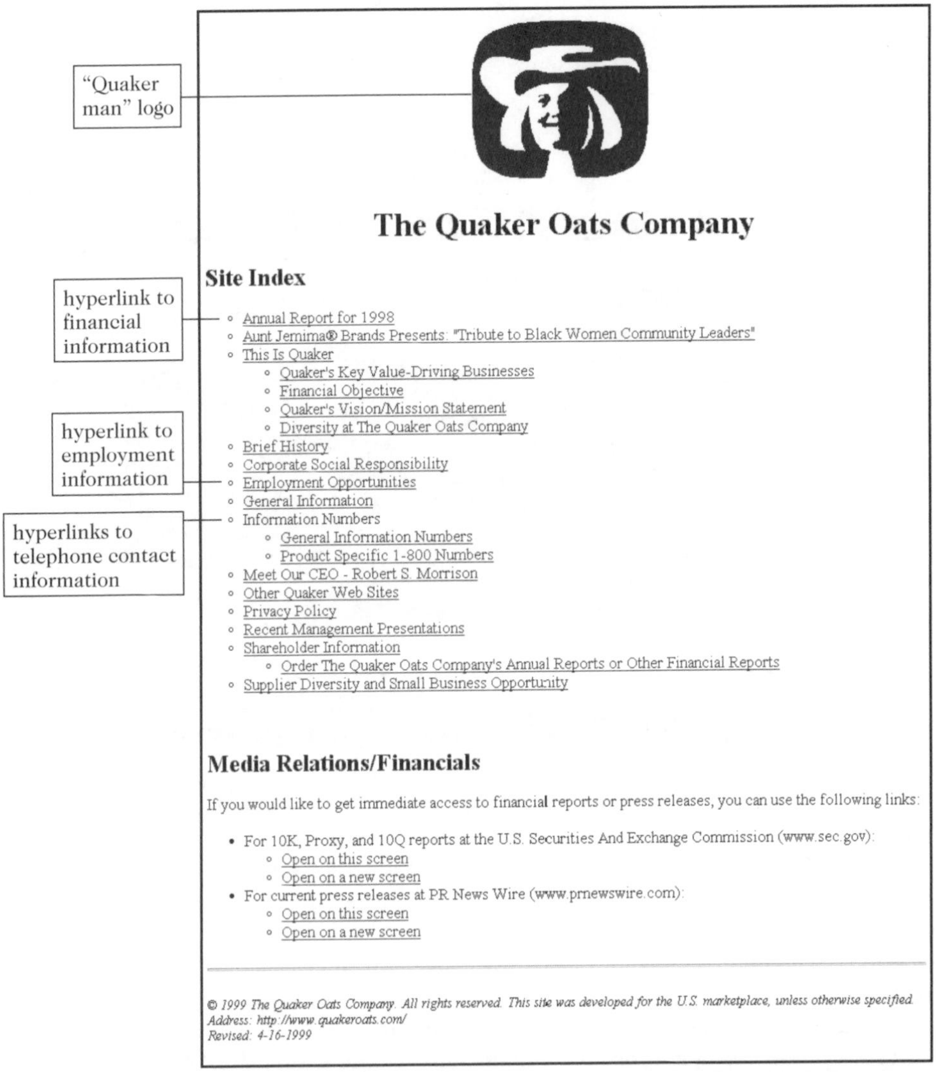
"Quaker man" logo
hyperlink to financial information
hyperlink to employment information
hyperlinks to telephone contact information
The Quaker Oats Company
Site Index
Annual Report for 1998
Aunt Jemima® Brands Presents: "Tribute to Black Women Community Leaders"
This Is Quaker
Quaker's Key Value-Driving Businesses
Financial Objective
Quaker's Vision/Mission Statement
Diversity at The Quaker Oats Company
Brief History
Corporate Social Responsibility
Employment Opportunities
General Information
Information Numbers
General Information Numbers
Product Specific 1-800 Numbers
Meet Our CEO - Robert S. Morrison
Other Quaker Web Sites
Privacy Policy
Recent Management Presentations
Shareholder Information
Order The Quaker Oats Company's Annual Reports or Other Financial Reports
Supplier Diversity and Small Business Opportunity
Media Relations/Financials
If you would like to get immediate access to financial reports or press releases, you can use the following links:
For 10K, Proxy, and 10Q reports at the U.S. Securities And Exchange Commission (www.sec.gov):
Open on this screen
Open on a new screen
For current press releases at PR News Wire (www.prnewswire.com):
Open on this screen
Open on a new screen
© 1999 The Quaker Oats Company. All rights reserved. This site was developed for the U.S. marketplace, unless otherwise specified.
Address: http://www.quakeroats.com/
Revised: 4-16-1999

Figure 3-14 *Quaker Oats old home page*

The site is a straightforward presentation of links to information about the firm. This page includes 24 links to financial information, employment opportunities, current press releases, and other information about the company. It includes links to contact information for the firm. It even has the corporate "Quaker man" logo. Although this Quaker Oats site provides access to much of the same information as the Toyota site, it offers the visitor a completely different experience and impression. In 1999, Quaker changed its Web page to include some pictures of its products and to improve its general appearance and user-friendliness. The overall impression of this new site is much more lively and fun than the original site. The updated Quaker Oats home page appears in Figure 3-15.

"Quaker man" logo

hyperlinks to telephone contact information

hyerlink to financial information

hyperlink to employment information

Figure 3-15   *Quaker Oats new home page*

Although the new site is more colorful and interesting, the basic information offered is essentially the same as that offered by the previous version. The new site offers this information more efficiently, using fewer links. In both the old and the new sites, Quaker Oats uses versions of its "Quaker man" logo to convey a highly recognizable image that is strongly associated with its company. Even though the new site is the result of a major overhaul, Quaker decided to include this image because it conveys the Quaker brand so well.

Not-for-Profit Organizations

The Toyota and Quaker Oats examples illustrate that the Web can integrate an opportunity for enhancing the image of a business with the provision of information. For some organizations, this integrated image-enhancement capability is a key goal of their Web presence efforts. Not-for-profit organizations are an excellent example of this. They can use their Web sites as a central resource for integrated communications with their varied and often geographically dispersed constituencies.

A key goal for many not-for-profit organizations is information dissemination. The Web allows them to integrate information dissemination with fund-raising in one location. Visitors who become engaged in the issues presented are usually just one or two clicks away from a page offering memberships or other opportunities to donate using a credit card. Web pages also provide a two-way contact channel with persons engaged in the organization's work who are not working directly for the organization—for example, many not-for-profits rely on volunteers and coordination with other organizations to accomplish their goals. This combination of information dissemination and a two-way contact channel is a key element in any successful electronic commerce Web site. Interestingly, not-for-profit organizations are ahead of many businesses in accomplishing this combination of elements in their Web presences. Figure 3-16 shows the home page of the **American Civil Liberties Union** (**ACLU**), which is devoted to the advocacy of individual rights in the United States.

Figure 3-16   *American Civil Liberties Union home page*

This page allows interested visitors to learn more about the ACLU and to join the organization if their interest is piqued by what they see. The Feedback link at the bottom of the page leads to a form that visitors can use to report a civil liberties violation, obtain assistance with legal research, ask questions about ACLU membership, or request permission to reprint ACLU publications. The page contains two links to a page that allows individuals to join the ACLU (and thus make a contribution to the organization).

The ACLU Web site is especially valuable for the ACLU because the organization serves many different constituencies, not all of whom agree with the ACLU or with each other on specific issues. If the ACLU were to create a print newsletter that contained interesting information for some of its supporters, that same information might offend other supporters. The links at the right side of the ACLU home page allow site visitors to select the issues in which they are interested—and only those issues.

Not-for-profit organizations can use the Web to stay in touch with existing stakeholders and identify new opportunities for serving them. Organizations as diverse as the **American Red Cross**, the **California Voter Foundation**, the **Public Broadcasting System**, and the **Union of Concerned Scientists** have created effective Web presences. Organizations such as **Amnesty International** and the **United Nations** also use the Web to create international communities of interested persons.

Political parties want to offer information about party positions on issues, recruit members, keep existing members informed, and provide communication links to visitors who have questions about the party. All the major United States political parties have Web sites, including the **Democratic**, **Green**, **Libertarian**, **Reform**, **Republican**, and **Socialist** parties. In addition, political organizations that are not affiliated with a specific party, such as the **Non-Partisan Center for Responsive Politics**, also accomplish similar goals with their Web presences.

Museums are finding that the Web offers a good way to introduce interested visitors to their collections and create a presence that is congruent with their image. The **Museum of Modern Art** (**MoMA**) in New York has created a page, which appears in Figure 3-17, that features a clean and functional design—two tenets of the modern art movement to which it is devoted.

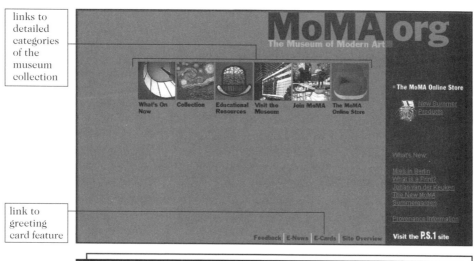

Figure 3-17   *Museum of Modern Art home page*

The MoMA page includes a directory of the entire museum collection, and, on the right side of the page, links to items offered by the museum's online store. The page includes a link to an E-Cards site feature that lets visitors send electronic cards with MoMA artwork images. The page succeeds at engaging the browsing visitor on several levels, yet provides clear paths to museum resources for visitors who know what they are seeking when they access the Web site.

How the Web Is Different

When firms first started creating Web sites in the mid-1990s, they often built simple sites that conveyed basic information about their businesses. Few firms conducted any market research to see what kinds of things potential visitors might want to obtain from these Web sites, and even fewer considered what business infrastructure adjustments would be needed to service the site. For example, few firms had e-mail address links on their sites. Those firms that did include an e-mail link often understaffed the department responsible for answering visitors' e-mail messages. Thus, many site visitors sent e-mail messages that were never answered.

This failure to understand how the Web is different from other presence-building media is one reason that so many businesses fail to achieve their Web objectives. To learn more about this issue, see Jakob Nielsen's **Failure of Corporate Websites** page in the Online Companion—an article that was written in 1998 but still accurately describes far too many Web sites.

Most Web sites that are designed to create an organization's presence in the Web medium include links to a fairly standard information set. The site should give the visitor easy access to the organization's history, a statement of objectives or mission statement, information about products or services, financial information, and a way to communicate with the organization. Sites achieve varying levels of success based largely on how they offer this information. Presentation is important, but so is realizing that the Web is an interactive medium.

A number of Web designers and consultants have taken firms to task for their uninspired use of the Web's interactive nature. Some of these criticisms appear in the print media, but many appear in online newsletters or e-zines. For example, Christopher Locke publishes **Entropy Gradient Reversals**, an e-zine that regularly discusses issues surrounding Web site design. He argues that large corporations should encourage their employees to engage in unrestricted online dialog with the firm's customers, suppliers, and other stakeholders. He believes that this dialog would be in the spirit of the Internet and would help companies create more honest and real personalities as part of their Web presences.

David Weinberger presents similar arguments in his online **Journal of the Hyperlinked Organization**. With two other consultants, Locke and Weinberger have published the **Cluetrain Manifesto** Web site, which presents 95 theses directed at major businesses and other organizations that use the Web. These theses include some short, direct statements, such as "Markets are conversations" and "Hyperlinks subvert hierarchy," and other more lengthy bits of advice for Web presence creators. The authors of the Web site have published a book, also titled *The Cluetrain Manifesto*, which includes additional material that does not appear on the Web site. The Cluetrain Manifesto Web site has generated considerable discussion and has been electronically "signed" by hundreds of visitors to the site, many of whom have added comments.

The main point that these consultants make is that large firms must acknowledge and use the Web's capability for two-way, meaningful communication with their customers. They further argue that use of this communication process is not optional; companies that fail to communicate effectively through this channel will lose customers to competitors that do.

Meeting the Needs of Web Site Visitors

Businesses that are successful on the Web realize that every visitor to their Web site is a potential customer. Thus, an important concern for businesses crafting a Web presence is the variation in important visitor characteristics. People who visit a Web site seldom arrive by accident; they are there for a reason.

Many Motivations of Web Site Visitors

Unfortunately, for the Web designer trying to make a site that will be useful for everyone, visitors arrive for many different reasons, including these:

- Learning about products or services that the company offers

- Buying the products or services that the company offers
- Obtaining information about warranty, service, or repair policies for products they have purchased
- Obtaining general information about the company or organization
- Obtaining financial information for making an investment or credit-granting decision
- Identifying the people who manage the company or organization
- Obtaining contact information for a person or department in the organization

Creating a Web site that meets the needs of visitors with such a wide range of motivations can be challenging. Not only do Web site visitors arrive with different needs that they hope to meet, they arrive with different experience and expectation levels. In addition to the problems posed by the diversity of visitor characteristics, technology issues can also arise. These Web site visitors will be connected to the Internet through a variety of communication channels that provide different bandwidths and data transmission speeds. They will also be using several different Web browsers. Even those who are using the same browser can have a variety of configurations. The wide array of browser add-in and plug-in software adds yet another dimension to visitor variability. Considering and addressing the implications of these many variations in visitor characteristics when building a Web site can help convert these visitors into customers.

Making Web Sites Accessible

One of the best ways to accommodate a broad range of visitor needs is to build flexibility into the Web site's interface. Many sites offer separate versions with and without frames and give visitors the option of choosing either one. Some sites offer a text-only version. As researchers at the **Trace Center** note, this can be an especially important feature for visually impaired visitors who use special browser software, such as the **IBM Home Page Reader**, to access Web site content. The **W3C Web Accessibility Initiative** site includes a number of useful links to information regarding these issues.

If the site uses graphics, it can give the visitor the option to select smaller versions of the images so that the page will load on a low-bandwidth connection in a reasonable amount of time. If the site includes streaming audio or video clips, it can give the visitor the option to specify a connection type so that the streaming media adjusts itself to the bandwidth for that connection.

A good site design lets visitors choose among information attributes, such as level of detail, forms of aggregation, viewing format, and downloading format. Many electronic commerce Web sites give visitors a selectable level of detail by presenting product information by product line. The site presents one page for each line of products. A product line page contains pictures of each item in that product line accompanied by a brief description. By using hyperlinked graphics for the product pictures, the site offers visitors the option of clicking the product picture, which opens a page of detailed specifications for that product.

One of the more controversial developments in Web site design is the use of animated graphics software, such as **Macromedia Flash** or **Adobe LiveMotion**, to create Web pages. These pages (or large portions of the pages) are not rendered in

HTML and are often very large files that take considerable time to download—especially for visitors who do not have broadband connections. Many Web site designers love these products because they provide them with an exciting creative design tool, but few major electronic commerce sites use these types of animated graphical pages. For an interesting discussion of the disadvantages of Flash and similar tools, see WebWord.com's **Flash Usability Challenge** pages. A good solution to the controversy is to give visitors an option to select a Flash version or a non-Flash version of the site on its home page.

Some specific tasks that customers want to perform do lend themselves to animated Web pages. For example, the **Lee FitFinder** is a series of Flash animation pages that can help Lee customers find the right size and style of jeans. One of the Lee Jeans FitFinder animation pages is shown in Figure 3-18.

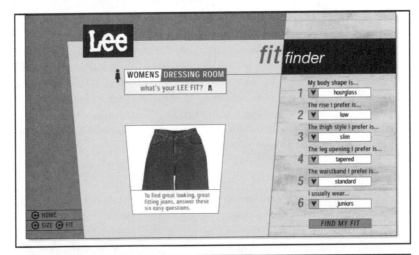

Figure 3-18   *Lee Jeans FitFinder Flash animation*

Web sites can also offer visitors multiple information formats by including links to files in those formats. For example, a page offering financial information could include links to an HTML file, an Adobe PDF file, and an Excel spreadsheet file. Each of these files would contain the same financial information in different formats; visitors can then choose the format that best suits their immediate needs. Visitors looking for a specific financial fact might choose the HTML file so that the information would appear in their Web browsers. Other visitors who want a copy of the entire annual report as it was printed would select the PDF file and either view it in their browsers or download and print the file. Visitors who want to conduct analyses on the financial data would download the spreadsheet file and perform calculations using the data in their own spreadsheet software.

To be successful in conveying an integrated image and in offering information to potential customers, businesses should try to meet the following goals when constructing their Web sites:

- Offer easily accessible facts about the organization
- Allow visitors to experience the site in different ways and at different levels

- Provide visitors with a meaningful, two-way (interactive) communication link with the organization
- Sustain visitor attention and encourage return visits
- Offer easily accessible information about products and services, and how to use them

Trust and Loyalty

When companies first started selling on the Web, many of them believed that their customers would use the abundance of information to find the best prices and disregard other aspects of the buying experience. For some products this may be true; however, most products include an element of service. When customers buy a product, they are also buying that service element. A seller can create value in a relationship with a customer by nurturing customers' trust and developing it into loyalty. Recent studies by business researchers have found that a 5 percent increase in customer loyalty measures (such as proportion of returning customers) can yield profit increases ranging from 25 percent to 80 percent.

Even when products are commodity items, the service element can be a powerful differentiating factor for which customers will pay extra. These services include such things as delivery, order handling, help with selecting a product, and after-sale support. Since many of these services are things that a potential customer cannot evaluate before purchasing a product, the customer must trust the seller to provide an acceptable level of service.

When a customer has an experience with a seller who provides good service, that customer begins to trust the seller. When a customer has multiple good experiences with a seller, that customer feels loyal to the seller. Thus, the repetition of satisfactory service can build customer loyalty that will prevent a customer from seeking out alternative sellers who offer lower prices.

Many companies that have begun doing business on the Web have spent large amounts of money to obtain customers. If they do not provide levels of customer service that lead customers to develop trust in and loyalty to the firm, the companies are unlikely to recover the money they spent to attract the customers in the first place, much less earn a profit.

Customer service is a problem for many electronic commerce sites. Recent research indicates that customers rate most retail electronic commerce sites to be average or low in customer service. A common weak spot for many sites is the lack of integration between the companies' call centers and their Web sites. As a result, when a customer calls with a complaint or problem with a Web purchase, the customer service representative does not have information about Web transactions and is unable to resolve the caller's problem.

A number of studies show that the e-mail responsiveness of electronic commerce sites has also been disappointing. Many major companies are slow to respond to e-mailed inquiries about product information, order status, or after-sale problems. A significant number of companies in these studies never acknowledged or responded to the e-mail queries.

Rating E-Business Web Sites

Two companies routinely review electronic commerce Web sites for usability, customer service, and other factors. Many people have found these review sites to be useful as they decide which sites to patronize. **BizRate.com** provides a comparison shopping service and offers links to sites with low prices and good service ratings for specific products. BizRate.com compiles its ratings by conducting surveys of sites' customers. When the customer places an order, a pop-up window appears asking for a rating of various aspects of his or her experience with the site. The customer is offered a chance at a prize in exchange for providing this information.

Gomez.com provides scorecards for electronic commerce sites in specific categories. Gomez.com computes each site's overall ranking based on assessments of the site's ease of use, customer confidence in the company, resources or features on the site, and quality of customer service. A Gomez.com scorecard for the sporting goods industry appears in Figure 3-19.

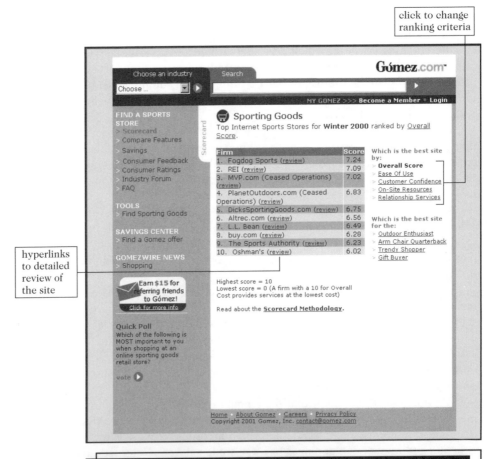

Figure 3-19 *Gomez.com scorecard for Web sporting goods stores*

Web Site Usability

Research indicates that few businesses accomplish all of the goals for a Web site in their current Web presences. Even sites that succeed in achieving most of these goals often fail to provide sufficient interactive contact opportunities for site visitors. Most firms' Web sites give the general impression that the firm is too important and its employees are too busy to respond to inquiries. This is no way to encourage visitors to become customers!

Usability Testing

In general, firms are only now starting to perform **usability testing** on their Web sites. As the practice of usability testing becomes more common, more Web sites will meet the goals outlined previously in this chapter. Companies that have done usability tests, such as **Eastman Kodak**, **T. Rowe Price**, and **Maytag**, have found that they can learn a great deal about meeting visitor needs by conducting focus groups and watching how different customers navigate through a series of Web site test designs. Analysts at **Cambridge Technology Partners** and **Forrester Research** agree that the cost of usability testing is so low compared to the total cost of a Web site design or overhaul that it should almost always be included in such projects. Two pioneers of usability testing are Ben Shneiderman and Jakob Nielsen. Dr. Shneiderman founded the **University of Maryland Human-Computer Interaction Lab** and has published a number of books on interface design. Dr. Nielsen's **Alertbox** Web site includes much information about how to conduct usability testing and use its results to improve Web site design and operation.

Major Web Site Redesigns

T. Rowe Price, a leading mutual fund, asset management, and securities brokerage firm, built its first Web site in 1996. Within two years, the site contained so much information that visitors had a difficult time getting to the data they were seeking. For example, users had to click through five or six pages to get to investor reports. The company rolled out a redesigned Web site in late 1998. One of the design team's goals was to ensure that customers would never need to click through more than two pages to get the information they were seeking. The new design was focused on function rather than visual appeal and included a greater number of links on the home page. Kodak also redesigned its site in response to user feedback. Kodak's redesign resulted in a Web site structure similar to that of T. Rowe Price, with a large number of links. Kodak has continued to improve its Web page; a recent version of the page appears in Figure 3-20.

links to free
picture
taking and
editing
advice

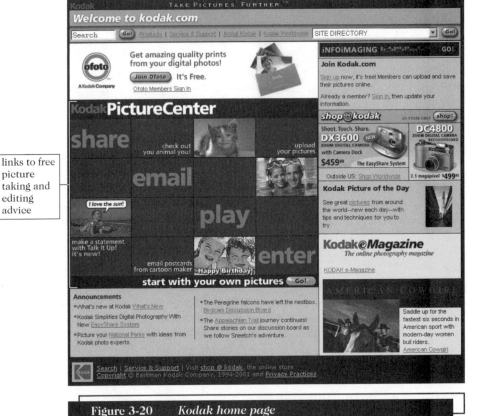

Figure 3-20 *Kodak home page*

Both Kodak and T. Rowe Price sell products that have a large intangible element. The benefits of high-quality photographs and sound investment plans are complex and can be hard for consumers to evaluate. Thus, both the Kodak and T. Rowe Price Web sites offer visitors a large number of links on their home pages so that visitors can explore and identify benefits in a different way.

In contrast to the experience of Kodak and T. Rowe Price, Maytag's usability research found that customers wanted fewer links on its home page. The tangible nature of Maytag's main product line is different from T. Rowe Price's intangible financial products and the intangible benefits of Kodak's cameras and photo finishing products. The new customer-oriented Maytag Web home page includes three main sections, one for each of the primary functions that Maytag's research showed site visitors wanted to perform: learning about Maytag products; getting help with a task that involves a Maytag product; and obtaining customer service. Each section has a drop-down list of links to detailed information in each category. The page also has two links below the main section of the page: One link leads to information about product innovations, and the other link leads to Maytag's online shopping service with delivery provided by local retail partners. The company's famous trademark "Maytag repair man" also appears in the lower-right corner of the page, as shown in Figure 3-21.

Figure 3-21 *Maytag home page*

Customer-Centric Web Site Design

A common theme in these site redesigns is the attention paid to creating a site that meets the needs of potential site visitors. Putting the customer at the center of all site designs is called a **customer-centric** approach to Web site design. A customer-centric approach implies some guidelines that Web designers can follow when creating any type of Web site. These guidelines include the following:

- Design the site around how visitors will navigate the links, not around the company's organizational structure
- Allow visitors to access information quickly
- Avoid using inflated marketing statements in product or service descriptions
- Avoid using business jargon and terms that visitors might not understand
- Build the site to work for visitors who are using the oldest browser software on the oldest computer connected through the lowest bandwidth connection—even if this means creating multiple versions of Web pages
- Be consistent in use of design features and colors
- Make sure that navigation controls are clearly labeled or otherwise recognizable
- Test text visibility on smaller monitors

- Check to make sure that color combinations do not impair viewing clarity for color-blind visitors
- Conduct usability tests by having potential site users navigate through several versions of the site

Web marketing consultant Kristin Zhivago of **Zhivago Marketing Partners** has a number of recommendations for Web sites that are designed for electronic commerce. She encourages Web designers to create sites focused on the customer's buying process rather than the company's perspective and organization. For example, she suggests that companies examine how much information their Web site provides and how useful that information is for customers. If the site does not provide substantial "content for your click" to visitors, they will not become customers.

Using these guidelines when you create your site can help make visitors' Web experiences more efficient, effective, and memorable—in a positive way. Usability is an important element of creating an effective Web presence. For an interesting look at Web design issues, you can visit the **Webby Awards** site. The Webby Awards are given to sites that "exemplify the kinds of sites that Internet users should visit every day for information and entertainment" as judged by a panel of Web designers, journalists, and industry leaders.

CONNECTING WITH CUSTOMERS

An important element of corporate Web presence is connecting with site visitors who are customers or potential customers. In this section, you will learn how a Web site can help firms identify and reach out to customers.

The Nature of Communication on the Web

Most businesses are familiar with two general ways of identifying and reaching customers: personal contact and mass media. These two approaches are often called **communication modes** because they each involve a characteristic way (or mode) of conveying information from one person to another (or communicating). In the **personal contact** model, the firm's employees individually search for, qualify, and contact potential customers. This personal contact approach to identifying and reaching customers is sometimes called **prospecting**. In the **mass media** approach, firms prepare advertising and promotional materials about the firm and its products or services. They then deliver these messages to potential customers by broadcasting them on television or radio, printing them in newspapers or magazines, posting them on highway billboards, or mailing them.

Some experts distinguish between broadcast media and addressable media. **Addressable media** are advertising efforts directed to a known addressee and include direct mail, telephone calls, and e-mail. Since few users of addressable media actually use address information in their advertising strategies, in this book we consider addressable media to be mass media. Many businesses use a combination of mass media and personal contact to identify and reach customers. For example, Prudential uses mass media to create and maintain the public's general awareness of its insurance products and reputation, while its salespersons use prospecting techniques to identify potential customers. Once an individual becomes a customer, Prudential maintains contact through a combination of personal contact and mailings.

The Internet is not a mass medium, even though a large number of people now use it and many companies seem to view their Web sites as billboards or broadcasts. Nor is the Internet a personal contact tool, although it can provide individuals the convenience of making personal contacts through e-mail and newsgroups. Jeff Bezos, founder of Amazon.com, has described the Web as the ideal tool for reaching what he calls "the hard middle"—markets that are too small to justify a mass media campaign, yet too large to cover using personal contact. Figure 3-22 illustrates the position of the Web as a customer-contact medium, between the large markets addressed by mass media and the highly focused markets addressed by personal contact selling and promotion techniques.

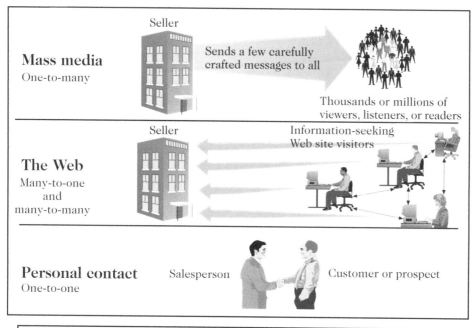

Mass media
One-to-many

Seller

Sends a few carefully crafted messages to all

Thousands or millions of viewers, listeners, or readers

The Web
Many-to-one
and
many-to-many

Seller

Information-seeking
Web site visitors

Personal contact
One-to-one

Salesperson

Customer or prospect

Figure 3-22   *Business communication modes*

To help you better understand the differences shown in Figure 3-22, read the following scenario. The scenario assumes that you have heard about a new book, but would like to learn more about it before buying it. Consider how your information acquisition process would vary, depending on the medium you used to gather the information.

- *Mass media.* You might have been exposed to general promotional messages from book publishers who have created impressions about quality associated with particular book brands. If your existing knowledge includes a brand identity for the book's publisher, these messages may influence your perceptions of the book. You may have been exposed to an ad for the title on television or radio, or in print. You may have heard the book's author interviewed on a radio program or might have read a review of the book in a publication such as *The New York Times Book*

Review or *Booklist* magazine. Notice that most of these process elements involve you as a passive recipient of information. This communication channel is labeled "Mass media" and appears at the top of Figure 3-22. Communication in this model flows from one advertiser to many potential buyers and thus is called a **one-to-many communication model**. The defining characteristic of the mass media promotion process is that the seller is active and the buyer is passive.

- *Personal contact.* Small-value items are not frequently sold through this medium because the costs of devoting a salesperson's efforts to a small sale are prohibitive. However, in the case of books, local bookshop owners and employees often devote considerable time and resources to developing a close relationship with their customers. Although each individual book sale is a small-value transaction, people who frequent local bookshops tend to buy large numbers of books over time. Thus, the bookseller's investment in developing personal contacts is often rewarded. In this scenario, you may visit your local bookshop and strike up a conversation with a knowledgeable bookseller. In the personal contact model, this would most likely be a bookseller with whom you have already established a relationship. The bookseller would offer an opinion on the book based on having read that book, books by the same author, or reviews of the book. This opinion would be expressed as part of a two-way conversational interchange. This interchange usually includes a number of conversational elements, such as discussions about the weather, local sports, or politics, that are not directly related to the transaction you are considering. These other interchanges are part of the trust-building and trust-maintaining activities that you undertake to develop the relationship element of the personal contact model. The underlying **one-to-one communication model** appears at the bottom of Figure 3-22 and is labeled "Personal contact." The defining characteristic of information gathering in the personal contact model is the wide-ranging interchange that occurs within the framework of an existing trust relationship. Both the buyer and the seller (or the seller's representative) actively participate in this exchange of information.

- *The Web.* To obtain information about a book on the Web, you could search for Web site references to the book, the author, or the subject of the book. You would likely identify a number of Web sites that offered such information. These sites might include those of the book's publisher, firms that sell books on the Web, independent book reviews, or discussion groups focused on the book's author or genre. **The New York Times Book Review** and **Booklist** magazine, both staples of mass media book promotion, now have online Web editions. Book review sites that did not originate in a print edition, such as the **BookBrowser**, have also appeared on the Web. Most Web-based booksellers maintain searchable space on their sites for readers to post reviews and comments about specific titles. If the author of the book is famous, there might even be independent Web fan sites devoted to him or her. If the book is about a notable person, incident, or time period, you might find Web sites devoted to those notable topics that include reviews of books related to the topic. You could examine any number of these resources to any extent you desired. You might encounter some advertising material created by the publisher while searching the Web. However, if you choose not to view the publisher's ads, you will find it as easy to click the Back button on your Web browser as it is to surf television channels with your remote control. The Web affords you many communication channels. Figure 3-22 shows only one of the communication models that can occur when using the Web to search for product information. The model labeled "The Web" in Figure 3-22 is the **many-to-one communication model**. The Web gives you the flexibility to use a one-to-one model (as in the personal contact model) in which you communicate over the Web with an individual working for the seller, or even to engage in **many-to-many communications** with other potential buyers. The defining characteristic of a product information search on the Web is that the buyer actively participates in the search and controls the length, depth, and scope of the search.

Summary

Businesses are using six main types of revenue-generating models on the Web: the Web catalog, information sales, advertising-supported, advertising-subscription mixed, fee-for-transaction, and fee-for-services models. You learned how these models work and what kinds of businesses use which models.

By understanding how the Web differs from other media and by designing a Web site to capitalize on those differences, companies can create an effective Web presence that delivers value to visitors. Every organization must realize that visitors to its Web site arrive with a variety of expectations, prior knowledge, and skill levels, and are connected to the Internet through different technologies. Knowing how these factors can affect the visitor's ability to navigate the site and extract information from the site can help organizations design better, more usable Web sites. Enlisting the help of users when building test versions of the Web site is also a good way to create a Web site that represents the organization well.

Firms must understand the nature of communication on the Web so they can use it to identify and reach the largest possible number of customers and qualified prospects. Using a many-to-one communications model enables Web sites to effectively reach potential customers.

Key Terms

Account aggregation
Advertising-subscription mixed revenue model
Advertising-supported revenue model
Addressable media
Bill presentment
Cannibalization
Catalog model
Channel conflict
Clipping service
Communication modes
Customer-centric
Demographic information
Disintermediation
E-zine
Fee-for-service revenue model
Fee-for-transaction revenue model
Mail order model
Many-to-many communications

Many-to-one communication model
Mass media
One-to-many communication model
One-to-one communication model
Personal contact
Portal (Web portal)
Presence
Prospecting
Reintermediation
Revenue model
Stakeholders
Stickiness
Sticky
Strategic alliance
Usability testing
Web catalog model
Web directory

Review Questions

1. Write a paragraph in which you describe the conditions under which a Web site can hope to become profitable if it relies exclusively on advertising revenue. In a second paragraph, provide an example of a company that is using the advertising-supported model and that is likely to be successful in the long run.

2. Describe two possible service-for-fee offerings that might become available on the Web. Write one paragraph for each service in which you outline the profit potential and risk of losses for each.

3. In one paragraph, define the term *presence*. Write an additional paragraph in which you explain why firms that do business on the Web should be more concerned about presence than firms that operate only in the physical world.

4. Write a brief paragraph in which you describe two Web design mistakes that firms often made when they created their first Web sites in the mid-1990s.

5. In an essay of about 500 words, explain how promoting products on the Web is different from using mass media promotion or personal contact.

Exercises

1. Page 117 includes a list of five things that Web sites can do to meet the needs of visitors. Create a table in which the first column is a list of the five needs of Web site visitors. Find three Web sites that meet three or more of the needs. Create a column for each Web site next to the first column (columns two through four) and rate how well each site meets each need in the first column. You may want to use the **Webby Awards** site as a starting point in your search, but do not use any of the award nominees or winners in your table.

2. You have been employed by Bob Drudge, the owner of **refdesk.com**, to sell space on his site to advertisers. Create a promotional press release of no more than 300 words in which you describe the advantages of advertising on refdesk.com. You may decide to promote space on the main page, other specific pages, or all pages. Be prepared to explain why your promotional strategy should work. You may find the **Art of Web Site Promotion**, **Promotion World**, or **Seltzer's "How to Publicize Your Web Site over the Internet"** Online Companion links helpful in your task.

3. A number of occupations are exposed to the risk of disintermediation posed by firms moving activities to the Web. Four of these occupations are: *auto salesperson*, *stockbroker*, *insurance agent*, and *travel agent*. For each occupation, write one paragraph in which you explain whether the employees will be able to find jobs in the new Web-based business activity and explain what those jobs would be. You may want to use **About.com** or your favorite Web search engine to find information on this topic.

For Further Study and Research

Baard, M. 2000. "Interacting With Customers," *Publish*, December, 50–52.

Barron, K. 1999. "Realtors 1, Web 0," *Forbes*, 163(3), February 8, 64–65.

Barsh, J., E. Kramer, D. Maue, and N. Zuckerman. 2001. "Magazines' Home Companion," *The McKinsey Quarterly*, April, 83–91.

Berner, R. 2000. "Going that Extra Inch," *Business Week*, September 18, 84.

Biggs, M. 2000. "E-Commerce Success Requires Commitment to Building a Proper Customer Community," *InfoWorld*, 22(14), April 3, 68.

Brown, E. and J. Fox. 1999. "Nine Ways to Win on the Web," *Fortune*, 139(10), May 24, 112–121.

Buckman, R. 1999. "Wall Street Is Rocked by Merrill's Online Plans," *The Wall Street Journal*, June 2, C1.

Byrnes, N. 2000. "Avon: The New Calling," *Business Week*, September 18, 136–143.

Carr, N. 2000. "Hypermediation: Commerce as Clickstream," *Harvard Business Review*, 78(1), January–February, 46–47.

Cavillini, S. 2001. "Stocks That Work: Job-Hunting Sites Are Among the Select Dot-Coms That Haven't Plunged," *The Industry Standard*, August 20–27, 78.

Chak, A. 2001. "Design for Buyers," *CNET Builder.com*, April 20. Available online at: (http://www.zdnet.com/ecommerce/stories/main/0,10475,2710578-1,00.html).

Champy, J. 1998. "Direct Sales or Electronic Channels? Give the Customer a Choice," *Computerworld*, 32(47), November 23, 64.

Chan, S. 2001. "Usability Guru Philosophizes on Web Subjects," *The Seattle Times*, April 7, E1.

Charski, M. 2001. "Games Could Be Good Bet for Advertisers," *Interactive Week*, 7(52), December 25, 14.

Christensen, C. and M. Overdorf. 2000. "Meeting the Challenge of Disruptive Change," *Harvard Business Review*, 78(2), March–April, 66–75.

Cox, R. 2001. "Wanted: E-recruits," *PC Magazine*, 20(9), May 8, 72.

Cuneo, A. 1999. "Lands' End Ads Pitch 'Ultimate Direct Merchant,'" *Advertising Age*, 70(17), April 19, 85.

Dahm, T. 2001. "Five Steps to Accessible Web Sites," *E-Commerce News*, June 29. Available online at: (http://www.zdnet.com/ecommerce/stories/main/0,10475,2781418,00.html).

Daly, J. 2000. "Sage Advice: Interview with Peter Drucker," *Business 2.0*, August 22, 134–144.

The Economist. 2000. "Shopping Around the Web," 354(8159), February 26, 5–6.

Eisenmann, T. 2002. *Internet Business Models*. New York: McGraw-Hill.

Evans, P. and T. Wurster. 1997. "Strategy and the New Economics of Information," *Harvard Business Review*, 75(5), September–October, 71–83.

Greco, S. 2001. "Fanatics!" *Inc. Magazine*, 23(5), April 1, 36–43.

Greenspun, P. 1999. *Philip and Alex's Guide to Web Publishing*. San Francisco: Morgan Kaufmann.

Gunderson, A. 2000. "Double, Double...Toil and Trouble: Online Shopping Is Still a Muddle," *Fortune*, September 4, 374.

Hagen, P., H. Manning, and Y. Paul. 2000. *Must Search Stink?* Cambridge, MA: Forrester Research. Available online at: (http://www.forrester.com/ER/Research/Report/0,1338,9412,FF.html).

Hamilton, A. 2001. "Net Addictions," *Time*, 157(12), March 26, 76.

Hicks, M. 1999. "Sold on the Simplicity of Web Sites," *PC Week*, June 7, 77.

Hill, A. 2001. "Stop Shopping Cart Abandonment: Top Five Reasons Customers Abandon Shopping Carts," *Smart Business*, February 13. Available online at: http://www.zdnet.com/smartbusinessmag/stories/all/0,6605,2677306,00.html).

Hogan, M. 1999. "The Internet Business 100," *PC Computing*, 12(7), July, 145–161.

Kaufman, L. and C. Deutsch. 2000. "Montgomery Ward to Close Its Doors," *The New York Times*, December 29, 1.

Kelsey, D. 2000. "Top E-Tail Web Sites' Customer Service Not Good," *BizReport*, August 8. Available online at: (http://www.bizreport.com/research/2000/08/20000809-1.htm).

Kemp, T. 2000. "Wal-Mart No Web Mart," *InternetWeek*, October 9, 1–2.

Kemp, T. 2000. "Lands' Ends Stays in Online Fashion," *InternetWeek*, December 18, 32.

Kling, J. 1999. "Seven Great Ways to Use Your Company's Web Site," *Harvard Management Update*, January, 3–4.

Lake, D. 2001. "Quick and Easy," *The Industry Standard*, March 5. Available online at: (http://www.thestandard.com/article/0,1902,22342,00.html).

Lee, L. 2000. "An E-Commerce Cautionary Tale," *Business Week*, March 20, 46–47.

Levine, R., C. Locke, D. Searle, and D. Weinberger. 2000. *The Cluetrain Manifesto: The End of Business as Usual*. Cambridge, MA: Perseus.

Locke, C. 1998. "Fear and Loathing on the Web," *The Industry Standard*, July 9.

Locke, C. 2000. "Smart Customers, Dumb Companies," *Harvard Business Review*, 78(6), November–December, 187–191.

Lorek, L. 2001. "Wrecks Pile Up in Online Auto Sales Sector," *Interactive Week*, April 2. Available online at: (http://www.zdnet.com/ecommerce/stories/main/1,10475,2702424,00.html).

Lorek, L. 2001. "Orbitz Takes Off," *Interactive Week*, 8(21), May 28, 16–19.

Managing Intellectual Property. 1999. "Developments," June, 90.

Manjoo, F. 2001. "Salon: Last One Standing," *Wired News*, June 15. Available online at: (http://www.wired.com/news/ebiz/0,1272,44464,00.html).

McHugh, J. 2001. "A Healthy Niche Market," *Forbes*, 167(5), Spring, 36–37.

McLaughlin, K. 2001. "Real Makes Gaming Play: Hopes Its Customer Base Will Pay for Online Diversions," *Business 2.0*, March 21. Available online at: (http://www.business2.com/ebusiness/2001/03/28783.htm).

Mowry, M. 2000. "Net Retail's Grim Reality," *The Industry Standard*, August 7, 154–155.

Mullaney, T. 2001. "Orbitz Doesn't Soar," *Business Week*, July 9. Available online at: (http://www.businessweek.com/magazine/content/01_29_/b3740621.htm).

Nash, K. 2000. "Little E-Engines That Could," *Computerworld*, 34(16), April 17, 46–47.

Neilsen, J. 1999. *Designing Websites With Authority: Secrets of an Information Architect*. Indianapolis: New Riders.

Neilsen, J. 2000. "End of Web Design," *Alertbox*, July 23. Available online at: (http://www.useit.com/alertbox/20000723.html).

Neilsen, J. 2000. "Flash: 99% Bad," *Alertbox*, October 29. Available online at: (http://www.useit.com/alertbox/20001029.html).

Neilsen, J. 2001. "Usability Metrics," *Alertbox*, January 21. Available online at: (http://www.useit.com/alertbox/20010121.html).

Neuborne, E. 2001. "Bridging the Loyalty Gap," *Business Week*, January 22, EB10.

Nielsen, J., K. Coyne, and M. Tahir. 2001. "Make It Usable," *PC Magazine*, 20(3), February 6, IPO1–IPO6.

Nunes, P., D. Wilson, and A. Kambil. 2000. " The All-in-One Market," *Harvard Business Review*, 78(3), May–June, 19–20.

Oh, J. 2001. "Looking Good," *The Industry Standard*, March 5. Available online at: (http://www.thestandard.com/article/0,1902,22337,00.html).

Oliva, R. 2000. "Brainstorm Your E-Business," *Marketing Management*, 9(1), Spring, 55–57.

Opdyke, J. 2001. "Tools to Trade By: Technology Is Giving Investors the Ability to Do All Kinds of New Things," *The Wall Street Journal*, June 11, R8.

Oreskovic, A. 2001. "Testing 1-2-3," *The Industry Standard*, March 5. Available online at: (http://www.thestandard.com/article/0,1902,22338,00.html).

Pack, T. 2001. "Use It or Lose It," *eContent*, June 1, 44.

Pollack, S. 2001. *Intelligent Customer Service: Improving Your Online Customer Interactions*. Multimedia Live White Paper. Petaluma, CA: Multimedia Live. Available online at: (http://www.mmlive.com/).

Pomerantz, D. 2001. "Shop 'Til You Flop," *Forbes*, 167(2), January 22, 56.

Ramsey, C. 2000. "Managing Web Sites as Dynamic Business Applications," *Intranet Design Magazine*, June. Available online at: (http://idm.internet.com/articles/200006/wm_index.html).

Rayport, J. and J. Sviokla. 1995. "Exploiting the Virtual Value Chain," *Harvard Business Review*, 73(6), November–December, 75–85.

Reed, S. 1999. "Can't Get No Satisfaction: Online Shopping Can Be as Bad as the Real Thing," *InfoWorld*, 21(21), May 24, 83.

Ross, P. 1999. "Web Winners," *Forbes*, 163(8), April 19, 226–228.

Roth, A. 2001. "Dimon Kills Wingspan; Cost of Brand Is Cited," *American Banker*, 166(125), June 29, 1–2.

Schaffer, E. 2000. "A Better Way for Web Design," *InformationWeek*, May 1, 194.

Schroeder, C. 2001. "Content Sites Will Work on the Web," *The Wall Street Journal*, April 16, A18.

Schwartz, E. 1999. *Digital Darwinism*. New York: Broadway Books.

Schwartz, E. 1997. *Webonomics*. New York: Broadway Books.

Shelton, B. 1999. "Building Customer Loyalty on the Web," 9(3), March 1, 24–29.

Shneiderman, B. 1997. *Designing the User Interface: Strategies for Effective Human-Computer Interaction*. Reading, MA: Addison-Wesley.

Sklar, J. 2000. *Principles of Web Design*. Cambridge, MA: Course Technology.

Slatalla, M. 2000. "Wonder How You'd Look in Yellow Eye Shadow?" *The New York Times*, August 18, G-4.

Surmacz, J. 2001. "Big Pipe Dreams: Evian Uses Broadband to Create a Web Site That Fits Its Image," *CIO E-Business Research Center*, January 31. Available online at: (http://www.cio.com/forums/ec/edit/013101_evian.html).

Sweat, J. 2001. "Well-Tailored E-Commerce," *InformationWeek*, April 16, 49–51.

Technology Review. 1998. "That Mess on Your Web Site," 101(5), September–October, 72–75.

Tedeschi, B. 2000. "Charitable Groups Discover New Revenue in Retailing Goods via Their Own Web Sites," *The New York Times*, March 27, C-11.

Tedeschi, B. 2000. "Retailers Are Letting their Right Hand Know What the Left Hand Is Up To for Better Customer Service," *The New York Times*, July 24, C-9.

Vickery, L. 2001. "Keeping the Cachet: Luxury-Goods Makers Want to Jump into the Chaos of Cyberspace Without Sullying Their Images," *The Wall Street Journal*, April 23, R28.

Waldman, A. 2000. "Vanguard Preps Site Overhaul," *Financial NetNews*, 5(46), November 13, 1–2.

Walsh, J. 1999. "Is Your Site Really Working?" *InfoWorld*, 21(10), March 8, 53, 56.

Walsh, T. 2000. "Four Wheel Drive," *Business 2.0*, August 22, 151–164.

Weinberger, D. 1999. "0:1 Marketing," *Journal of Hyperlinked Organizations*, May 20. Available online at: (http://www.hyperorg.com/backissues/joho-may20-99.html#01).

Weingarten, M. 2001. "Flash Backlash," *The Industry Standard*, March 5. Available online at: (http://www.thestandard.com/article/0,1902,22330,00.html).

Weiss, T. 2000. "Walmart.com Back Online After Four-Week Overhaul," *Computerworld*, 34(45), November 6, 24.

Wood, C. 2001. "Collusion in the Air," *PC Magazine*, 20(9), May 8, 199.

130

Yudkin, M. 2000. "Usability Matters," *Publish*, December. Available online at: (http://www.publish.com/features/0012/feature7.html).

Yudkin, M. 2001. "The Web's Pipsqueaks Stand Tall," *Business 2.0*, May 1. Available online at: (http://www.business2.com/magazine/2001/05/webs_pipsqueaks.htm).

Zellner, W. 2001. "Where the Net Delivers: Travel," *Business Week*, June 11. Available online at: (http://www.businessweek.com/magazine/content/01_24/b3736111.htm).

Zimmerman, A. 2000. "Wal-Mart Launches Web Site for a Third Time, This Time Emphasizing Speed and Ease," *The Wall Street Journal*, October 31, B12.

MARKETING ON
THE WEB

INTRODUCTION

In September 1997, a new gift shop opened for business on the Web. Of course, in 1997, there were already many gift shops on the Web. However, this store, named 911Gifts.com, carried items that were chosen specifically to meet the needs of last-minute gift shoppers. Using 911—the emergency telephone number used in most parts of the United States—in the store's name was intended to convey the impression of crisis-solving urgency. The company's two major strengths were its promise of next-day delivery on all items and its site layout, in which gift selections were organized by holiday rather than by product type. Thus, the harried shopper could simply click the Mother's Day gifts link and view a set of gift choices appropriate for that holiday that were ready for immediate delivery. The site also included a reminder service, called GiftAlert, so that its customers would not find themselves in an emergency gift situation on the next holiday.

By 1999, the company had 90,000 customers signed up for GiftAlert and was doing about $1 million in annual sales. It carried about 500 products, and each of the products was chosen to yield a gross margin of at least 40 percent. 911Gifts.com was a successful business, but without additional capital to build the company's brand, future growth would be difficult. To take the company to the next level, the company hired Hilary Billings, a retail marketing executive whose experience included building the

Pottery Barn catalog business at Williams-Sonoma. Billings undertook a complete reevaluation of the 911Gifts.com marketing plan and, after revising it, took it to investors and raised over $30 million for a rebranding and complete overhaul of the company's Web site. In October 1999, the new brand was born as **RedEnvelope**. In many Asian countries, gifts are sent in a simple, red envelope. The new brand was designed to create a sense of elegant simplicity to replace the sense of panic and emergency solutions conveyed by the old brand name.

The product line was revamped to fit the new image as well. About 300 products were dropped and replaced with different products that focus groups had judged to be more appealing. The new product line had an even higher average gross margin than the old line. Billings launched a massive brand-awareness campaign that included online advertising, buses in seven major cities painted red and festooned with large red bows, and print advertising in upscale publications. The most important change in advertising strategy was the launching of a print catalog. RedEnvelope catalogs are mailed to customers to coincide with major gift-giving holidays and serve as additional reminders. Because RedEnvelope sells a small set of products that are chosen for their visual appeal and for the status they are intended to convey, the full-color, lushly illustrated print catalogs are a powerful selling tool.

One year later, the results of this extensive makeover were clear. RedEnvelope had tripled its number of customers and had increased sales by over 400 percent. The company chose a specific part of the gifts market and targeted its offerings to meet the needs and desires of those customers. The company created a brand, a marketing plan, and a set of advertising and promotion strategies that would expose the company to the largest portion of that market it could afford to reach. The important point to remember is that RedEnvelope matched its inventory selection, delivery methods, and marketing efforts to each other and to the needs of its customers.

LEARNING OBJECTIVES

In this chapter, you will learn about:

- When to use product-based and customer-based marketing strategies
- Identifying and communicating with different market segments
- Customer relationship intensity and the customer relationship life cycle
- Using advertising on the Web
- Technology-enabled customer relationship management
- Creating and maintaining brands on the Web
- Affiliate marketing strategies
- Viral marketing strategies
- Search engine positioning
- Domain name selection

133

WEB MARKETING STRATEGIES

In this chapter, you will learn how companies are using the Web in their marketing strategies to advertise their products and services, and to promote their reputations. Increasingly, companies are classifying customers into groups and creating targeted messages for each group. The sizes of these targeted groups can be smaller when companies are using the Web—in some cases, just one customer at a time can be targeted. New research into the behavior of Web site visitors has even suggested ways in which Web sites can respond to visitors who arrive at a site with different needs at different times. This chapter will also introduce you to some of the ways companies are making money by selling advertising on their Web sites.

Most companies use the term **marketing mix** to describe the combination of elements that they use to achieve their goals for selling and promoting their products and services. When a company decides which elements it will use, it calls that particular marketing mix its **marketing strategy**. As you learned in Chapter 3, companies—even those in the same industry—try to create unique presences in their markets. A company's marketing strategy is an important tool that works with its Web presence to get the company's message across to its customers and prospective customers.

Most marketing classes organize the essential issues of marketing into the **four Ps of marketing**: product, price, promotion, and place. **Product** is the physical item or service that a company is selling. The intrinsic characteristics of the product are important, but customers' perceptions of the product, called the product's **brand**, can be as important as the actual characteristics of the product.

The **price** element of the marketing mix is the amount the customer pays for the product. In recent years, marketing experts have argued that companies should think of price in a broader sense, that is, the total of all financial costs that the customer pays (including transaction costs) to obtain the product. This total cost is subtracted from the benefits that a customer derives from the product to yield an estimate of the **customer value** obtained in the transaction. Later in this book, you will learn how the Web can create new opportunities for creative pricing and price negotiations through online auctions, reverse auctions, and group buying strategies. These Web-based opportunities are helping companies find new ways to create increased customer value.

Promotion includes any means of spreading the word about the product. On the Internet, new possibilities abound for communicating with customers and potential customers. In Chapter 2, you learned how companies are using the Internet to engage in meaningful dialogs with their customers using e-mail and other means. In this chapter, you will learn even more communication techniques that companies are using to promote their products.

For years, marketing managers dreamed of a world in which instant deliveries would give all customers exactly what they wanted when they wanted it. The issue of **place** (also called **distribution**) is the need to have products or services available in many different locations. The problem of getting the right products to the right places at the best time to sell them has plagued companies since commerce began. Although the Internet does not solve all of these logistics and distribution problems, it can certainly help. For example, digital products (such as software, music, and e-books) can be delivered almost instantly on demand through the Internet. Companies that sell products that must be shipped have found that the Internet gives them much better shipment tracking and control than did previous information technologies.

Product-Based Marketing Strategies

In Chapter 3, you learned about the importance of a company's Web presence and how this presence must integrate with the brand or other established images the company uses in its promotional activities. Most companies offer a variety of products that appeal to different groups. When creating a marketing strategy, managers must consider both the nature of their products and the nature of their potential customers.

Managers at many companies think of their businesses in terms of the products and services they sell. This is a logical way to think of a business, because companies spend a great deal of effort, time, and money to design and create those products and services. If you ask managers to describe what their companies are selling, they will usually provide you with a detailed list of the physical objects they sell or that they use to create the service. When customers are likely to buy items from particular product categories, or are likely to think of their needs in terms of product categories, this type of product-based organization makes sense. Most office supplies stores on the Web believe their customers organize their needs into product categories. The **Staples** home page shown in Figure 4-1 uses product categories as a very strong organizing theme.

tabs to product categories

lists of product category links

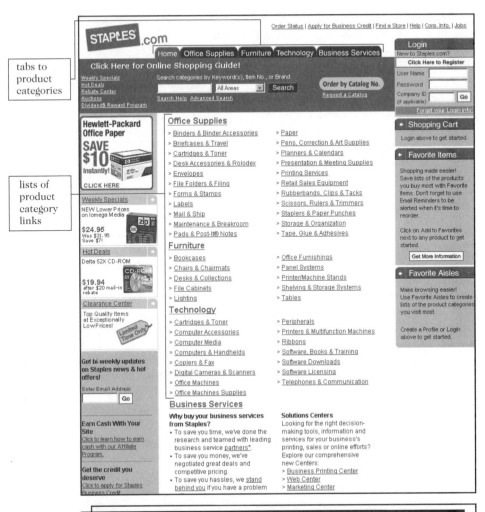

Figure 4-1   *Staples home page*

The Staples page has tabbed headings near the top of the page that are links to product categories. More detailed product category links fill the center of the page. Staples has designed its page to meet the needs of the customer who has a specific product category in mind. Even the search box near the top of the page includes a drop-down list of categories so that customers can narrow their searches within categories.

A company that sells to a very different market, but that uses a similar product-based marketing strategy, is **Service Merchandise**. Service Merchandise was originally a jewelry retailer that expanded into gifts and housewares. The company sold its products in physical stores and through catalogs for many years before opening its Web site. Most companies that used print catalogs in the past organized them by product category. As you can see in Figure 4-2, Service Merchandise has carried over its product-focused marketing strategy to its Web site.

links to
broad
product
categories

product
category
links

Figure 4-2   *Service Merchandise home page*

Both of these companies are using a product-based strategy. They have organized their Web sites from an internal viewpoint, that is, according to the way that they have organized their product design and manufacturing processes. If customers arrive at these Web sites looking for a specific type of product, this approach will work well. Alternatively, customers who are looking to fulfill a specific need, such as outfitting a new sales office or choosing a graduation gift, might not find these Web sites as useful. Many marketing researchers and consultants advise companies to think as if they were their own customers and to design their Web sites so that they become enabling experiences that can help customers meet their individual needs.

Customer-Based Marketing Strategies

In Chapter 3, you learned that the Web creates an environment that allows buyers and sellers to engage in complex communications modes. The communication structures on the Web can become much more complex than those in traditional mass media outlets such as broadcast and print advertising. When a company takes its business to the Web, it can create a Web site that is flexible enough to meet the needs of many different users. Instead of thinking of their Web sites as collections of products, companies can build their sites to meet the specific needs of various types of customers.

A good first step in building a customer-based marketing strategy is to identify groups of customers that share common characteristics. For example, the customers of an office supplies store might include individuals, small businesses, larger companies, and other types of organizations such as schools and government agencies. **Office Depot**, one of Staples' competitors in the office supplies business, has a Web site that directs customers into one of two branches, a Business Solutions Center designed to meet the needs of larger organizations and a product-based branch for everyone else. The Office Depot site appears in Figure 4-3.

branch of site for all other customers

branch of site for large business customers

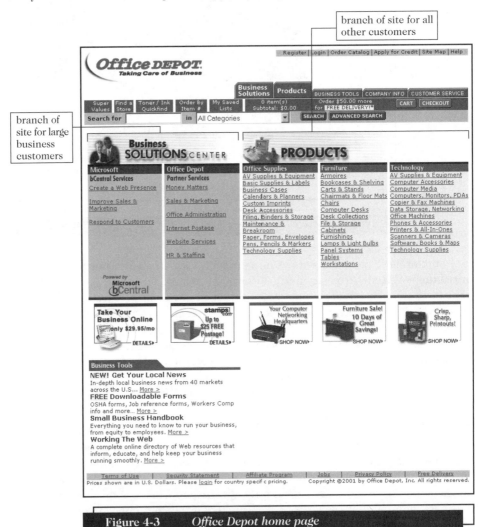

Figure 4-3   *Office Depot home page*

Although Office Depot's approach of breaking customers into two groups—large businesses and everyone else—is a good first step, several subgroups probably exist within each of those groups. Marketers can use their experience with selling in their industry to identify those subgroups and then develop marketing strategies and tactics that will reach customers in each subgroup effectively.

COMMUNICATING WITH DIFFERENT MARKET SEGMENTS

Identifying groups of potential customers is just the first step in selling to those customers. An equally important component of any marketing strategy is the selection of communication media to carry the marketing message.

In the physical world, companies can convey a large part of their message by the way they construct buildings and design their floor space. For example, banks have traditionally been housed in large, solid-looking buildings that provide passersby an ample view of the main safe and its thick, sturdy door. Banks use these physical manifestations of reliability and strength to convey an important part of their service offerings—that a customer's money will be safe and secure with the bank.

Media selection can be critical for an online firm because it does not have a physical presence. The only contact a potential customer might have with an online firm could well be the image it projects through the media and through its Web site. The challenge for online businesses is to convince customers to trust them even though they do not have an immediate physical presence.

Trust and Media Choice

The Web is an intermediate step between mass media and personal contact, but it is a very broad step. Using the Web to communicate with potential customers offers many of the advantages of personal contact selling and many of the cost savings of mass media. Figure 4-4 shows how these three information dissemination models compare on another important dimension: trust.

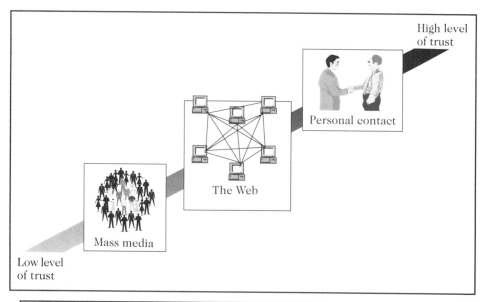

High level of trust

Personal contact

The Web

Mass media

Low level of trust

Figure 4-4   *Trust in three information dissemination models*

Although Mass Media offers the lowest level of trust, many companies continue to use it successfully. The cost of mass media advertising can be spread over the many people in its large audiences. For example, the cost of creating a television ad can be several hundred thousand dollars, but that ad will be viewed by millions of people. Thus, the cost of advertising per viewer is very low. Its low cost makes mass media advertising attractive to many companies.

After years of being barraged by television and radio commercials, many people have developed a resistance to the messages conveyed in the mass media. The impact on an audience of the shouted expression "New and improved!" has diminished to almost nothing in most cases. The overuse of superlatives has caused most people to distrust or ignore the messages contained in mass media. Television remote controls have a mute button and make channel surfing easy for a reason. Attempts to re-create mass media advertising on the Web are doomed to fail for the same reasons—many people will ignore or resist messages that lack content of any specific personal interest to them.

Companies can use the Web to capture some of the benefits of personal contact, yet avoid some of the costs inherent in that approach. Most experts agree that it is better to scale up the trust-based model of personal contact selling to the Web than to scale down the mass marketing approach. In a 1996 marketing report, the **GartnerGroup** concluded that customer-centered marketing strategies would be an excellent fit for the Internet marketplace that was then emerging. The GartnerGroup also noted that rising consumer expectations and reduced product differentiation had led to increased competition and a splintering of mass markets. Both of these results were reducing the effectiveness of mass media advertising.

Market Segmentation

Advertisers' response to this decrease in effectiveness was to identify specific portions of their markets and target them with specific advertising messages. This practice, called **market segmentation**, divides the pool of potential customers into **segments**. Segments are usually defined in terms of demographic characteristics such as age, gender, marital status, income level, and geographic location. Thus, for example, unmarried men between the ages of 19 and 25 might be one market segment.

In the early 1990s, firms began identifying smaller and smaller market segments for specific advertising and promotion efforts. This practice of targeting very small market segments is called **micromarketing**. However, the low cost per viewer of traditional mass media advertising campaigns becomes much higher when those methods are used to target very small market segments. This hampered the success of micromarketing strategies. Even though micromarketing was an improvement over mass media advertising, it still used the same basic approach and suffered from the weaknesses of that model.

Marketers have traditionally used three categories of variables to identify market segments. One variable is location. Firms divide their customers into groups by where they live or work. In this type of segmentation, called **geographic segmentation**, companies create different combinations of marketing efforts for each geographical group of customers. The grouping can be by nation, state (or province), city, or even by neighborhood. Alternatively, companies can develop one marketing strategy for urban customers, another for suburban customers, and yet a third for rural customers.

The second category uses information about age, gender, family size, income, education, religion, or ethnicity to group customers. This type of segmentation is called **demographic segmentation**. Demographic variables are frequently used by traditional marketers because research has shown that customer needs and usage of products is strongly related to these types of variables. Demographic segmentation also exists on the Web. For example, a number of sites are devoted to women's issues or are directed at specific age groups (such as teenagers) whose members tend to purchase music CDs and trendy clothing. Often, demographic and geographic segmenting strategies are combined. For example, an airline might target middle-income families living in Wisconsin and Michigan with mid-winter advertising for vacation trips to Florida.

In **psychographic segmentation**, marketers try to group customers by variables such as social class, personality, or their approach to life. For example, an auto company might direct advertising for a sports car to customers who are gregarious and have a high need for achievement. The use of psychographic segmentation has increased dramatically in recent years as marketers attempt to identify characteristic lifestyles and then design advertising to reach people who see themselves as having a particular lifestyle.

Companies that advertise on television often create messages designed to reach the likely audiences of various types of programs. These audiences represent one or more market segments. The market segments can be geographic, demographic, psychographic, or a combination of these. Figure 4-5 presents some examples from the television medium that show how companies do this.

Type of television program	Type of advertising
Children's cartoons	Children's toys and games
Daytime dramas	Household and laundry goods, pet foods
Late-night talk shows	Snack foods and nonprescription drugs
Golf tournaments	Golf equipment, investments, and life insurance
Baseball and football games	Snack foods, beer, autos
Documentary films	Books, CDs, educational videotapes

Figure 4-5 *Television advertising messages tailored to program audience*

Children's television shows are likely to feature advertising for products that will appeal to children. Ads on daytime dramas are directed at people who are home during the day and who thus might be interested in household and laundry care products. These people are more likely than others to own pets, so they will see ads for pet foods. Advertisers on late-night talk shows often direct their ads at people who might be having trouble getting to sleep. Advertisers also believe that this late-night audience is receptive to promotions for snack foods to eat while watching these programs or for nonprescription medications to overcome whatever might be keeping them up so late.

Advertisers use sports programming as a vehicle for two different market segments. Some sports shows, such as golf tournaments or tennis matches, appeal to higher income viewers. Other sports shows, such as baseball or football game broadcasts, appeal to viewers with more moderate incomes. As a result, programs that cover golf or tennis are more likely to include ads for investment and insurance products and luxury automobiles than are baseball or football programs. Also, since viewers of golf tournaments and tennis matches are likely to play the sport, these programs often include ads for game equipment.

Programs that feature documentaries (such as those on the History Channel or the Discovery Channel) will often carry ads for books, book clubs, CDs, or even educational videotapes. Advertisers believe that these types of products will appeal to the intellectual, arts-loving audiences of these programs.

Companies do much more than just match advertising messages to market segments. They also build a sales environment for their product or service that corresponds to the market segment they are trying to reach. In the physical world, store design and layout are often directed at specific market segments. If you walk through a shopping mall, you can observe that colors, displays, lighting, background music, and even the clothes worn by sales clerks will vary with the targeted segment. For example, a clothing store for young women will present a completely different experience to its customers than will a clothing store that sells expensive, conservative formalwear targeted toward more mature women with larger incomes.

Market Segmentation on the Web

The Web gives companies an opportunity to present different store environments online. Figures 4-6 and 4-7 show the home pages of **Old Navy** and **Eddie Bauer**, respectively.

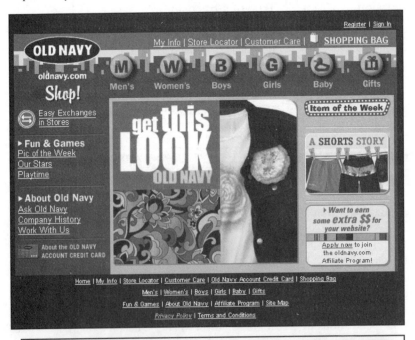

| Figure 4-6 | *Old Navy home page* |

Figure 4-7   *Eddie Bauer home page*

Both pages are well-designed and functional. However, you can see that they are each addressed to different market segments. The Old Navy site is targeted at young, fashion-conscious buyers. The site uses a wide variety of typefaces and bold graphics to convey its tone. The emphasis on the page is to "get this look" and, presumably, become the envy of your friends. In contrast, the Eddie Bauer page is rendered in a more muted, conservative style. The page is designed for older, more practical folks. The messages emphasized on the page are the price reductions—messages that appeal to a market segment full of people looking for classics instead of the latest trends.

In the physical world, retail stores have limited floor and display space. These limitations often force physical stores to decide on one particular message to convey. Exceptions do exist, such as a music store that has a separate room for classical CDs (with different background music than the rest of the store) or a large department store that can use lighting and display space differently in each department; however, smaller retail stores usually must choose the one image that will appeal to most of their customers. On the Web, retailers can provide separate virtual spaces for different market segments. Some Web retailers provide the ultimate in targeted marketing—they allow their customers to create their own stores, as you will learn in the next section.

Offering Customers a Choice on the Web

Dell Computer has done many things well in its online business. Its Web site offers customers a number of different ways to do business with the company. Its home page, shown in Figure 4-8, has links for each of the major groups of customers it has identified and also includes links to specific product categories.

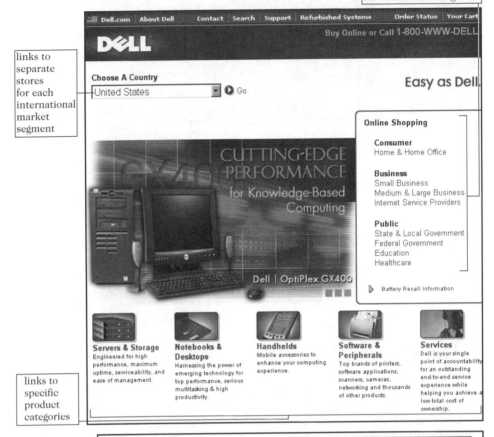

links to separate stores for each market segment

links to separate stores for each international market segment

links to specific product categories

Figure 4-8   *Dell Computer home page*

Two of Dell's major customer groups, Business and Public, are divided into subgroups. Dell has found that, for example, its Education customers have very different interests and needs than its Healthcare customers do. Dell's most highly specialized customer subgroup is Internet Service Providers (ISPs). Both the products offered for sale and the content provided to the ISPs is significantly different than the products and content offerings for other Dell customers. By providing a highly customized set of Web pages for this group, Dell can gain a competitive advantage over other sellers. The Dell Service Provider page, which appears in Figure 4-9, shows some of these customized Service Provider offerings.

Figure 4-9   *Dell's customized Service Provider page*

Dell Premier accounts give users the ultimate level of customer-based market segmentation. In these accounts, Dell offers each customer its own Dell Web site. Dell can customize a company's Premier account pages to show product selections for which price and terms have already been negotiated. Dell even allows individual employees of its customers to create their own personalized pages within their companies' Premier pages. This highly customized approach to offering products and services that match the needs of a particular customer is called **one-to-one marketing**. The Internet gives marketers the best opportunity for highly customized interactions with customers that they have had since the heyday of the door-to-door salesperson in the 1940s and 1950s.

BEYOND MARKET SEGMENTATION: CUSTOMER BEHAVIOR AND RELATIONSHIP INTENSITY

In the previous sections, you learned how companies can target groups of customers that are similar to each other as market segments. You also learned how one-to-one marketing gives companies a chance to create Web experiences that are unique to each individual customer. The next step—beyond market segmentation, even beyond one-to-one marketing—is when companies use the Web to target specific customers in different ways at different times.

Segmentation Using Customer Behavior

In the physical world, businesses can sometimes create different experiences for customers in response to their needs. For example, a company might decide that its mission is to sell prepared meals to hungry customers. A given potential customer will respond to hunger in different ways at different times. If a person is hungry in the morning, but late for work, that person might drive through a fast food restaurant or grab a quick cup of coffee at the train station. Lunch might be a sandwich ordered and delivered to the office, or it could require a nice restaurant if a client needs to be entertained. Dinner could be at a restaurant with friends, take-out food from a neighborhood Chinese restaurant, or a delivered pizza.

The point is that the same person will require different combinations of products and services depending on the occasion. In general, the creation of a separate experience for customers based on their behavior is called **behavioral segmentation**. When the behavioral segmentation is based on things that happen at a specific time or occasion, behavioral segmentation is sometimes called **occasion segmentation**.

Usually, businesses that operate in the physical world can meet only one or a few of customer's differing behavioral needs. For example, the Chinese restaurant mentioned in the preceding paragraph might offer dining room service and take-out service, but it probably would not offer a drive-through window or a morning coffee kiosk. Very few restaurants are able to offer everything from fast food through a five-course dinner. In the online world, it is much easier to design a single Web site that meets the needs of visitors who arrive in different behavioral modes. Thus, a Web site design can include elements that appeal to different behavioral segments.

Marketing researchers are just beginning to study how and why people prefer different combinations of products, services, and Web site features and how these preferences are affected by their mode of interaction with the site. Market researchers are finding that people want Web sites that offer a range of interaction possibilities from which they can select to meet their needs. Remember that a particular person might visit a particular Web site at different times and might be searching for different interactions each time. Customizing visitor experiences to match the site usage behavior patterns of each visitor or type of visitor is called **usage-based market segmentation**. Researchers have begun to identify common patterns of behavior and to categorize those behavior patterns. One set of categories that marketers use today includes browsers, buyers, and shoppers.

Browsers

Some visitors to a company's Web site are just surfing or browsing. Web sites intended to appeal to potential customers in this mode must offer them something that will pique their interest. The site should include words that are likely to jog the memory of visitors and remind them of something they want to buy on the site.

These key words are often called **trigger words** because they prompt a visitor to stay and investigate the products or services offered on the site. Links to explanations about the site or instructions for using the site can be particularly helpful to this type of customer. A site should include extra content related to the product or service the site sells. For example, a Web site that sells camping gear might offer reviews of popular camping destinations with photos and online maps. Such content can keep a visitor who is in browser mode interested long enough to stay at the site and develop a favorable impression of the company and buy on this visit or bookmark the site for a return visit.

Buyers

Visitors who arrive in buyer mode are ready to make a purchase right away. The best thing a site can offer to a buyer is to be certain that nothing will get in the way of the purchase transaction. For visitors who have chosen a product from a printed catalog, many Web sites include a text box control on their home pages that allow visitors to enter the catalog item number. This places that item in the site's shopping cart and takes the buyer directly to the shopping cart page.

The shopping cart page should offer a link that takes the visitor back into the shopping area of the site, but the primary goal is to get the buyer to the shopping cart as quickly as possible—even if the buyer is at the site for the first time. The shopping cart should allow the buyer to create an account and log in after placing the item into the cart. To avoid placing barriers in the way of customers who want to buy, the site should not ask visitors to log in until they near the end of the shopping cart procedure.

Perhaps the ultimate in shopping cart convenience is the 1-Click feature offered by **Amazon.com**, which allows customers to purchase an item with a single click. Any items that a customer purchases using the 1-Click feature within a 90-minute time period are aggregated into one shipment. Amazon.com has filed a patent application on the 1-Click feature and has successfully sued Barnes & Noble (which had implemented a similar feature on its Web site) to defend its exclusive right to use the feature.

Shoppers

Some customers will arrive at a Web site knowing that it offers items they are interested in buying. These visitors are motivated to buy, but they are looking for more information before they make a purchase decision. For the visitor who is in shopper mode, a site should offer comparison tools, product reviews, and lists of features. One site that offers many useful features for shopper-mode visitors is the Crutchfield consumer electronics Web store. Crutchfield's site includes features that allow customers to specify the level of detail presented for each product, to sort products by brand or price, and to compare products with each other side-by-side. A page from the Crutchfield site showing some of the shopper-friendly features appears in Figure 4-10.

short description
of product category

link to
customer
service pages
with
telephone
number for
immediate
help

controls that
allow the
visitor to
manipulate
information

product list

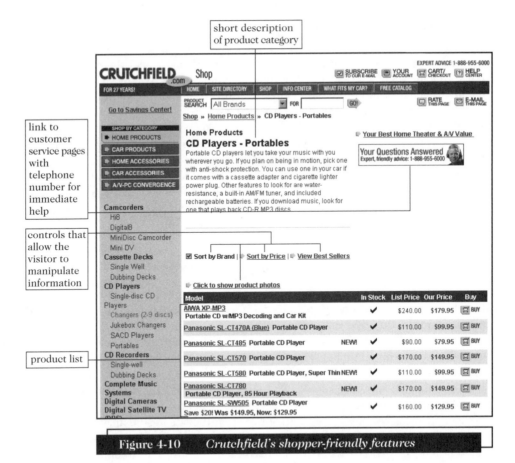

Figure 4-10   Crutchfield's shopper-friendly features

Remember that a person might visit a Web site one day as a browser, and then return later as a shopper or a buyer. People do not retain behavioral categories from one visit to the next—even for the same Web site.

Although many companies work with these three visitor modes, other researchers are exploring alternative models. Much of Web site visitor behavior is not yet well understood. One recent study conducted in 2000 by major consulting firm McKinsey & Co. examined the behavior of 50,000 users and identified six different groups of active Internet users. The following six behavior-based categories and their characteristic traits are:

- *Simplifiers* are users who like convenience. They are attracted by sites that make doing business easier, faster, or otherwise more efficient for them than is possible in the physical world.
- *Surfers* use the Web to find information, explore new ideas, and shop. They like to be entertained, and they spend far more time on the Web than other people. To attract surfers, sites must offer a wide variety of content that is attractive, well-displayed, and constantly updated.
- *Bargainers* are in search of a good deal. Although they make up less than 10 percent of the online population, they make up over half of all

visitors to the eBay auction site. They enjoy searching for the best price or shipping terms and are willing to move among sites to do that.

- *Connectors* use the Web to stay in touch with other people. They are intensive users of chat rooms, instant messaging services, electronic greeting card sites, and Web-based e-mail. Connectors tend to be new to the Web, are less likely than other people to purchase on the Web, and are actively trying to learn what the Web has to offer them.

- *Routiners* return to the same sites over and over again. They use the Web to obtain news, stock quotes, and other financial information. Routiners like the comfort of working with a user interface that they know well.

- *Sportsters* are similar to routiners, but they tend to spend time on sports and entertainment sites rather than at news and financial information sites. Since they view the Web as an entertainment vehicle, Sportsters are attracted by sites that are interactive and attractive.

The challenge for Web businesses today is to identify which of these (or similar) groups is visiting their sites and to formulate ways of generating revenue from each segment. As more researchers study Web site visitor behavior, perhaps we will learn how to recognize the various modes in which visitors arrive and then channel them into the appropriate sections of the site. Until then, many Web sites will use Dell's approach, in which visitors are asked to identify themselves as belonging to a particular category of customer when they enter the site.

Customer Relationship Intensity and Life-cycle Segmentation

One goal of marketing is to create strong relationships between a company and its customers. The reason that one-to-one marketing and usage-based segmentation are so valuable is that they help to strengthen companies' relationships with their customers. Good customer experiences can help create an intense feeling of loyalty toward the company and its products or services. Researchers have identified several stages of loyalty as customer relationships develop over time. A five-stage model of customer loyalty that is typical of these models appears in Figure 4-11.

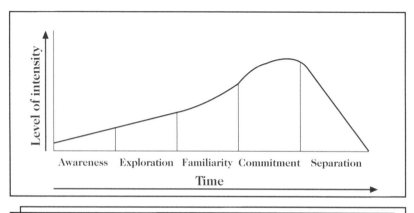

Figure 4-11   *Five stages of customer loyalty*

This model shows the increase in intensity of the relationship as the customer moves through the first four stages. In the fifth stage, a decline occurs and the relationship terminates. Not all customers go through the full five stages; some will stop at a stage and continue the relationship at that level of intensity or terminate the relationship at that point. As the figure shows, changes in the nature of the relationship do not occur suddenly as a customer moves from one stage to the next. Within each stage, the level of intensity changes gradually as the customer moves through that stage. The characteristics of the five stages are outlined in the next sections.

Awareness

Customers who recognize the name of the company or one of its products are in the awareness stage of customer loyalty. They know that the company or product exists, but have not had any interaction with the company. Advertising a brand or a company name is a common way for companies to achieve this level of relationship with potential customers.

Exploration

In the exploration stage, potential customers learn more about the company or its products. The potential customer might visit the company's Web site to learn more, and the two parties will often communicate by telephone or e-mail. A large amount of information interchange can occur between the parties at this stage.

Familiarity

Customers who have completed several transactions and are aware of the company's policies regarding returns, credits, and pricing flexibility are in the familiarity stage of their relationship with the company. In this stage, they are as likely to shop and buy from competitors as they are from the company.

Commitment

After having a considerable number of highly satisfactory experiences with a company, some customers develop a fierce loyalty or strong preference for the products or brands of that company. These customers have reached the commitment stage and are often willing to tell others about how happy they are with their interactions. To lure customers from the familiarity stage to the commitment stage, companies sometimes make concessions on price or terms. Usually, the value of the strong relationship is worth more to the company than these concessions.

Separation

Over time, the conditions that made the relationship valuable might change. The customer may be severely disappointed by changes in the level of service (either as provided by the company or as perceived by the customer) or product quality. The company can also evaluate the relationship and conclude that the loyal, committed customer is costing too much to maintain. As the intensity of the relationship fades, the parties enter a separation stage.

An important goal of any marketing strategy should be to move customers into the commitment stage as rapidly as possible and keep them there as long as possible.

Companies want to see customers move into the separation stage only if they are costing more to serve than they are worth.

Analyzing how customers' behavior changes as they move through the five stages can yield information about how they interact with the company and its products in each stage. The five stages are sometimes called the **customer life cycle** and using these stages to create groups of customers that are in each stage is called **life-cycle segmentation**. Two companies that undertake continuing research into market segmentation and how companies can use segment information to develop better relationships with their customers are **Claritas** and **Donnelley Marketing**.

Claritas created one of the first segment marketing databases, named PRIZM, in the early 1970s. Claritas built PRIZM to take advantage of people's tendency to live near other people with similar tastes and preferences. Thus, PRIZM identifies the demographic characteristics of people by neighborhood. Claritas has developed a number of other products that offer marketers databases with specific demographic, financial, and psychographic characteristics. Donnelley Marketing offers similar products, such as its Buyer Behavior Indicator and Affluence Models databases. Both Donnelley and Claritas have extended their research from traditional direct marketing to help firms sell online. You can learn more about these companies and their products by following their links on the Online Companion for this chapter.

In the rest of this chapter, you will learn some specific techniques that can be parts of successful Web marketing strategies. Remember that each of these techniques makes sense only when used in concert with another. Not all techniques will work well in all situations. For example, in the chapter's opening case, RedEnvelope found that a print catalog could be an integral part of promoting its online sales. RedEnvelope's success does not mean that printing catalogs is a good idea for all Web businesses (see the Kozmo Learning from Failures feature below). It is only a good idea if it provides customers with recognizable value and augments the rest of the company's marketing strategy.

L E A R N I N G   F R O M   FAILURES

KOZMO

Throughout New York City, people were in their homes late at night craving videos and snack foods. Kozmo was launched in 1998 to meet that need. With its orange-jacketed delivery persons riding bicycles or motor scooters, Kozmo promised delivery of most items within an hour of ordering. Kozmo did not offer as wide a range of items as most convenience stores sold, so its main competitive advantage was its delivery service. Kozmo attempted to become profitable by adding high-margin items such as DVD players and Sony PlayStations and by expanding its delivery areas to include higher-income neighborhoods. In addition to Manhattan, Kozmo opened operations for a short time in Houston and San Diego. In these cities, the higher average distances between deliveries made it even more difficult to cover costs.

(continued)

151

Despite its best efforts, Kozmo was unable to create an image that was much different from that of a convenience store on wheels. Kozmo found it difficult to convince customers that having snack food items and videos delivered made those items significantly more valuable. Most of Kozmo's product line consisted of items for which most people were accustomed to paying low prices.

In March 2001, just one month before closing operations, Kozmo announced a marketing plan that included spending $2.5 million to print and circulate 400,000 catalogs. The plan was a last-ditch attempt to increase brand awareness, gain new customers, and convince people who did not have an Internet connection to use Kozmo's phone order service. Unlike RedEnvelope, however, the Kozmo catalog was not a part of an integrated business plan and did not provide the same kind of added value that RedEnvelope's catalog provides—a bag of potato chips does not gain much appeal by appearing in a full-color catalog photo.

The lesson from Kozmo's experience is that using one element from a marketing strategy that has worked for another company is no guarantee that it will work for your company. Marketing techniques are effective only when implemented as part of an integrated strategy that fits the company's products and gives customers a compelling reason to buy.

ADVERTISING ON THE WEB

Advertising is all about communication. The communication might be between a company and its current customers, potential customers, or even former customers that the company would like to regain. To be effective, firms should send different messages to each of these audiences.

The five-stage customer life-cycle model shown in Figure 4-11 (in the previous section) can be helpful in creating the message to convey to each of these audiences. In the Awareness stage, the advertising message should inform. The message could describe a new product, suggest new uses for existing products, or describe specific improvements that have been made to a product. Audiences in the Exploration stage should receive messages that explain how a product or service works and encourage switching to that brand. In the Familiarity stage, the advertising message should be persuasive—convincing customers to purchase specific products or request that a salesperson call. Customers in the Commitment stage should be sent reminder messages. These ads should reinforce customers' good feelings about the brand and remind them to buy products or services. Companies generally do not target ads at customers who are in the Separation stage.

Most companies that launch an electronic commerce initiative will already have an advertising program in place. Online advertising should always be coordinated with existing advertising efforts. For example, print ads should include the company's URL.

Banner Ads

Most advertising on the Web uses banner ads. A **banner ad** is a small rectangular object on a Web page that displays a stationary or moving graphic and includes a hyperlink to the advertiser's Web site. Banner ads are versatile advertising vehicles—their graphic images can help increase awareness, yet users can click them to open the advertiser's Web site and learn more about the product. Thus, banner ads can serve both an informative function and a persuasive function.

Early banner ads used a simple graphic, usually in GIF format, that loaded with the Web page and remained on the page until the user moved to another page or closed the browser. Today, a variety of **animated GIFs** and **rich media objects** created using Shockwave, Java, or Flash are used to make attention-grabbing banner ads. These ads can be rotated so that each time the Web page is loaded into a browser, the ad changes. Banner ads come in a variety of sizes and shapes, as you can see in the Britannica.com home page shown in Figure 4-12.

Figure 4-12   *Banner ads on the Britannica.com home page*

Although Web sites can create banner ads in any dimensions they choose, advertisers decided early in the life of electronic commerce that it would be easier to standardize the sizes. The three banner ads on the Web page shown in Figure 4-12 are among the most common sizes: the **full banner** (468×60 pixels), the **half banner** (234×60 pixels), and the **square button** (125×125 pixels). The standard banner sizes that most Web sites have voluntarily agreed to use are called **interactive marketing unit (IMU) ad formats**. The **Interactive Advertising Bureau** (IAB) is a not-for-profit organization that promotes the use of Internet advertising and encourages effective Internet advertising (in April 2001, the IAB changed its name from the Internet Advertising Bureau). The IAB has established voluntary standards for IMUs. You can learn more about banner ads, including examples of the latest IAB-approved sizes, by following the Online Companion link to the IAB Web site.

Most advertising agencies that work with online clients can create banner ads as part of their services. Web site design firms also can create banner ads. Charges for creating banner ads range from about $100 to over $1500, depending on the complexity of the ad. You can make your own banner ads by using a graphics program or by using the tools provided by some Web sites. **ABCBanners** is an advertising-supported Web site that offers free downloadable graphics.

Banner Ad Placement

There are three different ways to arrange for other Web sites to display your banner ads. The first is to use a banner exchange network. A **banner exchange network** coordinates ad-sharing so that other sites run your ad while your site runs other exchange members' ads. Usually, the exchange requires you to accept two ads on your site for every one of your ads that appears on someone else's site. The exchange then makes its profit by selling the extra ad space to other businesses. Since banner exchanges are free, many Web businesses do use them; however, it is often difficult to find a group of other Web sites that have formed an exchange or that belong to an exchange that are not direct competitors of your business. This limitation prevents many businesses from using link exchange programs. Not-for-profit information Web sites are more likely to find a link exchange program suitable.

The second way that businesses can place their banner advertising is to find Web sites that appeal to one of the company's market segments and then pay those sites to carry the ads. This can take considerable time and effort. Smaller sites may not have an established pricing policy for advertising. Larger sites usually have high standard rates that they discount for larger customers. Smaller customers generally pay the standard rates. A company can hire an advertising agency to negotiate rates and help with ad placement. A full-service advertising agency can help design the ads, create the banners, and find appropriate Web sites. Agencies that do a lot of Internet work can often negotiate lower advertising rates with sites because the agencies can consolidate their clients' budgets and buy large blocks of advertising space at one time.

A third way to place banner advertising is to use a banner advertising network. A **banner advertising network** acts as a broker between advertisers and Web sites that carry ads. The larger banner advertising networks, such as **DoubleClick**, **LinkExchange**, and **ValueClick**, offer many of the same services as comprehensive ad agencies and often broker space primarily on larger Web sites (such as Yahoo!) that have high traffic rates and are, thus, more expensive. The smaller firms, on the other hand, often sell only leftover discounted space. Many of these smaller firms have fallen on hard times with the decline in advertising purchases that has occurred recently, and many have gone out of business.

Measuring Banner Ad Cost and Effectiveness

As more companies rely on their Web sites to make a favorable impression on potential customers, the issue of measuring Web site effectiveness has become important. Mass media efforts are measured by estimates of audience size, circulation, or number of addressees. When a company purchases mass media advertising, it pays a dollar amount for each thousand persons in the estimated audience. This pricing metric is called **cost per thousand** and is often abbreviated **CPM**.

Measuring Web audiences is more complicated because of the Web's interactivity and because the value of a visitor to an advertiser depends on how much information the site gathers from the visitor (for example, name, address, e-mail address, telephone number, and other demographic data). Since each visitor voluntarily chooses whether to provide these bits of information, all visitors are not of equal value. Internet advertisers have developed some Web-specific metrics for site activity, but these are not generally accepted and are currently the subject of considerable debate.

A **visit** occurs when a visitor requests a page from the Web site. Further page loads from the same site are counted as part of the visit for a specified period of time. This period of time is chosen by the administrators of the site and depends on the type of site. A site that features stock quotes might use a short time period, because visitors may load the page to check the price of one stock and reload the page 15 minutes later to check another stock's price. A museum site would expect a visitor to load multiple pages over a longer time period during a visit and would use a longer visit time window. The first time that a particular visitor loads a Web site page is called a **trial visit**; subsequent page loads are called **repeat visits**. Each page loaded by a visitor counts as a **page view**. If the page contains an ad, the page load is called an **ad view**.

Some Web pages have banner ads that continue to load and reload as long as the page is open in the visitor's Web browser. Each time the banner ad loads is an **impression**; and if the visitor clicks the banner ad to open the advertiser's page, that action is called a **click**, or a **click-through**. Banner ads are often sold on a CPM basis where the "thousand" is 1000 impressions. Rates vary greatly and depend on how much demographic information the Web site obtains about its visitors and what kinds of visitors the site attracts, but most rates range between $1 and $50 CPM. As recently as 1999, the range of online advertising rates was much higher, from about $5 to $100. Figure 4-13 shows a comparison of CPM rates for banner ads to CPM rates for advertising placed in other, traditional media outlets.

Medium	Description	Total cost	Audience size	Cost per thousand (CPM)
Network television	30-second commercial	$80,000 – $600,000	10 million - 20 million	$5–$30
Local television station	30-second commercial	$1,000 – $50,000	50,000 – 2 million	$3–$25
Cable television	30-second commercial	$3,000 – $10,000	100,000 – 500,000	$8–$20
Radio	60-second commercial	$200 – $1,000	50,000 – 2 million	$1–$18
Major metro newspaper	Full-page ad, single insertion	$20,000 – $80,000	100,000 – 600,000	$80–$130
Regional edition of a national magazine	Full-page ad, single insertion	$5,000 – $50,000	50,000 – 900,000	$40–$80
Local magazine	Full-page ad, single insertion	$2,000 – $10,000	3,000 – 80,000	$100–$140
Direct mail – coupon pack	Mailed in letter-sized envelope	$100 – $3,000	10,000 – 200,000	$15–$20
Billboard	Highway billboard (one-three months)	$5,000 – $25,000	100,000 – 3 million	$2–$5
World Wide Web	Banner ad (one month)	$200 – $30,000	10,000 – 50 million	$1–$50

Figure 4-13   *CPMs for advertising in various media*

One of the most difficult things for companies to do as they move onto the Web is to gauge the costs and benefits of advertising on the Web. Many companies have developed new metrics to evaluate the number of desired outcomes their advertising yields. For example, instead of comparing the number of click-throughs that companies obtain per dollar of advertising, they measure the number of new visitors to their site who buy for the first time after arriving at the site by way of a click-through. They can then calculate the advertising cost of acquiring one customer on the Web and compare that to how much it costs them to acquire one customer through traditional channels.

When banner ads first appeared on the Web in the mid-1990s, they provided a new experience for Web surfers. As users saw more ads, however, the ads lost their ability to attract attention. Click-through rates, which had been as high as 2 percent when banner ads were first introduced, have steadily dropped and now range from .3 percent to .5 percent, depending on the site's content.

To battle this decrease, banner ad designers first introduced animated GIFs with moving elements in the hopes that they might be more attractive to the user's eye than stationary graphics. When animated GIFs failed to halt the decline in click-through rates, designers created ads that displayed rich media effects, such as movie clips. They added interactive effects by writing Java programs that could respond to a user's click with some action (other than simply loading the advertiser's page into the browser). Some of these interactive ads even act like miniature video games. Designers also created banner ads that appear to be dialog boxes in the hope that confused users would click them. An example of one such banner ad is shown in Figure 4-14. This ad is designed to induce users to click the Correct button to fix the "error," but the banner actually links to a Web-based e-mail service.

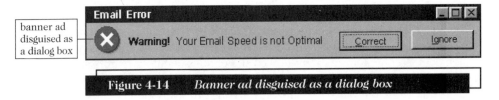

Figure 4-14   *Banner ad disguised as a dialog box*

Periodically, advertisers have tried new banner ad sizes and have placed the ads in page locations other than at the top or bottom. For example, some sites now use a large banner ad called a **skyscraper ad** that is designed to be placed on the side of a Web page and remain visible as the user scrolls down through the page.

Unfortunately for advertisers, none of these efforts has prevented the inexorable decline in click-through rates. In what some observers see as a last-ditch attempt to make advertising work on the Web, banner ad designers have turned to the alternative ad formats described in the next section.

Other Web Ad Formats

The steady decline in the effectiveness of banner ads has prompted advertisers to explore other formats for Web ads. One of these formats is the pop-up ad. A **pop-up ad** is an ad that appears in its own window when the user opens or closes a Web page. The window in which the ad appears does not include the usual browser controls. The only way to dismiss the ad is to click the small close control in the top right corner of the window's frame. Many users find pop-up ads to be extremely annoying. A particularly irritating variation on the pop-up ad technique occurs at Web sites that open more than one pop-up ad when a user leaves the site or closes the browser. If the user does not act quickly enough, the browser will spawn multiple windows and can even crash the computer.

Another type of pop-up ad is called the pop-behind ad. A **pop-behind ad** is a pop-up ad that is followed very quickly by a command that returns the focus to the original browser window. The result is an ad that is parked behind the user's browser waiting to appear when the browser is closed.

Despite user objections to pop-up ads (in all their variations), an increasing number of Web sites have started using them as a way of delivering a larger advertising image in a more forceful way. Some users have responded by using **ad-blocking software** that prevents banner ads and pop-up ads from loading (see the Online Companion's "Additional Information" section for links to Web sites that distribute or sell ad-blocking software).

Another intrusive ad format is the **interstitial ad**. When a user clicks a link to load a page, the interstitial ad opens in its own browser window, instead of the page that the user intended to load (the general meaning of the word "interstitial" is something that comes between two other things). Many interstitial ads close automatically, allowing the intended page to open in the existing browser window. Other interstitials require the user to click a button before they will close. Since they open in a full-sized browser window, interstitial ads offer the advertiser even more space than the pop-up ad format. These ads also completely cover the Web page that the user was trying to see. Many users find interstitials to be even more annoying than pop-up ads because they are larger and a more forceful interruption of the Web-browsing experience.

A fourth ad format is the active ad. **Active ads** generate graphical activity that "floats" over the Web page itself instead of opening in a separate window. One of the first active ads featured the figure of a little man who walked into the displayed Web page, unrolled a movie poster, and then pasted the poster onto the Web page (covering up part of the Web page content—content that a user might have been reading!). After about 10 seconds, the figure walked off the page and the poster disappeared. While it was open on the page, the poster was an active link to the movie's Web site.

Another active ad showed a Ford Explorer driving into the Web page. The Web page appeared to shake with the vibrations of the Explorer as it drove through. Active ads are certainly attention-getting and are even more intrusive than pop-ups or interstitials because they occur in the Web page itself and offer users no obvious way to dismiss them. Advertisers will likely continue to create new ad formats as users become accustomed to seeing active ads and they lose their effectiveness.

E-MAIL MARKETING

Sociologists and cultural anthropologists have proclaimed e-mail as one of the greatest tools for human communication to be developed in the 20th century. Since advertising is a process of communication, it is easy to see that e-mail can be a very powerful element in any company's advertising strategy. Many businesses would like to send e-mail messages to their customers and potential customers to announce new products, new product features, or sales on existing products. However, industry analysts have severely criticized some companies for sending e-mail messages to customers or potential customers. Some companies have even faced legal action after sending out mass e-mailings. Unsolicited e-mail is often considered to be spam, as you learned in Chapter 2. However, sending e-mail messages to Web site visitors who have expressly requested the e-mail messages is a completely different story. A key element in any e-mail marketing strategy is to obtain customers' approval before sending them any e-mail that includes a marketing or promotional message.

Permission Marketing

Many businesses are finding that they can maintain an effective dialog with their customers by using automated e-mail communications. Sending one e-mail message to a customer can cost less than one cent if the company already has the customer's e-mail address. Purchasing the e-mail addresses of persons who have asked to receive specific kinds of e-mail messages will add between a few cents and a dollar to the cost of each message sent. Another factor to consider is the conversion rate. The **conversion rate** of an advertising method is the percentage of recipients who respond to an ad or promotion. Conversion rates on requested e-mail messages range from 10 percent to over 30 percent. These are much higher than the click-through rates on banner ads, which are currently under .5 percent and decreasing.

The practice of sending e-mail messages to people who have requested information on a particular topic or about a specific product is called **opt-in e-mail** and is part of a marketing strategy called **permission marketing**. Seth Godin, the founder of YoYoDyne and later the vice president for direct marketing at Yahoo!, developed this marketing strategy and publicized it in a book he wrote with Don Peppers titled *Permission Marketing*.

Godin argues that, as the pace of modern life quickens, time becomes a valuable commodity. Most marketing efforts that traditional businesses use to promote their products or services depend on potential customers having enough time to listen to sales pitches and pay attention to the best ones. As time becomes more precious to everyone, people no longer wish to hear and evaluate advertising and promotional appeals for products and services in which they have no interest.

Thus, a marketing strategy that sends specific information only to persons who have indicated an interest in receiving information about the product or service being promoted should be more successful than a marketing strategy that sends general promotional messages through the mass media. One Web site that offers opt-in e-mail services is **yesmail.com**.

Combining Content and Advertising

One strategy for getting e-mail accepted by customers and prospects that many companies have found successful is to combine content with an advertising message. Articles and news stories that would interest specific market segments are good ways to increase acceptance of e-mail.

A good way to send content is to insert hyperlinks in the e-mail messages. The hyperlinks should take customers to the content, which is stored on the company's Web site. Once customers are viewing pages on the Web site, it is easier to induce them to stay on the site and to consider making purchases.

Outsourcing E-Mail Processing

Many companies find that the number of customers who opt-in to information-laden e-mails can grow rapidly. The job of handling e-mail lists and mass-mailing software can quickly outgrow the capacity of the company's information technology staff to handle it. A number of companies offer e-mail management services, and most small or medium-sized companies outsource their e-mail processing operation. The "Additional Information" section of the Online Companion pages for this chapter includes links to several companies that offer such services.

CUSTOMER RELATIONSHIP MANAGEMENT

The nature of the Web, with its two-way communication features and traceable connection technology, allows firms to gather much more information about customer behavior and preferences than they can gather using micromarketing approaches. Now, companies can measure a large number of things that are happening as customers and potential customers gather information and make purchasing decisions. The idea of technology-enabled relationship management has become possible when promoting and selling via the Web. **Technology-enabled relationship management** occurs when a firm obtains detailed information about a customer's behavior, preferences, needs, and buying patterns, *and* uses that information to set prices, negotiate terms, tailor promotions, add product features, and otherwise customize its entire relationship with that customer.

Although companies can use technology-enabled relationship management concepts to help manage relationships with vendors, employees, and other stakeholders, most companies currently use these concepts to manage customer relationships. Thus, technology-enabled relationship management is often called **customer relationship management (CRM)**, **technology-enabled customer relationship management**, or **electronic customer relationship management (eCRM)**. Figure 4-15 (on the next page) compares technology-enabled customer relationship management to traditional seller-customer interactions on seven dimensions.

Harvard Business School researchers Jeffrey Rayport and John Sviolka observed that firms today do business in both a physical world and in a virtual, information world. Rayport and Sviolka distinguish between commerce in the physical world, or marketplace, and commerce in the information world, which they term the **marketspace**. In the information world's marketspace, digital products and services can be delivered through electronic communication channels, such as the Internet.

In Chapter 1, you learned that the value chain model described the primary and support activities that firms use to create value. This value chain model is valid for activities in the physical world and in the marketspace. However, value creation requires different processes in the marketspace. By understanding that value creation in the marketspace is different, firms can identify value opportunities effectively in both the physical and information worlds.

For years, businesses have viewed information as a part of the value chain's supporting activities, but they have not considered how information itself might be a source of value. In the marketspace, firms can use information to create new value for customers. For example, **CDnow**, an online seller of music CDs, provides a number of valuable customer services that are derived purely from activities in the information world. CDnow will, if the customer so chooses, store an order history, provide recommendations based on previous purchases, and show current information on performing artists in whom the customer is interested.

Dimensions	Technology-enabled customer relationship management	Traditional relationships with customers
Advertising	Provide information in response to specific customer inquiries	"Push and sell" a uniform message to all customers
Targeting	Identify and respond to specific customer behaviors and preferences	Market segmentation
Promotions and discounts offered	Individually tailor to customer	Same for all customers
Distribution channels	Direct or through intermediaries; customer's choice	Through intermediaries chosen by the seller
Pricing of products or services	Negotiated with each customer	Set by the seller for all customers
New product features	Created in response to customer demands	Determined by the seller based on research and development
Measurements used to manage the customer relationship	Customer retention; total value of the individual customer relationship	Market share; profit

Figure 4-15 *Technology-enabled relationship management and traditional customer relationships*

Teenagers and young adults—people who often want more CDs than they can afford to buy right away—are important CDnow customers. Thus, CDnow's site offers a "Wish List" feature. Customers can store the names of CDs that they want to purchase later. Many CDnow customers order CDs as gifts for others, so the Web site provides a way to store gift recipient names and addresses to make repeat shopping easier. The CDnow site also has a "Gift Registry" in which customers can list CDs they would like to receive as gifts. Gift-givers can then obtain gift CD preferences for persons in the registry by entering the recipient's e-mail address.

When visitors register with the CDnow site, they are given a my CDnow page that is automatically configured to provide useful links. CDnow determines the links to include on the page by examining the visitor's history of purchases, Wish List and Gift Registry selections, and music interests that the visitor can expressly choose by following the link to Preferences and making selections. Figure 4-16 shows a my CDnow page that includes these marketspace features.

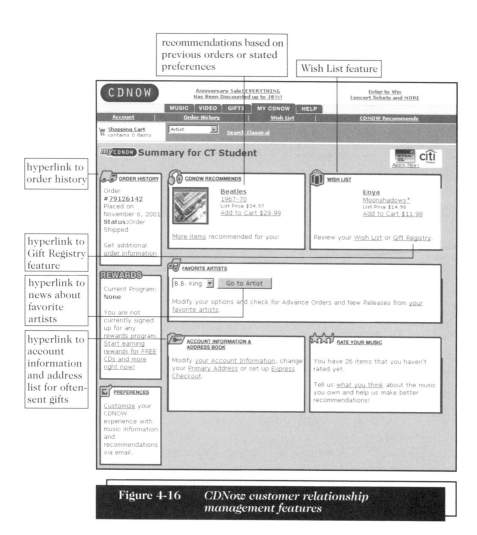

Figure 4-16 CDNow customer relationship management features

All the features that CDnow has built into its site add value for its customers. None of the features mentioned here occur in the physical marketplace; they all occur in the virtual information world of the marketspace. CDnow can provide these features only to customers who are willing to interact with the site or allow it to track their behaviors. Note that all of these marketspace site features are optional and must be actively selected by the individual customer. CDnow is implementing technology-enabled relationship management on five (advertising, targeting, promotions, distribution, and measurement) of the seven dimensions listed in Figure 4-15.

Successful new Web marketing approaches all involve enabling the potential customer to find information easily and customizing the depth and nature of that information; such approaches should encourage the customer to buy. Firms should track and examine the behaviors of their Web site visitors, and then use that information to provide customized, value-added digital products and services in the marketspace. Companies that use these technology-enabled relationship management tools to improve their contact with customers will be more successful on the Web than firms that adapt advertising and promotion strategies that were successful in the physical world but that are less effective in the virtual world. In Chapter 9, you will learn more about the specific software tools and other technologies that companies are using to implement CRM.

CREATING AND MAINTAINING BRANDS ON THE WEB

A known and respected brand name can present to potential customers a powerful statement of quality, value, and other desirable qualities in one recognizable element. Branded products are easier to advertise and promote, because each product carries the reputation of the brand name. Companies have developed and nurtured their branding programs in the physical marketplace for many years. Consumer brands such as Ivory soap, Walt Disney entertainment, Maytag appliances, and Ford automobiles have been developed over many years with the expenditure of tremendous amounts of money. However, the value of these and other trusted major brands far exceeds the cost of creating them.

Elements of Branding

The key elements of a brand, according to researchers at advertising agency Young & Rubicam, are differentiation, relevance, and perceived value. Product differentiation is the first condition that must be met to create a product or service brand. The company must clearly distinguish its product from all others in the market. This makes branding difficult for commodity products such as salt, nails, or plywood—difficult, but not impossible.

A classic example of branding a near-commodity product is Procter & Gamble's creation of the Ivory brand over 100 years ago. The company was experimenting with manufacturing processes and had accidentally created a bar soap that contained a high percentage of air. When one of the workers noted that the soap floated in water, the company decided to sell the soap using this differentiating characteristic in packaging and advertising by claiming "it floats". Thus was the Ivory soap brand born. **Procter & Gamble** maintains this brand differentiation on its Web site even today by listing the link to its **Ivory Soap** site under the heading "Beauty and Skin Care Products".

The second element of branding—relevance—is the degree to which the product offers utility to a potential customer. The brand will only have meaning to customers if they can visualize its place in their lives. Many people understand that **Tiffany & Co.** creates a highly differentiated line of jewelry and gift products, but very few people can see themselves purchasing and using such goods.

The third branding component—perceived value—is a key element in creating a brand that has value. Even if your product is different from others on the market and potential customers can see themselves using this product, they will not buy it unless they perceive value. Some large fast food outlets have well-established brands that actually work against them. People recognize these brands and avoid eating at these restaurants because of negative associations—such as low overall quality and high-fat-content menu items. Figure 4-17 summarizes the elements of a brand.

Element	Meaning to customer
Differentiation	In what significant ways is this product or service unlike its competitors?
Relevance	How does this product or service fit into my life?
Perceived value	Is this product or service good?

Figure 4-17   Elements of a brand

If a brand has established that it is different from competing brands and that it is relevant and inspires a perception of value to potential purchasers, those purchasers will buy the product and become familiar with how it provides value. Brands become established only when they reach this level of purchaser understanding and acceptance.

Emotional Branding vs. Rational Branding

Unfortunately, brands can lose their value if the environment in which they have become successful changes. A dramatic example is Digital Equipment Corporation (DEC). For years, DEC was a leading manufacturer of midrange computers. When the market for computing shifted to personal computers, DEC found that its branding did not transfer to the personal computers that it produced. The consumers in that market did not see the same perceived value or differentiation in DEC's personal computers that the buyers of midrange systems had seen for years. This is an important element of branding for Web-based firms to remember, because the Web is still evolving and changing at a rapid pace.

Companies have traditionally used emotional appeals in their advertising and promotion efforts to establish and maintain brands. One branding expert, Ted Leonhardt, has defined "brand" as "an emotional shortcut between a company and its customer". These emotional appeals work well on television, radio, billboards, and in print media, because the ad targets are in a passive mode of information acceptance. However, emotional appeals are difficult to convey on the Web because it is an active medium controlled to a great extent by the customer. Many Web users are actively engaged in such activities as finding information, buying airline tickets, making hotel reservations, and obtaining weather forecasts. These are busy people who will happily click away from emotional appeals.

Marketers are attempting to create and maintain brands on the Web by using **rational branding**. Companies that use rational branding offer to help Web users in some way in exchange for their viewing an ad. Rational branding relies on the cognitive appeal of the specific help offered, not on a broad emotional appeal. For example, Web e-mail services such as **Excite Mail**, **HotMail**, or **Yahoo! Mail** give users a valuable service—an e-mail account and storage space for messages. In exchange for this service, users see an ad on each page that provides this e-mail service.

Similarly, **MasterCard** promotes its brand name online through its **Shop Smart!** program. Shop Smart! is a third-party assurance mechanism. MasterCard ensures that any Web site displaying the Shop Smart! emblem (which happens to include a large MasterCard logo) is using what Master Card defines as a "safe" method of processing transactions. In exchange for this assurance on a Web shopping site, the Web user sees the MasterCard logo.

Brand Leveraging Strategies

Rational branding is not the only way to build brands on the Web. One method that is working for well-established Web sites is to extend their dominant positions to other products and services, a strategy called **brand leveraging**. **Yahoo!** is an excellent example of a company that has used brand leveraging strategies. Yahoo! was one of the first directories on the Web. It added a search engine function early in its development and has continued to parlay its leading position by acquiring other Web businesses and expanding its existing offerings. Yahoo! acquired GeoCities and Broadcast.com, and entered into an extensive cross-promotion partnership with a number of **Fox** entertainment and media companies. Yahoo! continues to lead its two nearest competitors, **Excite** and **Infoseek**, in ad revenue by adding features that Web users find useful and that increase the site's value to advertisers. Amazon.com's expansion from its original book business into CDs, videos, and auctions, is another example of a Web site leveraging its dominant position by adding features useful to existing customers.

Affiliate Marketing Strategies

Of course, this leveraging approach works only for firms that already have Web sites that dominate a particular market. As the Web matures, it will be increasingly difficult for new entrants to identify unserved market segments and attain dominance. A tool that many new, low-budget Web sites are using to generate revenue is affiliate marketing. In **affiliate marketing**, one firm's Web site—the affiliate firm's—includes descriptions, reviews, ratings, or other information about a product that is linked to another firm's site that offers the item for sale. For every visitor who follows a link from the affiliate's site to the seller's site, the affiliate site receives a commission. The affiliate site also obtains the benefit of the selling site's brand in exchange for the referral.

The affiliate saves the expense of handling inventory, advertising and promoting the product, and processing the transaction. In fact, the affiliate risks no funds whatever. CDnow and **Amazon.com** were two of the first companies to create successful affiliate programs on the Web. CDnow's Web Buy program, which includes more than 250,000 affiliates, is one of CDnow's main sources for new customers. The Amazon.com program has over 400,000 affiliate sites. Most of these affiliate sites are devoted to a specific issue, hobby, or other interest. Affiliate sites choose books or other items that are related to their visitors' interests and include links to the seller's site on their Web pages. Books and CDs are a natural for this type of shared promotional activity, but sellers of other products and services also use affiliate marketing programs to attract new customers to their Web sites.

One of the more interesting marketing tactics made possible by the Web is **cause marketing**, which is an affiliate marketing program that benefits a charitable organization (and, thus, supports a "cause"). In cause marketing, the affiliate site is created to benefit the charitable organization. When visitors click a link on the affiliate's Web page, a donation is made by a sponsoring company. The page that loads after the visitor clicks the donation link carries advertising for the sponsoring companies. Many companies have found that the click-through rates on these ads are much higher than the typical banner ad click-through rates.

Affiliate Commissions

Affiliate commissions can be based on several variables. In the **pay-per-click model**, the affiliate earns payment each time a site visitor clicks the link and loads the seller's page. This is similar to the click-through model of charging for banner advertising.

In the **pay-per-conversion model**, the affiliate earns payment each time a site visitor is converted from a visitor into either a qualified prospect or a customer. An example of a seller that might use the qualified prospect definition is a credit-card issuing bank. The bank might decide that its best strategy is to pay affiliates only when the visitor turns out to be a good credit risk. Alternatively, the bank may decide it wants to pay the affiliate only if the visitor is approved for the card and then accepts the card (completes the sale). A site that pays its affiliates on completed sales usually pays a percentage of the sale amount rather than a fixed amount per conversion. Some sites use a combination of these methods to pay their affiliates.

You can learn more about affiliate programs offered by different sites by going to each Web site and looking for a link to information about its affiliate program. Figure 4-18 shows the affiliate program information page for **Proflowers.com**, a top-rated online florist.

Figure 4-18 *Proflowers.com affiliate program information page*

Alternatively, you can visit an affiliate program broker site that offers affiliate program opportunities for a number of Web sites. An **affiliate program broker** is a company that serves as a clearinghouse or marketplace for sites that run affiliate programs and sites that want to become affiliates. These brokers also often provide software, management consulting, and brokerage services to affiliate program operators. For example, Proflowers.com uses affiliate program broker **Be Free** to manage its affiliate program. Be Free tracks affiliates' sales, calculates and pays affiliates' commissions, and handles any problems that arise. **Commission Junction** and **LinkShare** are two other popular affiliate program brokers.

Viral Marketing Strategies

Traditional marketing strategies have always been developed with an assumption that the company was going to communicate with potential customers directly or through an intermediary that was acting on behalf of the company, such as a distributor, retailer, or independent sales organization. Since the Web expands the types of communication channels available, including customer-to-customer communication, another marketing approach, viral marketing, has become popular on the Web. **Viral marketing** relies on existing customers to tell other persons—the company's prospective customers—about the products or services they have enjoyed using. Much like affiliate marketing uses Web sites to spread the word about a company, viral marketing approaches use individual customers to do the same thing. The number of customers increases the way a virus multiplies, thus the name.

Blue Mountain Arts, an electronic greeting card company, purchases very little advertising, but it has grown to become one of the most-visited sites on the Web. By late 1999, when the company was acquired by At Home Corporation for $780 million, Blue Mountain had more than 10 million people visiting its site each month. Electronic greeting cards are e-mail messages that include a link to the greeting card site. When people receive Blue Mountain Arts electronic greeting cards, they are taken to the Web site through that link and are likely to search for cards that they might like to send.

Brand Consolidation Strategies

Another way to leverage the established brands of existing Web sites was pioneered by Della & James, an online bridal registry that is now doing business as part of **WeddingChannel.com**. Although a number of national department store chains, such as **Macy's**, have established online registries for their own stores, Della & James created a single registry that connects to several local and national department and gift stores, including **Crate&Barrel**, **Gump's**, **Neiman Marcus**, **Tiffany & Co.**, and **Williams-Sonoma**. The logo and branding of each participating store are featured prominently on the WeddingChannel.com site. The founders had identified an opening for a market intermediary because the average engaged couple registers at three stores. Thus, WeddingChannel.com is providing a valuable consolidating activity for registering couples and their wedding guests that no store operating alone could provide. WeddingChannel.com also provides wedding planning services and every item that a bride and groom might need—from the bridal gown to the cake—all in one convenient Web location.

Costs of Branding

Transferring existing brands to the Web or using the Web to maintain an existing brand is much easier and less expensive than creating an entirely new brand on the Web. In 1998, a large number of companies began spending significant amounts of money to build new brands on the Web. According to studies by the **Intermarket Group**, the top 100 electronic commerce sites spent an average of $8 million each that year to create and build their online brands. Two of the top spenders included the battling Web sites **Amazon.com**, which spent $133 million, and **BarnesandNoble.com**, which spent $70 million. Most of this spending was for television, radio, and print media—not for online advertising. Online brokerages E*Trade and Ameritrade Holding were also among the top five in that first year of major brand-building on the Web, spending $71 million and $44 million, respectively.

Brand-building activity continued on the Web through 1999 and into the first months of 2000. In March 2000, the supply of money from lenders and venture capitalists began drying up, which resulted in smaller expenditures for most firms. By 2001, the peak of brand-building spending was long past for electronic commerce companies. Traditional firms had realized that an opportunity had opened for them to move their offline brands to the Web.

Promoting any company's Web presence should be an integral part of brand development and maintenance. The company's URL should always be included on product packaging and in mass media advertising on radio, television, and in print. Integrating the URL with the company logo on brochures can also be helpful in getting the word out about the Web site. Ensuring that the site appears in search engine listings is also very important, as you will learn in the next section.

SEARCH ENGINE POSITIONING AND DOMAIN NAMES

Potential customers find Web sites in many different ways. Some site visitors will be referred by a friend. Others will be referred by an affiliate marketing partner of the site. Some will see the site's URL in a print advertisement or on television. Others will arrive after typing a URL that is similar to the company's name. But many site visitors will be directed to the site by a search engine or directory Web site.

Search Engines and Web Directories

A **search engine** is a Web site that helps people find things on the Web. Search engines contain three major parts. The first part, called a **spider**, a **crawler**, or a **robot** (or simply **bot**), is a program that automatically searches the Web to find Web pages that might be interesting to people. When the spider finds Web pages that might interest visitors to the search engine site, it collects the URL of the page and information contained in the page. This information might include the page's title, key words included in the page's text, and information about other pages on that Web site. In addition to words that appear on the Web page, Web site designers can specify additional key words in the page that are hidden from the view of Web site visitors, but that are visible to spiders. These key words are enclosed in an HTML tag set called META tags. The word "META" is used for this tag set to indicate that the key words describe the content of a Web page and are not themselves a part of the content.

The spider returns this information to the second part of the search engine to be stored. The storage element of a search engine is called its **index** or **database**. The index checks to see if information about the Web page is already stored. If it is, it compares the stored information to the new information and determines whether to update the page information. The index is designed to allow fast searches of its very large amount of stored information.

The third part of the search engine is the search utility. Visitors to the search engine site provide search terms, and the **search utility** takes those terms and finds entries for Web pages in its index that match those search terms. The search utility is a program that creates a Web page that is a list of links to URLs that the search engine has found in its index that match the site visitor's search terms. The visitor can then click the links to visit those sites. You will learn more about the technologies used in search engines in later chapters of this book.

Some search engine sites also provide classified hierarchical lists of categories into which they have organized commonly searched URLs. Although these sites are technically called **Web directories**, most people refer to them as search engines. The most popular of these sites, such as Yahoo!, include a Web directory and a search engine. They give users the option of using the search engine to find categories of URLs as well as the URLs themselves. This combination of Web directory and search engine can be a very powerful tool for finding things on the Web. **Jupiter Media Metrix**, an online audience-measuring service, frequently posts lists of the most frequently visited search engine sites. The search engine and Web directory sites **AltaVista**, **AOL**, **Excite**, **Google**, **Lycos**, **MSN**, and **Yahoo!** are among the most popular sites listed by Jupiter Media Metrix.

Search Engine Positioning

Marketers want to make sure that when a potential customer enters search terms that relate to their products or services, their companies' Web site URLs appear among the first 10 returned listings. The combined art and science of having a particular URL listed near the top of search engine results is called **search engine positioning**, **search engine optimization**, or **search engine placement**.

For sites that obtain most of their visitors from search engines, good positioning is extremely important. An increasing number of search engine sites have started making the task easier—but for a price. These search engine sites offer companies a **paid placement** (also called a **sponsorship**), which is the option of purchasing a top listing in results listings for a particular set of search terms. The rates charged vary tremendously depending on the desirability of the search terms to potential sponsors.

Another option for companies is to buy banner ad space at the top of search results pages that include certain terms. For example, Chevrolet might want to buy banner ad space at the top of all search results pages that are generated by queries containing the words "new" and "car." Most search engine sites will sell banner ad space on this basis. An increasing number will sell space on results pages for the most desirable terms only to companies that agree to package deals that include paid placement and banner ad purchases.

Search engine positioning is a very complex subject. A number of consulting firms do nothing but advise companies on positioning strategy. Entire books have been written on the subject (one of the best currently available is Frederick Marckini's book, which is referenced in the "For Further Study and Research" section at the end of this chapter), and several major conferences are devoted to the subject each year. An excellent online resource is Danny Sullivan's **Search Engine Watch** Web site. Although much of the content on the site is limited to paid subscribers, the site does include a number of free resources and explanations that are very useful for learning about search engines and search engine optimization. The Search Engine Watch home page appears in Figure 4-19.

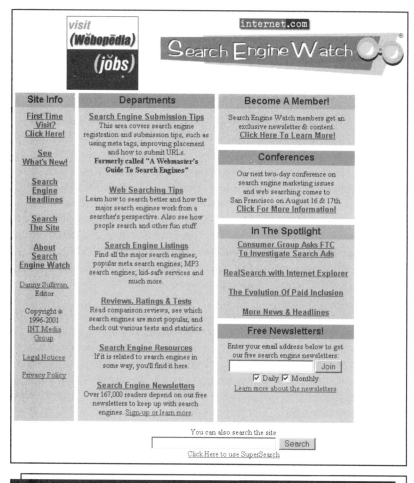

Figure 4-19   *Search engine watch home page*

Web Site Naming Issues

Companies that have a well-established brand name or reputation in a particular line of business usually want the URL for their Web sites to reflect that name or reputation. Obtaining identifiable names to use on the Web can be an important part of establishing a Web presence that is consistent with the company's existing image in the physical world.

Two airlines that started their online businesses with troublesome domain names have both purchased more suitable domain names. Southwest Airlines' domain name was www.iflyswa.com until it purchased www.southwest.com. Delta Air Lines' original domain name was www.delta-air.com. After several years of complaints from confused customers who could never remember to include the hyphen, the company purchased the domain name www.delta.com.

Buying and Selling Domain Names

In 1998, a poster art and framing company named Artuframe opened for business on the Web. With quality products and an appealing site design, the company was doing well, but it was concerned about its domain name, which was www.artuframe.com. After searching for a more appropriate domain name, the company's president found the Web site of Advanced Rotocraft Technology, an aerospace firm, at the URL www.art.com. After finding out that Advanced Rotocraft Technology's site was drawing 150,000 visitors each month who were looking for something art related, Artuframe offered to buy the URL. The aerospace firm agreed to sell the URL to Artuframe for $450,000. Artuframe immediately re-launched as **Art.com** and experienced a 30 percent increase in site traffic the day after implementing the name change.

The newly named site did not rely on the name change alone, however. It entered a joint marketing agreement with Yahoo! that placed ads for Art.com on art-related search results pages. Art.com also created an affiliate program with businesses that sell art-related products and not-for-profit art organizations. Although Art.com was ultimately unsuccessful in building a profitable business on the Web and liquidated in mid-2001, the domain name was snapped up immediately by already-profitable Allwall.com for an undisclosed amount. The new Allwall.com site, re-launched with the Art.com domain name, experienced a 100 percent increase in site visitors within the first month.

Another company that invested in an appropriate domain name was **Cars.com**. The firm paid $100,000 to the speculator who had originally purchased the rights to the name. Cars.com is a themed-portal site that displays ads for new cars, used cars, financing, leasing, and other car-related products and services. The major investors in this firm are newspaper publishers that wanted to retain an interest in automobile-related advertising as it moved online. As you learned in Chapter 3, classified ads are an important revenue source for many newspapers.

More recently, higher prices have prevailed in the market for domain names. Names such as Fruits.com, Question.com, Speaker.com, Tower.com, and Wisdom.com have each sold for prices over $100,000. Other names, including Cinema.com, Drugs.com, and ForSaleByOwner.com have sold for more than $500,000. Not long ago, eCompanies paid $7.5 million for the domain name Business.com. Although most domains that have high value are in the .com top-level domain, the name engineering.org sold at auction to the American Society of Mechanical Engineers, a not-for-profit organization, for just under $200,000. Figure 4-20 lists domain names that have sold for more than $1 million each.

Domain name	Price
Business.com	$ 7.5 million
Altavista.com	$ 3.3 million
Loans.com	$ 3.0 million
Wine.com	$ 3.0 million
Autos.com	$ 2.2 million
Express.com	$ 2.0 million
WallStreet.com	$ 1.0 million

Figure 4-20 *Domain names that have sold for more than $1 million*

URL Brokers

Several legitimate online businesses, known as **URL brokers**, are in the business of selling or auctioning domain names that they believe others will find valuable. Companies selling "good" (short and easily remembered) domain names include **Domains.com**, **DomainRace.com**, and **GreatDomains.com**. **Unclaimed Domains** sells a subscription to lists of recently expired domain names that it publishes periodically, and the **Netcraft** Web site has a URL search function to search for specific words in URLs.

Companies can also obtain domain names that have never been issued or that are currently unused from a domain name registrar. **ICANN** maintains a list of accredited registrars. Many of these registrars, such as **Network Solutions**, offer domain name search tools on their Web sites. A company can use these tools to search for available domain names that might meet their needs.

Summary

In this chapter, you learned how companies can use the principles of marketing strategy and the four Ps of marketing to achieve their goals for selling products and offering services on the Web. Some companies will use a product-based marketing strategy and some will use a customer-based strategy. The Web enables companies to mix these strategies and give customers a choice about which approach they would prefer.

Market segmentation using geographic, demographic, and psychographic information can work as well on the Web as it does in the physical world. The Web gives companies the powerful added ability to segment markets by customer behavior and life-cycle stage, even when the same customer exhibits different behavior during different visits to the company's site.

Online advertising has become more intrusive since it was introduced in the mid-1990s. You learned how companies are using various types of online ads, including banners, pop-ups, pop-behinds, and interstitials to promote their sites to potential customers. Permission marketing and opt-in e-mail show promise as viable alternatives to decreasingly effective Web page ads.

Many companies are using the Web to manage their relationships with customers in new and interesting ways. By understanding the nature of communication on the Web, companies can use it to identify and reach the largest possible number of qualified customers. Technology-enabled customer relationship management can provide better returns for businesses on the Web than the traditional unaided approaches of market segmentation and micromarketing.

Firms on the Web can use rational branding instead of the emotional branding techniques that work well in mass media advertising. Some businesses on the Web are sharing and transferring brand benefits through affiliate marketing and cooperative efforts among brand owners. Others are using brand leveraging and viral marketing approaches to increase their appeal and their customer bases.

Successful search engine positioning and domain name selection can be critical for many businesses in their quest for new online customers. The most important theme in this chapter is that companies must integrate the Web marketing tools they use into a cohesive and customer-sensitive overall marketing strategy.

Key Terms

Active ad

Ad-blocking software

Ad view

Affiliate marketing

Affiliate program broker

Animated GIF

Banner ad

Banner advertising network

Banner exchange network

Behavioral segmentation

Brand

Brand leveraging

Cause marketing

Click

Click-through

Conversion rate

Cost per thousand (CPM)

Crawler

Customer life cycle

Customer relationship management (CRM)

Customer value

Database

Demographic segmentation

Distribution

Electronic customer relationship management (eCRM)

Four Ps of marketing

Full banner

Geographic segmentation

Half banner

™

Thomson Learning (EMEA) Ltd
Distributed by: **Thomson Publishing Services** ,
Cheriton House, North Way, Andover , Hampshire, SP10 5BE, UK

IN CASE OF QUERY PLEASE MENTION		
ACCOUNT No.	DATE(TAX POINT)	INVOICE OR DOCUMENT No.
377888.0000	22-MAY-02	313501IT R

VAT Reg No GB198 9232 09

Payment to be received by	AREA
21-JUL-02	11320

.TON

I

PO4 8JF   I   I   *

DELIVER TO   PORTSMOUTH UNIV
*MRS J E BROWN
BUSINESS SCHL
LOCKSWAY RD MILTON
SOUTHSEA
HANTS

                                   PO4 8JF

TEXTBOOK ON 60 DAYS INSPECTION

N	Stock Location	Quantity	Title / Service / Author	Published Price	Discount %	Net Amount	Tax Code
114	22FU3C	1	Previous Ref. 191546IT 08-APR-02 MAR ELECTRONIC COMMERCE 3RD EDITI /Schneider PBN To be published on 23-MAY-02 IF YOU HAVE ANY QUERIES, OR ARE ADOPTING THIS BOOK AND WOULD LIKE INFORMATION ON ANY SUPPLEMENTS AVAILABLE TO ACCOMPANY THE BOOK, PLEASE CONTACT YOUR LOCAL REP TANYA CHAPMAN TEL: 0207 067 2655 FAX: 0207 067 2600 E-MAIL: tanya.chapman@itpuk.co.uk http://www.thomsonlearning.co.uk	27.99		27.99	

56196/300 Date 22-05-02 ICTA   SECURICOR B

TAX TOTAL	£ 0.00	VAT
TOTAL	£ 27.99	Due

COMPLETE THE QUESTIONNAIRE ON THE REVERSE

All Orders, Claims, Returns and Enquiries should be addressed to:

Thomson Publishing Services , Dept ICS , Cheriton House , North Way , Andover , Hampshire, SP10 5BE, UK

135010010

SOP 001/001

* Please be aware, your phone call may be recorded for training purposes.

Payment Methods

PLEASE ENSURE YOU QUOTE YOUR ACCOUNT NUMBER.

CHEQUE
Please make payable to
Thomson Publishing Services
and attach to this slip.

GIRO
Girobank A/C 1028456
Euro   A/C 62618489

CREDIT TRANSFER
Please quote :-
Thomson Publishing Services
and Invoice No.

Bankers:
The Royal Bank of Scotland Plc.
Corporate Banking Office
P O Box 450
5-10 Great Tower Street
London EC3P 3HX

Sterling
Sort Code: 16-04-00
Account No: 20121434

Euro
Sort Code 16-04-00
Account No: ITHPUSE EURC

SWIFT
Please quote SWIFT number.

Bankers:
The Royal Bank of Scotland Plc.
International & Wholesale Payments
P O Box 348
42 Islington High Street
London N1 8XL
Swift: RBOS GB 2L
Telex: 72130

CREDIT or CHARGECARD
Mastercard, Visa, Amex or Switch

Card number

Expiry date

Amount £

Signature

Account No.
377888.0000I

Invoice No..
313501IT

Amount.

£   27.99

PLEASE SEND YOUR PAYMENT TO

Thomson Publishing Services
PO Box 1633
Cheriton House, North Way
Andover, Hampshire
SP10 5QS, U.K

TLEAINSP IEI

Thank you for your interest the title named overleaf. Please complete the following questions as fully

Thomson Learning, Cheriton House, North Way, Andover, Hants, SP10 5BE, U.K.

☐ I have decided to use the above book for teaching purposes, and it will be :-

☐ Essential

☐ Recommended

☐ Supplementary

For use in the following course : -

Course Name _____ Course Level (eg. UG, PG, MBA) _

Course Start Date _____ Course End Date _____

No of Students per Year _____ I estimate _____ % of my students

If you expect 12 or more students to buy you may retain this copy free of charge.

Which other titles do you use on this course ? _____

_____

_____

Name(s) and Address(es) of your student's usual Bookseller(s) _____

_____

☐ Please tick here if you are buying this book. Return this form with your payment.

☐ Please tick here if you are returning this book. Please return in mint condition with this form.

What do you consider the main strengths and weaknesses of this book? (I may / may not be quoted) _____

_____

I can be contacted at work on telephone _____ extn _____ email _____

☐ Please tick here if you do not wish to receive information from Thomson Learning in the future.

Thank you for completing this form.

THOMSON

INVOICE TO PORTSMOUTH UNIV
BUSINESS SCHL
LOCKSWAY RD MI
SOUTHSEA
HANTS

Your Order Reference / Number	ISBN / ISS
BT / CUSTREC	0619063

OP MAR Batch

PLEASE

Impression

Index

Interactive marketing unit (IMU) ad format

Interstitial ad

Life-cycle segmentation

Market segmentation

Marketing mix

Marketing strategy

Marketspace

Micromarketing

Occasion segmentation

One-to-one marketing

Opt-in e-mail

Page view

Paid placement (sponsorship)

Pay-per-click model

Pay-per-conversion model

Permission marketing

Place (distribution)

Pop-behind ad

Pop-up ad

Price

Product

Promotion

Psychographic segmentation

Rational branding

Repeat visit

Rich media object

Robot (bot)

Search engine

Search engine optimization

Search engine placement

Search engine positioning

Search utility

Segments

Skyscraper ad

Spider

Sponsorship

Square button

Technology-enabled relationship management

Technology-enabled customer relationship
    management

Trial visit

Trigger words

URL brokers

Usage-based market segmentation

Viral marketing

Visit

Web directory

Review Questions

1. Assume you are a consultant to Fred's Sticks, a golf club manufacturer that sells its clubs direct to customers on the Web. Review Figure 4-5 in this chapter, which describes how advertisers select television programs that would be good hosts for their ads. Present a list of four magazines (other than golf magazines) in which Fred's Sticks should consider placing print advertising to support its Web sales effort. For each magazine, write one paragraph in which you explain why that magazine would be a good advertising outlet to reach potential customers of an online golf club store.

2. Write a paragraph in which you describe two Web site features that could help retain customers that are in the familiarity life-cycle stage of their relationship with your company.

3. Many people have strong negative reactions to pop-behind ads. Write a 500 word letter to the editor of an Internet industry magazine in which you explain, from the advertiser's viewpoint, why pop-behind ads are an effective advertising medium.

4. What are the key elements of technology-enabled customer relationship management? What advantages does technology-enabled customer relationship management have over traditional seller-customer interactions? Limit your answer to 250 words.

5. Describe two businesses, other than wedding planning and wedding gift registries, in which a brand consolidation Web site might provide enough value to customers to become profitable. Include specific examples of the value created.

Exercises

1. Visit the **RedEnvelope** Web site to examine how that company implements occasion segmentation. Write a report of no more than 200 words in which you describe two clear examples of occasion segmentation on the site.

2. You have been employed by Bob Drudge, the owner of **refdesk.com**, to sell space on his site to advertisers. Create a promotional press release of no more than 300 words in which you describe the advantages of advertising on refdesk.com. You may decide to promote space on the main page, specific other pages, or all pages. Be prepared to explain why your promotional strategy should work. You may find the **Art of Web Site Promotion**, **Promotion World**, **Sitelaunch**, or Seltzer's **"How to Publicize Your Web Site over the Internet"** Online Companion links helpful in your task.

3. Marti Baron operates a small Web business, The Cannonball, that sells parts, repair kits, books, and accessories to hobbyists who restore antique model trains. Many model train hobbyists and collectors have created Web sites on which they share photos and other information about model trains. Marti is interested in creating an affiliate marketing program that would allow those hobbyists to place links on their sites to The Cannonball and be rewarded with commissions on sales that result from visitors following those links. Examine the services offered by Be Free, Commission Junction, LinkShare, and any other affiliate program brokers you can find on the Web. Recommend at least one affiliate program broker that would be a good fit for Marti's business. In no more than 500 words, explain your recommendation. Be sure to consider the characteristics of Marti's business in your analysis.

For Further Study and Research

Agrawal, V., L. Arjona, and R. Lemmens. 2001. "E-Performance: The Path to Rational Exuberance," *The McKinsey Quarterly*, January, 30–43.

Balian, C. 2001. "Rolling Toward Ad Payoff," *eWeek*, July 1. Available online at: (http://www.xdnet.com/ecommerce/stories/main/0,10475,2780538,00.html).

Beardi, C. 2001. "Newspapers Work E-Beat to Home in on Consumers," *Advertising Age*, April 30, 26.

Bergert, S. and K. Kazimer-Shockley. 2001. "The Customer Rules," *Intelligent Enterprise*, 4(11), July 23, 31–34.

Blair, J. 2001. "Behind Kozmo's Demise: Thin Profit Margins," *The New York Times*, April 13. Available online at: (http://www.nytimes.com/2001/04/13/technology/13KOZM.html).

Blankenhorn, D. 2001. "Bigger, Richer Ads Go Online," *Advertising Age*, June 18, T10.

Boyd, H., O. Walker, J. Mullins, and J-C. Larréché. 2002. *Marketing Management*. Burr Ridge, IL: McGraw-Hill Irwin.

Buchwalter, C., M. Ryan, and D. Martin. 2001. *The State of Online Advertising*. Seattle, WA: AdRelevance (Jupiter Media Metrix).

Chuck, L. 1999. "On Being 'Consumer-ed': Marketing the User," *Searcher*, 7(5), May, 10–12.

Clarkson, B. 2001. "Got Brand?" *The Industry Standard*, March 19. Available online at: (http://www.thestandard.com/article/0,1902,22658,00.html).

Collins, S. 2001. "Affiliate Programs Growth Potential," *AdsGuide*, January 26. Available online at: (http://www.adsguide.com/news/2001/01/20010126-1.htm).

174

Compton, J. 2001. "Launch an E-Mail Ad Campaign," *Smart Business*, April 16. Available online at: (http://www.zdnet.com/ecommerce/stories/main/0,10475,2701095,00.html).

Dahlen, M. and Bergendahl, J. 2001. "Informing and Transforming on the Web: An Empirical Study of Response to Banner Ads for Functional and Expressive Products," *International Journal of Advertising*, 20(2), 189–206.

Delio, M. 2001. "Kozmo Kills the Messenger," *Wired News*, April 13. Available online at: (http://www.wired.com/news/business/0,1367,43025,00.html).

Dennis, S. 2001. "Online Advertisers Must Pick up on Multi-Channel Consumer," *BizReport*, April 5. Available online at: (http://www.bizreport.com/marketing/2001/04/20010405-1.htm).

E-Business. 2001. "Jupiter Takes a Stand on 'Pop-Up' Ads," July 13. Available online at: (http://www.zdnet.com/ecommerce/stories/main/0,10475,2784899,00.html).

Fitzgerald, K., A. Mand, and B. Tsui. 2001. "How 10 I-Shops Coped," *Advertising Age*, May 7, 6–7.

Forsyth, J., J. Lavoie, and T. McGuire. 2000. "Segmenting the E-market," *The McKinsey Quarterly*, 4. Available online at: (http://mckinseyquarterly.com/article_page.asp?tk=302016:939:24&ar=939&L2=24&L3=44).

Frangos, A. 2001. "Prescription for Change," *The Wall Street Journal*, April 23, R24.

Gardner, E. 1999. "Art.com," *Internet World*, March 15, 13. Also available online at: (http://www.iw.com/print/1999/03/15/).

Godin, S. and D. Peppers. 1999. *Permission Marketing: Turning Strangers into Friends, and Friends into Customers.* New York: Simon & Schuster.

Grebb, M. 2001. "The Art of Integration," *AdsGuide*, April 10. Available online at: (http://www.adsguide.com/article.php?id=13).

Grebb, M. 2001. "Sick of Banners? Try Immersion," *AdsGuide*, April 25. Available online at: (http://www.adsguide.com/article.php?id=15).

Green, H. and B. Elgin. 2001. "Do E-Ads Have a Future?" *Business Week*, January 22, EB46–EB50.

Halliday, J. 2001. "Edmunds.com Overhauls Auto Site: Ad Profits Needed to Cover Massive Expansion," *Advertising Age*, April 9, 28.

Helft, M. 2001. "Masters of the Rich Niche," *The Industry Standard*, July 16. Available online at: (http://www.thestandard.com/article/0,1902,27598,00.html).

Hilzenrath, D. 2001. "Saylor Firm Spent Millions Investing in Web Addresses," *Washington Post*, April 10, E1.

Hof, R. 2000. "Creative Coddling, Great Word of Mouth," *Business Week*, September 19, 57.

Hoffman, D. and T. Novak. 2000. "How to Acquire Customers on the Web," *Harvard Business Review*, 78(3), May–June, 179–188.

Holliday, H. 2001. "Drug Marketers Lead 1-on-1 March," *Advertising Age*, March 19, 8–9.

Howard, D. 2001. "Pay for Play," *Smart Business*, June 11. Available online at: (http://www.zdnet.com/smartbusinessmag/stories/all/0,6605,2764013,00.html).

Hwang, S. 2001. "Ad Nauseam," *The Wall Street Journal*, April 23, R8.

Jackson, J. 2001. "Are Clicks Through?" *eMarketer*, July 19. Available online at: (http://www.emarketer.com/analysis/eadvertising/20010719_ead.html).

Kaihla, P. 2001. "Want to Slim Down Customer Costs?" *eCompany*, March. Available online at: (http://www.ecompany.com/articles/mag/0,1653,9710,00.html).

Kaihla, P. 2001. "Five Battle-Tested Rules of Online Retail," *eCompany*, April. Available online at: (http://www.ecompany.com/articles/mag/0,1640,9599,00.html).

Kalin, S. 2001. "Brand New Branding," *Darwin*, July. Available online at: (http://www.darwinmag.com/read/070101/brand.html).

Khermouch, G. and T. Lowry. 2001. "The Future of Advertising," *Business Week*, March 26, 138.

Komenar, M. 1997. *Electronic Marketing.* New York: John Wiley & Sons.

Koprowski, G. 1998. "The (New) Hidden Persuaders: What Marketers Have Learned About How Consumers Buy on the Web," *The Wall Street Journal*, December 7, R10.

Kotler, P. and G. Armstrong. 1999. *Principles of Marketing*. Eighth Edition. Upper Saddle River, NJ: Prentice Hall.

Lawrence, S. 2001. "Finding a Future in Digital Ads," *The Industry Standard*, March 19. Available online at: (http://www.thestandard.com/article/0,1902,22689,00.html).

Lefton, T. 2001. "The Great Flameout," *The Industry Standard*, March 19. Available online at: (http://www.thestandard.com/article/0,1902,22685,00.html).

Livingston, B. 2001. "Got Spam? Ask JCPenney.com," *CNET News.com*, April 27. Available online at: (http://www.cnet.com/news/0-1278-210-5728743-1.html).

Lochridge, S. 2001. "Do You Really Know Your Customers?" *E-Business Advisor*, 19(4), April, 28–36.

Lowe, J. 2001. "Ignore Those E-Coupons and Miss Out on Savings," *Los Angeles Times*, January 18, T8.

Lytel, J. 2000. "Domain-Name Disputes Get Personal," *BizReport*, September 22. Available online at: (http://www.bizreport.com/marketing/2000/09/2000922-1.htm).

MacPherson, K. 2001. *Permission-Based E-Mail Marketing That Works!* Chicago: Dearborn Trade Press.

Maddox, K. 2001. "Tools of the Trade," *Advertising Age*, June 18, T1–T2.

Magid, L. 1999. "Web Ads Get Smart," *Upside Today*, June 9. Available online at: (http://www.upside.com/texis/mvm/larry_magid?id=375d863b0).

Marckini, F. 2001. *Search Engine Positioning*. San Antonio, TX: Republic of Texas Press.

Merlino, L. 2001. "Bull's Eye!" *Smart Business*, March 12. Available online at: (http://www.zdnet.com/smartbusinessmag/stories/all/0,6605,2688780,00.html).

Miles, S. 2001. "People Like Us: The Net Takes Customized Marketing to a Whole New Level," *The Wall Street Journal*, April 23, R30.

Naraine, R. 2001. "Ad Falloff Bites into New York Times Digital Revenue," *InternetNews*, July 19. Available online at: (http://www.internetnews.com/IAR/article/0,,12_804611,00.html).

Nelson, E. 2001. "The Soft Sell," *The Wall Street Journal*, April 23, R24.

Neuborne, E. 2001. "Pepsi's Aim Is True," *Business Week*, January 22, EB52.

Neuborne, E. and R. Hof. 1998. "Branding on the Net," *Business Week*, November 9, 76–81.

Norton, R. 2001. "The Bright Future of Web Advertising," *eCompany*, June. Available online at: (http://www.ecompany.com/articles/mag/0,1640,11620,FF.html).

Notess, G. 2000. "Internet Search Engine Update," *Online*, January–February, 13–14.

Ojala, M. 2000. "The Business of Domain Names," *Online*, May 1, 78.

Orenstein, S. 2000. "Boo.com: A Cautionary Tale," *The Industry Standard*, June 5, 106–113.

Osmer, N. 2001. "Avery Dennison Optimizes Customer Relationships with Permission Marketing," *Business 2.0*, February 16. Available online at: (http://www.business2.com/whatworks/entry/1,1981,1141,00.html).

Pastore, M. and C. Saunders. 2001. "Banners Can Brand, Honestly They Can," *Internet News*, July 19. Available online at: (http://www.internetnews.com/IAR/article/0,,12_804771,00.html).

Rewick, J. 2001. "Choices, Choices: A Look at the Pros and Cons of Various Types of Web Advertising," *The Wall Street Journal*, April 23, R12.

Rothenberg, R. 2001. "Famed Disasters: Hindenburg, Chicago Fire, and Banner Ads," *Advertising Age*, March 5, 19.

Rozanski, H., G. Bollman, and M. Lipman. 2001. "Seize the Occasion: Usage-Based Segmentation for Internet Marketers," *eInsights*, March 20, 1–13. Available online at: (http://www.strategy-business.com/enews/032001/032001.html).

176

Sandoval, G. 2001. "Kozmo to Shut Down, Lay Off 1,100," *News.com*, April 11. Available online at: (http://www.zdnet.com/ecommerce/stories/main/0,10475,5081050,00.html).

Saunders, C. 2001. "Nader Group Criticizes Pay for Placement Search Engines," *InternetNews*, July 17. Available online at: (http://www.internetnews.com/IAR/article/0,,12_803451,00.html).

Scheeres, J. 2001. "My Shoe Size? It'll Cost You," *Wired News*, June 11. Available online at: (http://www.wired.com/news/ebiz/0,1272,44278,00.html).

Shrake, S. 2000. "Start With the Right Brand Name," *Target Marketing*, April 1, 52.

Taylor, C. 2001. "E-Business Falls Back to Earth: Marketing Expenditures Continue to Drop," *Advertising Age Interactive*, May 21. Available online at: (http://www.adage.com/interactive/articles/20010521/article1.html).

Time, 2001. "Converting Web Surfers to Buyers," July 23, 46–47.

Underwood, R. 2001. "E-Marketing from a Blank Canvas," *Publish*, July 16. Available online at: (http://www.publish.com/ic_651108_4858_1-2742_138_05.html).

Upton, M. 2001. "eBusiness Executives Prefer Old Media to Attract Customers," *eBusiness Trends*, July 19. Available online at: (http://www.idc.com/ebusinesstrends/ebt20010719.htm).

Vickrey, L. 2001. "Keeping the Cachet," *The Wall Street Journal*, April 23, R28.

Weber, T. 2001. "Can You Say 'Cheese'? Intrusive Web Ads Could Drive Us Nuts," *The Wall Street Journal*, May 21, B1.

Williamson, D. 2001. "Is the Worst Yet to Come for Internet Shops?" *Advertising Age Interactive*, May 7. Available online at: (http://www.adage.com/interactive/articles/20010507/article2.html).

Wilson, T. 2001. "Salespeople Leverage the Net," *Internet Week*, June 1. Available online at: (http://www.internetwk.com/story/INW20010601S0003).

177

BUSINESS-TO-BUSINESS STRATEGIES: FROM ELECTRONIC DATA INTERCHANGE TO ELECTRONIC COMMERCE

INTRODUCTION

General Electric (**GE**) is one of the largest and most successful companies in the world. It engages in a wide range of businesses around the world, including the production of appliances and electrical and electronic products; broadcasting; and a variety of financial and insurance activities. One of its oldest lines of business is **GE Lighting**, which produces over 30,000 different kinds of lightbulbs in its 28 North American plants and other locations around the world. The raw materials used in making lightbulbs are fairly standard items: glass, aluminum, various insulating plastics, and filament materials. However, a major portion of each lightbulb's cost is the money that GE Lighting spends on indirect materials and parts for the machines used to fabricate and assemble the bulbs. These indirect materials and parts must conform to detailed specifications that GE stores on more than 3 million blueprints and other design drawings.

Since the technologies for making lightbulbs are mature and well-known, GE Lighting can solicit bids from a variety of suppliers for indirect materials and machinery replacement parts without worrying about the possible disclosure of trade secrets. Unfortunately, the bidding process at GE Lighting had become very slow and inefficient. Each transaction required the purchasing department to request the relevant blueprints, photocopy them, attach them to other material specification documents, and mail the whole package to suppliers who might be interested in bidding on the item. It would often take purchasing personnel more than four weeks to gather the information, send it to potential suppliers, obtain and evaluate suppliers' bids, negotiate with the chosen suppliers, and place an order. These long delays were limiting GE Lighting's flexibility and ability to respond to requests from its customers.

By applying the tools of electronic commerce to these purchase transactions, GE Lighting was able to make major improvements to the entire parts acquisition process. Today, purchasing personnel have access to a procurement system through their desktop computers. When they need to buy replacement parts for a machine, they create a new purchase file that includes basic quantity, delivery date, and delivery location information. Then, from a list generated by a continuously updated supplier database, they select suppliers from whom they will request quotes. Finally, they attach electronic copies of all necessary blueprints and engineering drawings, which are now digitized and stored in another database; with a mouse click, they send the entire bid package off in an encrypted format to all the selected suppliers. Assembling the bid package now takes hours instead of a week or more. Suppliers are asked to respond within a short time period—usually a week—through the Internet. The purchasing staff member can evaluate the returned bids and award a contract online, completing the entire process in about 10 days.

The most significant savings for GE Lighting were in process-time reduction—from four weeks or more to 10 days—and in the elimination of paper and the costs of paper handling. However, the company also realized other benefits. Because the online system made it easier to send out bid packages, purchasing was able to send out more bids to a wider range of suppliers. In particular, many foreign suppliers who had been difficult to reach with mailed bid packages were now being included in the solicitation for quotes. The increased competition drove prices down; GE Lighting has saved up to 20 percent on many of these items since moving the bid process online. Suppliers have welcomed the reduced time lag between submitting the bid and learning whether GE Lighting would award them the contract; this has made their production planning easier.

In this chapter, you will learn about:

- Strategies that businesses use to improve purchasing, logistics, and other support activities
- The ways that firms are creating network organizations that extend beyond traditional enterprise limits
- Electronic data interchange, how it works, and how businesses are moving it to the Internet
- Supply chain management and how businesses are using the Internet and Web technologies to improve it
- How businesses are creating electronic marketplaces that make purchase-sale negotiations easier and more efficient

180

PURCHASING, LOGISTICS, AND SUPPORT ACTIVITIES

In the previous two chapters, you learned about strategy issues that arise when businesses and other organizations provide information to potential customers. In terms of the value chain model described in Figure 1-9, you learned about primary activities: *identify customers*, *market and sell*, and *deliver*. You also became familiar with a number of business models for selling on the Web. Although most of these business models apply to both business-to-business and business-to-consumer electronic commerce, the emphasis in Chapters 3 and 4 was on business-to-consumer advertising, promotion, and sales activities.

In this chapter, you will learn how companies use electronic commerce to improve their primary activities of *purchasing* and *logistics*, and all of the support activities shown in Figure 1-9. At first glance, none of these business elements seems as exciting as creating a Web presence and selling to new customers on the Web; however, the potential for cost reductions and business process improvements in purchasing, logistics, and support activities is tremendous.

An emerging characteristic of purchasing, logistics, and support activities is that they need to be flexible. A purchasing or logistics strategy that works this year may not work next year. Fortunately, economic organizations are evolving from the hierarchical structures they have used since the Industrial Revolution to new, more flexible network structures. These network structures are made possible by the reductions in transaction costs caused by the emergence of the Internet and the Web.

Purchasing Activities

Purchasing activities include identifying vendors, evaluating vendors, selecting specific products, placing orders, and resolving any issues that arise after receiving the ordered goods or services. These issues might include late deliveries, incorrect quantities, incorrect items, and defective items. By monitoring all relevant elements of purchase transactions, purchasing managers can play an important role in maintaining and improving product quality and reducing cost. Many managers call this function procurement instead of purchasing to distinguish the broader range of responsibilities. The term **procurement** generally includes all purchasing activities, plus the monitoring of all elements of purchase transactions. It also includes managing and developing relationships with key suppliers. The business purchasing process is usually much more complex than most consumer purchasing processes. In Figure 1-2 (Chapter 1), you saw a six-step presentation of the buyer's purchase process that was a good description of a typical consumer purchase process. Figure 5-1 shows the steps in a typical business purchasing process.

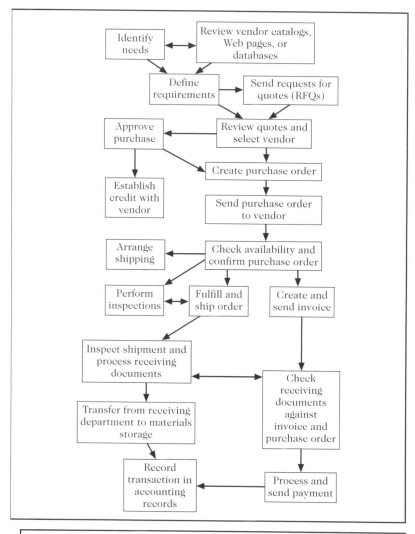

Figure 5-1   Steps in a typical business purchasing process

As you can see, the business purchasing process includes many more steps than the consumer purchasing process. In addition to having more steps, the business purchasing process requires a number of people to coordinate their individual activities as part of the process. In large companies, the procurement department that supervises the purchasing process may include hundreds of employees.

In many companies, procurement staff must have high levels of product knowledge to identify appropriate suppliers of products. The part of procurement activity devoted to identifying suppliers and determining the qualifications of those suppliers is called **sourcing**. Specialized Web purchasing sites can be particularly useful to procurement professionals responsible for sourcing.

Direct vs. Indirect Materials Procurement

Businesses make a distinction between direct and indirect materials. **Direct materials** are those materials that become part of the finished product in a manufacturing process. Steel manufacturers, for example, consider the iron ore that they buy to be a direct material. The procurement process for direct materials is an important part of any manufacturing business because the cost of direct materials is usually a very large part of the cost of the finished product. **Indirect materials** are all other materials that the company purchases, including factory supplies such as sandpaper, hand tools, and replacement parts for manufacturing machinery.

Large companies usually assign responsibility for purchasing these two kinds of materials to separate departments, one for direct materials and another for indirect materials. Most companies include the purchase of nonmanufacturing purchases—such as office supplies, computer hardware and software, and travel expenses—in the responsibilities of the indirect materials procurement department.

Many vendors that manufacture general industrial merchandise and standard machine tools for a variety of industries have created Web sites through which their customers can purchase. A number of customers buy these indirect materials products on a recurring basis, and many of them are commodities, that is, standard items that buyers usually select using price as their main criterion. These indirect materials items are often called **maintenance, repair, and operating (MRO)** supplies. Increasingly, procurement professionals are using the terms "indirect materials" and "MRO supplies" interchangeably.

By using a Web site to process orders, the vendors in this market can save the costs of printing and shipping catalogs and of handling telephone orders. They can also keep price and quantity information continually updated, which would be impossible to do in a printed catalog. Some industry analysts estimate that the cost to process an MRO order through a Web site can be less than one-tenth of the cost of handling the same order by telephone.

One of the largest MRO suppliers in the world is **W.W. Grainger**. Its Web site offers over 220,000 products for sale. Grainger's Web store, which appears in Figure 5-2, offers visitors a variety of ways to access information about and order Grainger products.

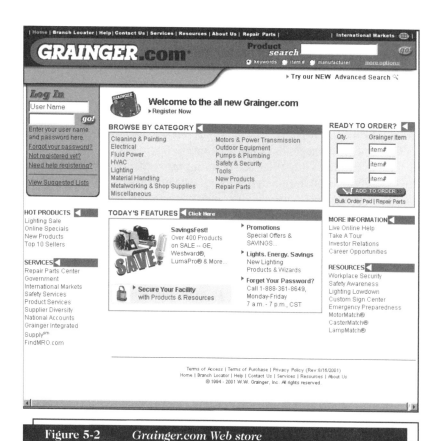

Figure 5-2   *Grainger.com Web store*

A visitor can enter the online catalog, use the product search box at the top of the page, or search by clicking a hyperlink to one of the categories listed in the middle of the page.

In 1999, **Milacron** opened its Milpro Web site to sell MRO items such as cutting tools, grinding wheels, and manufacturing fluids. Milacron, a diversified machinery manufacturer, had experienced difficulty in effectively reaching one of its key markets—over 100,000 widely scattered machine and job shops—and now reaches that market through its Web presence. Milacron is even accepting credit cards in payment for purchases on its Web site, which is highly unusual for an industrial products vendor. By accepting credit cards, Milacron gives both its smaller customers and its new customers (who do not have an established credit record with Milacron) a convenient way to buy.

Office equipment and supplies are also items that are used by a wide variety of businesses. Market leaders **Office Depot** and **Staples** each have well-designed Web sites devoted to helping business purchasing departments buy these routine items as easily as possible. On their business-to-business Web sites, **Digi-Key** and **Newark Electronics** sell electronic parts, and **Global Computer Supplies** sells computers and related items.

Logistics Activities

The classic objective of logistics has always been to provide the right goods in the right quantities in the right place at the right time. Logistics management is an important supporting activity for both the sales and the purchasing activities in a company. Businesses need to ensure that their products are delivered to customers on time and that the raw materials they use to create their products arrive when needed. The management of materials as they go from the raw materials storage area through production processes to become finished goods is also an important part of logistics.

Businesses have been increasing their use of information technology to achieve logistics objectives. Web-enabled automated warehousing operations are saving companies millions of dollars each year and major transportation companies such as **Schneider Logistics**, **Ryder System**, and **J.B. Hunt** now want to be seen by their customers as information management firms as well as freight carriers.

For example, the Schneider Track and Trace system delivers real-time shipment information to Web browsers on its customers' computers. This system shows the customer which freight carrier is transporting a shipment, where the shipment is, and when it should arrive at its destination. J.B. Hunt, which operates more than 60,000 trucks, trailers, and containers, implemented a Web site that lets its customers track their shipments themselves. With customers doing their own tracking, J.B. Hunt needs far fewer customer service representatives. Also, J.B. Hunt found that its customers could monitor their own shipments more effectively than the company could. Thus, this Web site saves more than $12,000 per week in labor and lost shipment costs.

FedEx has freight-tracking Web pages available to its customers, as does UPS. Firms that run their own trucking operations have implemented tracking systems that use global positioning satellite technology to monitor vehicle movements.

Logistics activities include managing the inbound movements of materials and supplies and the outbound movements of finished goods and services. Thus, receiving, warehousing, controlling inventory, scheduling and controlling vehicles, and distributing finished goods are all logistics activities. The Web and the Internet are providing an increasing number of opportunities to manage these activities better as they lower transaction costs and provide constant connectivity between firms engaged in logistics management.

Support Activities

Support activities include the general categories that appear in Figure 1-9: *finance and administration*, *human resources*, and *technology development*. Finance and administration includes activities such as making payments, processing payments received from customers, planning capital expenditures, and budgeting and planning to ensure that sufficient funds will be available to meet the organization's obligations as they come due. The operation of the computing infrastructure of the organization is also an administration activity. Human resource activities include hiring, training, and evaluating employees; administering benefits; and complying with government record-keeping regulations. Technology development can include a wide variety of activities, depending on the nature of the business or organization. It can include networking research scientists into virtual collaborative workgroups, posting research results, publishing research papers online, and providing connections to outside sources of research and development services.

A 1999 article in *Inc.* magazine described the challenges facing Allegiance Telecom. The company was growing very rapidly and hiring over 100 people each month to staff its sales offices throughout the United States. Each new hire had to receive a full briefing on medical, dental, and retirement benefits plans, and then he or she had to select from among several options for each. Since Allegiance was growing so rapidly, its human resources staff was spread thin and could not be in every sales office for every hire. The company turned to **Online Benefits**, a firm that duplicates its clients' human resource functions on a password-protected Web site that is accessible to clients' employees. The employees can then access their employers' benefits information, find the answers to frequently asked questions, and even perform complex benefit option calculations.

Other firms that offer support activities services include **TheTrip.com**, employee travel policies; **CyLex Systems**, document storage; **PayMaxx**, payroll processing; **Atrieva**, electronic file storage; and **HotOffice**, group calendar and communications. Larger firms are building these types of functions into their intranets. These larger firms are also including Web-enabled sales support and sales force automation functions in their extranets. For smaller firms, **DigitalWork** offers a number of business support activities, including debt collection, employee recruitment tools, and press release assistance.

Training and Knowledge Management

One common support activity that underlies multiple primary activities is training. In many companies, the human resources department handles training. Other companies may decentralize this function and have individual departments administer it. For example, insurance firms expend large amounts of resources on sales training. In most insurance companies, the sales and marketing department administers this training. By putting training materials on the company intranet, insurance companies can distribute the training materials to many different sales offices, yet coordinate the use of those materials in the corporate headquarters sales office.

In 1999, the Swedish telecommunications giant Ericsson launched an extranet for current and former employees, families of those employees, and employees of approved business partners. Ericsson has more than 100,000 employees scattered across the globe. One part of this extranet included a Web site that enabled current employees, retirees, and other recipients of payments from the company's medical and retirement plans to efficiently track their benefits. Another part of the extranet included a Web site that was designed to facilitate knowledge management. **Knowledge management** is the intentional collection, classification, and dissemination of information about a company, its products, and its processes. This type of knowledge is developed over time by individuals working for or with a company and is often difficult to gather and distill.

Ericsson managers hope that their knowledge network will generate new ideas, help solve problems, and improve business processes throughout the international organization. Designers of the system have identified their biggest challenge: to direct the information they collect in the extranet to projects and product development activities that will benefit from that information.

BroadVision, a software development and consulting firm, has installed an internal system called K-Net, or Knowledge Network, that organizes all information sources that its employees use regularly in their jobs. It found that many of its employees were visiting between 10 and 20 Web sites each day in the course of doing their jobs. K-Net brings together all of the information that each employee needs and combines it into one dashboard-style interface presented on a Web browser. Much of the interface is customized for employees, although some parts of the interface—such as health insurance, vacation days, and other human resources information—is standardized for all employees. BroadVision has found the K-Net system to be so useful that it is partnering with Bank of America, Hewlett-Packard, and Amadeus, a European travel services company, to develop a version of K-Net that it will sell to other companies.

E-Government

Although governments do not typically sell products or services to customers, they perform many functions for their stakeholders. Many of these functions can be enhanced by the use of electronic commerce. Governments also operate business-like activities; for example, they employ people, buy supplies from vendors, and distribute benefit payments of many kinds. They also collect a variety of taxes and fees from their constituents. The use of electronic commerce by governments and government agencies to perform these functions is often called **e-government**.

In 2000, the U.S. government's Financial Management Service (FMS) opened its **Pay.gov** Web site. The FMS is the agency responsible for receiving the government's tax, license, and other fee revenue (more than $2 trillion per year). It is also responsible for paying out more than $1.2 trillion in Social Security benefits, veterans benefits, tax refunds, and other disbursements. Federal agencies can link their Web sites to Pay.gov, which lets site visitors pay taxes and fees they owe to these agencies using their credit cards, debit cards, or various forms of electronic funds transfer and electronic cash.

The U.S. government's Bureau of Public Debt operates several Web sites that allow individuals and financial institutions to buy government bonds and other securities. The **TreasuryDirect** site offers the larger-denomination treasury bills, treasury bonds, and treasury notes; the **Savings Bonds Direct** site offers smaller-denomination savings bonds. The Bureau also operates the **SLGSafe** site, which provides state and local governments within the United States with a place to sell their bonds and notes to investors.

State governments are also creating their own Web sites for conducting business and interacting with their stakeholders. In 2001, the State of California opened its one-stop portal site, **my.ca.gov**, which appears in Figure 5-3.

link to business laws, regulations, and information about doing business with California

Figure 5-3 *State of California portal site my.ca.gov*

This site gives visitors access to virtually every California government agency and state operation. Site visitors can transact a wide array of business with the state, from renewing a driver's license to reserving a camping site. The goal of the site is to give constituents one site through which they can conduct all of their business with the State of California. For businesses, the site offers the full text of all California business laws and regulations. It also provides information about how to sell to and buy from the state and its agencies. Many other states have similar Web presences.

NETWORK MODEL OF ECONOMIC ORGANIZATION

In Chapter 1, you learned about the three different forms of economic organization: markets, hierarchies, and networks. One trend that is becoming clear in purchasing, logistics, and support activities is the shift away from hierarchical structures toward network structures. The traditional purchasing model had one hierarchically structured firm negotiating purchase terms with several similarly structured supplier firms and playing each supplier against the others. As is typical in a network organization, more businesses are now using their procurement departments to negotiate using a variety of tools and in a variety of economic forms, such as strategic alliances. For example, a buying firm might enter into a partnership with a supplier to develop a new technology that will reduce product costs overall. The technology development might be done by a third firm using research conducted by a fourth firm.

While reading the previous section, you may have noticed that companies can have other firms perform various support activities for them. Again, these are examples of firms moving to a network model of economic organization. Imagine a business that uses one supplier to manage its payroll, another to administer its employee benefits plans, and a third to handle its document storage needs. The document storage service supplier might store the documents of the payroll service supplier and the benefits administration firm. The payroll service supplier might handle the payroll for the benefits administration firm. A fourth firm might provide online backup storage for the files of the other three companies. Of course, the payroll firm and the employee benefits firm might form a marketing partnership to sell both of their services to particular market segments. The document storage firm and the online backup storage firm might form a similar strategic alliance.

Highly specialized firms can now exist and trade services very efficiently on the Web. The Web is enabling this shift from hierarchical forms of economic organization to network forms. These emerging networks of firms are more flexible and can respond to changes in the economic environment much more quickly than hierarchically structured businesses ever could. You can learn more about the economics of networked organizations at the **Network Economics** Web site maintained by the University of California, Berkeley. The roots of Web technology for business-to-business transactions, however, lie in a very hierarchically structured approach to interfirm information transfer: electronic data interchange.

ELECTRONIC DATA INTERCHANGE

You learned in Chapter 1 that electronic data interchange (EDI) is a computer-to-computer transfer of business information between two businesses that uses a standard format of some kind. The two businesses that are exchanging information are called **trading partners**. Firms that exchange data in specific standard formats are said to be **EDI-compatible**. The business information exchanged is often transaction data; however, it can also include other information related to transactions, such as price quotes and order-status inquiries. Transaction data in business-to-business transactions includes the information traditionally included on paper invoices, purchase orders, requests for quotations, bills of lading, and receiving reports. The data on these five types of forms accounts for over 75 percent of all information exchanged by trading partners in the United States. Thus, EDI was the first form of electronic commerce to be widely used in business—some 20 years before anyone used the term "electronic commerce" to describe anything!

Early Business Information Interchange Efforts

The emergence of large business organizations that occurred in the late 1800s and early 1900s brought with it the need to create formal records of business transactions. In the 1950s, companies began to use computers to store and process internal transaction records, but the information flows between businesses continued to be printed on paper; purchase orders, invoices, bills of lading, checks, remittance advices, and other standard forms were used to document transactions.

The process of using a person or a computer to generate a paper form, mailing that form, and then having another person enter the data into the trading partner's computer was slow, inefficient, expensive, redundant, and unreliable. By the 1960s, businesses that engaged in large volumes of transactions with each other had begun exchanging transaction information on punched cards or magnetic tape. Advances in data communications technology eventually allowed trading partners to transfer data over telephone lines instead of shipping punched cards or magnetic tapes to each other.

Although these information transfer agreements between trading partners increased efficiency and reduced errors, they were not an ideal solution. Since the data translation programs that one trading partner wrote usually would not work for other trading partners, each company participating in this information exchange had to make a substantial investment in computing infrastructure. Only large trading partners could afford this investment, and even those companies had to have a significant number of transactions to justify the cost. Smaller or lower-volume trading partners could not afford to participate in the benefits of these paper-free exchanges.

In 1968, a number of freight and shipping companies joined together to form the Transportation Data Coordinating Committee (TDCC), which was charged with exploring ways to reduce the paperwork burden that shippers and carriers faced. The TDCC created a standardized information set that included all the data elements that shippers commonly included on bills of lading, freight invoices, shipping manifests, and other paper forms. Instead of printing a paper form, shippers could convert information about shipments into a computer file that conformed to the TDCC standard format. The shipper could electronically transmit that computer file to any freight company that had adopted the TDCC format. The freight company translated the TDCC format into data it could use in its own information systems. The savings from not printing and handling forms, not entering the data twice, and not having to worry about error-correction procedures was significant for most shippers and freight carriers.

Although these early industry-specific data interchange efforts were very helpful, their benefits were limited to members of the industries that created standard-setting groups. In addition, most businesses that are in a particular industry buy goods and services from businesses that are in other industries. For example, a machinery manufacturer might buy from steel mills, paint distributors, electrical assembly contractors, and container manufacturers. Almost every business needs to buy office supplies and the services of freight and transportation companies. Thus, full realization of EDI's economies and efficiencies required standards that could be used by companies in all industries.

Emergence of Broader Standards

After a decade of fragmented attempts at setting broader EDI standards, a number of industry groups and several large companies decided to mount a major effort to create a set of cross-industry standards for electronic components, mechanical equipment, and other widely used items. The **American National Standards Institute (ANSI)** has been the coordinating body for standards in the United States since 1918. ANSI does not set standards itself, but it has created a set of procedures for the development of national standards and it accredits committees that follow those procedures.

In 1979, ANSI chartered a new committee to develop uniform EDI standards. This committee is called the **Accredited Standards Committee X12 (ASC X12)**. The committee meets three times each year to develop and maintain EDI standards. The committee and its subcommittees include information systems professionals from over 800 businesses and other organizations. Membership is open to organizations and individuals who have an interest in the standards. The administrative body that coordinates ASC X12 activities is the **Data Interchange Standards Association (DISA)**.

The ASC X12 standard has benefited from the participation of members from a wide variety of industries. The standard currently includes specifications for several hundred **transaction sets**, which are the names of the formats for specific business data interchanges. Figure 5-4 lists some of the more commonly used ASC X12 transaction sets.

104 - Air Shipment Information	829 - Payment Cancellation Request
110 - Air Freight Details and Invoice	840 - Request for Quotation
125 - Multilevel Railcar Load Details	841 - Specifications/Technical Information
151 - Electronic Filing of Tax Return Data Acknowledgement	842 - Nonconformance Report
170 - Revenue Receipts Statement	843 - Response to Request for Quotation
180 - Return Merchandise Authorization and Notification	846 - Inventory Inquiry/Advice
204 - Motor Carrier Shipment Information	847 - Material Claim
210 - Motor Carrier Freight Details and Invoice	850 - Purchase Order
213 - Motor Carrier Shipment Status Inquiry	853 - Routing and Carrier Instruction
214 - Transportation Carrier Shipment Status Message	854 - Shipment Delivery Discrepancy Information
304 - Shipping Instructions	855 - Purchase Order Acknowledgment
317 - Delivery/Pickup Order	856 - Ship Notice/Manifest
325 - Consolidation of Goods in Container	857 - Shipment and Billing Notice
350 - U.S. Customs Release Information	859 - Freight Invoice
404 - Rail Carrier Shipment Information	860 - Purchase Order Change Request–Buyer-Initiated
410 - Rail Carrier Freight Details and Invoice	861 - Receiving Advice/Acceptance Certificate
421 - Estimated Time of Arrival and Car Scheduling	865 - Purchase Order Change Acknowledgment/Request–Seller-Initiated
440 - Shipment Weights	867 - Product Transfer and Resale Report
466 - Rate Request	869 - Order Status Inquiry
511 - Requisition	870 - Order Status Report
810 - Invoice	879 - Price Change
812 - Credit/Debit Adjustment	893 - Item Information Request
813 - Electronic Filing of Tax Return Data	920 - Loss or Damage Claim–General Commodities
820 - Payment Order/Remittance Advice	924 - Loss or Damage Claim–Motor Vehicle
828 - Debit Authorization	997 - Functional Acknowledgment
	998 - Set Cancellation

Figure 5-4 *Commonly used ASC X12 transaction sets.*

Although the X12 standards were quickly adopted by major firms in the United States, in many cases businesses in other countries continued to use their own national standards. In the mid-1980s, the United Nations Economic Commission for Europe invited North American and European EDI experts together to build a common set of EDI standards based on the successful experiences of U.S. firms in using the ASC X12 standards. In 1987, the United Nations published its first standards under the title **EDI for Administration, Commerce**, and **Transport (EDIFACT**, or **UN/EDIFACT**). You can learn more about **UN/EDIFACT** in the Online Companion. As you can see from Figure 5-5, a number of the commonly used UN/EDIFACT standard transaction sets are similar to those in the ASC X12 standard.

```
AUTHOR  -  Authorization                          IFTCCA  -  Forwarding/Transport Shipment
BOPCUS  -  Balance of Payment Customer                        Charge Calculation
            Transaction Report                    IFTDGN  -  Dangerous Goods Notification
BOPDIR  -  Direct Balance of Payment Declaration   IFTFCC  -  International Transport Freight
BOPINF  -  Balance of Payment Information from                Costs/Other Charges
            Customer                              IFTMAN  -  Arrival Notice
COARRI  -  Container Discharge/Loading Report      INVOIC  -  Invoice
COHAOR  -  Container Special Handling Order        INVRPT  -  Inventory Report
CONAPW  -  Advice on Pending Works                 ORDCHG  -  Purchase Order Change Request
CONDPV  -  Direct Payment Valuation                ORDERS  -  Purchase Order
CONITT  -  Invitation to Tender                    ORDRSP  -  Purchase Order Response
CONPVA  -  Payment Valuation                       PAXLST  -  Passenger List
CONQVA  -  Quantity Valuation                      PAYMUL  -  Multiple Payment Order
COPRAR  -  Container Discharge/Loading Order       PAYORD  -  Payment Order
COREOR  -  Container Release Order                 PRODEX  -  Product Exchange Reconciliation
COSTCO  -  Container Stuffing/Stripping            QALITY  -  Quality Data
            Confirmation                          QUOTES  -  Quote
COSTOR  -  Container Stuffing/Stripping Order      RECADV  -  Receiving Advice
CREADV  -  Credit Advice                           REMADV  -  Remittance Advice
CUSDEC  -  Customs Declaration                     REQDOC  -  Request for Document
CUSRES  -  Customs Response                        REQOTE  -  Request for Quote
DEBADV  -  Debit Advice                            SSREGW  -  Notification of Registration of a
DELFOR  -  Delivery Schedule                                  Worker
HANMOV  -  Cargo/Goods Handling and Movement       STATAC  -  Statement of Account
IFCSUM  -  Forwarding and Consolidation            SUPRES  -  Supplier Response
            Summary
```

Figure 5-5   *Commonly used UN/EDIFACT transaction sets*

The ASC X12 organization and the UN/EDIFACT group agreed in late 2000 to develop one common set of international standards; however, no date for implementation of the common standards has been set. Both organizations created their transaction sets by extracting the information items from the paper forms used to document business transactions. Some critics of the current EDI standards argue that this reliance on forms has made it difficult for businesses to integrate EDI data flows into their business process-oriented information systems. Unfortunately, changing EDI transaction sets to follow business processes instead of paper transaction forms would require a complete redesign of standards that have become part of many organizations' computing infrastructures over the past 30 years.

How EDI Works

Although the basic idea behind EDI is straightforward, its implementation can be complicated, even in fairly simple business situations. For example, consider a company that needs a replacement for one of its metal cutting machines. This section describes the steps involved in making this purchase using a paper-based system, and then explains how the process would change using EDI. In both of these examples, assume that the vendor uses its own vehicles instead of a common carrier to deliver the purchased machine.

Paper-Based Purchasing Process

The buyer and the vendor in this example are not using any integrated software for business processes internally; thus, each information processing step results in the production of a paper document that must be delivered to the department handling the next step. Information transfer between the buyer and vendor is also paper based and can be delivered by mail, courier, or fax. The information flows that occur in the paper-based version of the purchasing process example are shown in Figure 5-6.

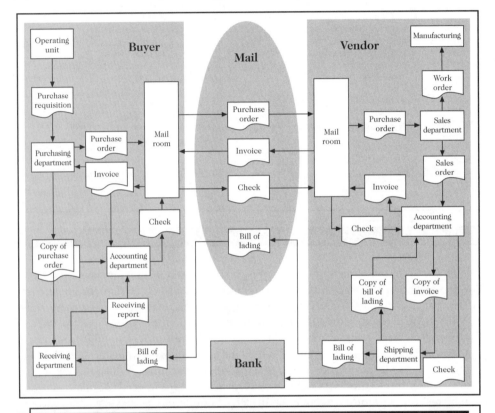

Figure 5-6 *Information flows in the paper-based purchasing process*

Once the production manager in the operating unit decides that the metal cutting machine needs to be replaced, the following process begins:

- The production manager completes a purchase requisition form and sends it to purchasing. This requisition describes the machine that is needed to perform the metal cutting operation.
- Purchasing contacts vendors to negotiate price and terms of delivery. When purchasing has selected a vendor, it prepares a purchase order and forwards it to the mail room.
- Purchasing also sends one copy of the purchase order to the receiving department so that receiving can plan to accept delivery when scheduled; purchasing sends another copy to accounting to advise it of the financial implications of the order.

- The mail room sends the purchase order it received from purchasing to the selected vendor by mail or courier.
- The vendor's mail room receives the purchase order and forwards it to its sales department.
- The vendor's sales department prepares a sales order that it sends to its accounting department and a work order that it sends to manufacturing. The work order describes the machine's specifications and authorizes manufacturing to begin work on it.
- When the machine is completed, manufacturing notifies accounting and sends the machine to shipping.
- The accounting department sends the original invoice to the mail room and a copy of the invoice to the shipping department.
- The mail room sends the invoice to the buyer by mail or courier.
- The vendor's shipping department uses its copy of the invoice to create a bill of lading and sends it with the machine to the buyer.
- The buyer's mail room receives the invoice at about the same time as its receiving department receives the machine with its bill of lading.
- The buyer's mail room sends one copy of the invoice to purchasing so the purchasing department knows that the machine has been received, and sends the original invoice to accounting.
- The buyer's receiving department checks the machine against the bill of lading and its copy of the purchase order. If the machine is in good condition and matches the specifications on the bill of lading and the purchase order, receiving completes a receiving report and delivers the machine to the operating unit.
- Receiving sends a completed receiving report to accounting.
- Accounting makes sure that all details on its copy of the purchase order, the receiving report, and the original invoice match. If they do, accounting issues a check and forwards it to the mail room.
- The buyer's mail room sends the check by mail or courier to the vendor.
- The vendor's mail room receives the check and sends it to accounting.
- Accounting compares the check to its copies of the invoice, bill of lading, and sales order. If all details match, accounting deposits the check in the vendor's bank and records the payment received.

EDI Purchasing Process

The information flows that occur in the EDI version of this sample purchasing process are shown in Figure 5-7. The mail service has been replaced with the data communications of an EDI network, and the flows of paper within the buyer's and vendor's organizations have been replaced with computers running EDI translation software.

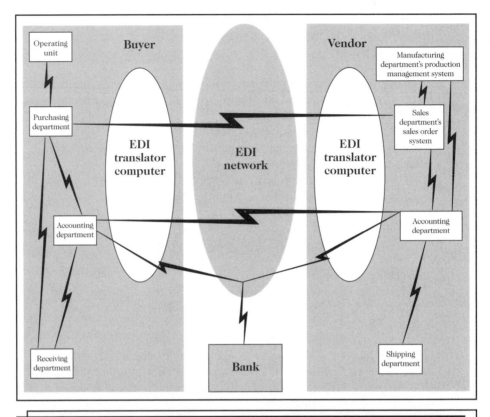

Figure 5-7   *Information flows in the EDI purchasing process*

In the EDI purchasing process, when the operating unit manager decides that the metal cutting machine needs to be replaced, the following process begins:

- The operating unit manager sends an electronic message to its purchasing department. This message will describe the machine that is needed to perform the metal cutting operation.
- Purchasing contacts vendors by telephone, e-mail, or through their Web sites to negotiate price and terms of delivery. After selecting a vendor, purchasing sends a message to the sales department announcing the selection.
- The buyer's EDI translator computer converts this message to a standard format purchase order transaction set, and then forwards the message through an EDI network to the vendor.
- Purchasing also sends one electronic message to the buyer's receiving department so it can plan to accept delivery when it is scheduled; purchasing sends to the buyer's accounting department another electronic message that includes details such as the agreed purchase price.
- The vendor's EDI translator computer receives the purchase order transaction set message and converts it to the file format that the vendor's information systems use.

- The converted purchase order details appear in the sales department's sales order system, and are automatically forwarded to the production management system in manufacturing and to the accounting system.
- The information that was automatically forwarded to manufacturing describes the machine's specifications and authorizes manufacturing to begin work on it.
- When the machine is completed, manufacturing notifies accounting and sends the machine to the vendor's shipping department.
- The vendor's shipping department sends an electronic message to its accounting department indicating that the machine is ready to ship.
- The vendor's accounting department sends a message to its EDI translator computer, which converts the message to the standard invoice transaction set and forwards it through the EDI network to the buyer.
- The buyer's EDI translator computer receives the invoice transaction set before its receiving department receives the machine. The computer then converts the invoice data to a format that the buyer's information systems can use. The invoice data becomes immediately available to both the buyer's accounting department and its receiving department.
- When the machine arrives, the buyer's receiving department checks the machine against the invoice information on its computer system. If the machine is in good condition and matches the specifications shown in the buyer's system, receiving sends a message to accounting confirming that the machine has been received in good order. It then delivers the machine to the operating unit.
- The buyer's accounting department system compares all details in the purchase order data, receiving data, and the decoded invoice transaction set from the vendor. If the details all match, the accounting system notifies its bank to reduce the buyer's account and increase the vendor's account by the amount of the invoice. The EDI network may provide services that perform this task.

Value-Added Networks

As you can see by comparing the paper-based purchasing process in Figure 5-6 to the EDI purchasing process in Figure 5-7, the departments are exchanging the same messages among themselves, but EDI reduces paper flow and streamlines the interchange of information among departments within a company and between companies. These efficiencies were responsible for the benefits described in the GE Lighting example presented in the introduction to this chapter. The three key elements shown in Figure 5-6 that alter the process so dramatically are the EDI network (instead of the mail service) that connects the two companies and the two EDI translator computers that handle the conversion of data from the formats used internally by the buyer and the vendor to standard EDI transaction sets. Trading partners can implement the EDI network and EDI translation processes in several ways. Each of these ways uses one of two basic approaches: direct connection or indirect connection.

Direct Connection Between Trading Partners

The first approach, called **direct connection EDI**, requires each business in the network to operate its own on-site EDI translator computer (as shown in Figure 5-7). These EDI translator computers are then connected directly to each other using modems and dial-up telephone lines or dedicated leased lines. The dial-up option becomes troublesome when customers or vendors are located in different time zones, and when transactions are time sensitive or high in volume. The dedicated leased-line option can become very expensive for businesses that must maintain many connections with customers or vendors. Trading partners that use different communications protocols can make either of the direct connection methods difficult to implement.

Indirect Connection Between Trading Partners

Instead of connecting directly to each of its trading partners, a company might decide to use the services of a value-added network. As you learned in Chapter 1, a value-added network (VAN) is a company that provides communications equipment, software, and skills needed to receive, store, and forward electronic messages that contain EDI transaction sets. To use the services of a VAN, a company must install EDI translator software that is compatible with the VAN. Often, the VAN will supply this software as part of its operating agreement.

To send an EDI transaction set to a trading partner, the VAN customer connects to the VAN using a dedicated or dial-up telephone line and then forwards the EDI-formatted message to the VAN. The VAN logs the message and delivers it to the trading partner's mailbox on the VAN computer. The trading partner then dials in to the VAN and retrieves its EDI-formatted messages from that mailbox. This approach is called **indirect connection EDI** because the trading partners pass messages through the VAN instead of connecting their computers directly to each other. Figures 5-8 and 5-9 show the differences between direct connection EDI and indirect connection EDI that uses a VAN.

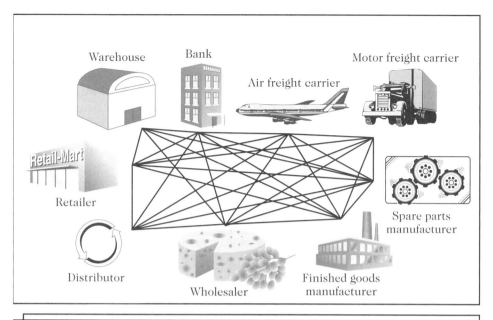

Figure 5-8 *Direct connection EDI*

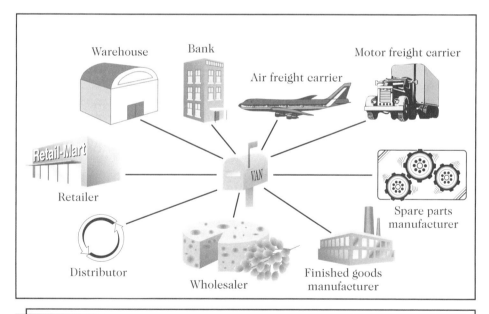

Figure 5-9 *Indirect connection EDI through a VAN*

Companies that provide VAN services include **Computer Associates**, **General Electric Information Services**, **GPAS**, **IBM Global Services**, **Kleinschmidt**, and

Peregrine Systems. Advantages of using a VAN are as follows:

- Users need to support only the VAN's one communications protocol instead of many possible protocols used by trading partners.
- The VAN records message activity in an audit log. This VAN audit log becomes an independent record of transactions, and this record can be helpful in resolving disputes between trading partners.
- The VAN can provide translation between different transaction sets used by trading partners (for example, the VAN can translate an ASC X12 set into a UN/EDIFACT set).
- The VAN can perform automatic compliance checking to ensure that the transaction set is in the specified EDI format.

VANs do have a number of disadvantages, however. One major issue is cost. Most VANs require an enrollment fee, a monthly maintenance fee, and a transaction fee. The transaction fee can be based on transaction volume, transaction length, or both. Trading partners that have few transactions often find it difficult to justify the high fixed costs of the enrollment fee and the monthly maintenance fee. For example, the up-front cost of implementing indirect connection EDI, including software, VAN enrollment fee, and hardware, can exceed $50,000. Other trading partners that have high transaction volumes find the VAN's ongoing transaction-based fees to be prohibitive.

In the past, many vendors were forced into bearing the high costs of participating in EDI to satisfy the needs of one or two large customers. This happened frequently to suppliers of the auto industry and the retail merchandising industry. Using VANs can become cumbersome and expensive for companies that want to do business with a number of trading partners, each using different VANs. Although some VANs do offer the service of exchanging messages with other VANs, the cost of this service can be unpredictable. Also, inter-VAN transfers do not always provide a clear audit trail for use in dispute resolution. Firms that had been precluded from adopting EDI by its high cost welcomed the Internet as a low-cost communications medium that could help them overcome some of the disadvantages of traditional EDI.

EDI on the Internet

As the Internet gained prominence as a tool for conducting business, trading partners that had been using EDI began to view the Internet as a potential replacement for the expensive leased lines and dial-up connections they had been using to support both direct and VAN-aided EDI. Companies that had been unable to afford EDI began to look at the Internet as an enabling technology that might get them back in the game of selling to large customers who demanded EDI capabilities of their suppliers.

The major roadblocks to conducting EDI over the Internet initially were general concerns about security and the Internet's general inability to provide audit logs and third-party verification of message transmission and delivery. As the basic TCP/IP structure of the Internet was enhanced with secure protocols such as secure HTTP (SHTTP), HTTP over a secure socket layer (HTTPS), and various other encryption schemes, businesses worried less about security issues, although concerns still existed. The lack of third-party verification continues to be an issue, since the Internet has no built-in facility for it. Because EDI transactions are business contracts and often involve large amounts of money, the issue of nonrepudiation is significant. **Nonrepudiation** is the ability to establish that a particular transaction actually occurred. It prevents each party from repudiating, or denying, the transaction's validity or existence. In the past, the nonrepudiation function was provided either by a VAN's audit logs for indirect connection EDI or a comparison of the trading partners' message logs for direct connection EDI.

Open Architecture of the Internet

A number of new firms, such as **Commerce One**, **DynamicWeb Enterprises**, the **EC Company**, **IPNet**, and **VanTree**, have begun providing EDI services on the Internet. Firms that had been providing traditional VAN services, such as **AT&T IP Services**, now offer EDI on the Internet. EDI on the Internet is also called **open EDI** because the Internet is an open architecture network, as you learned in Chapter 2. Many of the new EDI offerings go beyond traditional EDI and help trading partners accomplish information interchanges that are more complex than the EDI standard transaction sets.

The open architecture of the Internet allows trading partners virtually unlimited opportunities for customizing their information interchanges. New tools such as XML are helping trading partners be even more flexible in exchanging detailed information. Three EDI groups have recently met and charged a new ASC X12 task group with these broad objectives:

- Converting the ASC X12 EDI data elements and transaction set structures to XML in a way that retains a one-to-one mapping between the existing ASC X12 and the new XML data elements
- Developing XML data element names that allow people with existing ASC X12 transaction set expertise to continue to use that expertise when working with the new XML transaction sets
- Meeting the needs of application-to-application and human-to-application interfaces

Other firms are extending their internal networks (intranets) to their trading partners, which turns the intranets into extranets. Technologies such as virtual private networks (VPNs) are providing the security that makes such extranets increasingly attractive. For example, Nintendo USA has been using an EDI-based product registration system to prevent fraudulent returns. The system allows retailers to send directly to Nintendo USA the serial numbers of Nintendo systems that they have sold. This system has worked well for large retailers, but the benefits have not offset the costs for smaller toy stores. Therefore, in 1998, Nintendo expanded the registration system to include non-EDI adopters. Using an IPNet software package that captures serial number and other warranty information at the cash register and then sends it over the Internet to Nintendo, smaller retailers now enjoy all the benefits of EDI at a much lower cost than traditional EDI.

Financial EDI

Although Internet EDI is growing and offering new, flexible information interchange solutions for many trading partners, some elements of EDI remain difficult to transfer to the Internet. The EDI transaction sets that provide instructions to a trading partner's bank are called financial EDI (FEDI). All banks have the ability to perform electronic funds transfers (EFTs), which are the movement of money from one bank

199

account to another. You learned about EFTs in Chapter 1. The bank accounts involved in EFTs may be customer accounts or the accounts that banks keep on their own behalf with each other. When EFTs involve two banks, they are executed using a clearinghouse. In the United States, most EFTs are handled through the **Automated Clearing House (ACH)**. **EDI-capable banks** are those banks that are equipped to exchange payment and remittance data through VANs. Some banks also offer VAN services for nonfinancial transactions. These banks are called **value-added banks (VABs)**. Nonbank VANs that can translate financial transaction sets into ACH formats and transmit them to banks that are not EDI-capable are sometimes called **financial VANs (FVANs)**.

Many companies are reluctant to send over the Internet FEDI transaction sets that contain transfer instructions for large amounts of money—in some cases, millions of dollars—because of the perceived low level of security on the Internet. FEDI transaction sets are negotiable instruments—the electronic equivalent of checks. The reliability of FEDI itself is an issue, too. Since FEDI uses the Internet, it can be exposed to problems that are less likely to occur on the dedicated leased telephone lines used for connecting to a VAN. For example, if an Internet router outage delays an instruction to transfer $10 million, a trading partner could easily lose a day's interest on the funds. Thus, companies that have established indirect connection EDI through a VAN are likely to continue doing so to ensure added security for FEDI transaction sets, even though the cost is much higher than the cost of using the Internet for FEDI.

Hybrid EDI Solutions

Some firms are offering hybrid EDI solutions that use the Internet for part of the transaction. For example, **Bottomline Technologies**' PayBase package allows trading partners to send payment instructions over the secure ACH network and the detailed payment description information directly to the trading partner over the Internet. Since the payment description information is not a negotiable instrument, the concerns about security are less critical than with FEDI transactions. Other hybrid solutions include EDI-HTML translation services. These are offered by banks and other EDI service firms. One example of this kind of service is Northern Trust's **NetTransact** service, which provides an interface for the smaller business that is connected to the Internet but does not have EDI translation capability. Figure 5-10 shows how a service such as NetTransact works.

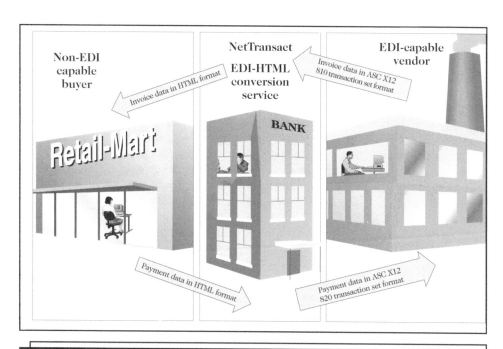

Non-EDI capable buyer

Invoice data in HTML format

NetTransact

EDI-HTML conversion service

Invoice data in ASC X12 810 transaction set format

EDI-capable vendor

BANK

Retail-Mart

Payment data in HTML format

Payment data in ASC X12 820 transaction set format

Figure 5-10 *NetTransact EDI-HTML conversion service*

For example, if an EDI-enabled vendor sends an invoice transaction set to a trading partner using NetTransact, that trading partner (the buyer) receives an HTML document instead of the ASC X12 810-formatted invoice data. The buyer can read the HTML document in a Web browser and can download the HTML document into a local accounting or purchase tracking system. NetTransact converts the buyer's response to the invoice into an ASC X12 820-formatted payment transaction set and forwards the set to the vendor or the vendor's VAN. To the vendor, this transaction appears to be all EDI. To the buyer, the transaction is no more difficult than using browser software to surf the Web. Increasingly, businesses will explore methods such as these that use the connectivity of the Internet to reduce the costs of EDI.

By 1999, the largest EDI service vendors—including GE Information Services, Peregrine's Harbinger division, and IBM Global Services—were all using the Internet, or at least the Internet's TCP/IP communications protocol. Since then, these companies have been working on ways to use XML as a way of transmitting EDI transaction sets.

EDI was the original form of electronic commerce, and it appears that it will continue to evolve and be a part of the electronic commerce boom on the Internet. Total EDI transaction volume was about $2 billion in 2001 and is expected to exceed $4 billion in 2002. The share of EDI carried by VANs has been declining. In 1997, VANs carried more than 95 percent of all EDI traffic. Experts predict that VANs' share of the transaction volume will decline to about 50 percent by 2003. This reflects a large expected increase in the number of smaller businesses that are willing to engage in EDI now that the use of an expensive VAN is no longer required.

SUPPLY CHAIN MANAGEMENT

In Chapter 1, you learned how companies can organize their strategic business unit activities using an industry value chain. The part of an industry value chain that precedes a particular strategic business unit is often called a **supply chain**. A company's supply chain for a particular product or service includes all the activities undertaken by every predecessor in the value chain to design, produce, promote, market, deliver, and support each individual component of that product or service.

For example, the supply chain of an automobile manufacturer would include every activity undertaken by every component supplier, including engine manufacturers, steel fabricators, glass manufacturers, wiring harness assemblers, and thousands of others. The purchasing department has traditionally been charged with buying all of these components at the lowest price possible. Usually, purchasing staff did this by identifying qualified vendors and asking them to prepare bids that described what they would supply and how much they would charge. The purchasing staff would then select the lowest bid that still met the quality standards for the component. This bidding process led to a very competitive environment with a large number of suppliers; this process focused excessively on the cost of individual components and ignored the total supply chain costs, including the cost to the manufacturing organization of dealing with such a large number of suppliers.

Value Creation in the Supply Chain

In recent years, businesses have realized that they can save money and increase product quality by taking a more active role in negotiations with suppliers. By engaging suppliers in cooperative, long-term relationships, companies have found that they can work together with these suppliers to identify new ways to provide their own customers with faster, cheaper, and better service. This process of taking an active role in working with suppliers to improve products and processes is called **supply chain management**. By coordinating the efforts of supply chain participants, firms that engage in supply chain management are reaching beyond the limits of their own organization's hierarchical structure and creating a new network form of organization among the members of the supply chain.

Supply chain management was originally developed as a way to reduce costs. It focused on very specific elements in the supply chain and tried to identify opportunities for process efficiency. Today, supply chain management is used to add value in the form of benefits to the ultimate consumer at the end of the supply chain. This requires a more holistic view of the entire supply chain than had been common in the early days of supply chain management.

Businesses that engage in supply chain management work to establish long-term relationships with a small number of very capable suppliers. These suppliers, called **tier one suppliers**, in turn, develop long-term relationships with a larger number of suppliers that provide components and raw materials to them. These **tier two suppliers** manage relationships with the next level of suppliers, called **tier three suppliers**, who provide them with components and raw materials. A key element of these relationships is trust between the parties. The long-term relationships that are created among participants in the supply chain are called **supply alliances**. The level of information sharing that must take place among the supply chain participants can be a major barrier to entering into these alliances. Firms are not used to disclosing detailed operating information and often perceive that information disclosure might hurt the firm by placing it at a competitive disadvantage.

In exchange for the stability of the closer, long-term relationships, buyers expect annual price reductions and quality improvements from suppliers at each stage of the supply chain. However, all supply chain participants share information and work together to create value. Ideally, the supply chain coordination creates enough value that each level of supplier can share the benefits of reduced cost and more efficient operations. Supply chain management has been gaining momentum during the past decade and is supported by major purchasing groups such as the **National Association for Purchasing Management** and the **Supply Chain Council**. By working together, supply chain members can reduce costs and increase the value of the product or service to the ultimate consumer.

One area in which differences in organizational goals often arise is described by Marshall Fisher in his 1997 *Harvard Business Review* article. He explains that firms often organize themselves to achieve either efficient process goals or market-responsiveness flexibility goals. Some companies structure themselves to be efficient producers, whereas others structure themselves to be flexible producers. The kinds of things that allow a firm to be an efficient, low-cost producer are exactly the things that prevent a firm from being flexible enough to respond to market changes. The efficient producer will invest in expensive machines that can stamp out large numbers of low-cost items. This investment drives the cost of production down, but makes it difficult for the producer to be flexible. For example, the large investment in specialized machinery prevents that producer from reconfiguring the plant layout. If even one member of the supply chain for a product that requires flexible production operates as an efficient producer (instead of as a flexible producer), every other firm in the supply chain will suffer. The efficient producer will create bottlenecks that will hamper the best efforts of all other supply chain members. Clear communication up and down the supply chain can keep each participant informed of what the ultimate consumer is demanding. The participants can then plot a strategy that will meet those demands.

Using Internet Technologies in the Supply Chain

Clear communications, and quick responses to those communications, are a key element of successful supply chain management. Technologies, and especially the technologies of the Internet and the Web, can be very effective communication enhancers. For the first time, firms can effectively manage the details of their own internal processes and the processes of other members of their supply chain. Software that uses the Internet can help all members of the supply chain review past performance, monitor current performance, and predict when and how much of certain products need to be produced. Figure 5-11 lists the advantages of using Internet technologies in supply chain management.

Suppliers can:
- Share information about customer demand fluctuations
- Receive rapid notification of product design changes and adjustments
- Provide specifications and drawings more efficiently
- Increase the speed of processing transactions
- Reduce the cost of handling transactions
- Reduce errors in entering transaction data
- Share information about defect rates and types

Figure 5-11 *Advantages of using Internet and Web technologies in supply chain management*

Increasing Efficiency in the Supply Chain

Many companies are using Internet and Web technologies to manage supply chains in ways that yield increases in efficiency throughout the chain. These companies have found ways to increase process speed, reduce costs, and increase manufacturing flexibility so that they can respond to changes in the quantity and nature of ultimate consumer demand.

For example, **Boeing**, the largest producer of commercial aircraft in the world, faces a huge task in keeping its production on schedule. Each airplane requires more than 1 million individual parts and assemblies, and each airplane is custom configured to meet the purchasing airline's exact specifications. These parts and assemblies must be completed and delivered on schedule or the production process comes to a halt.

In 1997, production and scheduling errors required Boeing to shut down two entire assembly operations for several weeks, costing the company over $1.5 billion. To prevent this from ever happening again, Boeing has invested in a number of new information systems that increase production efficiency by providing planning and control over logistics in every element of its supply chain. Using EDI and Internet links, Boeing is working with suppliers so that they can provide exactly the right part or assembly at exactly the right time. Even before starting an airplane into production, Boeing makes the engineering specifications and drawings available to its suppliers through secure Internet connections. As work on the airplane progresses, Boeing keeps every member of the supply chain continually informed of completion milestones achieved and necessary schedule changes.

By its second year of using these new systems, Boeing had cut in half the time needed to complete individual assembly processes. It has realized similar reductions in part defect costs. The combined effects of these increased efficiencies are helping Boeing do a much better job of meeting its customers' needs. Instead of waiting 36 months for delivery, customers can now have their new airplanes in 10 to 12 months.

To further benefit its customers, Boeing launched a spare parts Web site, **Boeing PART** (part analysis and requirements tracking). Over 500 airlines that are Boeing customers do not use EDI to order replacement parts. Boeing PART lets these customers register and then order parts using their Web browsers. The site is processing over 5000 transactions per day at a significantly lower cost to Boeing than if it were handling faxes, telephone calls, and mailed purchase orders. Boeing can deliver most parts ordered through Boeing PART on the same or next day.

Although **Dell Computer** has become famous for its use of the Web to sell custom-configured computers to individuals and businesses, it has also used technology-enabled supply chain management to give customers exactly what they want. It has reduced the amount of inventory it keeps on hand from three weeks' sales to six days' sales. Ultimately, Dell would like to see inventory levels measured in minutes. By increasing the amount of information it has about its customers, Dell has been able to dramatically reduce the amount of inventory it must hold. Dell has also shared this information with members of its supply chain.

Dell's top suppliers have access to a secure Web site that shows them Dell's latest sales forecasts along with other information about planned product changes, defect rates, and warranty claims. In addition, the Web site tells suppliers who Dell's customers are and what they are buying. All of this information helps these tier one suppliers plan their production much better than they could otherwise. The information sharing goes in both directions in Dell's supply chain: Tier one suppliers are required to provide Dell with current information on their defect rates and production problems. As a result, all members of the supply chain work together to reduce inventories, increase quality, and provide high value to the ultimate consumer. Much of this cooperative work requires a high level of trust. To enhance this trust and develop a sense of community, Dell maintains bulletin boards as an open forum in which its supply chain members can share their experiences in dealing with Dell and with each other.

For Boeing, Dell, and other firms such as **PPG Industries** and **BOC Gases**, the use of Internet and Web technologies in managing supply chains has yielded significantly increased process speed, reduced costs, and increased flexibility. All of these attributes combine to allow a coordinated supply chain to produce products and services that better meet the needs of the ultimate consumer.

Building and Maintaining Trust in the Supply Chain

The major issue that most companies must deal with in forming supply chain alliances is developing trust. Continual communication and information sharing are key elements in building trust. Since the Internet and the Web provide excellent ways to communicate and share information, they offer new avenues for building trust. Most procurement professionals have built trust on years of doing business with the same vendors. In many industries, vendors send sales representatives to call on buyers regularly. Vendors also participate actively in trade shows and conferences. By giving buyers frequent opportunities to interact with vendor representatives, vendors help build trust.

Vendors are finding that the Web gives them an opportunity to stay in contact with their customers more easily and less expensively. Although most buyers will still see sales representatives regularly, e-mail and the Web gives them nearly instant access to their sales representative and other vendor personnel. By providing comprehensive information at a moment's notice, vendors can build buyers' trust in the vendor's ability to deliver products and provide the personalized service that buyers need.

Using Technology to Create an Ultimate Consumer Orientation

One of the main goals of supply chain management is to help each company in the chain focus on meeting the needs of the consumer who is at the end of the supply chain. Companies in industries that have long supply chains have, in the past, often found it difficult to maintain this customer focus, which is often called an **ultimate consumer orientation**. Instead, companies have directed their efforts toward meeting the needs of the next member in the supply chain. This short-sighted approach can cause companies to miss opportunities to add value in subsequent steps of the chain.

One company that pioneered the use of Internet technology to go beyond the next step in its value chain is Michelin North America. Michelin has a highly respected brand name and reputation in the tire business. However, most consumers rely on local tire dealers to make specific recommendations when they need replacement tires for their vehicles. Michelin spends a great deal of money on direct advertising to their ultimate consumers. This advertising is directed at maintaining Michelin's powerful brand and convincing the consumer of the value of Michelin tires. The advertising and brand building effort can be wasted, however, if the consumer goes to a local tire dealer who recommends another brand.

Michelin launched an electronic commerce initiative in 1995 called BIB NET. The goal of this initiative was to sell more Michelin tires to consumers, but the initiative was directed at Michelin's tire dealers, not to the ultimate consumers. BIB NET was an extranet that allowed tire dealers to access tire specifications, inventory status, and promotional information about Michelin products through a simple-to-use Web browser interface. Before BIB NET, dealers calling Michelin for product information were sometimes placed on hold. A dealer who is talking to a customer cannot afford to wait on hold. By giving dealers the power to access Michelin product information directly and immediately, Michelin saved money (maintaining a Web page is much less expensive than answering thousands of phone calls) and gave dealers better service. Dealers using BIB NET were much less likely to recommend a competing tire product to their customers.

Since Internet technologies are tools that improve communications at a very low cost, they are ideal aids for enhancing the creation of a highly coordinated and effective supply chain. A number of polls and studies confirm that most information technology managers and most purchasing managers believe that information technology is helping to improve their firms' relationships with suppliers and supply chain management initiatives.

ELECTRONIC MARKETPLACES AND PORTALS

As the Web emerged in the mid-1990s, many business researchers and consultants believed that it would provide an opportunity for companies to establish information hubs for each major industry. These industry hubs would offer news, research reports, analyses of trends, and in-depth reports on companies in the industry—much as specialized industry trade magazines had provided in print format for years. In addition to information, these hubs would offer marketplaces and auctions in which companies in the industry could contact each other and transact business. Because these hubs would offer a doorway (or portal) to the Internet for industry members, and because these hubs would be vertically integrated (that is, each hub would offer services to just one industry), these planned enterprises were called **vertical portals**, or **vortals**. These are types of portals you learned about in Chapter 3.

As with many electronic commerce predictions, the prediction that vertical portals would change business forever did not turn out to be exactly correct. In this section, you will learn how B2B electronic marketplaces were conceived, developed, and operated as this sector of electronic commerce matured from 1997 through the present.

Industry Marketplaces

The first companies to launch industry hubs that followed the vertical portal model created trading exchanges that were focused on a particular industry. These vertical portals became known by various names that highlighted different elements of their collective nature, including **industry marketplaces** (focused on a single industry), **independent exchanges** (not controlled by a company that was an established buyer or seller in the industry), or **public marketplaces** (open to new buyers and sellers just entering the industry). **Ventro** opened its first industry marketplace, Chemdex, in early 1997 to trade in bulk chemicals. To leverage the high investment it had made in trading exchange technology, Ventro followed Chemdex with other Web marketplaces, including Promedix in specialty medical supplies, Amphire Solutions in food service, MarketMile in general business products and services, and a number of others.

Other companies were quick to follow in Ventro's chosen markets and many others. **SciQuest** founded an industry marketplace in life science chemicals. **VerticalNet** launched NECX, an independent exchange in the electronic components industry. The home page of **CheMatch.com**, which competed directly with Ventro in the bulk chemicals market, appears in Figure 5-12.

Figure 5-12   CheMatch.com home page

The number of new entrants into these businesses grew rapidly during the next two years. By mid-2000, there were more than 2200 independent exchanges in a wide variety of industries. For example, there were 200 exchanges operating in the metals industry alone (see "Learning from Failures: MetalSite"). As venture capital funding became scarce for companies that were not earning profits—and virtually all of these marketplaces were not earning profits—many of them closed. By late 2001, there were fewer than 180 industry marketplaces still operating. It simply did not make economic sense to have more than one or two independent marketplaces in any particular industry. Some of the industry pioneers who had closed their industry marketplace operations, such as Ventro and **E-Steel**, began selling the software and technology that they had developed to run their marketplaces. Their new customers were operators of other B2B marketplace models that arose to take business away from the independent marketplaces. You will learn about several of these models in the remainder of this section.

METALSITE

Although a number of small steel manufacturing plants (called mini-mills) have opened in the past 20 years, most of the world's steel is still produced in very large steel mills. In these steel mills, it is only economical to produce steel in large batches. Because of the high cost of reconfiguring machinery, a steel mill set up to create one type of steel (for example, rolled sheets) requires significant time and money to change over to produce another type of steel (for example, bar steel). To minimize these changeover costs, steel mills produce steel products in large batches to meet estimated demand rather than actual orders. Because production quantities are designed to meet estimated demand instead of actual demand, steel mills often have overproduction of some items.

Companies such as Bethlehem Steel, with revenues of over $4 billion and 14,000 employees, solved this problem in the past by sending faxes to potential buyers of their excess production. Buyers would respond with a bid on the product in which they were interested, and Bethlehem would negotiate with them to determine price and delivery terms.

In 1998, MetalSite was one of the first metal trading exchanges to begin doing business on the Web. These exchanges offered manufacturers such as Bethlehem an efficient way to reach a larger market for their excess production. By mid-2000, there were more than 200 metal exchanges operating on the Web. These exchanges were following a reintermediation strategy; that is, they were entering the supply chain of the steel industry to provide some added value that had not existed in the supply chain before. However, most industry analysts agreed that there was no need for more than one or two exchanges in the steel industry. In 2001, metal trading exchange sites began to fail.

MetalSite had grown rapidly. With over $35 million of investors' money, MetalSite was able to sign up 24,000 registered users and by mid-2001 was trading about $30 million worth of steel each month. However, its commissions of between 1 percent and 2 percent on each trade did not yield enough money to cover operating costs. The steel business was in a downturn along with the rest of the U.S. economy, and the downward pressure on commissions from competing exchanges was increasing rapidly. The major steel companies were discussing ways to form alliances to operate their own exchanges. After three years of operation and a desperate last-minute search for new investors, MetalSite closed in August 2001.

MetalSite had entered a business that could not support more than a handful of companies, and it was unable to become one of the survivors. The lesson from MetalSite's experience is that a reintermediation strategy must add significant value to the supply chain, and the company pursuing that strategy must be able to construct significant barriers that competitors must overcome to enter the business. MetalSite was unable to do either and thus failed. Many other B2B exchange sites that found themselves in similar competitive situations have also failed.

Private Stores and Customer Portals

As established companies in various industries watched new businesses open marketplaces, they became concerned that these independent operators would take control of transactions from them in supply chains—control that the established companies had spent years developing. Large companies that sell to many customers that are relatively small can exert great power in negotiating price, quality, and delivery terms with those customers. These sellers feared that industry marketplaces would dilute that power.

Many of these large sellers had already invested heavily in Web sites that they believed would meet the needs of their customers better than any industry marketplace would. For example, Cisco and Dell offer **private stores** for each of their major customers within their selling Web sites. A private store has a password-protected entrance and offers negotiated price reductions on a limited selection of products—usually products that the customer has agreed to purchase in certain minimum quantities. Other companies, such as Grainger and Milacron, provide additional services for customers on their selling Web sites. These **customer portal** sites offer private stores along with services such as part number cross-referencing, product use guidelines, safety information, and others that would be needlessly duplicated if the sellers were to participate in an industry marketplace.

Private Company Marketplaces

Similarly, large companies that purchase from vendors that are relatively small can exert similar power over those vendors in purchasing negotiations. The procurement departments of these companies can invest in procurement software from companies such as Ariba and CommerceOne (you will learn more about all types of electronic commerce software in Chapter 9). This software, generally referred to as **e-procurement software**, allows a company to manage its purchasing function through a Web interface. It automates many of the authorizations and other steps described in Figure 5-1 as part of business procurement operations. Although e-procurement software was originally designed to help manage the MRO procurement process, recent releases of this software have begun to include other marketplace functions, such as request for quote posting areas, auctions, and integrated support for purchasing direct materials. Most industry observers expect these features to be improved and expanded in future versions.

Companies that implement e-procurement software usually require their suppliers to bid on their business. For example, an office supplies provider would create a schedule of prices at which it would sell to the company. The company would then compare that pricing to bids from other suppliers. The selected supplier would provide product price and description information to the company, which would insert that information into its e-procurement software. This permits authorized employees to order office supplies at the negotiated prices through a Web interface.

When industry marketplaces opened for business, these larger companies were reluctant to abandon their investments in e-procurement software or to make the software work with industry marketplaces' software—especially in the early years of industry marketplaces when there were many of them in each industry. These companies use their power in the supply chain to force suppliers to deal with them on their own terms rather than negotiate with suppliers in an industry marketplace.

As marketplace software became more reliable, many of these companies purchased software and technology consulting services from companies, such as Ventro and e-Steel, that had abandoned their industry marketplace business and were offering the software they had developed to companies that wanted to develop private marketplaces. A **private company marketplace** is a marketplace that provides auctions, requests for quotes postings, and other features (many of which are similar to those of e-procurement software) to companies that want to operate their own marketplace. United Technologies, which sells more than $14 billion of high technology products and services to the aerospace and building systems industries, was one of the first major companies to open a private company marketplace in 1996. Since then, United Technologies has purchased more than $3 billion through its private marketplace and estimates that it has saved more than $400 million through lower prices and transaction cost savings on those purchases.

Industry Consortia-Sponsored Marketplaces

Some companies had relatively strong negotiating positions in their industry supply chains, but did not have enough power to force suppliers to deal with them through a private company marketplace. These companies began to form consortia to sponsor marketplaces. An **industry consortia-sponsored marketplace** is a marketplace formed by several large buyers in a particular industry.

One of the first such marketplaces was Covisint, which was created in 2000 by a consortium of DaimlerChrysler, Ford, and General Motors. Several thousand auto industry suppliers now belong to **Covisint**. In the hotel industry, Marriott, Hyatt, and three other major hotel chains formed a consortium to create the **Avendra** marketplace. These consortia-based marketplaces—along with private company marketplaces, private Web stores, and customer portals—have taken a large part of the market from the industry marketplaces that appeared to be so promising in the early days of B2B electronic commerce. Figure 5-13 summarizes the characteristics of five general forms of marketplaces that exist in B2B electronic commerce today, adapted from work presented by Warren Raisch, a Web marketplace consultant, in his book *The eMarketplace*.

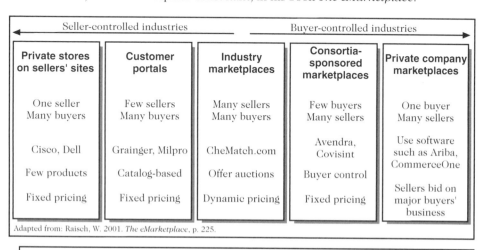

Seller-controlled industries			Buyer-controlled industries	
Private stores on sellers' sites	**Customer portals**	**Industry marketplaces**	**Consortia-sponsored marketplaces**	**Private company marketplaces**
One seller Many buyers	Few sellers Many buyers	Many sellers Many buyers	Few buyers Many sellers	One buyer Many sellers
Cisco, Dell	Grainger, Milpro	CheMatch.com	Avendra, Covisint	Use software such as Ariba, CommerceOne
Few products	Catalog-based	Offer auctions	Buyer control	Sellers bid on major buyers' business
Fixed pricing	Fixed pricing	Dynamic pricing	Fixed pricing	

Adapted from: Raisch, W. 2001. *The eMarketplace*, p. 225.

Figure 5-13 *Characteristics of B2B electronic marketplaces*

Summary

In this chapter, you learned that companies are using Internet and Web technologies in a variety of ways to improve their purchasing and logistics primary activities. Businesses are also making similar improvements in a wide range of support activities such as human resources, accounting, and technology development. Companies and other large organizations, such as government agencies, are finding it more important than ever to extend the reach of their enterprise planning and control activities beyond their organization's legal definition and to include parts of other organizations. This emerging network model of organization was introduced in Chapter 1 and is used in this chapter to describe the growth in interorganizational communication and coordination.

EDI, the first example of electronic commerce, was first developed by freight companies to reduce the paperwork burden of processing repetitive transactions. The spread of EDI to virtually all large companies over the past 30 years has led smaller businesses to seek an affordable way of participating in EDI. The Internet is now providing the inexpensive communications channel that EDI lacked for so many years.

The increase in communications capabilities offered by the Internet and the Web is, and will continue to be, an important force driving the adoption of supply chain management techniques in a variety of industries. Supply chain management incorporates several elements that can be implemented and enhanced through the use of the Internet and the Web. Increasingly, firms are connecting with their supply chain alliance partners to become more efficient and provide more value to the ultimate consumer of their value chain's products and services.

The emergence of industry electronic marketplaces in the mid-1990s gave way to the development of several different models for B2B electronic commerce, including private stores, customer portals, private marketplaces, and industry consortia-sponsored marketplaces. All four of these models continue to coexist with the original industry marketplace model today. It is not clear which of these models, if any, will dominate B2B electronic commerce in the future.

Key Terms

Accredited Standards Committee X12 (ASC X12)
American National Standards Institute (ANSI)
Automated Clearing House (ACH)
Customer portal
Direct connection EDI
Direct materials
E-government
E-procurement software
EDI for Administration, Commerce, and Transport (EDIFACT)
EDI-capable banks
EDI-compatible
Financial VANs (FVANs)
Independent exchange
Indirect connection EDI
Indirect materials

Industry marketplace
Industry consortia-sponsored marketplace
Knowledge management
Maintenance, repair, and operating (MRO)
Nonrepudiation
Open EDI
Private company marketplace
Private store
Procurement
Public marketplace
Sourcing
Supply alliances
Supply chain
Supply chain management
Tier one suppliers
Tier three suppliers

Tier two suppliers
Trading partners
Transaction sets

Ultimate consumer orientation
Value-added banks
Vertical portal (vortal)

Review Questions

1. In what ways can the Internet and the Web improve logistics management? Provide a list of at least five suggestions.

2. Which industries were the first to establish standard EDI transaction sets? State why, in your opinion, these industries were more interested in setting standards than other industries.

3. Define ASC X12 and briefly describe its operations.

4. Provide two examples of how a supply chain participant that has a different business strategy than other members of the supply chain can reduce the efficiency and value of the supply chain as a whole.

5. In under 300 words, describe the reasons a company might have for wanting to participate in an industry consortium marketplace instead of setting up its own private company marketplace.

Exercises

1. Use the **Thomas Register of American Manufacturers** Web site (note: free registration is required to access this site) to locate an industrial product with which you are completely unfamiliar. Note how many companies offer that product, how many of those companies have catalogs or Web sites, how many offer online ordering, and how many offer literature by fax. Summarize what you have learned about the product and its availability on the Web in a report of approximately 400 words.

2. Your boss, Andrew Wheeler, is the president of a small plastic parts fabricator. He wants you to look into improving client interactions through EDI. You know that a number of companies that provide VAN services for companies engaging in EDI, such as **General Electric Information Services**, **GPAS**, **IBM Global Services**, **Kleinschmidt**, and **Peregrine Systems**, have begun offering Internet EDI

services, too. Some of these VAN operators are targeting smaller businesses that, in the past, would not have been able to afford to implement EDI. Choose three of the VAN providers listed and examine their Web sites. For the three VAN providers that you choose, determine whether they are offering Internet EDI services. Also decide whether, in your opinion, their Web sites are targeting smaller businesses. In a memo to Andrew of approximately 200 words, summarize your findings for the three VAN providers you chose.

3. A number of standard-setting organizations offer memberships to business firms. You are working for Grace Crowley, chief information officer (CIO) of a medium-sized company that manufactures commercial seating products. These products include furniture for waiting rooms made in standard sizes and a wide range of customer-specified chair designs for auditoriums, music halls, and theaters. Grace has asked you to investigate the benefits of

joining two standard-setting organizations, the **Open Buying on the Internet Consortium** and **RosettaNet**. Prepare two memos for Grace, one for each organization. In your memos, outline the purposes of each organization, and the costs and benefits of becoming a member. Then, present your recommendation regarding whether your company should join.

For Further Study and Research

Adams, E. 2000. "Goodbye EDI, Hello XML?" *World Trade*, 13(2), February, 54–55.

Al-Kibsi, G., K. de Boer, M. Mourshed, and N. Rea. 2001. "Putting Citizens On-Line, not In Line," *The McKinsey Quarterly*, April, 64–73.

Ariba, Inc. 2000. *B2B Marketplaces in the New Economy*. Mountain View, CA: Ariba, Inc. Available online at: (http://www.ariba.com/com_plat/white_paper_form.cfm).

Ayers, J. 1999. "Supply Chain Strategies," *Information Systems Management*, 16(2), Spring, 72–80.

Baatz, E. 1999. "How Tech Tools Unlock Supply Value," *Purchasing*, 126(6), April 22, 28–34.

Black, J. 2001. "Build Lasting Partnerships: Collaboration Is the Name of the Game at General Motors," *Internet World*, August 1, 24.

Bort, J. 2000. "Checking the B2B Foundation, *Network World*, 17(37), September 11, 78–80.

Bovel, D. and M. Joseph. 2000. "From Supply Chain to Value Net," *Journal of Business Strategy*, 21(4), July–August, 24–28.

Brook, J. 1998. "Monitoring Purchases Via the Web," *InternetWeek*, November 23, 9.

Busch, J. and L. Reisman. 2000. "B-to-B Exchanges: Know Your Domain," *The Industry Standard*, July 24, 96.

Clark, P. 2001. "MetalSite Kills Exchange, Seeks Funding," *B to B*, 86(13), June 25, 3.

Cleary, M. 2001. "Metal Meltdown Doesn't Deter New Ventures," *Interactive Week*, 8(27), July 9, 29.

Cleary, M. 2001. "Volkswagen Steering Online Auctions In-House," *E-Business*, July 16. Available online at: (http://www.zdnet.com/ecommerce/stories/main/0,10475,2787112,00.html).

Colberg, T., N. Gardner, K. Horan, D. McGinnis, P. McLauchlin, and Y-H. So. 1995. *The Price Waterhouse EDI Handbook*. New York: John Wiley & Sons.

Computerworld. 2001. "Autopsy of a Dot Com: Chemdex," February 15. Available online at: (http://computerworld.com/cwi/story/0,1199,NAV65-665_STO57746,00.html).

Cronin, C. 2001. "Five Success Factors for Private Trading Exchanges," *E-Business Advisor*, July–August, 13–20.

Dalton, G., B. Violino, and J. Mateyaschuk. 1999. "E-Business Evolution," *Information Week*, June 7, 50–57.

Discount Store News. 1999. "EDI: Internet Revolutionizes EDI," 38(10), P3–P6.

Dobbs, J. 1999. *Competition's New Battleground: The Integrated Value Chain*. Cambridge, MA: Cambridge Technology Partners.

Ekoniak, J. 2001. "Economic Analysis: Automate Your Supply Chain," *Upside*, 13(5), May, 90–91.

Elliot, M. 2000. "Buyer's Guide: Supply Chain Software," *IIE Solutions*, 32(6), June, 39–44.

Esichaikul, V. and C. Chaichotiranant. 1999. "Selecting an EDI Third-Party Network," *Information Systems Management*, 16(1), Winter, 26–32.

Fickel, L. 1998. "MicroAge's Internet-Based Training Program," *CIO Magazine*, July 15, 72.

Fisher, M. 1997. "What Is the Right Supply Chain for Your Product?" *Harvard Business Review*, 75(2), March–April, 105–116.

Fisher, S. 1999. "EDI Vendors Spin a New Web," *Upside*, 11(5), May, 62–64.

Forbes. 2000. "B2B…to Be?" 166(5), August 22, 124–128.

Fox, P. 2001. "Boeing Shows How XML Can Help Business," *Computerworld*, 35(11), March 12, 28–29.

Frook, J. 1998. "Wal-Mart Opens its Arms to Internet EDI," *InternetWeek*, June 22, 16.

Hamel, G. 2000. "Waking Up IBM," *Harvard Business Review*, 78(4), July–August, 137–144.

Harreld, H. 2001. "Easier B-to-B Links Promised," *InfoWorld*, August 13, 16.

Hart, P. and C. Saunders. 1998. "Emerging Electronic Partnerships: Antecedents and Dimensions of EDI Use from the Supplier's Perspective," *Journal of Management Information Systems*, 14(4), Spring, 87–112.

Henriott, L. 1999. "Transforming Supply Chains Into E-Chains," *Supply Chain Management Review Global Supplement*, Spring, 15–18.

Hicks, M. 2001. "Survival Course," *eWeek*, August 14. Available online at: (http://www.zdnet.com/eweek/stories/general/0,11011,2804253,00.html).

Hill, S. 1999. "Supply Chain Management in the Age of E-Commerce," *Apparel Industry Magazine*, 60(3), March, 60–63.

Hoffman, C. 2000. "Run XBRL Right Now," *Journal of Accountancy*, 190(2), August, 28–29.

Hoffman, R. 1998. "Four Solutions to Rev Up Your E-Commerce Business," *Network Computing*, December 15, 75–86.

Ince, J. 2000. "To B2B, or Not to B2B," *Upside*, August, 130–136.

Jap, S. 2000. "Going, Going, Gone," *Harvard Business Review*, 78(6), November–December, 30.

Johnston, M. 2000. "Government Adopts Electronic Transactions," *InfoWorld*, 22(31), July 31, 5.

Kaplan, S. and M. Sawhney. 2000. "E-Hubs: The New B2B Marketplaces," *Harvard Business Review*, 78(3), May–June, 97–103.

Karpinski, R. 1999. "Web Links Supply Chain to Storefront," *InternetWeek*, June 21, 8–9.

Karpinski, R. 2001. "EDI-to-Web Move Promises ROI," *InternetWeek*, August 20, 9.

Kay, E. 2000. "From EDI to XML," *Computerworld*, 34(25), June 19, 84–85.

Kemp, T., D. Drucker, and C. Moozakis. 2001. "Reload: Slower Buy-In," *InternetWeek*, June 23, 10.

Kerwin, K., M. Stepanek, and D. Welch. 2000. "At Ford, E-Commerce Is Job 1," *Business Week*, February 28, 74–77.

King, J. 2000. "Quietly, Private E-Markets Rule," *Computerworld*, 34(36), September, 1, 16.

Konicki, S. 2001. "E-Business Key to Automotive Suppliers' Success, Survey Shows," *Information Week*, August 6. Available online at: (http://www.informationweek.com/story/IWK20010806S0005).

Lewis, W. 2000. "Pillar of the Community: XML Is Becoming the Standard Platform," *Intelligent Enterprise*, 3(13), August 18, 32–38.

Lin, B. and C. Hsieh. 2000. "Online Procurement: Implementation and Managerial Implications," *Human Systems Management*, 19(2), 105–110.

Marcella, A. and S. Chan. 1993. *EDI Security, Control, and Audit*, Norwood, MA: Artech House.

Massetti, B. and R. Zmud. 1996. "Measuring the Extent of EDI Usage in Complex Organizations: Strategies and Illustrative Examples," *MIS Quarterly*, 20(3), September, 331–345.

Mastony, C. and G. Button. 2000. "Maintenance, Repair & Operations," *Forbes*, 166(2), July 17, 174–176.

McAfee, A. 2000. "The Napsterization of B2B," *Harvard Business Review*, 78(6), November–December, 18–19.

McNatt, R. and D. Welch. 2000. "Oh, What a Feeling: B2B," *Business Week*, May 15, 14.

Meehan, M. 2001. "EDI, ebXML Groups Agree to Cooperate: B2B Standards Inch Forward," *Computerworld*, 35(27), July 20, 1–2.

Meehan, M. 2001. "Michelin Sees Long Road to B2B Adoption," *Computerworld*, 35(32), August 6, 10.

Miller, G. 2001. "Few Firms Using Net to Purchase Supplies, Study Shows," *Los Angeles Times*, January 23, C1.

Moozakis, C. 2001. "Ford Investment in Covisint Paid Off?" *InternetWeek*, July 20. Available online at: (http://www.internetwk.com/story/INW20010720S003).

Moozakis, C. 2001. "Large Auto Suppliers to Drive Web Supply Chain," *InternetWeek*, August 6. Available online at: (http://www.internetwk.com/story/INW20010806S0008).

215

Moozakis, C. and D. Joachim. 2001. "Auto Hub Revamps," *InternetWeek*, August 20, 9.

Ovans, A. 2000. "E-Procurement at Schlumberger," *Harvard Business Review*, 78(3), 21–22.

Pack, T. 2001. "Vortal Content: Eyes on the Emarketplace Prize," *EContent*, 24(5), July, 30–35.

Papazoglou, M. 2001. "Agent-Oriented Technology in Support of E-Business," *Communications of the ACM*, 44(4), April, 71–77.

Poirier, C. and M. Bauer. 2001. *E-Supply Chain: Using the Internet to Revolutionize your Business.* San Francisco: Barrett-Koehler.

Power, C. 1999. "Internet Systems Imperil EDI for Corporate Buying," *American Banker*, 164(73), April 19, 15.

Premkumar, G. 2000. "Interorganization Systems and Supply Chain Management: An Information Processing Perspective," *Information Systems Management*, 17(3), Summer, 56–69.

Purchasing. 2001. "MetalSite Shuts Operations While Seeking New Owner," July 5, 32.

Pushkin, A. and B. Morris. 1997. "Understanding Financial EDI," *Management Accounting*, 70(5), November, 42–46.

Radosevich, L. 1997. "The Once and Future EDI," *CIO Magazine*, January 1. Available online at: (http://www.cio.com/archive/ec_future_edi.html).

Raisch, W. 2001. *The eMarketplace: Strategies for Success in B2B Ecommerce.* New York: McGraw-Hill.

Rinat, Z. 2001. "Beyond Private Exchanges: The Private Business Network," *E-Business Advisor*, July–August, 20.

Roberti, M. 2001. "Commerce One Drops a $2 Billion Bombshell," *The Industry Standard*, July 19. Available online at: (http://www.thestandard.com/article/0,1902,28086,00.html)

Roberts, B. 1998. "Portals, You Say? This One's Private: Ericsson's Intranet Is a Give-and-Take Affair with Employees," *Intranet Design Magazine*, December 14. Available online at: (http://idm.internet.com/articles/200003/pt_03_15_00f.html).

Roberts-Witt, S. 2001. "Steel Gets Wired," *PC Magazine*, 20(11), June 12, 14.

Rosencrance, L. 2000. "Transporters Move to Deliver on E-Commerce," *Computerworld*, 34(27), July 3, 22.

Sancich, R. 2001. "Back 2 Business," *BizReport*, January 18. Available online at: (http://www.bizreport.com/insight/2001/0120010118-1.htm).

Schenecker, M., G. Desai, J. Patel, and J. Levitt. 1998. "Goodbye To Old-Fashioned EDI," *Information Week*, December 14, 73–80.

Schonfeld, E. 1999. "A Site Where Hospitals Can Click to Shop," *Fortune*, 139(7), April 12, 150.

Schwartz, E. 2000. "Exchange Evolution Points to Higher Savings," *InfoWorld*, 22(26), June 26, 12.

Senn, J. 1992. "Electronic Data Interchange," *Information Systems Management*, 9(1), Winter, 45–53.

Shinal, J. 2000. "A Web-Profit Prophet Spreads the Word," *Business Week*, September 18, 68.

Songini, M. 2001. "Group Maps RosettaNet to Supply-Chain Process," *Computerworld*, 35(16), April 16, 8.

Stein, T. and J. Sweat. 1998. "Killer Supply Chains," *Information Week*, November 9, 36–42.

Stundza, T. 1999. "A Tale of Two Commodities: Metals, Chemicals Buyers Eye E-Commerce," *Purchasing*, 126(6), April 22, S20–S27.

Sweat, J. 1999. "Integrated Enterprise," *Information Week*, June 14, 18–19.

Szygenda, R. 1999. "Executive Report: IT Value Chain," *Information Week*, February 8, 4–6.

Taylor, D. and A. Terhune. 2001. *Doing E-Business: Strategies for Thriving in an Electronic Marketplace.* New York: John Wiley & Sons.

Teschler, L. 2000. "New Role for B-to-B Exchanges: Helping Developers Collaborate," *Machine Design*, 72(19), October 5, 52–58.

Tie, R. 2000. "Comments Encouraged on Newly Named XBRL," *Journal of Accountancy*, 189(6), June, 14–15.

Thaler, M. 2001. "Private Exchanges: Are They All They're Cracked Up to Be?" *E-Business Advisor*, July–August, 16.

Tillet, S. 2001. "Feds Closing IT Gap: Agencies Aim to Capture E-Business During Downturn," *InternetWeek*, July 9, 1–2.

216

Tillett, S. 2001. "Medical Companies Track E-Learning," *InternetWeek*, August 20, 13.

Trommer, D. and S. Scheck. 1999. "Making the Connection: Automating Supply Chain Processes," *Electronic Buyer's News*, May 3, 50.

Tweney, D. 1999. "Microsoft Takes Aim at Language Barriers to Business Information," *InfoWorld*, 21(11), March 15, 59.

Upin, E. 2000. "On the Verge of the Next Great Wave," *Upside*, August, 170–172.

Vilar, A. 2001. "B2B: The Best Is Yet to Be," *Forbes*, September 10. Available online at: (http://www.forbes.com/global/2001/0917/070.html).

Vollmer, K. 2001. "VCML: The Right Choice to Bridge XML, EDI Data Formats," *InternetWeek*, August 20, 17.

Wagner, M. 2000. "Vendor, Dotcom Thyself: Sun Turns Net Focus Inward," *InternetWeek Online*, September 5. Available online at: (http://www.internetwk.com/lead/lead090500.htm).

Waugh, R. and S. Elliff. 1998. "Using the Internet to Achieve Purchasing Improvements at General Electric," *Hospital Material Management Quarterly*, 20(2), November, 81–83.

Weinberg, N. 2001. "B2B Grows Up," *Forbes*, September 10. Available online at: (http://www.forbes.com/best/2001/0910/018.html).

Weintraub, A. 2000. "The Be-All and End-All of B2B Sites?" *Business Week*, June 5, 56.

Wilson, T. 2000. "EDI Is Alive and Kicking, Study Says," *InternetWeek*, February 21, 15.

Wilson, T. 2001. "More than Cheap Supplies," *InternetWeek*, April 23, 1–2.

Zerega, B. 1998. "Extranet Lights a Fire Under Gas-Ordering Business," *InfoWorld*, 20(33), August 17, 46.

Ziemke, P. and P. Meadows. 2000. "Bridging Work and Home," *Upside*, 12(9), September, 105–110.

WEB AUCTIONS, VIRTUAL COMMUNITIES, AND WEB PORTALS

INTRODUCTION

In 1995, Pierre Omidyar was working as a Web programmer for General Magic, Inc. and, in his spare time, operating a small personal Web site that provided, among other things, updates on the Ebola virus. His girlfriend collected Pez candy dispensers, but had trouble finding other people who shared her interest. Omidyar decided to help her out by adding a page called AuctionWeb to his site. This page included a small auction function that allowed site visitors to trade Pez dispensers and other items. Interest in the site's auctions grew so rapidly that within a year, Omidyar had quit his job to devote his full energies to the Web auction business he had created. By the end of its second year in operation, Omidyar's Web site, which he had renamed **eBay**, had auctioned over $95 million worth of goods.

Inspired by this success, Omidyar obtained $5 million in funding from Benchmark Capital in 1997. Benchmark also helped him recruit a top-notch management team for the business. In September 1998, eBay offered its stock to the public and raised $63 million. Like many Internet companies, eBay experienced extremely rapid growth. In two years, sales grew from $372,000 to $47 million. Unlike many of the new dot-com companies of the time, eBay was profitable from its inception; its net income in 1998 was over $2 million. Because of its high growth rate and solid profitability, eBay was able to return to the stock market and raise an additional $765 million in April 1999. In just three years, eBay established itself as the dominant Web site for general consumer auctions with over 2 million registered buyers and sellers.

By 2001, eBay had increased its number of registered users to more than 34 million and it was hosting auctions for goods valued at more than $5 billion each year. These auctions were earning eBay a net income of more than $80 million on revenue that was growing at a rate of approximately 85 percent per year.

Since eBay was one of the first auction Web sites and because it has pursued an aggressive promotion strategy, it has become the first-choice site for many people who want to participate in auctions. Both buyers and sellers benefit from a large marketplace such as the one eBay has created. eBay's early advantage in the online auction business will be very difficult for competitors to overcome.

In Chapters 3 and 4, you learned how businesses are using the Web to create online identities, reach customers, and sell to them. In Chapter 5, you learned how businesses are using the Web to purchase goods and to work with their suppliers more effectively. In all three of these chapters, the focus was on how companies can use the Web to better do the things that they have been doing for years: buying and selling. In this chapter, you will learn how companies are using the Web to do things that they have never done before. These new things include running auctions, creating virtual communities, and operating Web portals.

LEARNING OBJECTIVES

In this chapter, you will learn about:
- Key characteristics of the six major auction types
- Strategies for general and specific consumer Web auction sites
- Strategies for business-to-business Web auction sites
- How businesses can use virtual communities to increase brand awareness and sales
- Strategies for Web portal sites

AUCTION BASICS

In many ways, online auctions provide a business opportunity that is perfect for the Web. An auction site can charge both buyers and sellers to participate, and it can sell advertising on its pages. People interested in trading specific items can form a market segment that advertisers will pay extra to reach. Thus, the same kind of targeted advertising opportunities that search engine sites generate with their results pages are available to advertisers on auction sites. This combination of revenue-generating characteristics makes it relatively easy to develop Web auction business models that yield profits early in the life of the project.

One of the Internet's strengths is that it can bring together people who share narrow interests but who are geographically dispersed. Web auctions can capitalize on that ability by either catering to a narrow interest or providing a general auction site that has sections devoted to specific interests.

Origins of Auctions

The earliest auctions for which we have written records are from Babylon in about 500 B.C. In those auctions, men bid against each other for the women they wished to marry. Roman soldiers used auctions to liquidate the property they took from their vanquished foes. In 193 A.D., the Praetorian Guard auctioned off the entire Roman Empire after killing the Emperor Pertinax. In later years, Buddhist temples held auctions to sell off the possessions of deceased monks.

Auctions became common activities in 17th-century England, where taverns held regular auctions of art and furniture. The 18th century saw the birth of two British auction houses—**Sotheby's** in 1744 and **Christie's** in 1766—that continue to be major auction firms today. The British settlers of the colonies that would become the United States brought auctions with them. Colonial auctions were used to sell farm equipment, animals, tobacco, and, sadly, human beings.

In an auction, a seller offers an item or items for sale, but does not establish a price. This is called "putting an item up for bid" or "putting an item on the (auction) block." Potential buyers are given information about the items or some opportunity to examine the item; they then offer **bids**, which are the prices they are willing to pay for the item. The potential buyers, or **bidders**, each have developed **private valuations**, or amounts they are willing to pay for the item. The whole auction process is managed by an **auctioneer**. In some auctions, people employed by the seller or the auctioneer can make bids on behalf of the seller. These people are called **shill bidders**. Shill bidders can artificially inflate the price of an item and may be prohibited from bidding by the rules of a particular auction.

English Auctions

Many different kinds of auctions exist. Most people who have attended or seen an auction on television have experienced only one type of auction, the **English auction**, in which bidders publicly announce their successive higher bids until no higher bid is forthcoming. At that point, the auctioneer pronounces the item sold to the highest bidder at that bidder's price. This type of auction is also called an **ascending-price auction**. An English auction is sometimes called an **open auction** (or **open-outcry auction**) because the bids are publicly announced; however, there are other types of auctions that use publicly announced bids that are also called open auctions.

In some cases, an English auction has a minimum bid, or reserve price. A **minimum bid** is the price at which an auction begins. If no bidders are willing to pay that price, the item is removed from the auction and not sold. In some auctions, a minimum bid is not announced, but sellers can establish a minimum acceptable price, called a **reserve price**, or simply **reserve**. If the reserve price is not exceeded, the item is withdrawn from the auction and not sold.

English auctions that offer multiple units of an item for sale and that allow bidders to specify the quantity they want to buy are called **Yankee auctions**. When the bidding in a Yankee auction concludes, the highest bidder is allotted the quantity bid.

If items remain after satisfying the highest bidder, those remaining items are allocated to successive lower (next-highest) bidders until all items are distributed. Although all successful bidders receive the quantity of items on which they bid, they only pay the price bid by the *lowest* successful bidder.

English auctions have drawbacks for both sellers and bidders. Since the winning bidder is only required to bid a small amount more than the next-highest bidder, winning bidders tend not to bid their full private valuations. Also, bidders risk becoming caught up in the excitement of competitive bidding and then bidding more than their private valuations. This psychological phenomenon, called the **winner's curse**, has been extensively documented by William Thaler and other behavioral economists.

Dutch Auctions

The **Dutch auction** is a form of open auction in which bidding starts at a high price and drops until a bidder accepts the price. Because the price drops until a bidder claims the item, Dutch auctions are also called **descending-price auctions**. Farmer's cooperatives in the Netherlands use this type of auction to sell perishable goods such as produce and flowers, which is how it came to be known as a "Dutch" auction.

In most Dutch auctions, the seller offers a number of similar items for sale. One common implementation of a Dutch auction uses a clock that drops the price with each tick. The first bidder to call out "stop," which stops the clock, becomes the winning bidder. The winning bidder can take all or any part of the auctioned items at that price. If any items remain, the clock is restarted and continues to run until all the items are taken by successive lower bidders. A Dutch auction is often better for the seller because the bidder with the highest private valuation will not let the bid drop much below that valuation for fear of losing the item to another bidder. Dutch auctions are particularly good for moving large numbers of commodity items quickly.

Sealed-Bid Auctions

In **sealed-bid auctions**, bidders submit their bids independently and are usually prohibited from sharing information with each other. In a **first-price sealed-bid auction**, the highest bidder wins. If multiple items are being auctioned, successive lower (next-highest) bidders are awarded the remaining items at the prices they bid. The **second-price sealed-bid auction** is the same as the first-price sealed-bid auction except that the highest bidder is awarded the item at the price bid by the *second*-highest bidder.

At first glance, one might wonder why a seller would even consider such an auction, because it gives the item to the winning bidder at a lower price. William Vickrey won the **1996 Nobel Prize in Economics** for his studies of the properties of this auction type. He concluded that it yields higher returns for the seller, encourages all bidders to bid the amounts of their private valuations, and reduces the tendency for bidders to collude. Since the winning bidder is protected from an erroneously high bid, all bidders tend to bid higher than they would in a first-price sealed-bid auction. Second-price sealed-bid auctions are commonly called **Vickrey auctions**.

Double Auctions

In a **double auction**, buyers and sellers each submit combined price-quantity bids to an auctioneer. The auctioneer matches the sellers' offers (starting with the lowest price and then going up) to the buyers' offers (starting with the highest price and then going down) until all the quantities offered for sale are sold to buyers. This type of auction works well only for items of known quality, such as securities or graded agricultural products, that are regularly traded in large quantities. Double auctions can be operated in either sealed-bid or open-outcry formats. The **New York Stock Exchange** conducts sealed-bid double auctions of stocks and bonds in which the auctioneer, called a specialist, manages the market using its own funds when necessary. The **Chicago Board of Trade** conducts open-outcry double auctions of commodity futures and stock options.

The six auction types described in this section are the most commonly used in business today. Figure 6-1 summarizes the key characteristics of each of these six major auction types.

Auction type	Key characteristics
English auction	Starting from a low price, bidding increases until no bidder is willing to bid higher.
Dutch auction	Starting from a high price, bidding automatically decreases until the bidder accepts the price.
First-price sealed-bid auction	Secret bidding process; the highest bidder pays the amount of the highest bid.
Second-price sealed bid auction (Vickrey auction)	Secret bidding process; the highest bidder pays the amount of the *second*-highest bid.
Double auction (open-outcry)	Buyers and sellers declare combined price-quantity bids. The auctioneer matches seller offers (lowest to highest) with buyer offers (highest to lowest). Buyers and sellers can modify bids based on knowledge gained from other bids.
Double auction (sealed-bid)	Buyers and sellers declare combined price-quantity bids. The auctioneer (specialist) matches seller offers (lowest to highest) with buyer offers (highest to lowest). Buyers and sellers cannot modify their bids.

Figure 6-1   *Key characteristics of six major auction types*

WEB AUCTION STRATEGIES

Web auctions are one of the fastest-growing segments of online business today. Millions of people buy and sell all types of goods on consumer auction sites each year. Industry analysts predict that Web auctions will account for 40 percent of all electronic commerce dollar volume by 2004. Although the online auction business is changing rapidly as it grows, three broad categories of auction Web sites have emerged: general consumer auctions, specialty consumer auctions, and business-to-business auctions. Some industry analysts consider the two types of consumer auctions to be business-to-consumer electronic commerce. Other analysts believe that a more appropriate term for the electronic commerce that occurs in general consumer auctions to be **consumer-to-consumer** or even **consumer-to-business** (since the bidders at a general consumer auction might be businesses). Their argument is that many sellers that participate in general consumer auctions are not really businesses but are ordinary people who use these auctions to sell personal items instead of holding, for example, a garage sale. Whether you prefer to think of the business model as business-to-consumer, consumer-to-consumer, or consumer-to-business, the largest number of online auctions occurs on general consumer auction sites.

General Consumer Auctions

One of the most successful consumer auction Web sites is **eBay**, the company described in the introduction to this chapter. The eBay home page appears in Figure 6-2.

The eBay home page includes links to categories of items. Alternatively, the potential bidder can use the Smart Search feature to find a specific item by entering descriptive terms. The bottom of the page includes a link to the **third-party assurance provider TRUSTe**. Organizations such as TRUSTe provide assurance that the privacy policies of the Web sites meet certain standards. You will learn more about TRUSTe and other assurance providers in Chapter 11.

Sellers and buyers must register with eBay and agree to the site's basic terms of doing business. Sellers pay eBay a listing fee and a sliding percentage of the final selling price. Buyers pay nothing to eBay. In addition to paying the basic fees, sellers can choose from a variety of enhanced and extra-cost services, including having their auctions listed in boldface type and included in lists of preferred auctions.

In an attempt to address buyer concerns about seller reliability, eBay has instituted a rating system. Buyers can submit ratings of sellers after doing business with them. These ratings are converted into graphics that appear with the seller's nickname on each auction in which that seller participates. Although this system is not without flaws, many eBay bidders feel that it affords them some level of protection from unscrupulous sellers. The converse is true also—sellers rate buyers, which provides some protection for sellers from unscrupulous buyers.

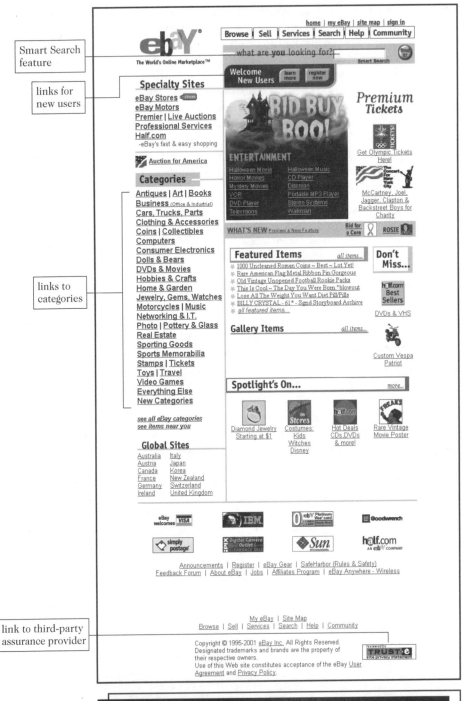

Smart Search feature

links for new users

links to categories

link to third-party assurance provider

Figure 6-2   *eBay home page*

The most common format used on eBay is a computerized version of the English auction. The eBay English auction allows the seller to set a reserve price. In eBay English auctions, the bidders are listed, but the bid amounts are not disclosed until after the auction is over. This is a slight variation on the in-person English auction, but since eBay always shows a continually updated high bid amount, a bidder who monitors the auction can see the bidding pattern as it occurs. The main difference between eBay and a live English auction is that bidders do not know who placed which bid until the auction is over. The eBay English auction also allows sellers to specify that an auction be made private. In an eBay private auction, the site never discloses bidders' identities and the prices they bid. At the conclusion of the auction, eBay notifies only the seller and the highest bidder.

Another auction type offered by eBay is an increasing-price format for multiple item auctions that eBay calls a Dutch auction. This format is not a true Dutch auction, but is instead a Yankee auction.

In either type of eBay auction, bidders must constantly monitor the bidding activity. All eBay auctions have a **minimum bid increment**, the amount by which one bid must exceed the previous bid, which is about 3 percent of the bid amount. To make bidding easier, eBay allows bidders to make a proxy bid. In a **proxy bid**, the bidder specifies a maximum bid. If that maximum bid exceeds the current bid, the eBay site will automatically enter a bid that is one minimum bid increment higher than the current bid. As new bidders enter the auction, the eBay site software continually enters higher bids for all bidders who have placed proxy bids. Although this feature is designed to make bidding require less bidder attention, if a number of bidders enter proxy bids on one item, the bidding rises rapidly to the highest proxy bid offered. This rapid rise in the current bid often occurs in the closing hours of an eBay auction.

EBay has been so successful because it was the first major Web auction site for consumers that did not cater to a specific audience and because it advertises widely—spending more than $100 million each year to market and promote its Web site. A significant portion of this promotional budget is devoted to traditional mass media outlets, such as television advertising. For eBay, such advertising has proven to be the best way to reach its main market: people who have a hobby or a very specific interest in items that are not locally available. Whether those items are jewelry, antique furniture, coins, first-edition books, or stuffed animals, eBay has created a place where people can become collectors, dispose of their collections, or trade out of their collections.

AUCTION UNIVERSE

One of the most promising new entrants into the general consumer auction business was Auction Universe. Times Mirror, the parent company of the *Los Angeles Times* newspaper, started Auction Universe in 1997 and then sold it in 1998 to a partnership of eight major newspaper companies (including Times Mirror itself) called **Classified Ventures**. These companies were concerned that classified advertising on the Web posed a threat to their newspapers' classified advertising, which is one of the most profitable elements in the newspaper business. Through their Classified Ventures partnership, these newspaper companies started their own Web sites for classified ads such as Apartments.com, Cars.com, and NewHomeNetwork.com. These sites earn revenue by charging for running ads, by selling advertising on their pages, or both. Classified Ventures believed that the Auction Universe site could become an important and profitable part of its Web presence.

Auction Universe closed in August 2000. Classified Ventures' classified ad sites continue to operate. The Auction Universe site was modeled on eBay and offered similar types of auctions and services for buyers and sellers. Some critics believed that the Auction Universe interface was more intuitive than eBay's and included a better search engine; however, the site failed to mount a sustained challenge to eBay's dominance. Even with major corporate sponsorship and a $10 million advertising campaign behind it, Auction Universe was unable to displace the advantage eBay obtained as the first Web auction site for general consumers.

Since one of the major determinants of Web auction site success is attracting enough buyers and sellers to create markets in many different items, some Web sites that already have a large number of visitors have entered this business. Portal sites such as **Yahoo!** and **Excite** have created general consumer auctions patterned after eBay. The Yahoo! Auctions home page appears in Figure 6-3.

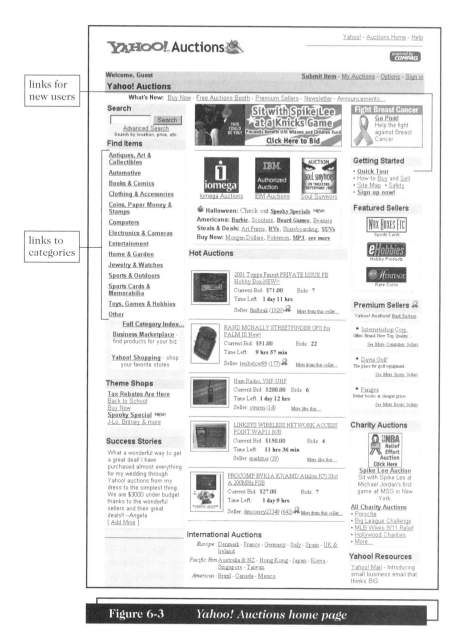

Figure 6-3   *Yahoo! Auctions home page*

As you examine the home page in Figure 6-3, you will notice that it includes many of the same features as the eBay home page. For example, it has links to categories of auction items and a search function.

Yahoo! had some early success in attracting large numbers of auction participants, in part because it offered its auction service to sellers at no charge. Yahoo! had been less successful in attracting buyers, resulting in less bidding action in each auction than generally occurs on eBay. In January 2001, Yahoo! began charging sellers in the face of dropping ad revenues in its other Web operations. Within one month, Yahoo! had lost about 80 percent of its auction listings; however, the percentage of listed items that ended in a sale increased sixfold, and the dollar amount of completed auctions remained constant. Since Yahoo! draws a large number of visitors every month, it is possible that Yahoo! will be able to further increase participation in its auctions and attract some of the sellers who left in reaction to the fees. Yahoo! is one of the most-visited sites on the Web; it consistently appears at or near the top of the Jupiter Media Metrix monthly report of most-visited Web sites.

Amazon.com, the pioneering Web bookseller, has also expanded its business to include auctions. Although its number of auctions is still small, Amazon is aggressively marketing its auction business. Unlike eBay, Amazon has not yet earned profits. Some industry observers note that Amazon might earn more by charging a commission on the auction of a used book than it could earn by selling the same title as a new book. With over 30 million registered users, Amazon is well-positioned to challenge eBay. Unlike visitors to portal sites such as Yahoo!, visitors to Amazon are already accustomed to purchasing goods on the Web. This may give Amazon an edge over portal sites in converting its visitors into auction participants.

One of the aggressive marketing positions that Amazon took to promote its auctions business is its "Auctions Guarantee." This guarantee directly addresses concerns raised in the media by eBay customers about being cheated by unscrupulous sellers. When Amazon opened its Auctions site, it agreed to reimburse any buyer for merchandise purchased in an auction that was not delivered or that was "materially different" from the seller's representations. Amazon limited its guarantee to items costing $250 or less; however, buyers of more expensive items generally protect themselves by using a third-party **escrow service**, which holds the buyer's payment until he or she receives and is satisfied with the purchased item. eBay responded immediately by offering its customers a similar guarantee, but not before Amazon had capitalized on media coverage of its guarantee to establish itself as a serious competitor in the Web auction business.

Amazon also has undertaken a joint online venture with Sotheby's, the famous British auction house, to hold Web auctions of fine art, antiques, jewelry, and other high-value collectibles. In general, it is difficult to sell these types of items on the Web because of the importance of direct, in-person inspection. Such inspections help establish the item's authenticity and condition. Sotheby's and its international network of dealers obtain the items for their online auctions and guarantee the authenticity and condition of items, just as at a Sotheby's in-person auction. Again, Amazon is addressing a serious concern of some of eBay's most prized customers, those who participate in auctions of high-value items. The Sotheby's joint venture and Amazon's recent press releases suggest that Amazon is trying very hard to differentiate its auction site from eBay's as a more attractive home for the upper end of the auction market.

Most general consumer Web auction sites invite independent sellers to auction their items on the site. Only a few general consumer Web auction sites offer items for sale from their own inventory. One of the more unusual and interesting of these sites is the **Klik-Klok Dutch Auction**. Klik-Klok sells general consumer items from its inventory using a true Dutch auction format. Figure 6-4 shows an in-progress Klik-Klok auction of a table clock.

Table Clock

Suggested retail price: $ 90.00
Starting price: $ 75
Shipping and handling: 9.50

Polished brass case featuring single columns, floating dial on a thick mineral glass panel. This handsome clock also has an alarm. Size 6 7/8" x 7 3/8"

quantity
selector

price drops as
timer counts
down to zero

Dutch Auction Home
Department Store
Home

Current Price: $68.80
Time Remaining: 0:31
Pieces Remaining: 5

Buy Now!!

Figure 6-4 *Klik-Klok Dutch auction in progress*

These Dutch auctions occur in a very short time period, usually a few minutes. The price begins at a stated starting price and drops every few seconds as the timer counts down to zero. Bidders can enter a quantity up to the number in the "Pieces Remaining" indicator and click the "Buy Now!" button to enter a bid at the price shown in the "Current Price" indicator.

The premier general consumer Web auction site is still eBay. Although it appears that portal sites or well-established Web merchants, such as Amazon.com, might be able to establish successful auction sites, they must overcome the strong advantage that eBay has built. Any challenger to eBay will find that the economic structure of markets is biased against new entrants. Since markets become more efficient (yielding fairer prices to both buyers and sellers) as the number of buyers and sellers increases, new auction participants are inclined to patronize established marketplaces. Thus, existing auction sites, such as eBay, are inherently more valuable to customers than new auction sites. This basic economic fact, which economists call a **lock-in effect**, will make the task of creating other successful general consumer Web auction sites even more difficult in the future.

A somewhat ironic example of the lock-in effect exists in the Japanese general consumer auction market. In this market, unlike in the United States, Yahoo! was the first major company to offer Web auctions. At the time (early 1999), Yahoo! did not charge fees to sellers. When eBay entered the Japanese market five months later, it charged fees and found few people interested in its services. Although Yahoo! has since instituted fees for its auctions, the lock-in effect has preserved its strong lead in Japan—in 2001, Yahoo! Auctions held 95 percent of the $1.6 billion market, while eBay's market share was 3 percent.

Specialty Consumer Auctions

Rather than struggle to compete with a well-established rival such as eBay in the general consumer auction market, a number of firms have decided to identify special interest market targets and create specialized Web auction sites that meet the needs of those market segments. Several early Web auction sites started by featuring technology items such as computers, computer parts, photographic equipment, and consumer electronics.

Doug Salot started an auction site, **Haggle Online**, in September 1996. Salot had been buying and selling computer equipment on the Internet's Usenet newsgroups before the Web existed. He saw the potential for the Web's graphical user interface in creating auctions. Haggle has officially branched out to include items unrelated to computers; however, technology products continue to be its mainstay. The site even includes an online computer museum! Other sites devoted to computers and related products include the **CNET.com** technology portal site and the Ziff-Davis computer publishing companies' **ZDNet** site.

Unlike the CNET.com and ZDNet sites, which auction only items provided by independent sellers, other Web auction sites sell their own inventory of closeouts and refurbished computers and computer-related items. Creative Computers, a direct-mail catalog marketer of computer hardware and software, owns one such site, **uBid**. The uBid auction site targets the same technology market as ZDNet and CNET.com. Although uBid has added other auction categories to its site, its true strength is in the computer and technology markets.

Although computers and technology were obvious early market segments that would find Web auctions appealing in the first wave of electronic commerce, a number of other specialized Web auction sites emerged as the Web matured. Although their operations are much smaller than those of general consumer auction sites, some companies that operate specialty consumer auctions have succeeded in building loyal followings. **PotteryAuction.com** and **JustBeads.com** are two examples of auction sites that cater to buyers and sellers who are geographically dispersed but who share highly focused interests. The JustBeads.com auction site appears in Figure 6-5.

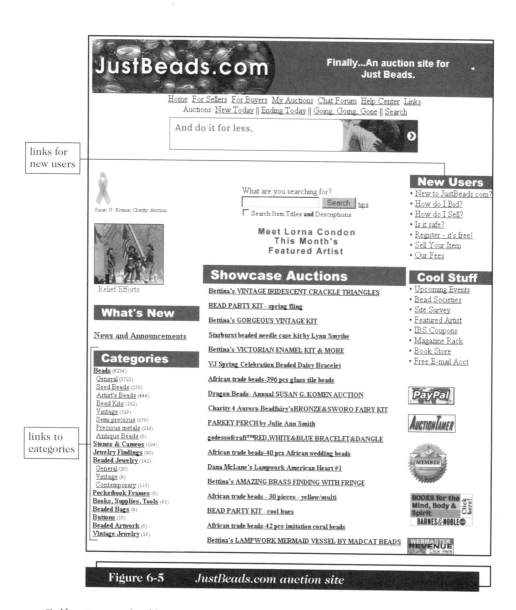

links for new users

links to categories

Figure 6-5 *JustBeads.com auction site*

Golfers in search of bargains on new and used clubs can find them at the **Golf Club Exchange**. Wine enthusiasts can bid on their favorite vintages at **Winebid**. Tobacco fanciers can find the perfect smoke at **Cigarbid.com**. Antique and art lovers can bid on dealer's offerings at **eHammer**. Investors can bid on the rates they wish to earn on USABancShares certificates of deposit at its Web auction site, **CDenergy**.

All of these sites gain an advantage by identifying a strong market segment that has readily identifiable products that are desired by persons with relatively high levels of disposable income. Golf clubs, wine, and technology products all meet these requirements very well. As other Web auction site developers identify similar market segments, these specialized consumer auctions may become profitable niches that can successfully coexist with large general consumer sites, such as eBay. One company that has identified a completely different way to segment the Web auction market is **CityAuction**. CityAuction lists its auctions by the location of an item's seller. The site's assumption is that people are more willing to bid on items offered for sale by a local person. In response to sites such as CityAuction, eBay has added a feature to its site that groups auctions by the seller's geographic area.

Business-to-Business Web Auctions

Unlike consumer Web auctions, business-to-business Web auctions evolved to meet a very specific need. Many manufacturing companies periodically need to dispose of unusable or excess inventory. Despite the best efforts of procurement and production management, businesses occasionally buy more raw materials than they need. Many times, unforeseen changes in customer demand for a product can saddle manufacturers with excess finished goods or spare parts.

Depending on its size, a firm will typically use one of two methods to distribute excess inventory. Large companies sometimes have liquidation specialists who find buyers for these unusable inventory items. Smaller businesses often sell their unusable and excess inventory to **liquidation brokers**, which are firms that find buyers for these items. Web auctions are the logical extension of these inventory liquidation activities to a new and more efficient channel, the Internet.

Two of the three main business-to-business Web auction models that are emerging are direct descendants of these two traditional methods for handling excess inventory. In the large-company model, the business creates its own auction site that sells excess inventory. In the small-company model, a third-party Web auction site takes the place of the liquidation broker and auctions excess inventory listed on the site by a number of smaller sellers. The third business-to-business Web auction model resembles consumer Web auctions. In this model, a new business entity enters a market that lacked efficiency and creates a site at which buyers and sellers who have not historically done business with each other can participate in auctions. An alternative implementation of this model occurs when a Web auction replaces an existing sales channel.

One of the earliest examples of the large company model is Ingram Micro's Auction Block site, which Ingram Micro started in 1997. Ingram Micro is a major distributor of computers and related equipment to value-added resellers (VARs), which are companies that configure computer hardware and software, such as network servers, for business users. Because computer technology changes rapidly, Ingram Micro often finds itself with outdated disk drives, computer chips, and other items that it formerly turned over to liquidation brokers.

Ingram Micro now auctions those items to its established customers through the Auction Block site. In 2000, Ingram Micro auctioned over $6 million worth of obsolete inventory through its Auction Block site; it is now expanding the site to include automated ordering, automatic bidding, and previews of future auctions. The VARs that are Ingram Micro's main customers now have the option of putting the Auction Block program on their own sites, which will allow their customers to participate in the bidding.

Ingram Micro estimates that the auction prices it receives on the site average about 60 percent of the items' costs. This percentage compares favorably to the average of 10 percent to 25 percent of cost that Ingram Micro had been obtaining from liquidation brokers. In effect, large companies such as Ingram Micro are removing the liquidation brokers from the value chain and claiming the brokers' intermediary profits. Recall that this process is called disintermediation.

Another large computer technology company that decided to build its own auction site to dispose of obsolete inventory is CompUSA. Although CompUSA sells to individuals, a significant portion of its sales are to corporate customers. Instead of selling through liquidation brokers, CompUSA decided to let midsized and smaller businesses bid directly on its technology inventory. Its Web auction site, **CompUSA Auctions**, appears in Figure 6-6.

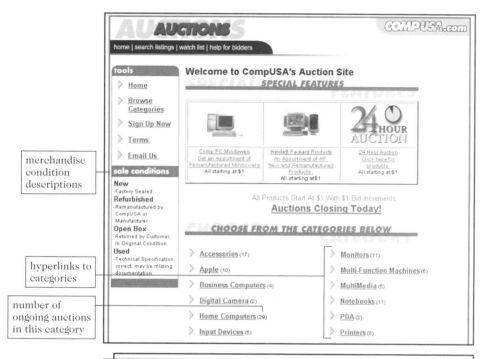

Figure 6-6   *CompUSA auctions home page*

In the second business-to-business auction model, smaller firms sell their obsolete inventory through an independent third-party auction site. In some cases, these Web auctions are conducted by the same liquidation brokers that have always handled the disposition of obsolete inventory. These brokers have adapted to the changed environment and have implemented electronic commerce to stay in business. One example is the **DoveBid** site established by the Ross-Dove Company, which has been a traditional liquidation broker for many years.

Gordon Brothers, another liquidation broker, has been selling the inventory of failed retailers since 1903. The company has used its expertise to launch or help others launch a number of Web sites that liquidate retailer inventories, including **RetailExchange.com** and **SmartBargains.com**. As many dot-com companies began to fail, the savvy liquidation company identified yet another business opportunity.

Gordon Brothers created the **Website Recycling Company** site, which sells entire Web sites, software, hardware, and even the intellectual property left in the wake of failed Web ventures.

Other third-party auction sites have been started by newcomers or even by companies that want to liquidate their inventory and are willing to do the same for other companies in their industry. Examples of third-party Web auction sites include **FastParts.Com** for electronic components, **CheMatch** for bulk petrochemicals, and **ChemConnect** for the entire chemicals market. In some industries, new auction markets on the Web are replacing older ways of doing business. For example, telecommunications companies can buy or sell time on their networks to each other through the **Band-X** Web auction site. Sellers list the number of minutes they have available, and the price of airtime minutes fluctuates in response to buyers' bids on those minutes.

Established securities trading organizations such as the New York Stock Exchange (NYSE) and the Chicago Board of Trade (CBOT) are facing an electronic challenge to their time-honored ways of doing business. In 1998, a new venture called the **International Securities Exchange (ISE)** was funded by electronic brokers E*Trade and Ameritrade Holdings, with contributions from several other brokerage firms. This new exchange is the first to be registered in the United States since 1973. In May 2000, the ISE began trading 82 of the most actively traded stock options contracts. The ISE's goal is to provide a low-cost trading forum in which the exchange will ultimately be able to trade 600 different stock options. The ISE home page appears in Figure 6-7.

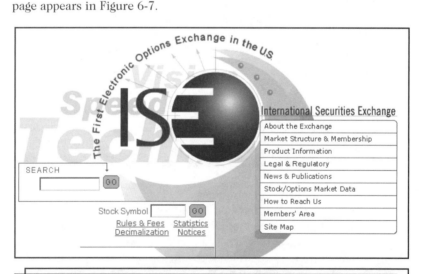

Figure 6-7 *International Securities Exchange home page*

This new electronic securities exchange poses a threat to all existing securities exchanges, because its lower fees might attract the most lucrative large trades of active issues from existing exchanges. Once the ISE becomes better established and expands its list of securities traded, industry analysts question whether traditional exchanges such as the NYSE and the CBOT can continue to exist.

Another online auction innovation is the new approach to bidding pioneered by **FreeMarkets**. Instead of using a public Web auction site, FreeMarkets uses its software and hardware tools to coordinate private online auctions that allow businesses to solicit bids from suppliers. Instead of undergoing the laborious process of sending out request for proposal packages to many suppliers, a business can list its request for proposals with FreeMarkets. Companies that have used FreeMarkets report savings of 10 percent to 20 percent in their procurement costs. FreeMarkets gives firms a way to begin procuring electronically without first investing in their own integrated system. In effect, FreeMarkets has moved the traditional first-price sealed-bid auction form onto the Internet.

Auction-Related Services

A common concern among people bidding in Web auctions is the reliability of the sellers. Recent surveys indicate that as many as 15 percent of all Web auction buyers have either not received the items they purchased, or have found the items to be different from the seller's representation. About half of those buyers were unable to resolve their disputes to their satisfaction. When purchasing high-value items, buyers can use an escrow service to protect their interests.

You learned earlier in this chapter that an escrow service is an independent party that holds a buyer's payment until the buyer receives the purchased item and is satisfied that the item is what the seller represented it to be. Some escrow services will take delivery of the item from the seller and inspect it for the buyer. Escrow services do, however, charge fees ranging from 1 percent to 5 percent of the item's cost, subject to a minimum fee. The minimum fee provision can make escrow services too expensive for small purchases. Escrow services that will handle Web auction transactions include **Tradenable** and **SafeBuyer.com**. Some of these escrow firms also sell auction buyer's insurance, which can protect buyers from nondelivery risks and also some quality risks.

Another service offered by some firms on the Web is a directory of auctions. Sites such as **Auctionguide.com** offer guidance for new auction participants and helpful tips on bidding strategies and other auction fine points for more experienced buyers and sellers. The **AuctionWatch** site is an auction directory site that provides links to auctions sorted by category, advice for auction participants, and other information services for both buyers and sellers. The AuctionWatch Universal Search functions provides a powerful search tool and a comprehensive categories list, which appears in Figure 6-8.

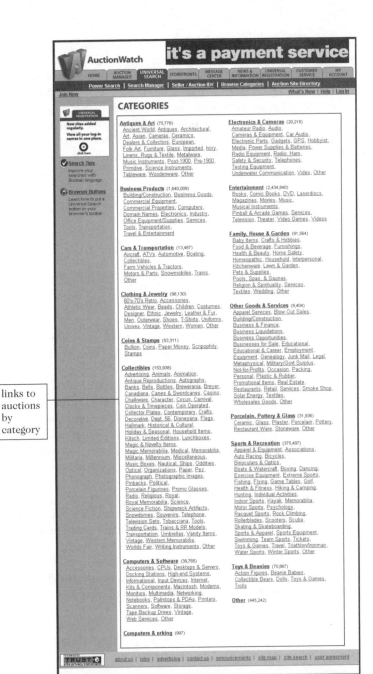

links to auctions by category

Figure 6-8 *Auction Watch Universal Search categories list*

Web sites that offer auction search and price-monitoring services include **AuctionBeagle**, **AuctionHawk**, and **BidXS**. These sites allow users to monitor bidding action on specific items or categories of items as it occurs on some of the more popular Web auction sites. **Price Watch** is an advertiser-supported site on which those advertisers post their current selling prices for computer hardware, software, and consumer electronics items. Although this monitoring is a retail pricing service designed to help shoppers find the best price on new items, Web auction participants find it can help them with their bidding strategies. **PriceSCAN** is a similar price-monitoring service that also includes prices on books, movies, music, and sporting goods, in addition to the types of items monitored by Price Watch.

Both auction buyers and sellers purchase software to help them manage their Web auctions. Sellers often run many auctions at the same time. Companies such as **Andale** sell auction management software that helps sellers create auction pages and manage the details of their auctions, such as payment and shipment tracking.

For buyers, a number of companies sell auction sniping software. **Sniping software** observes auction progress until the last second or two of the auction clock. Just as the auction is about to expire, the sniping software places a bid high enough to win the auction (unless that bid exceeds a limit set by the sniping software's owner). Because the software synchronizes its internal clock to the auction site clock, the software will almost always win out over a human bidder. The first sniping software, named Cricket Jr., was written by David Eccles in 1997. He sells the software on his **Cricket Sniping Software** site. A number of other sniping software sellers have entered the market—each claiming that its software will outbid other sniping software.

Another service that auction bidders and sellers use is electronic payments. Electronic payments services are provided by a number of companies such as PayPal and by many of the auction sites themselves. Although these payment services are currently used most often on auction sites, more and more Web sites are accepting them. You will learn more about electronic payments and how they work in Chapter 11.

Seller-Bid (Reverse) Auctions and Group Purchasing Sites

An interesting variation on the traditional auction model has come to the Web on sites such as **Respond.com.** In this variation, sellers bid the prices at which they are willing to sell. These types of auctions are sometimes called **reverse auctions**, because the role of the bidder as buyer is reversed to bidder as seller. For example, at the Respond.com site, a site visitor fills out a form that describes the item or service in which he or she is interested. The site then routes the visitor's request to a group of participating merchants who reply to the visitor by e-mail with offers to supply the item at a particular price. This type of offer is called a **reverse bid**. The buyer can then accept the lowest offer or the offer that best matches the buyer's criteria.

Many people view **Priceline.com** as a seller-bid auction site. Priceline.com allows site visitors to state a price they are willing to pay for airline tickets, car rentals, hotel rooms, and a few other services. If the price is sufficiently high, the transaction is completed. However, Priceline.com completes many of its transactions from an inventory that it has purchased from airlines, car rental agencies, and hotels. To the extent that Priceline sells out of its inventory, it operates more as a liquidation broker than as a true reverse auction. The Priceline.com home page appears in Figure 6-9.

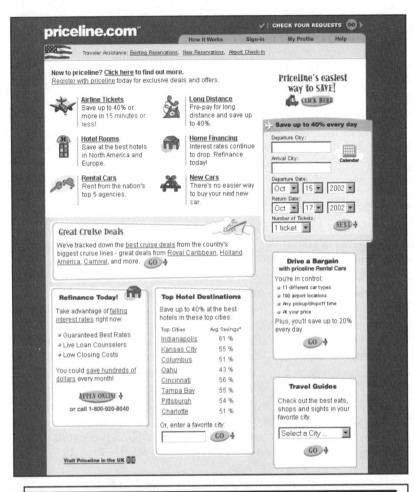

Figure 6-9 *Priceline.com home page*

In Chapter 5, you learned how businesses are creating various types of electronic marketplaces to conduct B2B transactions. Many of these marketplaces include auctions and reverse auctions. In 2001, glass and building materials producer Owens Corning held more than 200 reverse auctions for a variety of items including chemicals, conveyors, pipe fittings, and even bottled water. Having its suppliers bid to supply items has reduced the cost of those items by an average of 10 percent. Since Owens Corning buys about $3.4 billion each year, the potential for future cost savings is significant. Both the U.S. Navy and the federal government's General Services Administration are experimenting with reverse auctions to acquire a small part of the billions of dollars they spend each year on materials and supplies.

Not all companies are enthusiastic about reverse auctions. Some purchasing executives argue that reverse auctions cause suppliers to compete on price alone, which can lead suppliers to cut corners on quality or miss scheduled delivery dates. Others argue that reverse auctions can be useful for non-strategic commodity items that have established quality standards. However, as R. Gene Richter noted in a

2001 interview published in *Purchasing*, "Everything is strategic to somebody. Talk about ballpoint pens. A secretary has spots all over her brand new blouse because the pen you bought for a cent and a half is leaking." With compelling arguments on both sides, the extent to which reverse auctions will be used in the B2B sector is not yet clear.

Another new Web business is the group purchasing site. On a **group purchasing site**, the seller posts an item with a price. As individual buyers enter bids on an item (these bids are agreements to buy one unit of that item, but no price is specified), the site can negotiate a better price with the item's provider. The posted price ultimately decreases as the number of bids increases, but only if the number of bids increases. The types of products that are ideal for group purchasing sites are branded products that have a well-established reputation. This allows buyers to be confident that they are getting a good bargain and are not trading off price for reduced quality. The products should also have a high value-to-size ratio and should not be perishable.

Mercata was the first major group purchasing site on the Web for consumers, but it closed its doors in January 2001. Some sites still offer group purchasing, such as the European site **LetsBuyIt.com**, but consumer group purchasing sites have had difficulty attracting sellers' interest. The companies selling products that are well-suited to group purchasing efforts—such as computers, consumer electronics, and small appliances—have not been very interested in working with the group purchasing sites. These sellers have not found any compelling advantage in offering reduced prices on their merchandise to the group purchasing sites. Most of these sellers believe that these sites cannibalize product sales in their existing sales channels and are reluctant to offend the current distributors of their products by selling through group purchasing sites.

VIRTUAL COMMUNITY AND PORTAL STRATEGIES

Consider the following scenario:

Fran Dennison has arrived in Paris one day early for a series of business meetings. She is hoping to recover from her jet lag and enjoy a little French food before her work begins. She has found a lovely café and, using her basic knowledge of French, has successfully ordered lunch. Fran is reading the business section of *Le Monde*, a local newspaper. She begins reading an article about one of the business partners she will meet tomorrow, but her French is not good enough to completely understand the article. Fran opens her notebook computer and enters a request for translation services. She specifies that she needs immediate real-time translation of up to 500 words and is willing to pay up to 20 cents per word. She notes that the material to be translated is an article in today's *Le Monde*; she also enters the title of the article. Her computer, which contains a cellular link to her office network, launches an immediate search of online communities and marketplaces for this exact service. Two minutes later, a message appears on her computer from a French graduate student in the United States, Philippe Desmarest. His message indicates that he is willing to provide an immediate translation at Fran's quoted rate and that his computer has found the article on the *Le Monde* Web site. Five minutes later, an English translation appears in Fran's mailbox and $94.20 has been moved from her checking account to Philippe's. Fran has time to read the article and think about how she will adjust her presentation at tomorrow's meeting before her salad arrives.

This scenario could not occur today, but it is very close to becoming possible. Three key elements are required to make things such as Fran's on-demand translation a reality: cellular-satellite communications technology, electronic marketplaces, and software agents. All three of these elements exist today, but they have yet to be completely integrated.

Cellular-satellite communications technology capable of linking Fran to the Internet already exists and can be packaged with notebook computers, personal digital assistants (PDAs), and mobile phones. In Chapter 5, you learned that electronic marketplaces have grown in the B2B sector. As wireless and satellite data transmission technologies become integrated with marketplaces, these marketplaces can serve people who want to buy and sell a wide range of products and services. Firms such as **AvantGo** already provide downloads of Web site contents, news, restaurant reviews, and maps to PDAs. AvantGo's home page appears in Figure 6-10.

Figure 6-10 *AvantGo home page*

The **MIT Media Lab Software Agents Group** and other researchers have developed a number of intelligent software agents. These **intelligent software agents** (also called **software robots**, or **bots**) are programs that search the Web and find items for sale that meet a buyer's specifications. In addition to finding product item matches, software agents such as Simon can find the lowest price for an item. Simon is one of the best shopping agents currently available. You can find Simon at the **mySimon** Web site, which appears in Figure 6-11.

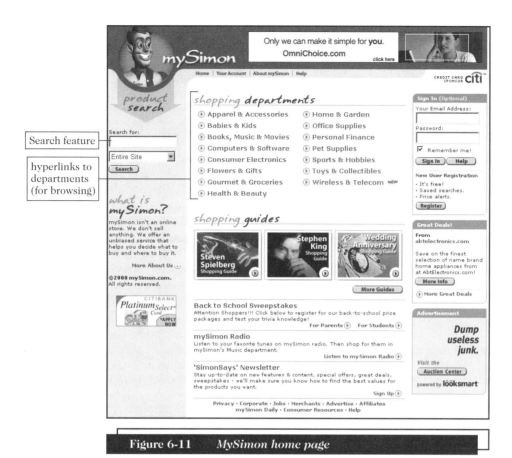

Search feature

hyperlinks to departments (for browsing)

Figure 6-11   *MySimon home page*

Some software agents are focused on a particular category of product, such as **Best Book Buys**, which searches more than 20 online bookstores for the best prices on books. In addition to obtaining price information, researchers are developing other software agents that track ratings of buyer and seller reputations. In much the same way that eBay makes reputation reports available to its bidders and sellers about each other, more general software agents can create and search databases of all kinds of buy-sell transactions on the Web. A Web site called the **BotSpot**, which is maintained by internet.com, contains information about a number of software agents, including a list of more than 40 shopping bot sites. You will learn more about software agents in Chapter 8.

In the next section, you will learn how companies are creating virtual communities on the Web for people like Fran, who engage in the business processes you learned about earlier in this book: identifying customers, promoting and selling goods and services to these customers, purchasing raw materials and supplies, and managing both the firm's internal operations and its relationships with other participants in the firm's supply chains.

Virtual Communities

A **virtual community**, also called a **Web community** or an **online community**, is a gathering place for people and businesses that does not have a physical existence.

Howard Rheingold described the characteristics of these communities in his 1993 book, *The Virtual Community*, which has become recognized as the definitive book on the subject. Virtual communities exist on the Internet today in various forms, including Usenet newsgroups, chat rooms, and Web sites. In addition to fulfilling the social interaction needs of individuals, virtual communities can help companies, their customers, and their suppliers plan, collaborate, transact business, and interact in ways that benefit all of them.

Although most Web communities are business-to-consumer strategy implementations, some successful business-to-business virtual communities have also emerged. Milacron's **Milpro** site, which you learned about in Chapter 5, is a good example of a business-to-business site that includes virtual community features. Milacron manufactures machine tools and sells them to a wide variety of industrial customers. Some of these customers are large manufacturers that use the site to purchase their MRO (maintenance, repair, and operations) supplies. These larger customers have in-house engineering departments that provide technical help and advice to the purchasing department as it selects the Milacron products the company needs. However, many of Milacron's customers are small, geographically isolated job shops that do not have in-house engineering expertise. Job shops manufacture custom metal parts and specialized machinery. These job shops buy MRO items from Milacron, but they rely on Milacron to provide technical help and advice regarding manufacturing processes.

Milacron decided that the most efficient and least expensive way to provide help and advice to these customers was to create a virtual community in which customers could interact with Milacron and with each other. Milacron hoped that this virtual community would also help forge a stronger relationship between these customers and Milacron. The Milpro site, shown in Figure 6-12, includes a Milacron product catalog, machine ordering and quotation services, and links to customer account information.

Figure 6-12   *Milpro home page*

In addition to providing these Milacron-focused services, the site also includes a number of links to interactive features that are focused on customers and their exchanges with each other. For example, the site offers a machinery flea market in which customers can trade used machinery and spare parts with each other. The site also features the Milpro Wizard, which offers free technical help to job shop workers. The Wizard can recommend specific products for specialized needs and offers suggestions for solving process problems. By providing this kind of free assistance, Milacron is using a rational branding strategy to attract existing customers and potential customers to the site and entice them into becoming regular visitors.

The lower border of the Web page even includes drawings of drill bits and grinding wheels. Milacron uses these graphics and the trademarked subtitle "Your Internet Tool Crib" to convey a feeling that this is "the" site for people who know machine tools. In this way, Milacron combines a small bit of emotional branding appeal with its rational branding strategy. The sense of community that Milacron hopes its Milpro site will provide for its customers has great potential value for Milacron. For example, if the site encourages customers to check in regularly and stay longer, those customers will gradually develop a high opinion of Milacron and its products. This highly interactive method of using a sticky Web site to improve a firm's image with its customer can be a powerful brand-building strategy. For business-to-consumer sites, high levels of stickiness are even more important, since they can be used to justify higher advertising rates.

Early Web Communities

One of the first Web communities was the **WELL.** The WELL, which is an acronym for "whole earth 'lectronic link," predates the Web. It began as a series of dialogs among the authors and readers of the *Whole Earth Review* in 1985. Most WELL members were originally from the San Francisco Bay area, and the influence of that area's counterculture heritage is a significant part of the WELL's ambiance. Members of the WELL pay a monthly fee to participate in its forums and conferences. The WELL has been home to many important researchers and participants in the growth of the Internet and the Web. Its membership also includes noted writers and artists. In 1999, the e-zine publisher **Salon.com** bought the WELL and has maintained the sense of community that had existed there for 14 years.

As the Web emerged in the mid-1990s, its potential for creating new virtual communities was quickly exploited. In 1995, Beverly Hills Internet opened a virtual community site that featured two Webcams aimed down Hollywood streets and had links to entertainment information Web sites. The theme of this community was the formation of digital cities around the focus of the Webcams. The founders of Beverly Hills Internet wanted to create a sense of community and felt that the Webcams would help accomplish that goal. Their hope was that people would be attracted by the Webcam images and want to add their own contributions, thus becoming members of a virtual neighborhood. Members were given free space on the Web site to create pages within these virtual cities to add their contributions. As it turned out, the Webcams never did attract much traffic, but the offer of free Web space did. The first of these digital cities were created around Webcams in the Los Angeles area and therefore were named for Los Angeles-area communities. As the site grew to include more geographic areas, it changed its name to GeoCities. GeoCities earned revenue

243

by selling advertising on members' Web pages and on pop-up pages that appeared whenever a visitor accessed a member's site. GeoCities grew rapidly and was purchased in 1999 by Yahoo! for $5 billion.

Other similar sites became virtual communities. Tripod was founded in 1995 in Massachusetts and offered its participants free Web page space, chat rooms, news and weather updates, and health information pages. Like GeoCities, Tripod sold advertising on its main pages and on participants' Web pages. The search engine site **Lycos** purchased Tripod in 1998 for $58 million.

Theglobe.com, also started in 1995, was the outgrowth of a class project at Cornell University. The students who created the site included bulletin boards, chat rooms, discussion areas, and personal ads. They then sold advertising to support the site's operation. Later additions included news feeds, an online art gallery, and shopping pages. Although Theglobe.com offered free Web page space, it did not emphasize that feature to the same extent that competing virtual communities had. Theglobe.com turned down several offers to purchase its community during its lifetime. The company experienced declines in its advertising revenues during the economic slowdown of 2000 and finally closed in 2001.

Web Community Consolidation

Virtual communities for consumers can continue as money-making propositions—or at least as organizations that earn enough to cover expenses—if they offer something sufficiently valuable to justify a charge for membership. For example, people joining the WELL community obtain access to a very interesting set of existing members who frequent the WELL's discussion areas. These areas are open only to members. Thus, the WELL has been able to charge a monthly membership fee. As you learned in the previous section, most virtual communities have been unable to support themselves and have either closed or been sold to companies such as Yahoo! or sites that have other revenue-generating activities that they can provide to the purchased community. Strategies that build on a combination of virtual communities and other activities are called Web portal strategies, discussed in the next section.

Web Portal Strategies

By the late 1990s, virtual communities were selling advertising to generate revenue. Search engine sites and Web directories were also selling advertising to generate revenue. Beginning in 1998, a wave of purchases and mergers occurred among these sites. The new sites that emerged still used an advertising-only revenue generation model and included all the features offered by virtual community sites, search engine sites, Web directories, and other information-providing and entertainment sites. These portals, which you first learned about in Chapter 3, are so named because their goal is to be every Web surfer's doorway to the Web.

Advertising-Supported Web Portals

Some Web observers believe that Web portal sites could be the great revenue-generating businesses of the future and that adding portal features to their existing sites or converting those sites to portals is a wise business strategy. They think that incorporating the sense of belonging developed in Web communities with the handy tools of search engine and Web directory sites will yield Web sites with high degrees of stickiness that will be extremely attractive to advertisers.

One rough measure of stickiness is how long each user spends at the site. Figure 6-13 lists 10 of the most popular sites on the Web based on the number of home users who accessed the sites during the week of October 7, 2001. The information in Figure 6-13 is adapted from a **Nielsen//NetRatings** report and shows sites grouped by owner. For example, the numbers for AOL Time Warner include activity on all sites operated by AOL Time Warner. Figure 6-14 shows the same rankings for people who accessed the sites from their computers at work. (Note: People who have broadband access at work and not at home often use their at-work computers for personal business during nonwork hours.)

Name	Minutes per unique visitor	Millions of unique visitors
eBay	35	7
Yahoo! sites	26	31
MSN	22	29
AOL Time Warner	16	39
Disney	15	7
Excite@Home	13	7
Lycos Network	8	9
Google	6	6
About.com	5	6
Microsoft	4	10

Adapted from reports for October 7, 2001 published by Nielsen//NetRatings at http://www.nielsennetratings.com/

Figure 6-13 *Stickiness of popular Web sites accessed from at-home computers*

Name	Minutes per unique visitor	Millions of unique visitors
eBay	52	6
Yahoo! sites	42	19
MSN	32	18
AOL Time Warner	27	20
Disney	16	6
Amazon.com	11	5
Google	9	6
Microsoft	6	9
Lycos Network	6	6
About.com	6	5

Adapted from reports for October 7, 2001 published by Nielsen//NetRatings at http://www.nielsennetratings.com/

Figure 6-14 *Stickiness of popular Web sites accessed from at-work computers*

245

Nielsen//NetRatings determines site popularity by measuring the number of unique visitors to a site. These leading sites had between 5 and 39 million unique visitors during the week measured. In Figures 6-13 and 6-14, the sites are not ranked by popularity, but by the average number of minutes that users spent on the sites. Since Web portals ask their members to provide demographic information about themselves, the potential for targeted marketing is very high. Industry observers predicting success for Web portals may be correct. More than half of the top 10 most-visited Web sites shown in Figures 6-13 and 6-14 are Web portals. High visitor counts can yield high advertising rates for these sites. In the past, Web portals were able to obtain up-front cash payments from advertisers, which is very unusual for any kind of advertising sale. For example, Excite paid Netscape (now a part of AOL) a $70 million advance fee for two-year rights to a prominent advertising location on its Netcenter Web portal site. Other portal sites have negotiated advertising deals that included a percentage of sales generated from sales leads on the portal site.

The companies that run Web portals certainly believe in the power of portals. They have been aggressively adding sticky features such as chat rooms, e-mail, and calendar functions—often by purchasing the companies that create those features. In addition to buying the virtual community site Tripod, Lycos purchased the online directory WhoWhere? for $133 million. In 1999 alone, Yahoo! spent over $10 billion in cash and stock to expand the range of services available on its Web portal site.

Of course, advertising-supported Web portals rely on a consistent stream of advertising revenue. Advertising spending on the Web has declined precipitously since mid-2000 and this downturn has seriously damaged the business prospects of even the largest and most well-known Web portals, such as Yahoo! and Excite. Many smaller portals have closed. The future of the advertising-supported Web portal is uncertain at this time.

Mixed-Model Web Portals

One of the most successful Web portals is AOL, which has always charged a fee to its users and which has always run advertising on its site. Many Web portals that are now struggling with their advertising-supported revenue models have been moving gradually towards AOL's strategy. Yahoo! now charges for the Internet phone service that had been free. It still offers free e-mail accounts, but now sells other features, such as more space to store messages and attached files, to members who pay for the "premium" e-mail service.

Other advertising-supported Web portals are following the lead of Yahoo! in a strategy called monetizing eyeballs or monetizing visitors. **Monetizing** refers to the conversion of existing regular site visitors seeking free information or services into fee-paying subscribers or purchasers of services. Many of the portals that are conducting these monetizing campaigns are worried about visitor backlash. They are unsure how many existing visitors will stay and pay for services they have previously been receiving at no cost.

Recently, more and more industry analysts are predicting the end of the "free Web." Although the largest portal sites should be able to survive using a mixed revenue model, it is unclear how smaller portals will fare.

Summary

Companies are now using the Web to do things that they have never done before, such as operating auction sites, creating virtual communities, and serving as Web portals. You learned about the key characteristics of the six major auction types (English, Dutch, first-price sealed-bid, second-price sealed-bid, double auction open-outcry, and double auction sealed-bid) and how firms are using Web auction sites to sell goods to their customers and generate advertising revenue.

Newspaper companies and other businesses have created Web sites that sell classified advertising. Escrow services have begun providing protection to Web auction participants by acting as independent, trusted third parties that hold both goods and payments until the buyer is satisfied with the transaction. Buyers have become sellers by participating in reverse auctions, in which service providers bid for the right to provide lowest-cost services. Businesses are also conducting Web auctions to liquidate excess inventories instead of selling those inventories to liquidation brokers.

A number of businesses offer ancillary services to Web users who participate in auction sites. These businesses include auction directories, auction information sites, escrow services, and auction management software for both sellers and bidders.

New companies have formed to take advantage of the Web's ability to bring together people and organizations that share narrow interests but that are geographically dispersed. Businesses are exploiting this characteristic of the Web by creating virtual communities with their customers and suppliers and by using these communities to sell goods and services. The major Web search engine sites have evolved into Web portals by adding virtual communities and related features to their sites' offerings, but the future of these businesses is in question following rapid declines in advertising revenue.

247

Key Terms

Ascending-price auction

Auctioneer

Bid

Bidder

Consumer-to-business

Consumer-to-consumer

Descending-price auction

Double auction

Dutch auction

English auction

Escrow service

First-price sealed-bid auction

Group purchasing site

Intelligent software agent (software robot or bot)

Liquidation broker

Lock-in effect

Minimum bid

Minimum bid increment

Monetizing

Online community

Open auction (open-outcry auction)

Private valuation

Proxy bid

Reserve

Reserve price

Reverse auction

Reverse bid

Sealed-bid auction

Second-price sealed-bid auction

Shill bidder

Sniping software

Third-party assurance provider

Vickrey auction

Virtual community

Web community

Winner's curse

Yankee auction

Review Questions

1. In one paragraph, name and briefly describe the key characteristics of the English auction format. In a second paragraph, explain how those characteristics are implemented in an English format Web auction.

2. In approximately 200 words, define the term *reserve price* and explain how the use of a reserve price can affect the progress and outcome of an auction.

3. Write three paragraphs in which you name and briefly describe each of the three models that are emerging for business-to-business Web auctions.

4. Describe the services that an escrow service company offers. Name one advantage and one disadvantage of using an escrow service. Limit your answer to 300 words.

5. Describe at least four sticky features that Web sites use to attract and keep visitors. For each feature, explain what makes it sticky. Explain why stickiness is important to companies operating Web sites. Limit your answer to 500 words.

Exercises

1. Use the Online Companion to examine the projects list at the **MIT Software Agents Group Projects** site. Choose a software agent technology used in one of these projects and, in approximately 200 words, describe how you could use it in an electronic commerce application.

2. Use **mySimon, Best Web Buys**, or another Web pricing robot of your choice to find sources for a book or CD that you would like to buy. Evaluate the results provided by the robot in terms of how useful the robot was in helping you plan for your purchase. Summarize your findings in a report of approximately 200 words.

3. You have been hired as an electronic commerce consultant to Oyster Bay, Inc., a dealer in ocean-going yachts. Oyster Bay maintains offices and marinas in major U.S. East Coast ports. The typical purchaser of an Oyster Bay yacht is a high-income business executive, a retiree, or a person of significant inherited wealth. Oyster Bay salespersons have noted that their customers are increasingly aware of the Web. Prepare a proposal for an Oyster Bay Web portal site. You do not need to design the Web pages, but your proposal should include a detailed list of features that should be included in the site design. Describe each feature in detail, and explain why you believe it should be included. For each feature, note whether it will be supplied by Oyster Bay personnel or purchased from an outside supplier. To learn more about existing yacht sales sites, you may wish to consult the **Carver Yachts, Yacht Sales International**, and **YachtWorld.com** links in the Online Companion.

For Further Study and Research

Anders, G. 1998. "Search for Pez Launched Web's Newest Mogul," *The Wall Street Journal*, September 25, B1.

Atanasov, M. 2001. "Going Out of Business Since 1903," *Smart Business*, 14(7), July, 72–76.

Barron, J. 2000. "For 1776 Copy of Declaration, A Record in an Online Auction," *The New York Times*, June 30, B1.

Belson, K., R. Hof, B. Elgin. 2001. "How Yahoo! Japan Beat eBay at Its Own Game," *Business Week*, June 4, 58.

Borden, M. 1999. "And Perhaps a Furby to Go with your Fuhrer?" *Fortune*, 139(9), May 10, 28.

Borrego, A. 2000. "How I Learned to Stop Worrying and (Almost) Love EBay," *Inc.*, May 16, 32–36.

Carrns, A. 1999. "www.doctorsmedicinesdiseases-galore.com—Today's Cybercraze Is Any Web Site Devoted To Health or Maladies," *The Wall Street Journal*, June 10, B1.

Cassady, R. 1967. *Auctions and Auctioneering*, Berkeley, CA: University of California Press.

Cleary, M. 2001. "Infighting May Delay Fed Reverse Auctions," *Interactive Week*, 8(19), May 14, 22.

Cohen, A. 2001. "The Sniper King," *On Magazine*, May. Available online at: (http://www.onmagazine.com/on-mag/magazine/article/1,9985,105315-1,00.html).

Consumer's Research Magazine. 1999. "Fraud Surge," 82(3), March, 7–9.

Dobrzynski, J. 2000. "F.B.I. Opens Investigation of EBay Bids," *The New York Times*, June 7, 1.

Dodge, J. 1999. "Making a Web Site Work: Three Companies Illustrate How Comfortable Internet Homes Can Be," *The Wall Street Journal*, June 14, 16.

Draenos, S. 2000. "Bidding for Auction Success," *Upside*, May 1, 41–44.

The Economist. 1997. "Going, Going..." May 31, 61.

The Economist. 2001. "We Have Lift-Off" February 3, 69–71.

Enos, L. 2001. "Fees Cause Yahoo! Auctions to Drop More Than 80 Percent," *E-Commerce Times*, February 13. Available online at: (http://www.ecommercetimes.com/perl/story/7444.html).

Fisher, S. 2000. "Web Auctions Open Doors for Art Collectors," *InfoWorld.com*, September 15. Available online at: (http://www2.infoworld.com/articles/hn/xml/00/09/18/000918hnetend.xml).

Freeman, L. 2000. "Blue Mountain Arts: Mark Rinella," *Advertising Age*, June 26, S20.

Furchgott, R. and R. McNatt. 1999. "When Bidders Are Ringers," *Business Week*, May 17, 8.

Gilbert, J. and A. Kerwin. 1999. "Newspapers Carve Slice of Auction Pie," *Advertising Age*, 70(26), June 21, 32–34.

Gomes, L. 2000. "EBay's Earnings Beat Forecasts as Sales, Number of Users Surge," *The Wall Street Journal*, July 26, B6.

Grimes, A. 2000. "Hoaxes on EBay Raise Questions on Auction Web Site's Safeguards," *The Wall Street Journal*, January 7, B6.

Gross, N. 1999. "Building Global Communities: How Business Is Partnering with Sites that Draw Together Like-Minded Consumers," *Business Week*, March 22, EB42.

Guernsey, L. 2000. "A New Caveat for EBay Users: Seller Beware," *New York Times*, August 3, G1.

Hannon, D. 2001. "Navy Supply Uncovers Benefits of Auctions," *Purchasing*, 130(12), June 21, 17.

Hof, R. 2000. "Will Auction Frenzy Cool?" *Business Week*, September 18, 140.

Jones, K. 1999. "Two B-to-B Auction Models Emerge," *Interactive Week*, 5(11), March 23, 44.

Kelsey, D. 2001. "EBay Sellers Making 25% Less Than Year Ago," *BizReport*, May 3. Available online at: (http://www.bizreport.com/article.php?art_id=1294).

Kenczyk, M. and V. Reitz. 2001. "Reverse Auctions Are Risky Models for Buying Custom Parts," 73(6), March 22, 148.

Kennedy, J. 1998. "Radio Daze," *MIT's Technology Review*, 101(6), November–December, 68–71.

Kollock, P. 1998. "Design Principles for Online Communities," *PC Update*, 15(5), June, 58–60.

Lloyd, J. 2001. "EBay Founder Pierre Omidyar: His Devotion to Community Created a Global Auction House," *Investor's Business Daily*, August 20, A4.

249

Ma, M. 1999. "Agents in E-Commerce," *Communications of the ACM*, 42(3), March, 78–80.

Machlis, S. 1999. "Should My Company Care About Portals?" *Computerworld*, 33(9), March 1, 44–45.

McGuire, D. 2000. "Auction Sites Stay Popular Despite Fraud Warnings—Study," *Newsbytes*, September 25. Available online at: (http://www.nbnn.com/).

McLean, B. 1999. "Sothebys.com," *Fortune*, 139(3), February 15, 200.

Merlino, L. 2000. "Auction Anxiety," *Upside*, October. Available online at: (http://www.upside.com/texis/mvm/story?id=39ac0ec70).

Moozakis, C. 2001. "Bargain Basement Bids," *InternetWeek*, July 23, 19.

Nardi, B. and V. O'Day. 1999. *Information Ecologies: Using Technology with Heart*, Cambridge, MA: MIT Press.

Newsweek. 1999. "Spend Time, Get a Dime," March15, 84.

Olsen, S., J. Hu, and M. Yamamoto. 2001. "Free Web: Its Days Are Numbered?" *ZDNet News*, June 4. Available online at: (http://www.zdnet.com/zdnn/stories/news/0,4586,2768421,00.html).

Petersen, A. 1999. "Some Places to Go When You Want to Feel Right at Home: Communities Focus on People Who Need People," *The Wall Street Journal*, January 6, B6.

Petrecca, L. and B. Snyder. 1998. "Auction Universe Puts in $10 Mil Bid for Customers," *Advertising Age*, 43(8), October 26, 8.

Plotkin, H. 1998. "Art Net," *Forbes*, 161(7), April 6, 29–31.

Priluck, J. 1999. "MIT: E-Commerce Just Beginning," *Wired News*, February 24, Available online at: (http://www.wired.com/news/news/technology/story/18104.html).

Purchasing. 2001. "What Top Supply Execs Say About Auctions," 130(12), June 21, S2–S3.

Quan, J. 1999. "Risky Business," *Rolling Stone*, March 4, 91–92.

Razzi, E. 1999. "Online Lenders: Bidding for You," *Kiplinger's Personal Finance Magazine*, 53(5), May, 56–58.

Regan, K. 2001. "Should EBay Be Nervous?" *E-Commerce Times*, February 16. Available online at: (http://www.ecommercetimes.com/perl/story/7584.html).

Rheingold, H. 1993. *The Virtual Community: Homesteading on the Electronic Frontier*, New York: HarperCollins.

Robins, W. 2000. "Auctions.com Now a Dot-Goner," *Editor & Publisher*, August 28, 6.

Roth, D. 2000. "Meet EBay's Worst Nightmare," *Fortune*, June 26, 199–202.

Sanborn, S. 2001. "Reverse Auctions Make a Bid for Business," *InfoWorld*, 23(12), March 19, 32.

Schiffman, B. 2001. "A Community That Stays Together, Pays Together," *Forbes*, August 28. Available online at: (http://www.forbes.com/technology/ecommerce/2001/08/28/0828yahoo.html).

Schuyler, N. 2000. "Going… Going… Gotcha!" *PC World*, October 1, 181.

Shellenbarger, S. 1999. "New Technologies Can Help the Elderly Stay More Connected," *The Wall Street Journal*, April 21, B1.

Solomon, S. 1999. "Go Sell It on the Mountain," *Inc.*, December, 48–62.

Stone, A. 2001. "EBay: Bidding for Web Domination," *Business Week*, January 11. Available online at: (http://www.businessweek.com/ebiz/0101/es0111.htm).

Stone, M. 2001. "Court Dismisses Class Action Against EBay," *BizReport*, January 19. Available online at: (http://www.bizreport.com/daily/2001/01/20010119-4.htm).

Tedeschi, B. 2000. "Creating Marketplaces for Business-to-Business Transactions," *The New York Times*, January 24, C10.

Thaler, R. 1994. *The Winner's Curse: Paradoxes and Anomalies of Economic Life*, Princeton, NJ: Princeton University Press.

Vickrey, W. 1961. "Counterspeculation, Auctions, and Competitive Sealed Tenders," *Journal of Finance*, 16(1), March, 8–37.

The Wall Street Journal. 2000. "Art of Antiquing Turned on Its Head by Online Auctions," August 8, B12B.

The Wall Street Journal. 2001. "Yahoo! Inc. Auction Listings Drop Off Since Fees Were Introduced," February 28, C13.

Wamsley, D. 1999. "Online Ad Auctions Offer Sites More than Bargains," *Advertising Age*, 70(13), March 29, 43.

Wang, S. 1999. "Analyzing Agents for Electronic Commerce," *Information Systems Management*, 16(1), Winter, 40–48.

Ward, E. 1999. "How to Build Community on Your Site and Participate in Others," *Business Marketing*, June 1, 24.

Weber, T. 2000. "To Build Virtual Trust, Web Sites Develop 'Reputation Managers,'" *The Wall Street Journal*, July 17, B1.

Wingfield, N. 2001. "EBay Watch: Corporate Sellers Put the Online Auctioneer on Even Faster Track," *The Wall Street Journal*, June 1, A1.

Wingfield, N. 2001. "Andale Hitches Its Wagon to EBay's Fortunes," *The Wall Street Journal*, September 29, B11.

THE ENVIRONMENT OF ELECTRONIC COMMERCE: INTERNATIONAL, LEGAL, ETHICAL, AND TAX ISSUES

INTRODUCTION

In 1999, **Dell Computer** and Micron Electronics (now doing business as **MicronPC**), two companies that sell personal computers through their Web sites, agreed to settle U.S. Federal Trade Commission (FTC) charges that they had disseminated misleading advertising to their customers and potential customers. The advertising in question was for computer leasing plans that both companies had offered on their Web sites. The ads stated the price of the computer along with a monthly payment. Unfortunately for Dell and Micron, stating the monthly payment without disclosing full details of the lease plan is a violation of the Consumer Leasing Act of 1976. This law is implemented through a federal regulation that was written and that is updated periodically by the Federal Reserve Board. This regulation, called Regulation M, was designed to require banks and other lenders to fully disclose the terms of leases so that consumers would have enough information to make informed financing choices when leasing cars, boats, furniture, and other goods.

Both Dell and Micron had included the required information on their Web pages, but FTC investigators noted that important details of the leasing plans, such as the number of payments and the fees due at the signing of the lease, were placed in a small typeface at the bottom of a long Web page.

A consumer who wanted to determine the full cost of leasing a computer would need to scroll through a number of densely filled screens to obtain enough information to make the necessary calculations.

In the settlement, both companies agreed to provide consumers with clear, readable, and understandable information in their lease advertising. The companies also agreed to record-keeping and federal monitoring activities designed to ensure their compliance with the terms of the settlement.

Dell and Micron are computer manufacturers. It apparently did not occur to them that they needed to become experts in Regulation M, generally considered to be a banking regulation. Companies that do business on the Web expose themselves, often unwittingly, to liabilities that arise from the environment of business today. That environment includes laws and ethical considerations that may be different from those with which the business is familiar. In the case of Dell and Micron, they were unfamiliar with the laws and ethics of the banking industry. The banking industry has a different culture than that of the computer industry—it is unlikely that a bank advertising manager would have made such a mistake.

As you will learn in this chapter, Dell and Micron are by no means the only Web businesses that have run afoul of laws and regulations. As companies expand their markets on the Web, they encounter unfamiliar laws and different ethical frameworks, especially when they begin doing business in other countries activities.

LEARNING OBJECTIVES

In this chapter, you will learn about:
- International electronic commerce
- Laws that govern electronic commerce activities
- Ethics issues that arise for companies conducting electronic commerce
- Conflicts between companies' desire to collect and use data about their customers and the privacy rights of those customers
- Taxes that are levied on electronic commerce activities

INTERNATIONAL NATURE OF ELECTRONIC COMMERCE

Since the Internet connects computers all over the world, any business that engages in electronic commerce instantly becomes an international business. When companies use the Web to create a corporate image, build a brand, sell or purchase products or services, conduct auctions, or build a community, they are automatically operating in a global environment. A now-famous cartoon that appeared in *The New*

Yorker magazine is shown in Figure 7-1. It illustrates a kind of anonymity that extends to all aspects of a Web presence.

"*On the Internet, nobody knows you're a dog.*"

| Figure 7-1 | *This cartoon from* The New Yorker *illustrates anonymity on the Web* |

For example, a U.S. bank can establish a Web site that offers services throughout the world. No potential customer visiting the site will know by browsing through the site's pages just how large or well-established the bank is. Since Web site visitors will not become customers unless they trust the company behind the site, a plan for establishing credibility is essential. Sellers on the Web cannot assume that visitors will know that the site is operated by a trustworthy business.

Customers' inherent lack of trust in "strangers" on the Web is logical and to be expected; after all, people have been doing business with their neighbors—not strangers—for thousands of years. When businesses grew to become large corporations with multinational operations, their reputations grew commensurately. Before a company could do business in dozens of countries, it had to prove its trustworthiness by satisfying customers for many years as it grew. Businesses on the Web must find ways to overcome this well-founded tradition of distrusting strangers, because today a company can incorporate one day and, through the Web, be doing business the next day with people in almost every country in the world. For businesses to succeed on the Web, they must find ways to quickly generate the trust that traditional businesses took years to develop.

An important element of business trust is anticipating how the other party to a transaction will act in specific circumstances. That is one reason why companies that have an established brand can build an online business more quickly and easily than a new company without a reputation—as you learned in Chapter 4. The brand conveys some expectations about how the company will behave. For example, a potential buyer might like to know how the seller would react to a claim by the buyer that the seller has misrepresented the quality of the goods sold. Part of this knowledge derives from the buyer and seller sharing a common language and common customs. Another part derives from having a common legal structure for resolving disputes. The combination of language and customs is often called **culture**. Most researchers agree that culture varies across national boundaries and, in many cases, varies across regions within nations. Businesses engaging in electronic commerce must be aware of the differences in language and customs that make up the culture of any region in which they intend to do business. The barriers to international electronic commerce include language, culture, and infrastructure issues.

Language Issues

Most companies have realized that the only way to do business effectively in other cultures is to adapt to those cultures. The phrase "think globally, act locally" is often used to describe this approach. The first step that a Web business usually takes to reach potential customers in other countries, and thus in other cultures, is to provide local language versions of its Web site. This may mean translating the Web site into another language or regional dialect. Researchers have found that customers are far more likely to buy products and services from Web sites in their own language, even if they can read English well. Only 370 million of the world's 6 billion people learned English as their native language.

Researchers estimate that about 75 percent of the content available on the Internet today is in English, but more than 46 percent of current Internet users do not read English. International Data Corporation predicts that by 2004, more than two-thirds of Internet users will be outside the United States and that 65 percent of Web use and 52 percent of electronic commerce sales will involve at least one party located outside the United States.

The non-English languages used most frequently by U.S. companies on their Web sites are Spanish, German, Japanese, and Chinese. Following closely behind is a second tier of languages that includes Italian, French, Korean, Portuguese, Dutch, Russian, and Swedish. These language choices by U.S. businesses closely match the native languages of Internet users. Some differences exist between non-English languages used on the Internet and language choices by U.S. companies because "Internet use" includes other activities besides electronic commerce and because many non-English speaking persons do not conduct business with U.S. companies.

As of September 2001, there were about 500 million people using the Internet for e-mail, Web surfing, electronic commerce, and other activities. Other than English, the languages most used by these people on the Internet were Chinese (9.2 percent),

Japanese (9.2 percent), German (6.7 percent), Spanish (6.7 percent), Korean (4.4 percent), Italian (3.8 percent), French (3.3 percent), and Portuguese (2.5 percent). English is the native language of fewer than 40 percent of the people on the Internet. The Web site of **Global Reach**, a consulting firm that offers Web site globalization services, maintains current information about language use on the Web.

Some languages require multiple translations for separate dialects. For example, the Spanish spoken in Spain is different from that spoken in Mexico, which is different from that spoken elsewhere in Latin America. People in parts of Argentina and Uruguay use yet a fourth dialect of Spanish. Many of these dialect differences are spoken inflections, which are not important for Web site designers (unless, of course, their sites include audio or video elements); however, a significant number of differences occur in word meanings and spellings. You may be familiar with these types of differences, since they occur in the U.S. and British dialects of English. The U.S. spelling of *gray* becomes *grey* in Great Britain, and the meaning of *bonnet* changes from a type of hat in the United States to an automobile hood in Great Britain. Chinese has two main systems of writing: one used in mainland China, and another used in Hong Kong and Taiwan.

Most companies that translate their Web sites translate all of their pages. However, some sites have thousands of pages with much targeted content, and the cost of translating all pages might be prohibitive for some firms. The decision whether to translate a particular page should be made by the corporate department responsible for each page's content. The home page should have versions in all supported languages, as should all first-level links to the home page. Beyond that, pages that are devoted to marketing, product information, and establishing brand should be given a high translation priority. Some pages, especially those devoted to local interests, might be maintained only in the relevant language. For example, a weekly update on local news and employment opportunities at a company's plant in Frankfurt probably needs to be maintained only in German.

In Chapter 2, you learned how Web browsers and Web servers communicate with each other. Recall that the request message that a Web browser (client) sends to the Web server when it establishes a connection can include a request header. That request header can contain information about the browser software, including the browser's default language setting. The Web server can thus detect the default language setting of the browser and automatically redirect the browser to the set of Web pages created in that language.

Another approach is to include links to multiple language versions on the Web site's home page. The Web site visitor must select one of the languages by clicking the appropriate link. Web sites that use this approach must make sure that the links are identifiable to visitors who read only those languages. Thus, the links should show the name of each language in that language. Many Web sites use country flags to indicate language. This practice can lead to errors and unintentional ill will. For example, a Bolivian visitor who must click the flag of Spain to navigate to the Spanish language pages might have difficulty identifying the Spanish flag and might resent that the Bolivian flag was not presented as a choice. The **Europages** home page, shown in Figure 7-2, includes identifiable links to pages in German, English, Spanish, French, Italian, and Dutch.

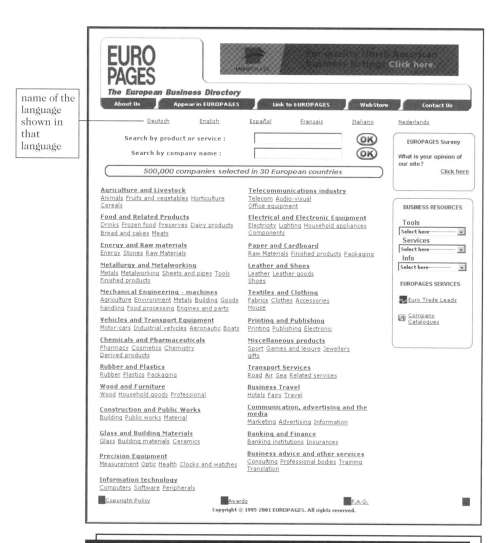

name of the language shown in that language

EURO PAGES

The European Business Directory

For quality North American business listings, Click here.

INFOBASE

| About Us | Appear in EUROPAGES | Link to EUROPAGES | WebStore | Contact Us |

Deutsch   English   Español   Français   Italiano   Nederlands

Search by product or service : [   ] OK

Search by company name : [   ] OK

500,000 companies selected in 30 European countries

EUROPAGES Survey

What is your opinion of our site?   Click here

Agriculture and Livestock
Animals Fruits and vegetables Horticulture
Cereals

Food and Related Products
Drinks Frozen food Preserves Dairy products
Bread and cakes Meats

Energy and Raw materials
Energy Stones Raw Materials

Metallurgy and Metalworking
Metals Metalworking Sheets and pipes Tools
Finished products

Mechanical Engineering - machines
Agriculture Environment Metals Building Goods
handling Food processing Engines and parts

Vehicles and Transport Equipment
Motor-cars Industrial vehicles Aeronautic Boats

Chemicals and Pharmaceuticals
Pharmacy Cosmetics Chemistry
Derived products

Rubber and Plastics
Rubber Plastics Packaging

Wood and Furniture
Wood Household goods Professional

Construction and Public Works
Building Public works Material

Glass and Building Materials
Glass Building materials Ceramics

Precision Equipment
Measurement Optic Health Clocks and watches

Information technology
Computers Software Peripherals

Telecommunications industry
Telecom Audio-visual
Office equipment

Electrical and Electronic Equipment
Electricity Lighting Household appliances
Components

Paper and Cardboard
Raw Materials Finished products Packaging

Leather and Shoes
Leather Leather goods
Shoes

Textiles and Clothing
Fabrics Clothes Accessories
House

Printing and Publishing
Printing Publishing Electronic

Miscellaneous products
Sport Games and leisure Jewellery
gifts

Transport Services
Road Air Sea Related services

Business Travel
Hotels Fairs Travel

Communication, advertising and the media
Marketing Advertising Information

Banking and Finance
Banking institutions Insurances

Business advice and other services
Consulting Professional bodies Training
Translation

BUSINESS RESOURCES

Tools
[Select here---------- ▼]
Services
[Select here---------- ▼]
Info
[Select here---------- ▼]

EUROPAGES SERVICES

Euro Trade Leads

Company Catalogues

Copyright Policy   Awards   F.A.Q.

Copyright © 1995-2001 EUROPAGES. All rights reserved.

Figure 7-2   *Europages home page*

257

Firms that provide Web page translation services and translation software for companies include **Alis Technologies**, **Berlitz**, **LexFusion**, **Lernout & Hauspie**, **Rubric, Ltd.**, **Transparent Language**, and **Worldpoint Interactive**. These firms will translate Web pages and maintain them for a fee that is usually between 25 and 90 cents per word for translations done by skilled human translators. Languages that are complex or that are spoken by relatively few people are generally more expensive to translate than other languages.

Different approaches can be appropriate for translating the different types of text that appear on an electronic commerce site. For key marketing messages, the touch of a human translator can be essential to capture subtle meanings. For more routine transaction processing functions, automated software translation may be an acceptable alternative. Software translation, also called **machine translation**, can reach speeds of 400,000 words per hour, so even if the translation is not perfect, businesses might find it preferable to a human who can translate between 400 and 600 words per hour. Many of the companies in this field are working to develop software and databases of previously translated material that can help human translators work more efficiently and accurately.

Idiom Technologies sells software that automates the process of maintaining Web pages in multiple language versions. Idiom's WorldServer software tracks text that needs to be translated and inserts the translations into all sites that include that language; however, a human translator must still perform the translation. By automating the process, companies can reduce the cost of maintaining multiple-language Web sites. WorldServer places XML tags in the text of each translated Web site. These tags identify each text element with a corresponding text element at the company's main Web site. When the text in a page at the main Web site changes, the software sends a notification of the change to the translation team. The notification shows which pages need to be updated and tracks the exact location of the change in each page that needs updating. When the human translation work is complete, the software automatically inserts the translated text into the correct locations.

The translation services and software manufacturers that work with electronic commerce sites do not generally use the term "translation" to describe what they do. They prefer the term **localization**, which means a translation that considers multiple elements of the local environment, such as business and cultural practices, in addition to local dialect variations in the language. The cultural element is very important, since it can affect—and sometimes completely change—the user's interpretation of text.

Culture Issues

Managers at Virtual Vineyards (now a part of **Wine.com**), a company that sells wine and specialty food items on the Web, were perplexed. The company was getting an unusually high number of complaints from customers in Japan about short shipments. Virtual Vineyards sold most of its wine in case (12 bottles) or half-case quantities. Thus, to save on operating costs, it stocked shipping materials only in case, half-case, and two-bottle sizes. After investigation, the company determined that many of its Japanese customers ordered only one bottle of wine, which was shipped in a two-bottle container. To these Japanese customers, who consider packaging to be an important element of a high-quality product such as wine, it was inconceivable that anyone would ship one bottle of wine in a two-bottle container. They were e-mailing to ask where the other bottle was, notwithstanding the fact that they had ordered only one bottle.

Some errors stemming from subtle language and cultural standards have become classic examples that are regularly cited in international business courses and training sessions. For example, General Motors' choice of name for its Chevrolet Nova automobile amused people in Latin America—*no va* means "it will not go" in Spanish. Pepsi's "Come Alive" advertising campaign fizzled in China because its message came across as "Pepsi brings your ancestors back from their graves."

Another story that is widely used in these training sessions is about a company that sold baby food in jars adorned with the picture of a very cute baby. The jars sold well everywhere they had been introduced except in parts of Africa. The mystery was solved when the manufacturer learned that food containers in those parts of Africa always carry a picture of their contents. This story is particularly interesting because it never happened. However, it illustrates a potential cultural issue so dramatically that it continues to appear in marketing textbooks and in international business training materials.

On the Web, designers must be very careful when choosing icons that represent common actions. For example, in the United States, a shopping cart is a good symbol to use when building an electronic commerce site. However, many Europeans use shopping *baskets* when they go to a store and may never have seen a shopping *cart*. In Australia, people would recognize a shopping cart image, however, they call them shopping *trolleys*. In the United States, people often form a hand signal (the index finger touching the thumb to create a circle) that indicates "OK" or "everything is just fine." A Web designer might be tempted to use this hand signal as an icon to indicate that the transaction is completed or the credit card is approved, unaware that in countries such as Brazil, this hand signal is a very offensive gesture.

The cultural overtones of simple design decisions can be dramatic. In India, for example, it is inappropriate to use the image of a cow in a cartoon or other comical setting. Potential customers in Muslim countries can be offended by an image that shows human arms or legs uncovered. Even colors or Web page design elements can be troublesome. A Web page that is divided into four segments can be offensive to a Japanese visitor because the number four is a symbol of death in that culture.

The design of a Web site built to attract customers from outside the United States should do more than avoid offending those visitors; it should also entice them. Web designers can do this by reflecting the visual preferences of the culture in which the site will operate. A site that strongly reflects cultural design preferences is **Bol.com**, a company that resulted from the mergers of a number of online bookstores from 12 countries around the world, including China, Finland, Germany, Japan, Malaysia, Sweden, and Switzerland. If you explore the Bol.com site, you can see a number of different design approaches used in the home pages for each of the different countries. For example, the Bol.com Japan home page (a portion of which appears in Figure 7-3) has very few graphics and a large amount of text. Japanese culture values factual information over broad visual appeal. Also, the graphics are considerably smaller than those that appear on Web pages directed at U.S. consumers. Very few Japanese consumers have broadband connections to the Internet, and this Web page is designed with that in mind.

only a few,
small
graphics
appear on
page

Figure 7-3 *Bol.com Japan home page*

You can explore the other Bol.com pages to find other characteristics that reflect the cultures of the target countries. For example, the pages for Finland and Sweden are less cluttered and have a crisp, clean design that is typical of Scandinavian graphic arts, furniture design, and architecture. These pages also have many fewer hyperlinks and much more white space than the typical U.S. retail Web site. Alternatively, the United Kingdom site is similar to what you would expect to see on a U.S. site. The Singapore site resembles the United Kingdom site, but it includes only one featured product on the home page. These Bol.com page designs are not necessarily better; they are just different in ways that are likely to make customers in each country feel more comfortable using the pages that are directed to them.

Softbank, a major Japanese firm that invests in Internet companies, has devised a way to introduce electronic commerce to a reluctant Japanese population. Japanese shoppers have resisted the U.S. version of electronic commerce because they generally prefer to pay in cash or by cash transfer instead of by credit card, and they have a high level of apprehension about doing business online. In 1999, Softbank created a joint venture with **7-Eleven**, **Yahoo! Japan**, and Tohan (a major Japanese book distributor) to sell books and CDs on the Web. This venture, called eS-Books, allows customers to order items on the Internet, and then pick them up and pay for them in cash at the local 7-Eleven convenience store. By adding an intermediary—the exact opposite of the strategy used by U.S. firms—that satisfies the needs of the Japanese customer, Softbank has been highly successful in bringing business-to-consumer electronic commerce to Japan.

Nike, a major U.S.-based maker of sports products, realized that it had to create special Web pages to attract the millions of its customers who live outside the United States. One such effort is the **Nike Football** site shown in Figure 7-4.

link to other-language versions of the site

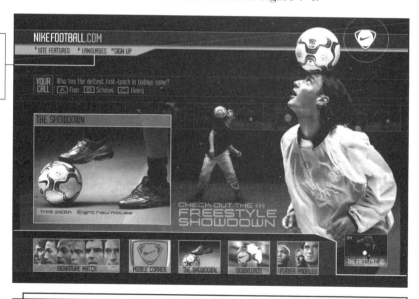

Figure 7-4   *Nike football site*

This soccer imagery is not what most U.S visitors would expect to see when visiting a "football" site! Since Nike already had a site that was designed for its U.S. audience (and which includes coverage of the U.S. game of "football"), it used the Nike Football site to appeal to European visitors by capitalizing on its involvement with the Euro 2000 soccer championships. Nike promoted these pages in television and newspaper advertising campaigns related to the Euro 2000 games. The site was so successful that it quickly was expanded to include pages devoted to Brazilian soccer. Today, the site is available in eight languages and covers soccer events around the world. Nike plans to develop even more Web pages directed at unique combinations of particular sports in particular countries in the appropriate languages for those sport-country combinations.

Some parts of the world have cultural environments that are extremely inhospitable to electronic commerce initiatives. For example, a report issued in 1999 by **Human Rights Watch** stated that many countries in the Middle East and North Africa do not allow their citizens free access to the Internet. The report notes that many governments in this part of the world regularly prevent free expression by their citizens and have taken specific steps to prevent the exchange of information outside of state controls. For example, Saudi Arabia, Yemen, and the United Arab Emirates all make efforts to filter the Web content that is available in their countries. Jordan has imposed taxes that put the cost of Internet access beyond the means of most Jordanians. Jordan also has laws that prohibit publications in any media that conflict with the values of an Islamic nation. An organization devoted to the international promotion of democracy and civil liberties, **Freedom House**, offers a number of downloadable publications on its site that include in-depth reports on Internet censorship activities of governments throughout the world.

In most North African and Middle Eastern countries, officials have publicly denounced the Internet for carrying materials that are sexually explicit, anti-Islam, or that cast doubts on the traditional role of women in their societies. In many of these countries, uncontrolled use of Internet technologies is so at odds with existing traditions, cultures, and laws that electronic commerce is unlikely to exist in these countries at any significant level in the near future. In contrast, Algeria, Morocco, and the Palestinian Authority have not limited online access or content.

Other countries, such as the People's Republic of China and Singapore, are wrestling with the issues presented by the growth of the Internet as a vehicle for doing business. These countries have a tradition of controlling their citizens' access to information from outside the country, but they want their economies to reap the benefits of electronic commerce. China has created a complex set of registration requirements and regulations that govern any business that engages in electronic commerce. These regulations are enforced by the Public Security Bureau, which is a branch of the state police, not an independent administrative agency. For example, ISPs must register all of their customers with the Public Security Bureau and must retain copies of all e-mail messages and chat room conversations for 60 days. Chinese citizens entering a chat room at **Sohu.com**, one of China's leading portal sites ("sohu" means "search fox" in Chinese), are greeted with a Web page containing the following text (translated here from the original Chinese):

Warning! Please take note that the following issues are prohibited according to Chinese law: 1) Criticism of the People's Republic of China Constitution. 2) Revealing State secrets, and discussion about overthrowing the Communist government. 3) Topics which damage the reputation of the State.

In July 2001, the Chinese government shut down 2000 Internet cafes for failing to keep adequate records and required another 6000 to suspend operations while they implemented the required electronic record keeping procedures. Singapore has also adopted a number of restrictive rules and policies. Only time will tell whether these attempts will effectively control Internet activity, yet allow business transactions to be conducted successfully.

Some countries, although they do not ban electronic commerce entirely, have strong cultural requirements that have found their way into the legal codes that govern business conduct. In France, an advertisement for a product or service must be in French. Thus, a business in the United States that advertises its products on the Web and that is willing to ship goods to France must provide a French version of its pages if it intends to comply with French law. Many U.S. electronic commerce sites include in their Web pages a list of the countries from which they will accept orders through their Web sites.

The official language of the Canadian province of Quebec is French. Quebec provincial law requires street signs, billboards, directories, and advertising created by Quebec businesses to be in French. In 1999, the government of Quebec fined photographer Michael Calomiris and ordered him either to remove his English-language Web site or to add a French translation of the pages to the site. Calomiris had been advertising his photographs for sale on his Quebec-based Web site and had targeted his ads to the U.S. market. He paid the fine and has appealed the government's decision. He is maintaining his Web site in English while his appeal is pending.

Infrastructure Issues

Businesses that successfully meet the challenges posed by language and culture issues still face the challenge posed by variations and inadequacies in the infrastructure that supports the Internet throughout the world. Internet infrastructure includes the computers and software connected to the Internet and the communications networks over which the message packets travel. In many countries other than the United States, the telecommunications industry is either government owned or heavily regulated by the government. In many cases, regulations in these countries have inhibited the development of the telecommunications infrastructure or limited the expansion of that infrastructure to a size that cannot reliably support Internet data packet traffic.

Local connection costs through the existing telephone networks in many countries are very high compared to U.S. costs for similar access. This can have a profound effect on the behavior of electronic commerce participants. For example, in Europe, where Internet connection costs can be quite high, few businesspeople would spend time surfing the Web to shop for a product. They will use a Web browser only to navigate to a specific site that they know will offer the product they want to buy. Thus, to be successful in selling to European businesses, a company must advertise its Web presence in traditional media instead of relying on high placement on search engine results pages.

The Organization for Economic Cooperation and Development's (OECD) Directorate of Science, Technology, and Industry has issued a number of **Statements on Information and Communications Policy** that deal with telecommunications infrastructure development issues throughout the world. These OECD statements have provided guidance for businesses and governments as they have begun building the technology capabilities that will support international electronic commerce in the future.

In recent years, business and government leaders in an increasing number of European countries have been pushing for flat-rate telephone line Internet access charges. Most Europeans pay for the time they are using the telephone line, including time for local calls, even in countries that have deregulated their telecommunications industries. In a **flat-rate access** system, the consumer or business pays one monthly fee for unlimited telephone line usage. Activists in these countries have argued that flat-rate access has been the key to the success of electronic commerce in the United States. Although many factors contributed to the rapid rise of U.S. electronic commerce, many industry analysts agree that flat-rate access has been one of the most important of those factors.

The paperwork and often-convoluted processes that accompany international transactions are targets for technological solutions. Most firms that conduct business internationally rely on a complex array of freight forwarding companies, customs brokers, international freight carriers, and importers to navigate the maze of paperwork that must be completed at every step of the transaction to satisfy government and insurance requirements. The multiple flows of information and transfers of physical objects that occur in a typical international trade transaction are illustrated in Figure 7-5.

As you can see in Figure 7-5, the information flows can be very complex. Domestic transactions usually include only the seller, the buyer, their respective banks, and one freight carrier. International transactions almost always require physical handling of goods by several freight carriers, storage in a freight forwarder's facility before international shipment, and storage in a port or bonded warehouse facility in the destination country. This handling and storage requires monitoring by government customs offices in addition to the monitoring by seller and buyer that occurs in domestic transactions. International transactions usually require the coordinated efforts of customs brokers and freight forwarding agencies because the regulations and procedures governing international transactions are so complex.

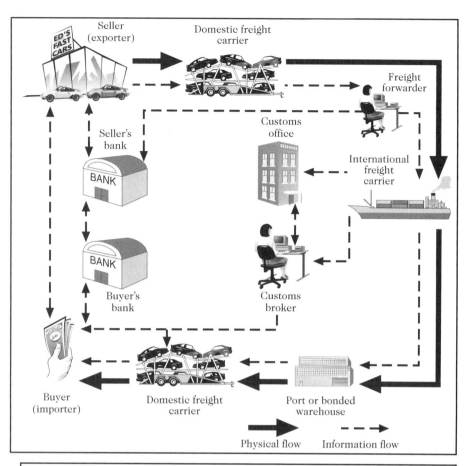

Figure 7-5 *Parties involved in a typical international trade transaction*

Analysts estimate that more than half of all businesses on the Web turn away all international orders because they do not have the processes in place to handle such orders. Some of these companies are losing millions of dollars worth of international business each year. This problem is global; not only are U.S. businesses having difficulty reaching their international markets, but businesses in other countries are having the same difficulty reaching the U.S. market.

The United Nations estimated that the cost of handling paperwork for international transactions is $540 billion, or approximately 6 percent of the total $9 trillion spent in worldwide international trade. Companies such as **ClearCross** and **NextLinx** sell software designed to automate much of the international trade process. This is a difficult task, because each country has its own paper-based forms and procedures with which international shippers must comply. To further complicate matters, some countries that have automated some of the procedures use computer systems that are incompatible with those of other countries. A consortium of 120 banks and logistics firms has founded **Bolero.net**, an association dedicated to replacing the paper maze with a set of interoperable electronic commerce applications. In many ways, these software vendors and consortia are working to repeat the success that EDI advocates had when they replaced common business documents with electronic data flows.

THE LEGAL ENVIRONMENT OF ELECTRONIC COMMERCE

Businesses that operate on the Web must comply with the same laws and regulations that govern the operations of all businesses. If they do not, they face the same set of penalties—fines, reparation payments, court-imposed dissolution, and even jail time for officers and owners—that any business faces.

Businesses operating on the Web face two additional complicating factors as they try to comply with the law. First, the Web extends a company's reach beyond traditional boundaries. As you learned in the previous section, a business that uses the Web immediately becomes an international business. Thus, a company can become subject to many more laws more quickly than a traditional brick-and-mortar business based in one specific physical location. Second, the Web increases the speed and efficiency of business communications. As you learned in Chapters 3 and 4, customers often have much more interactive and complex relationships with the companies they buy from on the Web than they do with traditional merchants. Further, the Web creates a network of customers who often have significant levels of interaction with each other. Web businesses that violate the law or breach ethical standards can face rapid and intense reactions from many customers and other stakeholders who become aware of the businesses' activities.

Borders and Jurisdiction

Territorial borders in the physical world serve a useful purpose in traditional commerce: They mark the range of culture and reach of applicable laws very clearly. When people travel across international borders, they are made aware of the transition in many ways. For example, exiting one country and entering another usually requires a formal examination of documents, such as passports and visas. In addition, both the language and the currency usually change upon entry into a new country. Each of these experiences, and countless others, are manifestations of the differences in legal rules and cultural customs in the two countries. In the physical world, geographic boundaries almost always coincide with legal and cultural boundaries. The limits of acceptable ethical behavior and the laws that are adopted in a geographic area are the result of the influences of the area's dominant culture. The relationships among a society's culture, laws, and ethical standards appear in Figure 7-6.

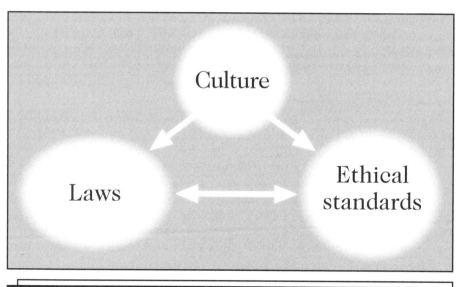

Figure 7-6   Culture determines laws and ethical standards

The geographic boundaries on culture are logical—for most of our history, we humans have been unable to travel great distances to learn about other cultures. In recent years, some countries have decided that times have indeed changed and that people can travel easily from one country to another within a geographic region. One example is the European Union (EU), which allows free movement within the EU for citizens of member countries. The EU has even adopted a common currency, the euro, to replace the currencies of the individual member countries. Legal scholars define the relationship between geographic boundaries and legal boundaries in terms of four elements: power, effects, legitimacy, and notice.

Power

Power, in the form of control over physical space and the people and objects that reside in that space, is a defining characteristic of statehood. For laws to be effective, a government must be able to enforce them. Effective enforcement requires the power both to exercise physical control over residents, if necessary, and to impose sanctions on those who violate the law. The ability of a government to exert control over a person or corporation is called **jurisdiction**.

Laws in the physical world do not apply to people who are not located in or who do not own assets in the geographic area that created those particular laws. For example, the United States cannot enforce its copyright laws on a citizen of Japan who is doing business in Japan and who owns no assets in the United States. Any assertion of power by the United States over such a Japanese citizen would conflict with the Japanese government's recognized monopoly on using force with its citizens. Japanese citizens who bring goods into the United States to sell, however, would be subject to applicable U.S. copyright laws. A Japanese Web site that offered delivery of goods into the United States would, similarly, be subject to applicable U.S. laws.

The level of power asserted by a government is limited to that which is accepted by the culture that exists within its geographic boundaries. Ideally, geographic boundaries, cultural groupings, and legal structures will all coincide. When they do not, internal strife and civil wars can erupt.

Effects

Laws in the physical world are grounded in the relationship between physical proximity and the effects of a person's behavior. Personal or corporate actions have stronger effects on people and things that are nearby than on those that are far away. Government-provided trademark protection is a good example of this. For instance, the Italian government can provide and enforce trademark protection for a business named Caruso's Ristorante located in Rome. The effects of another restaurant using the same name are strongest in geographic areas in Rome, close to Rome, and in other parts of Italy. If someone were to open a restaurant in Kansas City and call it Caruso's Ristorante, the restaurant in Rome would experience few, if any, negative effects from having this restaurant use its trademarked name. Thus, the effects of the trademark violation are controlled by the law in Italy because of the limited range within which such a violation has an effect.

The characteristics of laws are determined by the local culture's acceptance of or reluctance to various kinds of effects. For example, certain communities in the United States require that houses be built on lots that are at least five acres. Other communities prohibit outdoor advertising of various kinds. The local cultures in these communities make the effects of such restrictions acceptable.

When businesses begin operations online, the traditional measures of effects—and the laws that have been developed using those measures over many years—do not work very well. For example, France has a law that prohibits the sale of Nazi memorabilia. The people of France have considered this to be a reasonable law for many years. The United States does not have a similar prohibition in its laws. When U.S.-based online auction sites began offering auctions of Nazi memorabilia, the auction sites were in compliance with U.S. laws. However, because of the international nature of the Web, these auctions were available to people around the world, including residents of France. The French government ordered Yahoo! Auctions to stop these auctions. Yahoo! argued that it was in compliance with U.S. law, but the French government insisted that the effects of those Yahoo! auctions extended to France and thus violated French law. To avoid protracted legal actions over the jurisdiction issue, Yahoo! decided that it would no longer carry such auctions.

Legitimacy

Most people agree that the legitimate right to create and enforce laws derives from the mandate of those who are subject to those laws. In 1970, the **United Nations** passed a resolution that affirmed this idea of governmental legitimacy. The resolution made clear that the people residing within a set of recognized geographic boundaries are the ultimate source of legitimate legal authority for people and actions within those boundaries. Thus, **legitimacy** is the idea that those subject to laws should have some role in formulating them.

Some cultures allow their governments to operate with a high degree of autonomy and unquestioned authority. China and Singapore are countries in which national culture permits the government to exert high levels of unchecked authority. Other cultures, such as those of the Scandinavian countries, place strict limits on governmental authority.

The levels of authority and autonomy with which governments of various countries operate varies significantly from one country to another. Online businesses must be ready to deal with a wide variety of regulations and levels of enforcement of those regulations as they expand their businesses to other countries. This can be difficult for smaller businesses that operate on the Web.

Notice

Physical boundaries are a convenient and effective way to announce the ending of one legal or cultural system and the beginning of another. The physical boundary, when crossed, provides notice that one set of rules has been replaced by a different set of rules. People can obey and perceive a law or cultural norm as fair only if they are notified of its existence. Borders provide this notice in the physical world. The legal systems of most countries include a concept called constructive notice. Persons receive **constructive notice** that they have become subject to new laws and cultural norms when they cross an international border, even if they are not specifically warned of the changed laws and norms by a sign or a border guard's statement. Thus, ignorance of the law is not a sustainable defense, even in a new and unfamiliar jurisdiction.

This presents particular problems for online businesses, since they may not know that customers from another country are accessing their Web sites. Thus, the concept of notice—even constructive notice—does not translate very well to online business.

Jurisdiction on the Internet

Defining, establishing, and asserting jurisdiction are much more difficult on the Internet than they are in the physical world, mainly because traditional geographic boundaries do not exist. For example, a Swedish company that engages in electronic commerce may have a Web site that is entirely in English and a URL that ends in ".com," thus not indicating to customers that it is a Swedish firm. The server that hosts this company's Web page could be in Canada, and the people who maintain the Web site might work from their homes in Australia.

If a Mexican citizen buys a product from the Swedish firm and is unhappy with the goods received, that person might want to file a lawsuit against the seller firm. However, the world's physical border–based systems of law and jurisdiction do not help this Mexican citizen determine where to file the lawsuit. The Internet does not provide anything like the obvious international boundary lines in the physical world. Thus, the four considerations that work so well in the physical world—power, effects, legitimacy, and notice—do not translate very well to the virtual world of electronic commerce.

Governments that want to enforce laws regarding business conduct on the Internet must establish jurisdiction over that conduct. A **contract** is a promise or set of promises between two or more legal entities—persons or corporations—that provides for an exchange of value (goods, services, or money) between or among them. A **tort** is an intentional or negligent action taken by a legal entity that causes harm to another legal entity. People or corporations that wish to enforce their rights based on either contract or tort law must file their claims in courts with jurisdiction to hear their case. A court has sufficient jurisdiction in a matter if it has both subject-matter jurisdiction and personal jurisdiction.

Subject-Matter Jurisdiction

Subject-matter jurisdiction is a court's authority to decide a particular type of dispute. For example, in the United States, federal courts have subject-matter jurisdiction over issues governed by federal law (such as bankruptcy, copyright, patent, and federal tax matters), and state courts have subject-matter jurisdiction over issues governed by state laws (such as professional licensing and state tax matters). If the parties to a contract are both located in the same state, a state court has subject-matter jurisdiction over disputes that arise from the terms of that contract. The rules for determining whether a court has subject-matter jurisdiction are very clear and easy to apply. Very few disputes arise over subject-matter jurisdiction.

Personal Jurisdiction

Personal jurisdiction is, in general, determined by the residence of the parties. A court has personal jurisdiction over a case if the defendant is a resident of the state in which the court is located. In such cases, the determination of personal jurisdiction is straightforward. However, an out-of-state person or corporation can also voluntarily submit to the jurisdiction of a particular state court by agreeing to do so in writing or by taking certain actions in the state.

One of the most common ways that people voluntarily submit to a jurisdiction is by signing a contract that includes a statement, known as a **forum selection clause**, that the contract will be enforced according to the laws of a particular state. That state then has personal jurisdiction over the parties who signed the contract regarding any enforcement issue that arises from the terms of that contract.

In the United States, individual states have laws that can create personal jurisdiction for their courts. The details of these laws, called **long-arm statutes**, vary from state to state, but generally create personal jurisdiction over nonresidents who transact business or commit tortious acts in the state. For example, suppose that an Arizona resident drives recklessly while in California and, as a result, causes a collision with another vehicle that is driven by a California resident. Due to the driver's tortious behavior in the state of California, the Arizona resident can expect to be called into a California court. In other words, California courts have personal jurisdiction over the matter.

Businesses should be aware of jurisdictional considerations when conducting electronic commerce over state and international lines. In most states, the extent to which these laws apply to companies doing business over the Internet is unclear. Since these procedural laws were written before electronic commerce existed, their application to Internet transactions continues to evolve as more and more disputes arise from online commercial transactions. The trend in this evolving law is that the more business activities a company conducts in a state, the more likely it is that a court will assert personal jurisdiction over that company through the application of a long-arm statute.

One exception to the general rule for determining personal jurisdiction occurs in the case of tortious acts. A business can commit a tortious act by selling a product that causes harm to a buyer. The tortious act can be negligent, in which the seller unintentionally provides a harmful good, or it can be an intentional tort, in which the seller knowingly or recklessly causes injury to the buyer. The most common examples of business-related intentional torts are defamation, misrepresentation, fraud, and theft of trade secrets. Although case law is rapidly developing in this area also, courts tend to invoke their respective states' long-arm statutes much more readily in the case of tortious acts than in other business cases. If the matter involves an intentional tort or a criminal act, courts will more liberally assert jurisdiction.

Jurisdiction in International Commerce

Jurisdiction issues that arise in international business are even more complex than the rules governing personal jurisdiction across state lines within the United States. The exercise of jurisdiction across international borders is governed by treaties between the countries engaged in the dispute. In general, U.S. courts determine personal jurisdiction for foreign companies and persons in much the same way that these courts interpret the long-arm statutes in domestic matters. Non-U.S. corporations and individuals can be sued in U.S. courts if they conduct business or commit tortious acts in the United States. Similarly, foreign courts can enforce decisions against U.S. corporations or individuals through the U.S. court system if those courts can establish jurisdiction over the matter.

Jurisdictional issues are complex and change rapidly. Any business that intends to conduct electronic commerce should consult an attorney who is well-versed in these procedural issues. However, there are a number of resources online that can be useful to non-lawyers who want to do preliminary investigation of a legal topic such as jurisdiction. The **UCLA Online Institute for Cyberspace Law and Policy** has been operating since 1995 to provide legal news updates and reference materials online. The **John Marshall Law School's Center for Information Technology and Privacy Law** Web page includes links to current cases, law review articles, and other updated resources related to electronic commerce legal issues. The Center provides a collection of materials related to cyberspace law, as shown in Figure 7-7.

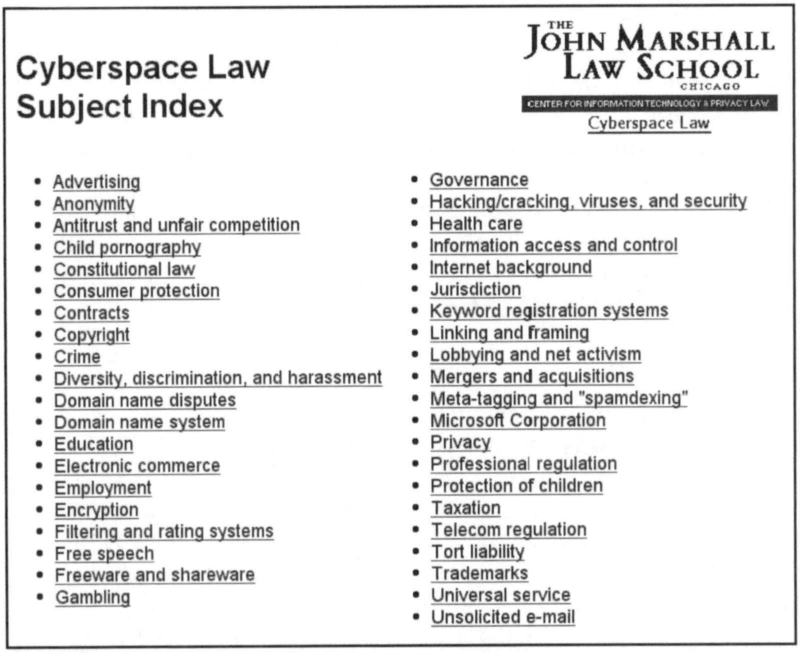

Cyberspace Law Subject Index

THE JOHN MARSHALL LAW SCHOOL
CHICAGO
CENTER FOR INFORMATION TECHNOLOGY & PRIVACY LAW
Cyberspace Law

- Advertising
- Anonymity
- Antitrust and unfair competition
- Child pornography
- Constitutional law
- Consumer protection
- Contracts
- Copyright
- Crime
- Diversity, discrimination, and harassment
- Domain name disputes
- Domain name system
- Education
- Electronic commerce
- Employment
- Encryption
- Filtering and rating systems
- Free speech
- Freeware and shareware
- Gambling

- Governance
- Hacking/cracking, viruses, and security
- Health care
- Information access and control
- Internet background
- Jurisdiction
- Keyword registration systems
- Linking and framing
- Lobbying and net activism
- Mergers and acquisitions
- Meta-tagging and "spamdexing"
- Microsoft Corporation
- Privacy
- Professional regulation
- Protection of children
- Taxation
- Telecom regulation
- Tort liability
- Trademarks
- Universal service
- Unsolicited e-mail

Figure 7-7 *John Marshall Law School Cyberspace Law site*

Contracting and Contract Enforcement in Electronic Commerce

Any contract includes three essential elements: an offer, an acceptance, and consideration. The contract is formed when one party accepts the offer of another party. An **offer** is a commitment with certain terms made to another party, such as a declaration of willingness to buy or sell a product or service. An offer can be revoked as long as no payment, delivery of service, or other consideration has been accepted. An **acceptance** is the expression of willingness to take an offer, including all of its stated terms. **Consideration** is the bargained-for exchange of something valuable, such as money, property, or future services. When a party accepts an offer based on the exchange of valuable goods or services, a contract has been created. An **implied contract** can also be formed by two or more parties that act as if a contract exists, even if no contract has been written and signed.

People enter into contracts on a daily, and often hourly, basis. Every kind of agreement or exchange between parties, no matter how simple, is a type of contract. For example, every time a consumer buys an item at the supermarket, the elements of a valid contract are met:

- The store offers an item at a stated price.
- The consumer accepts this offer by indicating a willingness to buy the product for the stated price.
- The store exchanges its product for another valuable item: the consumer's payment.

Contracts are a key element of traditional business practice, and they are equally important on the Internet. Offers and acceptances can occur when parties exchange e-mail messages, engage in electronic data interchange (EDI), or fill out forms on Web pages. These Internet communications can be combined with traditional methods of forming contracts, such as the exchange of paper documents, faxes, and verbal agreements made over the telephone or in person. An excellent resource for many of the laws concerning contracts, especially as they pertain to U.S. businesses, is the Cornell Law School Web site, which includes the full text of the **Uniform Commercial Code** (UCC).

When a seller advertises goods for sale on a Web site, that seller is not making an offer, but is inviting offers from potential buyers. If a Web ad were a legal offer to form a contract, the seller could easily become liable for the delivery of more goods than it has available to ship. When a buyer submits an order, which is an offer, the seller can accept that offer and create a contract. If the seller does not have the ordered items in stock, the seller has the option of refusing the buyer's order outright or counteroffering with a decreased amount. The buyer then has the option to accept the seller's counteroffer.

Making a legal acceptance of an offer is quite easy to do in most cases. When enforcing contracts, courts tend to view offers and acceptances as actions that occur within a particular context. If the actions are reasonable under the circumstances, courts tend to interpret those actions as offers and acceptances. For example, courts have held that actions—including mailing a check, shipping goods, shaking hands, nodding one's head, taking an item off a shelf, or opening a wrapped package—are all, in some circumstances, legally binding acceptances of offers. Although the case law is limited regarding acceptances made over the Internet, it is reasonable to assume that courts would view clicking a button on a Web page, entering information in a Web form, or downloading a file to be legally binding acceptances.

Written Contracts on the Web

In general, contracts are valid even if they are not in writing or signed. However, certain categories of contracts are not enforceable unless the terms are put into writing and signed by both parties. In 1677, the British Parliament enacted a law that specified the types of contracts that had to be in writing and signed. Following this British precedent, every state in the United States today has a similar law, called a **Statute of Frauds**. Although these state laws vary slightly, each Statute of Frauds specifies that contracts for the sale of goods worth over $500 and contracts that require actions that cannot be completed within one year must be created by a signed writing. Fortunately for businesses and people who want to form contracts using electronic commerce, a writing does not require either pen or paper.

Most courts will hold that a **writing** exists when the terms of a contract have been reduced to some tangible form. An early court decision in the 1800s held that a telegraph transmission was a writing. Later courts have held that tape recordings of spoken words, computer files on disks, and faxes are writings. Thus, the parties to an electronic commerce contract should find it relatively easy to satisfy the writing requirement. Courts have been similarly generous in determining what constitutes a signature. A **signature** is any symbol executed or adopted for the purpose of authenticating a writing. Courts have held names on telegrams, telexes, faxes, and Western

Union Mailgrams to be signatures. Even typed names or names printed as part of a letterhead have served as signatures. It is reasonable to assume that a symbol or code included in an electronic file would constitute a signature. As you will learn in Chapter 10, the United States now has a law that explicitly makes digital signatures legally valid for contract purposes.

Firms conducting international electronic commerce do not need to worry about the signed writing requirement in most cases. The main treaty that governs international sales of goods, Article 11 of the **United Nations Convention on Contracts for the International Sale of Goods**, requires neither a writing nor a signature to create a legally binding acceptance.

Warranties

Most firms conducting electronic commerce have little trouble fulfilling the requirements needed to create enforceable, legally binding contracts on the Web. One area that deserves attention, however, is the issue of warranties. Any contract for the sale of goods includes implied warranties. A seller implicitly warrants that the goods that it sells are fit for the purposes for which they are normally used. If the seller knows specific information about the buyer's requirements, acceptance of an offer from that buyer may result in an additional implied warranty of fitness, which suggests that the goods are suitable for the specific uses of that buyer. Sellers can also create explicit warranties by providing a specific description of the additional warranty terms. It is also possible for a seller to create explicit warranties, often unintentionally, by making general statements in brochures or other advertising materials about product performance or suitability for particular tasks.

Sellers can avoid some implied warranty liability by making a warranty disclaimer. A **warranty disclaimer** is a statement that the seller will not honor some or all implied warranties. Any warranty disclaimer must be conspicuously made in writing, which means it must be easily noticed in the body of the written agreement. On a Web page, sellers can meet this requirement by putting the warranty disclaimer in larger type, a bold font, or a contrasting color. To be legally effective, the warranty disclaimer must be stated obviously and must be easy for a buyer to find on the Web site.

Authority to Form Contracts

As explained previously in this section, a contract is formed when an offer is accepted for consideration. Problems can arise when the acceptance is issued by an imposter or someone who does not have the authority to bind the company to a contract. In electronic commerce, the online nature of acceptances can make it relatively easy for identity forgers to pose as others.

Fortunately, the Internet technology that makes forged identities so easy to create also provides the means to avoid being deceived by a forged identity. In Chapter 10, you will learn how companies and individuals can use digital signatures to establish identity in online transactions. If the contract is for any significant amount, the parties should require each other to use digital signatures to avoid identity problems. In general, courts will not hold a person or corporation whose identity has been forged to the terms of the contract; however, if negligence on the part of the person or corporation contributed to the forgery, a court may hold the negligent party to the terms of the contract. For example, if a company was careless about protecting passwords and allowed an imposter to enter the company's system and accept an offer, a court might hold that company responsible for fulfilling the terms of that contract.

Determining whether an individual has the authority to commit a company to an online contract is a greater problem than forged identities in electronic commerce. This issue, called **authority to bind**, can arise when an employee of a company accepts a contract and the company later asserts that the employee did not have such authority. For large transactions in the physical world, businesses check public information on file with the state of incorporation or ask for copies of corporate certificates or resolutions to establish the authority of persons to make contracts for their employers. These methods are available to parties engaged in online transactions; however, they can be time consuming and awkward. You will learn about some good electronic solutions, such as digital signatures and certificates from a certification authority, in Chapter 10.

Terms of Service Agreements

Many Web sites have stated rules that site visitors must follow, although very few visitors are aware of these rules. If you examine the home page of a Web site, you will often find a link to a page titled "Terms of Service," "Conditions of Use," "User Agreement," or something similar. If you follow that link, you will find a page full of detailed rules and regulations, most of which are intended to limit the Web site owner's liability for what you might do with information you obtain from the site. These contracts are often called **terms of service (ToS)** agreements even when they appear under a different title. In most cases, a site visitor is held to the terms of service even if that visitor has not read the text or clicked a button to indicate agreement with the terms. The visitor is bound to the agreement by simply using the site. The first few sections of the Amazon.com terms of service agreement appear in Figure 7-8, which shows the top of Amazon's "Conditions of Use" page.

275

amazon.com. | ☒ VIEW CART | WISH LIST | YOUR ACCOUNT | HELP

WELCOME | YOUR STORE | BOOKS | ELECTRONICS | TOYS & GAMES | SOFTWARE | CELL PHONES & SERVICE | COMPUTER & VIDEO GAMES | ▶ SEE MORE STORES

 Help > Privacy & Security > Privacy Notice > Conditions of Use

Conditions of Use

Welcome to Amazon.com. Amazon.com and its affiliates provide their services to you subject to the following conditions. **If you visit or shop at Amazon.com, you accept these conditions.** Please read them carefully. In addition, when you use any current or future Amazon.com service (e.g., Friends & Favorites, e-Cards, Auctions, and Honor System) or visit or purchase from any business affiliated with Amazon.com, whether or not included in the Amazon.com Web site, you also will be subject to the guidelines and conditions applicable to such service or business.

> explains that these terms of service apply to all site visitors

PRIVACY

Please review our Privacy Notice, which also governs your visit to Amazon.com, to understand our practices.

ELECTRONIC COMMUNICATIONS

When you visit Amazon.com or send e-mails to us, you are communicating with us electronically. You consent to receive communications from us electronically. We will communicate with you by e-mail or by posting notices on this site. You agree that all agreements, notices, disclosures and other communications that we provide to you electronically satisfy any legal requirement that such communications be in writing.

COPYRIGHT

All content included on this site, such as text, graphics, logos, button icons, images, audio clips, digital downloads, data compilations, and software, is the property of Amazon.com or its content suppliers and protected by United States and international copyright laws. The compilation of all content on this site is the exclusive property of Amazon.com and protected by U.S. and international copyright laws. All software used on this site is the property of Amazon.com or its software suppliers and protected by United States and international copyright laws.

TRADEMARKS

AMAZON.COM; AMAZON.COM BOOKS; EARTH'S BIGGEST BOOKSTORE; IF IT'S IN PRINT, IT'S IN STOCK; MOODMATCHER; 1-CLICK; and other marks indicated on our site are registered trademarks of Amazon.com, Inc. or its subsidiaries, in the United States and other countries. EARTH'S BIGGEST SELECTION, PURCHASE CIRCLES, SHOP THE WEB, ONE-CLICK SHOPPING, AMAZON.COM ASSOCIATES, AMAZON.COM MUSIC, AMAZON.COM VIDEO, AMAZON.COM TOYS, AMAZON.COM ELECTRONICS, AMAZON.COM e-CARDS, AMAZON.COM AUCTIONS, zSHOPS, CUSTOMER BUZZ, AMAZON.CO.UK, AMAZON.DE, BID-CLICK, GIFT-CLICK, AMAZON.COM ANYWHERE, AMAZON.COM OUTLET, BACK TO BASICS, BACK TO BASICS TOYS, NEW FOR YOU, AMAZON HONOR SYSTEM, PAYPAGE, UNPAY, and other Amazon.com graphics, logos, page headers, button icons, scripts, and service names are trademarks or trade dress of Amazon.com, Inc. or its subsidiaries. Amazon.com's trademarks and trade dress may not be used in connection with any product or service that is not Amazon.com's, in any manner that is likely to cause confusion among customers, or in any manner that disparages or discredits Amazon.com. All other trademarks not owned by Amazon.com or its subsidiaries that appear on this site are the property of their respective owners, who may or may not be affiliated with, connected to, or sponsored by Amazon.com or its subsidiaries.

PATENTS

One or more United States and international patents apply to this site, including without limitation: U.S. Patent Nos. 5,715,399; 5,727,163; 5,826,258; 5,960,411; 5,963,949; and 5,999,924.

LICENSE AND SITE ACCESS

Amazon.com grants you a limited license to access and make personal use of this site and not to download (other than page caching) or modify it, or any portion of it, except with express written consent of Amazon.com. This license does not include any resale or commercial use of this site or its contents; any collection and use of any product listings, descriptions, or prices; any derivative use of this site or its contents; any downloading or copying of account information for the benefit of another merchant; or any use of data mining, robots, or similar data gathering and extraction tools. This site or any portion of this site may not be reproduced, duplicated, copied, sold, resold, visited, or otherwise exploited for any commercial purpose without express written consent of Amazon.com. You may not frame or utilize framing techniques to enclose any trademark, logo, or other proprietary information (including images, text, page layout, or form) of Amazon.com and our affiliates without express written consent. You may not use any meta tags or any other "hidden text" utilizing Amazon.com's name or trademarks without the express written consent of Amazon.com. Any unauthorized use terminates the permission or license granted by

276

Figure 7-8   *Amazon.com Conditions of Use page*

Web Site Content Issues

A number of legal issues can arise regarding the Web page content of electronic commerce sites, including deceptive trade practices, regulation of advertising claims, defamation, product disparagement, and violations of intellectual property rights. Intellectual property rights include the protections afforded to individuals and companies by the government through its granting of copyrights, patents, and through its registration of trademarks and service marks.

Copyright Infringement

A **copyright** is a right granted by a government to the author or creator of a literary or artistic work. The right is for the specific length of time provided in the copyright law and gives the author or creator the sole and exclusive right to print, publish, or sell the work. Creations that can be copyrighted include virtually all forms of artistic or intellectual expression—books, music, artworks, recordings (audio and video), architectural drawings, choreographic works, product packaging, and computer software. In the United States, works created after 1977 are protected for the life of the author plus 70 years. Works copyrighted by corporations or not-for-profit organizations are protected for 95 years from the date of publication or 120 years from the date of creation, whichever is earlier.

The idea contained in the expression is not copyrightable. The particular form of expression of an idea creates a work that can be copyrighted. If an idea cannot be separated from its expression in a work, that work cannot be copyrighted. For example, mathematical calculations cannot be copyrighted. A collection of facts can be copyrighted, but only if the collection is arranged, coordinated, or selected in a way that causes the resulting work to rise to the level of an original work. For example, the Yahoo! Web Directory is a collection of links to URLs. These facts existed before Yahoo! selected and arranged them into the form of its directory. However, most copyright lawyers would argue that the selection and arrangement of the links into categories probably makes the directory copyrightable.

In the past, many countries (including the United States) required the creator of a work to register that work to obtain copyright protection. U.S. law still allows registration, but registration is no longer required. A work that does not include the words "copyright" or "copyrighted" or the copyright symbol (©), and that was created after 1977, is copyrighted automatically by virtue of the copyright law unless the creator has specifically released the work into the public domain.

Most Web pages are protected by copyright because they arrange the elements of words, graphics, and HTML tags in a way that creates an original work. This creates a potential problem because of the way the Web works. As you learned in Chapter 2, when a Web client requests a page, the Web server sends an HTML file to the client. Thus, a copy of the HTML file (along with any graphics or other files needed to render the page) resides on the Web client computer. Most legal experts agree that this copying is a fair use of the copyrighted Web page. The U.S. copyright law includes an exemption from infringement actions for fair use of copyrighted works. The **fair use** of a copyrighted work includes copying it for use in criticism, comment, news reporting, teaching, scholarship, or research. The law's definition of fair use is intentionally broad and can be difficult to interpret. When you make fair use of a copyrighted work, you must be careful to provide a citation to the original work to avoid charges of plagiarism.

Copyright law has always included elements, such as the fair use exemption, that make it difficult to apply. The Internet has made this situation worse because it allows the immediate transmission of exact digital copies of many materials. In the case of digital music, the Napster site provided a network that millions of people used to trade music files that they had copied from their CDs and compressed into MPEG version 3 format, commonly referred to as MP3. This constituted copyright violation on a grand scale, and a group of music recording companies sued Napster for facilitating the violations. Napster argued that it had only provided the "machinery" used in the copyright violations—much as electronics companies manufacture and sell VCRs that might be used to make illegal copies of videotapes—and had not itself infringed on any copyrights. Both the U.S. District Court and the Federal Appellate Court held that Napster was guilty of "vicarious infringement" by failing to monitor its network and by profiting from its operation even though Napster itself did not transfer any copies. The courts ordered that Napster be shut down. In late 2001, Napster agreed to pay $26 million in damages for copyright infringement to a group of music publishing associations and was working on re-launching the site with agreements in place to pay copyright holders for the music that would be downloaded in the future.

Patent Infringement

A **patent** is an exclusive right to make, use, and sell an invention that a government grants to the inventor. In the United States, patents on inventions protect the inventor's rights for 20 years. A patent on the design for an invention provides protection for 14 years. To be patentable, an invention must be genuine, novel, useful, and not obvious given the current state of technology. In the early 1980s, companies began obtaining patents on software programs that met the terms of the U.S. patent law. However, most firms that develop software to use in Web sites and for related transaction processing have not found the patent law to be very useful. The process of obtaining a patent is expensive and can take several years. Most developers of Web-related software believe that the technology in the software could become obsolete before the patent protection is secured.

One type of patent has been of interest to companies engaging in electronic commerce. A U.S. Court of Appeals ruled in 1998 that patents could be granted on "methods of doing business." The **business process patent**, which protects a specific set of procedures for conducting a particular business activity, is quite controversial. In addition to the Amazon.com patent on its 1-Click purchasing method (which you read about in Chapter 4), other Web businesses have obtained business process patents. The Priceline.com "name your own price" price-tendering system, About.com's approach to aggregating information from many different Web sites, and Cybergold's method of paying people to view its Web site have each received business process patents. Many legal experts and business researchers believe that the issuance of business process patents grants the recipients unfair monopoly power and is an inappropriate extension of patent law. The U.S. Supreme Court has not yet ruled on any cases involving business process patents.

Trademark Infringement

A **trademark** is a distinctive mark, device, motto, or implement that a company affixes to the goods it produces for identification purposes. A **service mark** is similar to a trademark, but it is used to identify services provided. In the United States, trademarks and service marks can be registered with state governments, the Federal government, or both. The name (or a part of that name) that a business uses to identify itself is called a **trade name**. Trade names are not protected by trademark laws unless the business name is the same as the product (or service) name. They are protected, however, under common law. **Common law** is the part of British and U.S. law that is established by the history of court decisions that has built up over many years. The other main part of British and U.S. law, called **statutory law**, arises when elected legislative bodies pass laws, which are also statutes.

The owners of registered trademarks have often invested a considerable amount of money in the development and promotion of their trademarks. Web site designers must be very careful not to use any trademarked name, logo, or other identifying mark without the express permission of the trademark owner. For example, a company Web site that included a photograph of its president who happened to be holding a can of Pepsi could violate Pepsi's trademark rights. Pepsi can argue that the appearance of its trademarked product on the Web site implies an endorsement of the president or the company by Pepsi.

Defamation

A **defamatory** statement is a statement that is false and that injures the reputation of another person or company. If the statement injures the reputation of a product or service instead of a person, it is called **product disparagement**. In some countries, even a true and honest comparison of products may give rise to product disparagement. Since the difference between justifiable criticism and defamation can be hard to determine, commercial Web sites should avoid making negative, evaluative statements about other persons or products.

Web site designers should be especially careful to avoid potential defamation liability by altering a photo or image of a person in a way that depicts the person unfavorably. In most cases, a person must establish that the defamatory statement caused injury. However, most states recognize a legal cause of action, called **per se defamation**, in which a court deems some types of statements to be so negative that injury is assumed. For example, the court will hold inaccurate statements alleging conduct potentially injurious to a person's business, trade, profession, or office as defamatory per se—the complaining party need not prove injury to recover damages. Thus, online statements about competitors should always be carefully reviewed before posting to determine whether they contain any elements of defamation.

Deceptive Trade Practices

The ease with which they can edit graphics, audio, and video files allows Web site designers to do many creative and interesting things. Manipulations of existing pictures, sounds, and video clips can be very entertaining. If the objects being manipulated are trademarked, however, these manipulations can violate the trademark holder's rights. Fictional characters can be trademarked or otherwise protected.

Many personal Web pages include unauthorized use of cartoon characters and scanned photographs of celebrities; often, these images are altered in some way. An electronic commerce Web site that uses an altered image of Mickey Mouse speaking in a modified voice is likely to hear from the Disney legal team.

Web sites that include links to other sites must be careful not to imply a relationship with the companies sponsoring the other sites unless such a relationship actually exists. For example, a Web design studio's Web page may include links to company Web sites that show good design principles. If those company Web sites were not created by the design studio, the studio must be very careful to state that fact. Otherwise, it would be easy for a visitor to assume that the linked sites were the work of the design studio.

In general, trademark protection prevents another firm from using the same or a similar name, logo, or other identifying characteristic in a way that would cause confusion in the minds of potential buyers of the trademark holder's products or services. For example, the trademarked name *Visa* is used by one company for its credit card services and another company for its type of synthetic fiber. This use is acceptable because the two products are very different. However, the use of very well-known trademarks can be protected for all products if there is a danger that the trademark might be diluted. Various state laws define **trademark dilution** as the reduction of the distinctive quality of a trademark by alternative uses. Trademarked names such as *Hyatt*, *Trivial Pursuit*, and *Tiffany*, and the shape of the Coca-Cola bottle have all been protected from dilution by court rulings. A Web site that sells gift-packaged seafood and claims to be the "Tiffany of the Sea" risks a lawsuit from the famous jeweler claiming trademark dilution.

Advertising Regulation

In the United States, advertising is regulated primarily by the **Federal Trade Commission**. The FTC publishes regulations and investigates claims of false advertising. Its Web site, shown in Figure 7-9, includes helpful guidelines for businesses that want to comply with the law and avoid such claims.

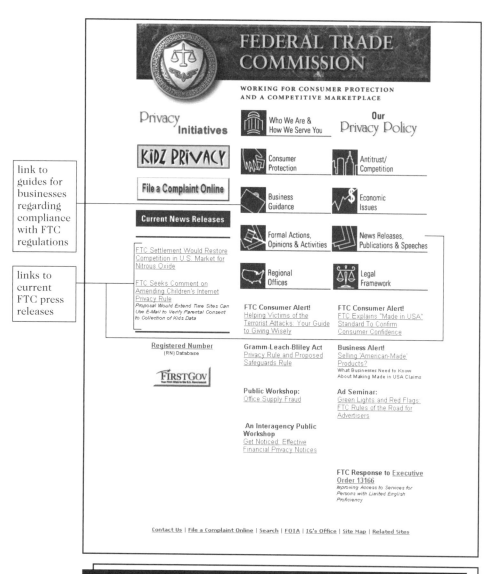

281

Figure 7-9 *Federal Trade Commission home page*

Any advertising claim that can mislead a substantial number of consumers in a material way is illegal under U.S. law. In addition to conducting its own investigations, the FTC accepts referred investigations from organizations such as the Better Business Bureau. The FTC provides policy statements that can be helpful guides for designers creating electronic commerce Web sites. These policies include information on what is permitted in advertisements, and covers specific areas such as these:

- Bait advertising
- Consumer lending and leasing
- Endorsements and testimonials

- Energy consumption statements for home appliances
- Guarantees and warranties
- Prices

Other federal agencies have the power to regulate online advertising in the United States. These agencies include the Food and Drug Administration (FDA), the Bureau of Alcohol, Tobacco, and Firearms (BATF), and the Department of Transportation (DOT). The FDA regulates information disclosures for food and drug products. In particular, any Web site that is planning to advertise pharmaceutical products will be subject to the FDA's drug labeling and advertising regulations. The BATF works with the FDA to monitor and enforce federal laws regarding advertising for alcoholic beverages and tobacco products. These laws require that every ad for such products includes statements that use very specific language. Many states also have laws that regulate advertising for alcoholic beverages and tobacco products. The state and federal laws governing advertising and the sale of firearms are even more restrictive. Any Web site that plans to deal in these products should consult with an attorney who is familiar with the relevant laws before posting any advertising for them online. The DOT works with the FTC to monitor the advertising of companies over which it has jurisdiction, such as bus lines, freight companies, and airlines.

Web-Based Crime, Terrorism, and Warfare

The Internet has opened up many possibilities for people to communicate and get to know each other better—no matter where in the world they live. The Internet has also opened doors for businesses to reach new markets and create opportunities for economic growth. It is sad that some people in our world have found the Internet to be a useful tool for perpetrating crimes, advocating terrorism, and waging war.

Crime on the Web includes online versions of crimes that have been undertaken for years in the physical world, including theft, stalking, distribution of pornography, and gambling. Other crimes, such as commandeering one computer to launch attacks on other computers, are new.

Law enforcement agencies have difficulty combating many types of online crime. The first obstacle they face is the issue of jurisdiction. As you learned earlier in this chapter, determining jurisdiction can be tricky on the Internet. Consider the case of a person living in Canada who uses the Internet to commit a crime against a person in Texas. It is unclear which elements of the crime could establish sufficient contact with Texas to allow police there to proceed against a citizen of a foreign country. It is possible that the actions that are considered criminal under Texas and U.S. law might not be considered so in Canada. If the crime is theft of intellectual property (such as computer software or computer files), the questions of jurisdiction become even more complex.

Enforcing laws against distribution of pornographic material has also been difficult because of jurisdiction issues. The distinction between legal adult material and illegal pornographic material is, in many cases, subjective and often difficult to make. Courts have tended to allow state and local courts to draw the line based on local community standards. This creates problems for Internet sales. For example, consider a case in which questionable content is sold on a Web site located in Oregon to a customer who downloads the material in Georgia. A difficult question arises regarding which community standards might apply to the sale.

A similar jurisdiction issue arises in the case of online gambling. Many gambling sites are located outside the United States. If persons in California use their computers to connect to an offshore gambling site, it is unclear where the gambling activity occurs. Several states have passed laws specifically outlawing Internet gambling, but the jurisdiction of those states to enforce laws that limit Internet activities is not yet clear.

Another problem facing law enforcement officers is the difficulty of applying laws that were written before the Internet became prevalent to criminal actions carried out on the Internet. For example, most states have stalking laws that provide criminal penalties to persons who harass, annoy, or alarm another person in a way that presents a credible threat. Many of these laws are triggered by physical actions, such as physically following the person targeted. The Internet gives a stalker the opportunity to use e-mail or chat room discussions to create the threatening situation. Laws that require physical action on the part of the stalker are not effective against online stalkers. Only a few states have passed laws that address specifically the problem of online stalking.

Many Internet security experts believe that we are at the dawn of a new age of terrorism and warfare that could be carried out or coordinated through the Internet. A considerable number of Web sites currently exist that openly support or are operated by hate groups and terrorist organizations. The Internet provides an effective communications network on which many people and businesses have become dependent. Although the Internet was designed from its inception to continue operating while under attack, a sustained effort by a well-financed terrorist group or rogue state could slow down the operation of major transaction processing centers. As more business communications traffic moves to the Internet, the potential damage that could result from this type of attack increases. You will learn more about security threats and countermeasures for those threats in Chapters 10 and 11.

ETHICAL ISSUES

Companies using Web sites to conduct electronic commerce should adhere to the same ethical standards that other businesses follow. If they do not, they will suffer the same consequences that all companies suffer: the damaged reputation and long-term loss of trust that can result in loss of business. In general, advertising on the Web should include only true statements and should not omit any information that could mislead potential purchasers or wrongly influence their impressions of a product or service. Even true statements have been held to be misleading when the ad omits important related facts. Any comparisons to other products should be supported by verifiable information.

Ethical considerations are important in determining advertising policy on the Web. Recall from Chapter 3 that buyers on the Web often communicate with each other. Reports of an ethical lapse that is rapidly passed among customers can seriously affect a company's reputation. In 1999, *The New York Times* ran a story that disclosed Amazon.com's arrangements with publishers for book promotions. Amazon.com was accepting payments of up to $10,000 from publishers to give their books editorial reviews and placement on lists of recommended books as part of a cooperative advertising program. When this news broke, Amazon.com issued a statement that it had done nothing wrong and that such advertising programs were a standard part of publisher-bookstore relationships. The outcry on the Internet in newsgroups and mailing lists was overwhelming. Two days later—before most mass media outlets had even reported the story—Amazon.com announced that it would end the practice and offer unconditional refunds to any customers who had purchased a promoted book. Amazon.com had done nothing illegal, but the practice appeared to be unethical to many of its customers and potential customers.

In early 1999, eBay faced a similar ethical dilemma. Several newspapers had begun running stories about sales of illegal items, such as assault weapons and drugs, on the eBay auction site. At this point in time, eBay was listing about 250,000 items each day. Although eBay would investigate claims that illegal items were up for auction on its site, eBay did not actively screen or filter listings before the auctions were placed on the site.

Even though eBay was not legally obligated to screen the items auctioned, and even though the screening would be fairly expensive to do, eBay's executive team decided that screening for illegal and copyright-infringing items would be in the best long-run interest of eBay. The team decided that such a decision would send a signal about the character of the company to its customers and the public in general. The eBay executive team also decided to remove an entire category—firearms—from the site. Not all of eBay's users were happy about this decision—the sale of firearms on eBay, when done properly, was legal. However, the eBay executive team again decided that presenting an overall image of an open and honest marketplace was so important to the future success of eBay that it chose to ban all firearm sales.

Privacy Rights and Obligations

The issue of online privacy is continuing to evolve as the Internet and the Web grow in importance as tools of communication and commerce. Many legal and privacy issues remain unsettled and are hotly debated in various forums. The **Electronic Communications Privacy Act of 1986** is the main law governing privacy on the Internet today. Of course, this law was enacted before the general public began its wide use of the Internet. A more recent law, the **Children's Online Privacy Protection Act of 1998**, provides restrictions on data collection that must be followed by electronic commerce sites aimed at children.

In recent years, a number of legislative proposals have been advanced that specifically address online privacy issues, but, thus far, none have withstood constitutional challenges. In July 1999, the FTC issued a report that examined how well Web sites were respecting visitors' privacy rights. Although it found a significant number of sites without posted privacy policies, the report concluded that companies operating Web sites were developing privacy practices with sufficient speed and that no federal laws regarding privacy were required at that time. Privacy advocacy groups responded to the FTC report with outrage and calls for legislation. Thus, the near-term future of privacy regulation in the United States is unclear. The Direct Marketing Association (DMA), a trade association of businesses that advertise their products and services directly to consumers by using mail, telephone, Internet, and mass media outlets, has established a set of privacy standards for its members.

However, critics note that past efforts by the DMA to regulate its members' activities have been less than successful.

Ethics issues are significant in the area of online privacy because laws have not kept pace with the growth of the Internet and the Web. The nature and degree of personal information that Web sites can record when collecting information about visitors' page-viewing habits, product selections, and demographic information can threaten the privacy rights of those visitors. Differences in cultures throughout the world have resulted in different expectations about privacy in electronic commerce. In Europe, for example, most people expect that information they provide to a commercial Web site will be used only for the purpose for which it was collected. Many European countries have laws that prohibit companies from exchanging consumer data without the express consent of the consumer. In 1998, the European Union adopted a **Directive on the Protection of Personal Data**. This directive codifies the constitutional rights to privacy that exist in most European countries and applies them to all Internet activities. In addition, the directive prevents businesses from exporting personal data outside the European Union unless the data will continue to be protected in accordance with provisions of the directive.

Until the legal environment of privacy regulation becomes more clear, electronic commerce sites should be conservative in their collection and use of customer data. Mark Van Name and Bill Catchings, writing in *PC Week* in 1998, outlined four principles for handling customer data that provide a good outline for Web site administrators. These principles are as follows:

- Use the data collected to provide improved customer service.
- Do not share customer data with others outside your company without the customer's permission.
- Tell customers what data you are collecting and what you are doing with it.
- Give customers the right to have you delete any of the data you have collected about them.

LEARNING FROM FAILURES

DOUBLECLICK

As you learned in Chapter 4, **DoubleClick** is one of the largest banner advertising networks in the world. DoubleClick arranges the placement of banner ads on Web sites. Like many other Web sites, DoubleClick uses cookies, which are small text files placed on Web client computers, to identify visitors who are returning.

For visitors to most sites, the privacy risk posed by cookies is not great. Visitors to Amazon.com, for example, will have Amazon.com cookies placed on their computers so that the Web server at Amazon.com will recognize them when they return. This can be useful, for example, when a visitor who has placed several items in a shopping cart before being interrupted can return to Amazon.com

(continued)

later in the day and find the shopping cart intact—the Web server can read the client's Amazon.com cookie and find the shopping cart from the client's previous session. The Amazon.com server can read only its own cookies; it cannot read the cookies placed on the client computer by any other Web server.

There are two important differences between the Amazon.com scenario and what happens when DoubleClick serves a banner ad. First, the visitor usually does not know that the banner ad is coming from DoubleClick (and thus, does not know that the DoubleClick server could be writing a cookie to the client computer). Second, DoubleClick serves ads through thousands of different Web sites. As a visitor moves from one Web site to another, that visitor's computer can collect many DoubleClick cookies. The DoubleClick server can read all of its own cookies and gather from each one information about which ads were served and the sites through which they were served. Thus, DoubleClick can compile a tremendous amount of data about where a visitor has been on the Web.

Even this amount of information collection would not trouble most people. DoubleClick can use the cookies to track a particular computer's connections to Web sites, but it does not record any identity information about the owner of that computer. Therefore, DoubleClick accumulates a considerable record of Web activity, but cannot connect that activity with a person.

In 1999, DoubleClick arranged a $1.7 billion merger with Abacus Direct Corporation. Abacus had developed a way to link information about people's Web behavior (collected through cookies such as those placed by DoubleClick's banner ad servers) to the names, addresses, and other information about those people that had been collected in an offline consumer database.

The reaction from online privacy protection groups was immediate and substantial. The FTC launched an investigation, the Internet's privacy issues e-mail lists and chat rooms buzzed with discussions and, in the end, DoubleClick abandoned its plans to integrate its cookie-generated data with the identity information in the Abacus database. Although DoubleClick is still one of the largest banner advertising networks, it has not met its profitability targets. DoubleClick had been counting on generating additional revenue by using the information in the combined database that it was unable to create.

When the FTC probe concluded two years later, DoubleClick was not charged with any violations of laws or regulations. The lesson here is that a company violates the Internet community's ethical standards at its own peril, even if the transgression does not break any laws.

TAXATION AND ELECTRONIC COMMERCE

Companies that do business on the Web are subject to the same taxes as any other company. However, even the smallest Web businesses can become instantly subject to taxes in many states and countries because of the Internet's worldwide scope. Traditional businesses may operate in one location and be subject to only one set of tax laws for years. By the time those businesses are operating in multiple states or countries, they have developed the internal staff and record-keeping infrastructure needed to comply with multiple tax laws. Firms that engage in electronic commerce must comply with these multiple tax laws from their first day of existence.

A government acquires the power to tax a business when that business establishes a connection with the area controlled by the government. For example, a business that is located in Kansas has a connection with the state of Kansas and is subject to Kansas taxes. If that company opens a branch office in Arizona, it forms a connection with Arizona and becomes subject to Arizona taxes on the portion of its business that occurs in Arizona. This connection between a taxpaying entity and a government is called **nexus**. The concept of nexus is similar in many ways to the concept of personal jurisdiction discussed earlier in this chapter. The activities that create nexus vary from state to state. Nexus issues have been frequently litigated, and the resulting common law is fairly complex. Determining nexus can be difficult when a company conducts only a few activities in or has minimal contact with the state. In such cases, it is advisable for the company to obtain the services of a professional tax advisor.

An online business is potentially subject to several types of taxes, including income taxes, transaction taxes, and property taxes. Income taxes are levied by national, state, and local governments on the net income generated by business activities. Transaction taxes, which include sales taxes, use taxes, and customs duties, are levied on the products or services that the company sells or uses. Customs duties are taxes levied by the United States and other countries on certain commodities when they are imported into the country. Property taxes are levied by states and local governments on the personal property and real estate used in the business. In general, the taxes that cause the greatest concern for Web businesses are income taxes and sales taxes.

Income Taxes

The **Internal Revenue Service** (**IRS**) is the U.S. government agency charged with administering the country's tax laws. A basic principle of the U.S. tax system is that any verifiable increase in a company's wealth is subject to federal taxation. Thus, any company whose U.S.-based Web site generates income is subject to U.S. federal income tax. Further a Web site maintained by a company in the United States must pay federal income tax on income generated outside of the United States. To reduce the incidence of double taxation of foreign earnings, U.S. tax law provides a credit for taxes paid to foreign countries. The IRS Web site appears in Figure 7-10.

link to Taxpayer
Advocate Service

link to
downloadable tax
forms and IRS
publications

Figure 7-10   *Internal Revenue Service home page*

The IRS has been subject to criticism in recent years for its heavy-handed tactics and intimidating auditors. The agency is making attempts to improve its operations and its image with taxpayers. Its Web site is a good example of this effort. The site uses a friendly newspaper motif to help calm visitors who might be desperate to find a tax form or ruling as they prepare their tax returns at the last minute. The site includes links to downloadable tax forms, copies of tax regulations and IRS publications, and the Taxpayer Advocate Service.

Most states levy an income tax on business earnings. If a company conducts activities in several states, it must file tax returns in all of those states and apportion its earnings in accordance with each state's tax laws. In some states, the individual cities, counties, and other political subdivisions within the state also have the power to levy income taxes on business earnings. Companies that do business in multiple local jurisdictions must apportion their income and must file tax returns in each locality that levies an income tax. The number of taxing authorities in the United States exceeds 30,000.

Companies that sell through their Web sites do not, in general, establish nexus everywhere their goods are delivered to customers. Usually, a company can accept orders and ship from one state to many other states and avoid nexus by using a contract carrier such as FedEx or United Parcel Service to deliver goods to customers.

Sales Taxes

Most states levy a sales tax on goods sold to consumers. Businesses that establish nexus with a state must file sales tax returns and remit the sales tax they collect from their customers. If a business ships goods to customers in other states, it is not required to collect sales tax from those customers unless the business has established nexus with the customer's state. However, the customer in this situation is liable for payment of a use tax in the amount that the business would have collected as sales tax if it had been a local business. Few consumers file use tax returns and few states enforce their use tax laws with regularity.

Larger businesses use complex software to manage their sales tax obligations. Not only are the sales tax rates different in the 7500 U.S. sales tax jurisdictions (which include states, counties, cities, and other sales tax authorities), but the rules about which items are taxable differ. For example, New York's sales tax law provides that large marshmallows are taxable (because they are "snacks"), but small marshmallows are not taxable (because they are "food").

Some purchasers are exempt from sales tax, such as certain charitable organizations and businesses buying items for resale. Thus, to determine whether a particular item is subject to sales tax, a seller must know where the customer is located, what the laws of that jurisdiction say about taxability and tax rate, and the taxable status of the customer.

Summary

Businesses face many challenges posed by differences in language, culture, and infrastructure when conducting electronic commerce across international borders. Due to the speed and vast availability of online information, companies must quickly establish credibility with online customers to succeed in new and different markets.

Translation of Web pages by companies familiar with the culture of the target country helps to avoid some of the problems that can arise from doing business across international borders. Some countries require that companies doing business within their borders make their Web pages available in the local language.

Ethics issues can arise even when no laws have been broken. Since laws and ethics standards derive from local cultures, and local cultures vary significantly around the world, businesses must work hard to become aware of those cultural, ethical, and legal differences—especially those that exist in their target markets. Strategies and systems that would fail in the United States may be prerequisites for electronic commerce success in other countries.

Variations and inadequacies of the infrastructure that supports the Internet worldwide can make it challenging to conduct electronic commerce in certain countries. Even in countries that have adequate infrastructure, government or control of the communication networks can impair visitors' ability to access the Internet.

The relationship between geography and culture is historically and legally intertwined. For most of our history, people have not been able to travel great distances to learn about other cultures. The relationship between geographic boundaries and legal boundaries is based on four elements: power, effects, legitimacy, and notice.

These four elements have helped governments create the legal concept of jurisdiction in the physical world. Since the four elements exist in somewhat different form on the Internet, the jurisdiction rules that have worked so well in the physical world do not always work well in the online world. In many areas, the legal concept of jurisdiction on the Internet is still unclear and ill-defined.

As in traditional commerce, contracts are a part of doing business on the Web and are established through various types of offers and acceptances. Any contract for the electronic sale of goods or services includes implied warranties. Many companies include contracts or rules on their Web sites in the form of terms of service agreements. Contracts can be invalidated when one of the parties to the transaction is an imposter; however, forged identities are becoming easier to detect through electronic security tools.

Seemingly innocent inclusion of photographs, whether manipulated or not, and other elements of a Web page can lead to infringement of trademarks, copyrights, defamation, patents, and violation of intellectual property rights. Electronic commerce sites must be careful not to imply relationships that do not actually exist. Negative evaluative statements about entities, even when true, are best avoided given the subjective nature of defamation and product disparagement.

Collecting information and tracking consumer habits raises questions of ethics regarding online privacy. Some countries are far more restrictive than others in terms of what type of information collection is acceptable and legal.

Unfortunately, some people in our world have found the Internet to be a useful tool for perpetrating crimes, advocating terrorism, and waging war. Law enforcement agencies have difficulty combating many types of online crime, and governments are working to create adequate defenses for online war and terrorism.

Companies that conduct electronic commerce are subject to the same laws and taxes as other companies, but the nature of doing business on the Web can expose companies to a large number of laws and taxes sooner than traditional companies usually face them. Although some legal issues are straightforward, others are difficult to interpret and follow because of the newness of electronic commerce and the unsettled nature of applicable law. The large number of government agencies that have jurisdiction and the power to tax makes it essential that companies doing business on the Web understand the potential liabilities of doing business with customers in those jurisdictions.

Key Terms

Acceptance	Nexus
Authority to bind	Offer
Business process patent	Patent
Common law	Per se defamation
Consideration	Personal jurisdiction
Constructive notice	Product disparagement
Contract	Service mark
Copyright	Signature
Culture	Statute of Frauds
Defamatory	Statutory law
Fair use	Subject-matter jurisdiction
Flat-rate access	Terms of service (ToS)
Forum selection clause	Tort
Implied contract	Trade name
Jurisdiction	Trademark
Legitimacy	Trademark dilution
Localization	Warranty disclaimer
Long-arm statute	Writing
Machine translation	

Review Questions

1. Explain the difference between language translation and language localization in fewer than 200 words.

2. In a paragraph, describe the advantages of a flat-rate telecommunications access system for countries that want to encourage electronic commerce.

3. What is the difference between subject-matter jurisdiction and personal jurisdiction? Keep your explanation under 300 words.

4. Define product disparagement. Describe a situation that would be an example of product disparagement. Limit your answer to two paragraphs.

5. In 300 words or fewer, explain nexus. Why is it an important concept in state taxation? In what ways is it similar to jurisdiction?

Exercises

1. Use the **AltaVista Translation** Web site to translate the following business messages from English to one of the foreign languages available on that site. Translate each message back into English. Write a short memo that summarizes the problems you think an electronic commerce Web site that uses translation software might experience. Translate the following messages:

   - The flight has been delayed for several hours and your shipment of components will not arrive as scheduled.

- We would be happy to bid on your proposal; however, we will need the drawings of subassembly #24 and the supervising mechanical engineer's quality control report by next Thursday.
- Our company offers the latest and greatest hot deals on wheels. We would love to send you a brochure that explains why our brakes, wheels, and suspension components will do the job for you effectively and economically.

2. Use **Northern Light** or your favorite Web search engine to obtain a list of Web pages that include the word "warranty." Visit the Web pages on the search results list until you find a page that includes a warranty or guarantee statement for products that the site is offering for sale. Print the page and turn it in with your answers to the following questions:

- Is the warranty statement on a separate page?
- Could you read the entire warranty statement without scrolling your browser window?

- Does the statement include a warranty disclaimer? If so, is the disclaimer conspicuous? List the main items disclaimed.
- Does the statement make any explicit claims about the suitability of the product for specific purposes? List those claims.
- Does the statement deny warranty coverage if the product is used in specific ways? List the ways specified.
- Write one paragraph in which you evaluate the clarity of the warranty statement.

3. Visit the Better Business Bureau's **BBBOnLine** Web site and the **TRUSTe** Web site. Examine each site to determine what types of privacy policies a member company must have to qualify for these two programs. Evaluate the two programs from the standpoint of a consumer who is interested in privacy protection—that is, determine which privacy program you would prefer to see in a Web site with which you are doing business. Summarize your findings in a memo of about 200 words.

For Further Study and Research

Angwin, J. 2001. "Are Domain Panels the Hanging Judges of Cyberspace?" *The Wall Street Journal*, August 20, B1.

Ardito, S., P. Eiblum, and R. Daulong. 1999. "Realistic Approaches to Enigmatic Copyright Issues," *Online*, 23(3), May–June, 91–95.

Balkin, R. 1999. "AltaVista's Automatic Translation Program," *Database*, 22(2), April–May, 56–57.

Beckman, D. and D. Hirsch. 2000. "Web Worries of Dot-Com Lawyers," *ABA Journal*, June, 82.

Berkeley, S. 2000. "Web Attack," *Harvard Business Review*, September–October, 20.

Betts, M. 2001. "Report: Global E-Commerce Still Faces Big Challenges," *Computerworld*, May 3.

Available online at: (http://www.computerworld.com/cwi/story/0,1199,NAV47_STO60164.00.html).

Betts, M., C. Sliwa, and J. DiSabatino. 2000. "Global Web Sites Prove Challenging," *Computerworld*, 34(34), August 21, 17.

Bingi, P., A. Mir, and J. Khamalah. 2000. "The Challenges Facing Global E-Commerce," *Information Systems Management*, 17(4), Fall, 26–34.

Bond, R. and C. Whiteley. 1998. "Untangling the Web: A Review of Certain Secure E-Commerce Legal Issues," *International Review of Law, Computers & Technology*, 12(2), July, 349–370.

Bonisteel, S. 2001. "WIPO Rejects Hotel's Bid for Casino Domain," *BizReport*, October 2. Available online at: (http://www.bizreport.com/article.php?art_id=2264).

Boyle, M., J. Peterson, W. Sample, T. Schottenstein, and G. Sprague. 1999. "The Emerging International Tax Environment for Electronic Commerce," *Tax Management International Journal*, 28(6), June 11, 357–382.

Brandweek. 2000. "Trends in Trademarks 2000," 41(24), June 12, 68–70.

Brilmayer, L. 1989. "Consent, Contract, and Territory," *Minnesota Law Review*, 74(1), 11–12.

Brown, W. 2000. "Modern Technology Speaks with a Global Tongue; and It Certainly Isn't French," *The Daily Telegraph*, February 2, 10.

Carney, D. 2000. "E-Exchanges May Keep Trustbusters Busy," *Business Week*, May 1, 52.

Clark, P. 2000. "E-Hub Will Court Power of Attorneys," *B to B*, 85(13), August 28, 3.

Clark, P. 2001. "Doubts Cloud DoubleClick's Repositioning," *B to B*, 86(15), August 28, 1–2.

Clausing, J. 1999. "Study Says Most Children's Web Sites Are Lax on Privacy," *The New York Times*, July 20. Available online at: (http://www.nytimes.com/library/tech/99/07/cyber/articles/20privacy-day.html).

Cohn, M. 2001. "China Seeks to Build the Great Firewall," *The Toronto Star*, July 21, A1.

Corgel, J. 2000. "International Business: E-Business in Europe," *Vital Speeches of the Day*, 66(20), August 1, 637–640.

Computerworld. 2001. "Special Report: Globalization," January 11. Available online at: (http://computerworld.com/cwi/story/0,1199,NAV63_STO56162,00.html).

Crane, E. 2000. "Double Trouble," *Ziff Davis Smart Business*, 13(10), October, 62.

Creed, A. 2001. "E-Trade Swallows $90,000 Fine," *BizReport*, July 10. Available online at: (http://www.bizreport.com/article.php?id=1692).

Davenport, T. 2000. "E-Commerce Goes Global," *CIO*, 13(20), August 1, 52–54.

Delio, M. 2001. "Does Media Fuel Buyers' Fears?" *Wired News*, June 22. Available online at: (http://www.wired.com/news/business/0,1367,44895,00.html).

Dempsey, G. and R. Sussman. 1999. "A Hands-On Guide for Multilingual Web Sites," *Communication World*, 16(6), June–July, 45–47.

DePalma, A. 2000. "Getting There Is Challenge for Latin America E-Tailing," *The New York Times*, August 17, 4.

Digital Millennium Copyright Act. 1998. Public Law No. 105–304, 112 Statutes 2860.

DiLodovico, A., W. Lewis, V. Palmade, and S. Sankhe. 2001. "India—From Emerging to Surging," *The McKinsey Quarterly*, September, 28–65.

Direct Marketing. 2001. "FTC Closes DoubleClick Investigation," 63(12), April, 18.

DiSabatino, J. 2000. "Globalization," *Computerworld*, 34(28), July 10, 46.

DiSabatino, J. 2001. "Privacy Groups Want FTC Action on Passport," *Computerworld*, October 23. Available online at: (http://www.computerworld.com/itresources/rcstory/0,4167,STO65003_KEY11,00.html).

Echikson, W., C. Matlack, and D. Vannier. 2000. "American E-Tailers Take Europe by Storm," *Business Week*, August 7, 54–56.

The Economist. 2000. "Business Ethics: Doing Well by Doing Good," 355(8167), April 22, 65–67.

The Economist. 2000. "The Internet's Chastened Child," 357(8196), November 11, 80.

Einhorn, B., A. Webb, and P. Engardio. 2000. "China's Tangled Web: Will Beijing Ruin the Net by Trying to Control It?" *Business Week*, July 17, 28–30.

Emond, M. 2000. "Preparing for the E-Planet," *E-com*, 2(4). Available online at: (http://www.e-commag.com/printresources/v2n4/v2n4037.htm).

Federal Trade Commission (FTC). 1999. *Self-Regulation and Privacy Online: A Report to Congress*. Washington: FTC.

Flynn, L. 2000. "Whose Name Is It Anyway? Arbitration Panels Favoring Trademark Holders in Disputes Over Web Names," *The New York Times*, September 4, C3.

Friedman, M. 1999. "Photographer Fights Quebec Language Law," *Computing Canada*, 25(24), June 18 1, 4.

293

Gleckman, H. 2000. "The Tempest Over Taxes: The 'Too Complex' Excuse Won't Last for Long," *Business Week*, February 7, EB32–EB33.

Gleckman, H. and D. Carney. 2000. "Watching Over the World Wide Web: The Internet and the Rise of Globalization Are Creating New Pressure to Develop a Commercial Code That's Recognized from Kuala Lumpur to Kansas City," *Business Week*, August 28, 195–196.

Goldstein, E. 1999. *The Internet in the Mideast and North Africa: Free Expression and Censorship.* Washington: Human Rights Watch.

Gurley, W. 2000. "Like It or Not, Every Startup Is Now Global," *Fortune*, June 26, 324.

Guttman, R. 2000. "Erkki Liikanen: European Commissioner for Enterprise and the Information Society," *Europe*, May, 11–13.

Haddock, F. 2000. "European E-volution," *Global Finance*, 14(4), April, 39–40.

Hardesty, D. 1999. *Electronic Commerce Taxation and Planning.* Boston: Warren, Gorham & Lamont.

Harmon, A. 2001. "As Public Records Go Online, Some Say They're Too Public," *The New York Times*, August 24, A1.

Harvard Law Review. 1999. "The Criminalization of Copyright Infringement in the Digital Era," 112(7), May, 1705–1722.

Heckman, J. 2000. "Trademarks Protected Through New Cyber Act," *Marketing News*, 34(1), January 3, 6–7.

Heilemann, J. 2000. "David Boies: The Wired Interview," *Wired*, October. Available online at: (http://www.wired.com/wired/archive/8.10/boies.html).

Hemphill, T. 2000. "DoubleClick and Consumer Online Privacy: An E-Commerce Lesson Learned," *Business & Society Review*, 105(3), Fall, 361–372.

Hirschman, C. 2001. "Prosecuting in the Name of Privacy," *Telephony*, 241(7), August 13, 82.

Hong, V. 2000. "'Brussels 1' Angers EC Businesses," *The Industry Standard*, December 1. Available online at: (http://www.thestandard.com/article/display/0,1151,20531,00.html).

Hurt, E. 2000. "FTC Wins Internet's Respect," *Business 2.0*, October 13. Available online at: (http://www.business2.com/content/channels/technology/2000/10/13/21123).

Janal, D. 1999. "Thirty Essential Steps to Take Right Now to Prevent Online Crime," *Communication World*, 16(4), March, 34–36.

Jensen, M. 2001. "The African Internet: A Status Report," May. Available online at: (http://www3.sn.apc.org/africa/afstat.htm).

Jones, J. 2000. "Protecting Privacy," *InfoWorld*, 22(18), May 1, 40–41.

Kahin, B. and C. Nesson (eds.). 1997. *Borders in Cyberspace.* Cambridge, MA: MIT Press.

Keeler, D. 2000. "Taxation Slips Through the Net," *Global Finance*, 14(6), June, 60–61.

King, J. 1999. "Idiom App Speaks Your Language," *Computerworld*, 33(22), May 31, 66.

Lapres, D. 2000. "Legal Do's and Don'ts of Web Use in China," *China Business Review*, 27(2), March–April, 26–28.

Leo, A. 2001. "The World Wide Translator," *Technology Review*, September 21. Available online at: (http://www.techreview.com/web/leo/leo092101.asp).

Le Seac'h, M. and A. Klotz. 1999. "Corporate Translating: Handle with Care," *Business and Economic Review*, 45(2), January–March, 12–14.

Lessig, L. 2000. *Code and Other Laws of Cyberspace.* New York: Basic Books.

Levaux, J. 2001. "Adapting Products and Services for Global E-Commerce: The Next Frontier is Beyond Localization," *World Trade*, 14(1), January, 52–54.

Loro, L. 1999. "Marketers Look to Stave Off Net Taxes," *Advertising Age's Business Marketing*, 84(4), April, 1, 31.

Manjoo, F. 2001. "Fine Print Not Necessarily in Ink," *Wired News*, April 6. Available online at: (http://www.wired.com/news/business/0,1367,42858,00.html).

McCarthy, B. 2000. "All E-Business Is Global," *Informationweek*, June 5, 204.

McClintock, M., N. Maguire, J. Kilby, and D. Barlow. 2000. "Electronic Commerce," *International Tax Review*, July–August, 9–13.

294

McCune, J. 1999. "English Written Here," *Management Review*, 88(2), February, 12.

Meller, P. 2000. "Europe Passes Stiff E-Commerce Law," *The Industry Standard*, December 1. Available online at: (http://www.thestandard.com/article/display/0,1151,20526,00.html).

Messmer, E. 1999. "Teaching the Web to Speak to Everyone," *Network World*, 16(2), May 24, 29–30.

Miller, R. and G. Jentz. 2002. *Law for E-Commerce*. Cincinnati: West.

Morrow, J. 2000. "Study: Fraud No Threat to E-Commerce," *E-Commerce Times*, September 20. Available online at: (http://www.ecommercetimes.com/news/articles2000/000920-1.shtml).

Moschella, D. 1999. "Consumers Being Forgotten in the Copyright Debate," *Computerworld*, 33(25), June 21, 35.

Murray, J. 2000. "E-Contracts Present Courts with Special Legal Challenges," *Purchasing*, 129(3), August 24, 119–120.

Olin, J. 2001. "Reducing International E-Commerce Taxes," *World Trade*, 14(3), March, 64–66.

Oliva, R. and S. Prabakar. 1999. "Copyright Perils Can Lurk on the Business Web," *Marketing Management*, 8(1), Spring, 54–57.

Pantazis, A. 1999. "Zeran v. America Online, Inc.: Insulating Internet Service Providers from Defamation Liability," *Wake Forest Law Review*, 34(2), Summer, 531–555.

Porter, K. and S. Bradley. 1999. *eBay, Inc.* Case #9-700-007. Cambridge, MA: Harvard Business School.

Posch, R. 1999. "What Is Fair Use?" *Direct Marketing*, 62(1), May, 26–28.

Radcliffe, D. 2001. "E-Merchant Beware," *Computerworld*, 35(25), June 18, 42.

Reagle, J. 1999. "The Platform for Privacy Preferences," *Communications of the ACM*, 42(2), February, 48–51.

Retsky, M. 1999. "Protect Your Property Intellectually," *Marketing News*, 33(13), June 21, 10–11.

Rewick, J. 2000. "DoubleClick Finds Its Abacus Unit Nettlesome," *The Wall Street Journal*, October 19, B6.

Rich, J. 2000. "Latin America Is a Difficult—But Attractive—Region for E-Marketplaces that Are Trying to Enlist Small Businesses," *The New York Times*, September 25, C4.

Roberts, B. 2000. "Ready, Fire, Aim," *Electronic Business*, 26(7), July, 80–88.

Samborn, H. 2000. "Nibbling Away at Privacy," *ABA Journal*, 86(2), June, 26–27.

Samuelson, P. 1999. "Good News and Bad News on the Intellectual Property Front," *Communications of the ACM*, 42(3), March, 19–24.

Schwartau, W. 2000. "Safe Passage," *Network World*, 17(9), February 28, 95–97.

Sclafane, S. 1998. "Web Sites Create World Wide Exposures," *National Underwriter*, 102(6), February, 9, 14, 31.

Segal, J. 2000. "Cybertraps for HR," *HRMagazine*, 45(6), June, 217–231.

Shaller, D. 2000. "E-mail, the Internet, and Other Legal and Ethical Nightmares," *Strategic Finance*, August, 82(2), 48–52.

Shannon, P. 2000. "Including Language in your Global Strategy for B2B E-Commerce," *World Trade*, 13(9), September, 66–68.

Shari, M. 2000. "Cutting Red Tape in Singapore," *Business Week*, September 18, 92.

Skipton, C. 1999. "Think Globally, Act Locally," *New Media*, 9(6), June, 58.

Sliwa, C. 2000. "Boo.com Makeover Draws Skeptical Reactions at Unveiling," *Computerworld*, 34(32), August 7, 6.

Smedinghoff, T. (ed.). 1996. *Online Law: The SPA's Legal Guide to Doing Business on the Internet*. Reading, MA: Addison-Wesley Developers Press.

Steinberg, J. 2000. "An Unusual Commute Leads to a Different Kind of Partnership," *The New York Times*, September 20, H4.

Stone, M. 2001. "Court Dismisses Class Action Against eBay," *BizReport*, January 19. Available online at: (http://www.bizreport.com/daily/2001/01/20010119-4.htm).

Swire, P. and R. Litan. 1998. *None of your Business: World Data Flows, Electronic Commerce, and the European Privacy Directive*. Washington: Brookings Institution Press.

295

Thornton, J. 2000. "Should Drug Ads Be Legal?" *Marketing*, May 4, 28–29.

Towle, H. 2000. "No Guiding Light," *CIO*, 13(21), August 15, 72–74.

Tynan, D. 2000. "Privacy 2000: In Web We Trust?" *PC World*, 18(6), June, 103–111.

United Nations. 1970. "Declaration on Principles of International Law Concerning Friendly Relations and Cooperation Among States in Accordance with the Charter of the United Nations," *General Assembly Resolution*, #2625, 35th Session.

Van Name, M. and B. Catchings. 1998. "Practical Advice about Privacy and Customer Data," *PC Week*, 15(27), July 6, 38.

Vijayan, J. and K. Ohlson. 2000. "Standards Issue Mars E-Signatures," *Computerworld*, 34(28), July 10, 1, 16.

Walker, P. 2000. "Watch Out for the Web," *Credit Management*, March, 24–25.

Wallraff, B. 2000. "What Global Language?" *The Atlantic Monthly*, 286(5), 52–66.

Whitaker, B. 2000. "The Web Makes Going Global Easy, Until You Try to Do It," *The New York Times*, H20.

Wiley, L. 1999. "Proposed Revisions to European Copyright Laws Cause a Stir," *E Media Professional*, 12(4), April, 16–17.

Wilke, J. 2001. "Twenty States Oppose Airlines' Proposal for Joint Venture in Online Reservations," *The Wall Street Journal*, January 11, A10.

Williams, J., J. Clark, C. Clark, and J. Noe. 1999. "What a Tangled Web: The Legal and Public Relations Dangers of Operating a Web Site," *Information Strategy*, 15(3), Spring, 6–12.

Wilson, T. 2001. "Spotty Infrastructure Impairs World View," *InternetWeek*, March 26, 1–3.

Winston, J. "Copyright on the Internet," *Target Marketing*, 22(6), June, 40.

Wood, C. 2001. "Collusion in the Air," *PC Magazine*, 20(9), May 8, 199.

Wright, B. 1995. *The Law of Electronic Commerce: EDI, E-Mail, and Internet: Technology, Proof, and Liability*. Boston: Little, Brown.

Zuckerman, A. 2001. "Somebody Out There Wants to Cheat You!" *World Trade*, 13(3), March, 36–38.

296

WEB SERVER HARDWARE AND SOFTWARE

INTRODUCTION

As you learned in earlier chapters, **Lands' End** is one of the most successful clothing retailers on the Web. The company has been a leader in adding features that attract customers to the site and that keep these customers coming back. Behind the scenes at Lands' End, a team of experienced technology professionals implements new Web page features and performs many regular maintenance tasks that are necessary to keep the Web site running.

Lands' End closely monitors the performance of its Web site to make sure that customers have a consistent experience when they return to the site. The Web site's technical team works hard to make sure that site visitors do not notice the Web site's operating characteristics. Since volume on the site has doubled or tripled each year since the site opened for business and the company regularly makes major improvements to the site, this goal has not always been easy to attain.

Lands' End's specific goals for performance change as Web technologies improve. For example, the site management team has a target for the time it takes a Web page from the site to load on a visitor's computer. In the early days of the site, that target was 15 seconds. Today, the target is under 5 seconds. The Web site's technical team has always taken a conservative approach to operating the site so that the site can meet its performance goals more easily. For example, the technical team specifies the maximum and average sizes of Web pages and graphics files that the content team can use. In addition, the technical team must complete all major changes to the site (including thorough testing) before November 1 each year, prior to the holiday selling season. Lands' End makes over 40 percent of its total annual sales in November and December and does not want to take any chances with Web site changes during that time period.

The server hardware at Lands' End is a mix of **Sun** and **IBM** computers that are managed by another computer that allocates incoming Web traffic. Some of the Web site's advanced features, such as the graphics-intensive My Virtual Model, are created on a separate set of computers. These computers are all located at the company's headquarters in a small town near Madison, Wisconsin. The computers run a UNIX-based operating system from Sun called Solaris and a version of the Apache Web server software that you will learn more about in this chapter. Although the technical team writes some of the software that it uses to monitor the Web site's performance, the company also uses the services of Keynote Systems. Keynote can measure how fast particular pages load or how rapidly transactions are completed at various times of the day. Keynote can make these measurements at a number of locations around the world.

By paying close attention to the details, the technical team at Lands' End keeps the Web site operating at or above expected levels. When customers become so absorbed in the shopping experience that they do not notice the operation of the site, the technical team has done its job.

LEARNING OBJECTIVES

In this chapter, you will learn about:
- Web server hardware considerations
- Measuring the performance of Web server hardware
- How individual computers are combined to provide large-scale Web services
- Web server software, including Apache, Microsoft Internet Information Server, and iPlanet Web Server
- Other software that works with Web server software to accomplish the basic operations of a Web site

This chapter provides background information on the basic technical requirements of a Web site that can support electronic commerce operations. You will learn about software that accomplishes specific electronic commerce functions in later chapters.

WEB SERVER HARDWARE AND PERFORMANCE EVALUATION

When corporate Web sites first appeared in the mid-1990s, they were often curiosities that were far less important to businesses than the software used to do their accounting or production planning. Today, a Web site may be the first place customers go to conduct business with traditional companies, while electronic commerce sites have become the main business focus for many organizations.

As you learned in Chapter 2, a Web server responds to requests from clients. The two main ingredients in a Web server are its hardware—the computers and related components—and its Web server software. In this chapter, you will learn about the hardware requirements you need to host a Web site. You will learn about specific software features later in the chapter.

A popular site with many visitors must have much greater Web server capacity than a less popular site. Most companies begin developing their Web sites by determining which Web server software and which operating system software they want to run; different software provides different capabilities. Then, the company decides which hardware it should obtain to support the operation of the chosen software combination.

Types of Web Sites

An important first step in planning a Web server is to determine what the company wants to accomplish with the server. The company must estimate how many visitors will be connecting to the Web site and what types of files (graphics, multimedia, or text) will be delivered through the site. The company must also assess its existing information technology staff. Some companies will have a large staff with a depth of experience, while others will have a small or relatively inexperienced staff. Companies create Web sites for a wide variety of reasons and in a wide variety of forms including: simple development (testing) sites, intranets, information-only sites for customers, business-to-business portals, storefront sites, or content delivery sites. Each has a different purpose, requires different computer hardware and software, and requires different monetary and personnel resources. Decisions about server hardware and software should be driven by the volume and type of Web activities expected.

Development Sites

The simplest Web site—and the least costly to implement and maintain—is a development site. Companies can use a development site to experiment with and evaluate different Web designs with little initial investment. A development site can reside on an existing PC and can be developed with low-cost Web site building tools such as **Microsoft FrontPage** or **Macromedia Dreamweaver**. Testers can access the site through their PCs on the existing LAN or directly through the workstation housing the Web site.

Intranets

Corporate intranets house internal memos, corporate policy handbooks, expense account worksheets, budgets, newsletters, and a variety of other corporate documents. Because intranets are shielded from the Internet, they do not require additional security software to protect them against threats from outside the company. However, many companies do create safeguards to protect their intranet servers from ill-intentioned employees working within the company.

Transaction-Processing Sites

Transaction-processing sites, such as business-to-business and business-to-consumer electronic commerce sites, must be available 24 hours a day, seven days a week; that is, they must be **high reliability servers**. These sites require reliable and robust servers. Transaction-processing sites also need to have spare server computers for handling high traffic volumes that occur periodically; that is, they must be **high availability servers**. In addition to requiring fast and reliable hardware, commerce sites must run Web and commerce software that is efficient and easily upgraded when site traffic increases. These sites must also run security software, which is important whenever a server is connected to the Internet. You will learn more about security software in Chapter 11.

Content-Delivery Sites

Content-delivery sites like *The Wall Street Journal*, *The New York Times*, and C-NET sell and deliver content such as news, histories, summaries, and other digital information. The content must be presented rapidly on the visitor's screen. Hardware requirements for content sites are similar to those of business-to-business and business-to-commerce sites. Visitors must be able to locate articles quickly with a fast and precise search engine.

Web Hosting Choices

When companies need to incorporate electronic commerce components, they may opt to run servers in-house; this is called **self-hosting**. However, the company may decide that a third-party Web hosting service provider is a better choice than self-hosting. As you will learn in Chapter 9, many small Web stores use a third-party host provider for both Web services and electronic commerce functions, particularly when the Web site is small or the company sells a limited number of products.

As you learned in Chapter 2, a number of companies, called Internet service providers (ISPs), are in the business of providing Internet access to companies and individuals. Many of these companies offer Web hosting services as well. To distinguish themselves from companies that provide only Internet access services, these hosting service firms sometimes call themselves something other than ISPs. Because the hosting services they offer are designed to help companies conduct electronic commerce, these hosting service firms sometimes call themselves **commerce service providers (CSPs)**. These firms often offer Web server management and the renting of application software (such as databases, shopping carts, and content management programs); thus, these companies also sometimes call themselves **managed service providers (MSPs)** or **application service providers (ASPs)**. Despite the increasing variety of acronyms, many companies that provide some or all of these additional services still call themselves ISPs.

Service providers offer clients hosting arrangements that include shared hosting, dedicated hosting, and co-location. **Shared hosting** means that the client's Web site is on a server that hosts other Web sites simultaneously and is operated by the service provider at its location. With **dedicated hosting**, the service provider makes a Web server available to the client, but the client does not share the server with other clients of the service provider. In both shared hosting and dedicated hosting, the service provider owns the server hardware and leases it to the client. The service provider is responsible for maintaining the Web server hardware and software, and provides the connection to the Internet through its routers and other network hardware. In a **co-location** (also spelled **collocation** and **colocation**) service, the service provider rents a physical space to the client to install its own server hardware. The client installs its own software and maintains the server. The service provider is responsible only for providing a reliable power supply and a connection to the Internet through its routers and other networking hardware. You can find service providers by looking in your local telephone directory or by using a Web directory such as **The List**, which appears in Figure 8-1.

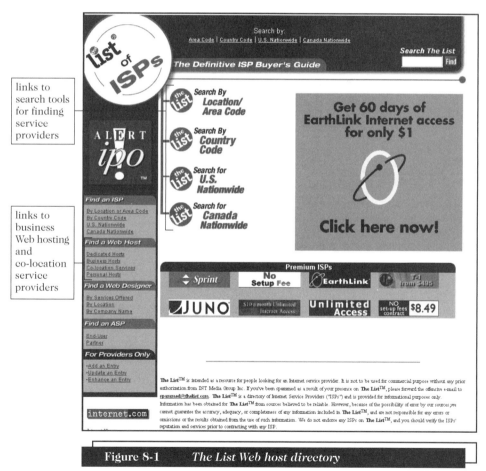

links to search tools for finding service providers

links to business Web hosting and co-location service providers

Figure 8-1 *The List Web host directory*

The **HostIndex** site provides a convenient collection of Web pages that compare Web hosts with one another. **TopHosts.com** provides one of the most comprehensive link collections available to aid in researching Web hosting alternatives and services.

When making Web server hosting decisions, a company should ask whether the hardware platform and software combination can be upgraded when the traffic on its Web site increases. A company's Web server requirements are directly related to its electronic commerce transaction volume and Web site traffic. The best hosting services provide Web server hardware and software combinations that are **scalable**, which means that they can be adapted to meet changing requirements when their clients grow.

Using a service provider's shared or dedicated hosting services instead of building an in-house server or using a co-location service means that the staffing burden shifts from the company to the Web host. **EZ Webhost** and **Interland.com** are examples of Web hosting companies. Because these companies offer a variety of services, they might be called ISPs, CSPs, MSPs, or ASPs by different users, depending on the service they are seeking. Figure 8-2 shows the dedicated hosting page from the Interland Web site.

links to information about dedicated hosting services offered by Interland.com

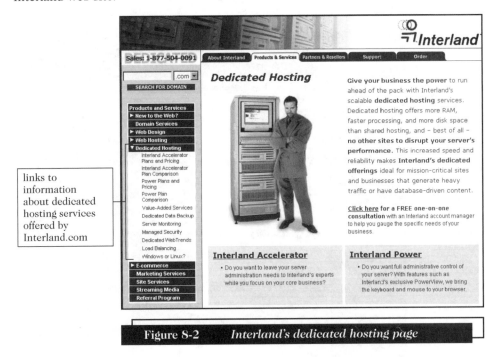

Figure 8-2   *Interland's dedicated hosting page*

Operating Systems for Web Servers

Companies must be forward thinking when making Web hardware choices. Obviously, a fast server is always better than a slower one. A company must also think about which server is a good choice for the present, while the business is small, and for the future, when the business has grown. Another important factor in server hardware choices is the internal (intranet) and external traffic or transactions that are likely to occur on the server. A small organization just starting out might get a few thousand visits, or hits, per hour initially. Larger, more well-known organizations employing a server for the first time might expect tens of thousands of "eyeballs" at the Web site each hour. Careful planning and thorough testing will help companies to determine the best combination for their needs. Because some visitors to a site might be using expensive client machines on high-speed connections and others might be using PC clients on low-speed dial-up connections, a company might want more than one machine to serve the two different client types.

One of the most important variables in the server decision is whether the server hardware is scalable; that is, whether it can be upgraded or even connected to additional servers seamlessly. At some point when the server traffic is sufficiently high, more computing power will need to be added to the site. Running a large Web server on a personal computer is not feasible. Similarly, purchasing a $50,000 application server for a small site is excessive. Running database software such as Microsoft SQL Server or Oracle on the same computer as the Web server or the electronic commerce application software is not a good idea, because database products have large processing and memory requirements and can slow down Web server response times. Operating system tasks include running programs and allocating computer resources such as memory and disk space to programs. Operating system software also provides input and output services to devices connected to the computer, including the keyboard, monitor, and printers. A computer must have an operating system to run programs. For large systems, the operating system has even more responsibilities, including keeping track of multiple users logged on to the system and ensuring that they do not interfere with one another.

Most Web servers run on computers that use one of the following operating systems: Microsoft Windows NT Server, Microsoft Windows 2000 Server Edition, Linux, or one of several UNIX-based operating systems, such as Solaris or FreeBSD. Each operating system has distinct advantages and disadvantages. For example, many people find the **Microsoft server products** simpler to learn and use than the somewhat arcane UNIX-based systems. However, UNIX-based machines are more popular, and many users claim that they are more robust machines on which to run a Web server. **Linux** is a free operating system that is easy to install, very fast, and efficient. **IBM** sells many of its Web server hardware products with the Linux operating system installed. Although Linux is available at no cost, most companies buy the operating system through a commercial distributor. The commercial distributions of Linux include useful additional software such as installation utilities. The distributor provides support for the operating system. Commercial Linux distributors that sell versions of the operating system with utilities for Web servers include **Caldera**, **Mandrake**, **Red Hat**, and **SuSE**. **Sun Microsystems** sells Web server hardware along with its UNIX-based operating system, Solaris.

The best way to choose Web server hardware and operating system configurations is to run tests on various combinations, remembering to consider the system's scalability. **Mindcraft**, one of a handful of companies that is an independent testing lab, tests software, hardware systems, and network products for users. Its site contains reports and statistics comparing combinations of application server platforms, operating systems, and Web server software products. Figure 8-3 shows Mindcraft reports on various Web servers.

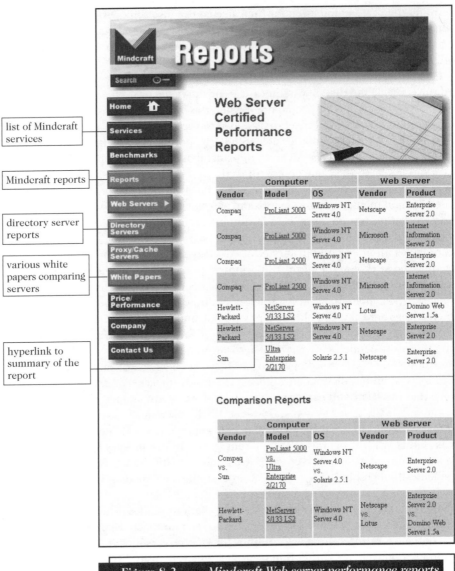

list of Mindcraft services

Mindcraft reports

directory server reports

various white papers comparing servers

hyperlink to summary of the report

Figure 8-3 *Mindcraft Web server performance reports*

Web Server Performance Evaluation

Benchmarking Web server hardware and software combinations can help in making informed decisions for a system. **Benchmarking** is testing that is used to compare the performance of hardware and software. Because technology is continually changing, any suggestion in this book about particular Web server hardware or software could become obsolete; but guidelines to help make good choices are helpful, so some basic rules are presented below.

Elements affecting server performance include hardware, operating system software, connection speed, user capacity, and type of Web pages being delivered. When evaluating Web server performance, a company should know exactly what factors are being measured and ensure that these are important factors relative to the expected use of the Web server.

Hardware and operating systems are key areas for benchmarking. A PC with a midrange CPU, small hard drive, and 128 megabytes of memory will perform poorly compared with a high-end workstation or a powerful UNIX-based computer. Operating systems yield different performance values under various Web benchmark tests.

Another factor that can affect a Web server's performance is the speed of its connection. A server on a T3 connection can deliver Web pages to clients much faster than it could on a T1 connection.

The number of users the server can handle is also important. This can be difficult to measure, because results are affected by the server's line speed, the clients' line speeds, and the sizes of the Web pages that are delivered. Two factors to evaluate when measuring a server's Web page delivery capability are throughput and response time. **Throughput** is the number of HTTP requests that a particular hardware and software combination can process in a unit of time. **Response time** is the amount of time a server requires to process one request. These values should be well within the anticipated loads a server can experience, even during peak load times.

Finally, the mix and type of Web pages a system is likely to deliver in response to client requests greatly affect performance. A **dynamic page** is a Web page whose content is shaped by a program in response to user requests, whereas a **static page** is an unchanging page retrieved from disk. A server delivering mostly static Web pages will perform better than the same server delivering dynamic Web pages, because static page delivery requires far less computing power than dynamic page delivery. The largest performance differences between competing Web server products appear when servers deliver dynamic pages.

SPECweb99 and **WebStone** are among the Web server benchmarking programs that are available. They make a number of measurements that can help companies decide which server to employ. Benchmarking programs can range in price from free to a few hundred dollars.

WebStone is the original Web server benchmarking program and is still very popular. The information WebStone collects is similar to information collected by other Web benchmark programs. WebStone works by measuring the response of Web servers to a workload it creates. The workload simulates multiple Web clients (users connecting to a Web site with their clients) accessing the Web server. WebStone can simulate over 100 Web clients on a single computer. Webmaster, a program that controls all the testing done by WebStone, runs on one of the client computers and distributes the Web client software and test files to client computers.

After Webmaster starts the execution of a benchmark, it waits for client computers to report the performance that each client measured. When all client performance information is available, Webmaster consolidates the information into a summary report. The files used by the Web client computers determine the performance measured by WebStone. WebStone supplies a standard set of files so administrators can compare the evaluation results fairly for different Web servers. Because of the way that WebStone benchmark tests are structured, evaluation results measure the performance of the combination of the Web server's operating system, Web server software, network connection speed, and CPU speed.

WebStone uses three tests to measure performance: HTML, CGI, and API. The HTML test measures server performance when the client asks the server to retrieve and send it an HTML-encoded file (a static Web page). The CGI test causes the Web server to run another program, using the CGI protocol. The **common gateway interface (CGI)** protocol is a common way for Web servers to interact dynamically with clients. Web pages that contain forms with text boxes, option buttons, and list boxes can collect information from users that CGI programs can use to manipulate databases, store information, or retrieve data. The third test, called API, tests the Web server's ability to pass information from a Web client to the server's **application program interface (API)**, which is a set of protocols, routines, and tools for building application code blocks. The API request launches another program that locates information for the Web server and passes it back. An example of an API request is a Web client request for information found in a database on another computer.

SPECweb99 is a benchmark program from the Standard Performance Evaluation Corporation, a nonprofit standards organization. SPECweb99 provides system workloads that stress-test Web servers. These workloads originate from sample Web sites and include Web files from 1 KB to 1000 KB in size.

Anyone contemplating purchasing a server that will have to handle heavy traffic should compare standard benchmarks for a variety of hardware and software configurations. Customized benchmarks can give Web managers guidelines for modifying file sizes, cache sizes, and other parameters. Web managers should run benchmarks regularly. Benchmarks are not as meaningful for small Web sites with much smaller numbers of daily visitors. In the latter case, a focus on Web design and site navigation can maximize clients' satisfaction.

Besides testing Web servers' raw performance, it is important to test server software features for efficiency and usability. These tests will reveal whether a particular feature is easy to use and whether it performs well. Web server software features are described in the next section.

DESIRABLE FEATURES OF WEB SERVERS

Web server software responds to requests from client programs. Electronic commerce support, backend programs, and databases are all managed by the server. Backend program and database responses are formatted and passed back to the server, which sends the formatted Web page to the requesting client.

Web server software program features can range from basic to extensive. Web server programs provide a core feature set that typically includes core capabilities, site management, site development, application construction, and dynamic content. Not all Web server programs' features fit precisely into one of these categories. First, we will examine the core capabilities any Web server program should have.

Core Capabilities

Recall that the most fundamental duty of a Web server is to process and respond to Web client requests that are sent using the HTTP protocol. For a client request for a Web page, the server program finds and retrieves the page, creates an HTTP header, and appends the HTML document to it. For dynamic pages, the server uses an architecture with three or more tiers that invokes other programs, receives the results from the backend process, formats the response, and sends the pages and other objects to the requesting client program. IP-sharing, or a virtual server, is a feature that allows different groups to share a single Web server's Internet protocol (IP) address. A **virtual server** or **virtual host** is a feature that maintains more than one server on one machine. This means that different groups can each have their own domain name, but all domain names refer to the same physical Web server. For example, ABC Corporation's marketing department could have the domain name www.marketing.abc.com, while sales could have the domain name www.sales.abc.com—both names referring to the same ABC Corporation Web server.

Indexing and Searching

Search engines and indexing programs are important elements of many Web servers. Search engines or search tools search either a specific site or the entire Web for requested documents. An indexing program can provide full-text indexing that generates an index for all documents stored on the server. When a browser requests a Web site search, the search engine compares the index terms to the requester's search term to see which documents contain matches for the requested term or terms. Many Web server software products contain indexing software. Indexing software can often index documents stored in many different file formats.

Data Analysis

Web servers can capture visitor information, including data about who is visiting a Web site (the visitor's URL), how long the visitor's Web browser viewed the site, the date and time of each visit, and which pages were displayed. This data is placed into a Web **log file**. As you can imagine, the file grows very quickly—especially for popular sites with thousands of visitors each day. Careful analysis of the log file can be fruitful and reveal many interesting facts about site visitors and what they like. To make sense of a log file, you must run third-party Web log file analysis programs. These programs summarize log file information by querying the log file and either returning gross summary information or accumulating details that reveal how many visitors came to the site per day, hour, or minute, or which hours of the day were peak loading times. Two of the most popular Web log file analysis programs are the **Analog** Web server log file analyzer and the **WebTrends** Web server log file analyzer. Figure 8-4 shows part of WebTrends log file analysis report.

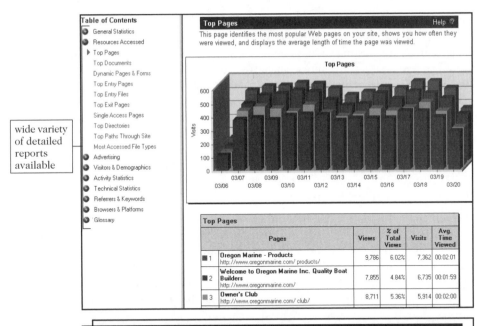

wide variety of detailed reports available

Top Pages   Help ?

This page identifies the most popular Web pages on your site, shows you how often they were viewed, and displays the average length of time the page was viewed.

Top Pages

	Pages	Views	% of Total Views	Visits	Avg. Time Viewed
▪1	**Oregon Marine - Products** http://www.oregonmarine.com/ products/	9,786	6.02%	7,362	00:02:01
▪2	**Welcome to Oregon Marine Inc. Quality Boat Builders** http://www.oregonmarine.com/	7,855	4.84%	6,735	00:01:59
▪3	**Owner's Club** http://www.oregonmarine.com/ club/	8,711	5.36%	5,914	00:02:00

Figure 8-4   *WebTrends log file analysis*

Site Management Tools

Comprehensive Web site development software includes products such as Macromedia Dreamweaver and Microsoft Front Page. These comprehensive programs also include tools for managing Web sites once they have been created. Other programs provide functions that help Web site managers keep their sites running smoothly. Macromedia's **HomeSite** Web page design software validates graphics, computes page download times, validates links, and validates HTML code.

Link Checking

Dedicated site management tools include a standard set of features, starting with link checking. A **link checker** examines each page on the site and reports on any URLs that are broken, that seem to be broken, or that are in some way incorrect. It can also identify orphan files. An **orphan file** is a file on the Web site that is not linked to any page. Other important site management features include script checking and HTML validation. Some management tools can locate error-prone pages and code, list broken links, and e-mail maintenance results to site managers.

On the company Web site, it is important to regularly check links that point to pages both within and outside the corporate Web site. Some Web server software does contain link-checking features. A **dead link**, when clicked, displays an error message rather than a Web page. Maintaining a site that is free of dead links is vital, because too many dead links on a site will cause people to jump to another site. Web-browsing customers are just a click away from going to a competitor's site if they become annoyed with an errant Web link.

Free link-checking and Web site validation programs, such as **Elsop LinkScan**, can be launched by entering the address of a Web site's home page and checking a few boxes. The results of the link checker are either displayed automatically or e-mailed to a recipient. Besides checking links, Web site validation programs sometimes check spelling and other structural components of Web pages.

Checkers that run on a company's own Web site are also available. Commercial site checkers such as **Big Brother** software from **Watchfire** produce more comprehensive results and more detailed site analyses than do the free products. The Watchfire Linkbot products include an enterprise version that is a complete Web site analysis tool; it scans a site for broken links and many other potential problems and then generates reports detailing any errors it finds. Figure 8-5 shows a report produced by the Watchfire Linkbot enterprise product.

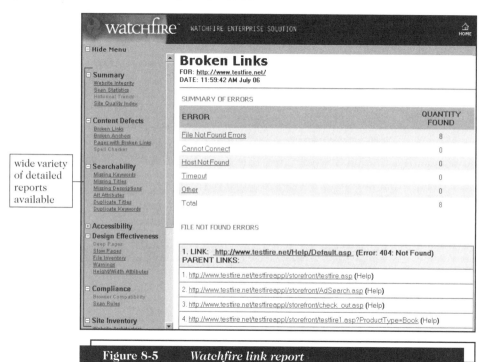

wide variety of detailed reports available

Figure 8-5 *Watchfire link report*

LinxCop is one of several reverse link checkers available. A **reverse link checker** checks on sites with which a company has entered a link exchange program (which you learned about in Chapter 4) and ensures that link exchange partners are fulfilling their obligation to include a link back to the company's Web site.

Remote Server Administration

With **remote server administration**, a Web site administrator can control a Web site from any Internet-connected computer. Although all Web sites provide administrative controls—most through a workstation computer on the same network as the server computer or through a Web browser—it is convenient for an administrator to be able to fix the server from wherever he or she happens to be. For example, an administrator can install **Web Site Garage** on any Internet-connected Windows machine and monitor and change anything on the Web site from that computer. The **NetMechanic** site, which appears in Figure 8-6, offers a variety of link checking, HTML troubleshooting, site monitoring, and other programs that can be useful in managing the operation of a Web site.

Web site trouble-shooting tools

Web site performance monitoring tools

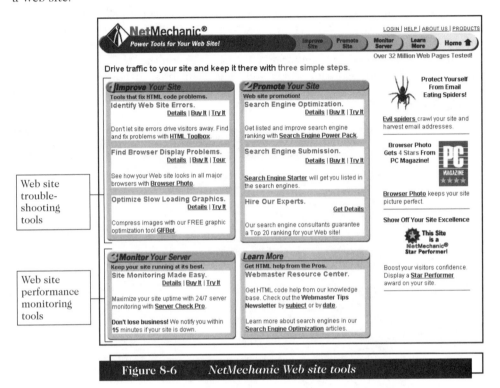

Figure 8-6   *NetMechanic Web site tools*

Dynamic Content

Dynamic content is nonstatic information constructed in response to a Web client's request. For example, if a Web client inquires about the status of an existing order by entering a unique customer number or order number into a form, the Web server will search the customer information (or will send a query to the backend database in a higher tier) and generate a dynamic Web page based on the customer information it found (or that the database management software provided), thus fulfilling the client's request. Assembled from backend databases and internal data on the Web site, a dynamic page is a specific response to the requester's query.

In a Web site that is a collection of HTML pages, the content on the site can be changed only by editing the HTML in the pages. This is cumbersome and does not allow customized pages to be produced in response to specific queries from site visitors. Most Web sites today that provide dynamic Web pages use an approach called server-side scripting. In **server-side scripting** (also called **server-side includes** or, more generally, **server-side technologies**), programs running on the Web server create the Web pages before sending them back to the requesting Web clients as parts of response messages. Web sites can accomplish this page-generation task in a number of ways. Microsoft has developed a dynamic page-generation technology called **active server pages (ASP)**. Sun Microsystems has developed a similar technology called **Javaserver pages (JSP)**, and the open-source Apache Software Foundation has sponsored a third alternative called **PHP: Hypertext Preprocessor (PHP)**. In these approaches, the server-side scripts are mixed with HTML-tagged text to create the dynamic Web page. For example, ASP allows Web programmers to use their choice of programming languages, such as VBScript, Jscript, and Perl. Java, a programming language created by Sun, can be used to produce dynamic pages. Such server-side programs are called **Java servlets**.

The Future of Dynamic Web Page Generation

Many critics of the server-side scripting approaches note that these approaches do not really solve the problem of dynamic Web page generation. They argue that it merely shifts the task of creating dynamic pages from HTML code writers to ASP (or JSP or PHP) code writers. Several initiatives are under way that are directed at a more comprehensive solution to the dynamic Web page creation problem. The **Apache Cocoon Project** is one of these initiatives, using the XML (extensible markup language) technology that you learned about in Chapter 2.

In this approach, the content is stored with XML tags that describe the semantics (the meaning) of each content item. The information request is handled by a Java servlet (called the producer) that can read the XML file and select the requested content items using the XML tags in the content file. Instead of creating a Web page, Cocoon can produce a response tailored to the request by applying a style sheet to the data. If a site visitor requests, for example, an Adobe Portable Document Format (PDF) file or a Wireless Markup Language (WML) file for display on a wireless hand-held device, a Web site using Cocoon technology can generate the results in those file formats from its XML content files. Business logic rules also can be implemented in the Java program that processes the XML content files. Many industry experts believe that the Apache Cocoon Project or similar development efforts by Microsoft and Oracle may provide a better way to generate dynamic Web pages in the future.

311

WEB SERVER SOFTWARE

The Web server market includes two distinct areas: intranet servers and public Web servers. Some Web server software runs on only one computer operating system, while others run on several operating systems. This section describes the three most popular Web server programs; Apache HTTP Server, Microsoft Internet Information Server (IIS), and iPlanet Enterprise Server (often called by its former name, Netscape Enterprise Server). These popularity rankings were accumulated through surveys done by **Netcraft**, a networking consulting company in Bath, England, known throughout the world for its Web Server Survey. Netcraft conducts continual surveys to tally the number of Web sites in existence and to measure the relative popularity of Internet Web server software.

Figure 8-7 shows the market share of Apache, Microsoft, and iPlanet along with the National Center for Supercomputing Applications (NCSA) Web server and a ranking that includes all other Web servers. The NCSA Web server was one of the first Web servers developed in the United States. Because it was developed with U.S. government research funds, it is available at no cost.

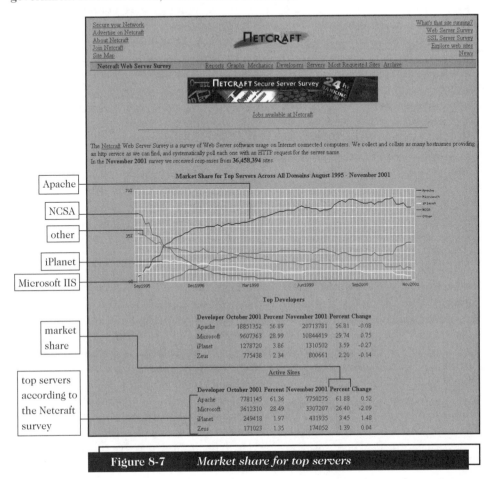

Figure 8-7   *Market share for top servers*

Server software from Apache and Microsoft accounts for the majority of the installed sites. Although these numbers will be somewhat different as you read this, the three packages highlighted in this section are likely to be popular for some time. Click the **Netcraft surveys** link on the Online Companion to check out the latest survey results.

According to a *PC Magazine* survey (see the Alwang reference listed in the "For Further Study and Research" section at the end of the chapter), the market share percentages for intranet Web servers are quite different than for public Web servers. Although the Web server software packages described in this chapter are all top selections among intranet servers, Microsoft IIS and iPlanet Enterprise Servers together account for 75 percent of installed intranet server programs.

Recall from the previous section that the performance of one Web server differs from that of another based on workload, operating system, and the size and type of

Web pages being loaded. *PC Magazine* evaluates computer products regularly. In these tests, some Web software fared well when delivering static HTML pages, but other Web server software performed better when delivering dynamic Web page content. The differences among servers can be significant: picking the right server for each different business need is critical (see the Machrone reference listed in the "For Further Study and Research" section).

The sections that follow contain descriptions of Apache HTTP Server, Microsoft Internet Information Server, and iPlanet Enterprise Server. Each description examines the program's general characteristics, its configuration and management tools, log files and reporting tools, security and directory support, applications development, and database connectivity.

Apache HTTP Server

Apache is an ongoing group software development effort. Rob McCool developed Apache while he was working at the University of Illinois at the **National Center for Supercomputing Applications** (**NCSA**) in 1994. Several Webmasters from around the world created their own extensions to the server and formed an e-mail group so that they could coordinate their changes (known as "patches") to the system. The system became known as Apache because it consisted of the original core system with a lot of patches—thus, it became known as "a patchy" server, or simply, "Apache."

Apache HTTP Server dominates the Web, in part, because it is free and performs very efficiently. It is powerful enough that IBM has licensed it for its own WebSphere application server package. In the period from 1996 through 2000, Apache has enjoyed the highest increase in Internet Web sites of all Web servers, according to a Netcraft survey. Currently, Apache is more widely used than all the other Web servers combined. Apache runs on many operating systems (including AIX, FreeBSD-UNIX, HP-UX, Linux, Microsoft Windows, SCO-UNIX, and Solaris) and the hardware that supports them. Apache has a built-in search engine and HTML authoring tools, and supports FTP.

Apache has wizards available to create new sites and directories, and the server provides for multiple logs that can be automatically cycled or archived. **Cycling** a log means replacing the oldest log with the newest, thus recycling the space it occupies. **Archiving** a log means saving it, perhaps on a backup storage device. The log entries conform to the established, standard NCSA common log format to which many servers adhere.

Apache's application development tools support CGI and several proprietary APIs. Once the API blocks are built, programmers can invoke them to perform their duties by using the common API interface. Apache supports server-side technologies for generating dynamic Web pages. Apache supports the ODBC standard and can access Oracle, Sybase, Microsoft SQL Server, and IBM's DB2 databases. Figure 8-8 shows Apache's home page.

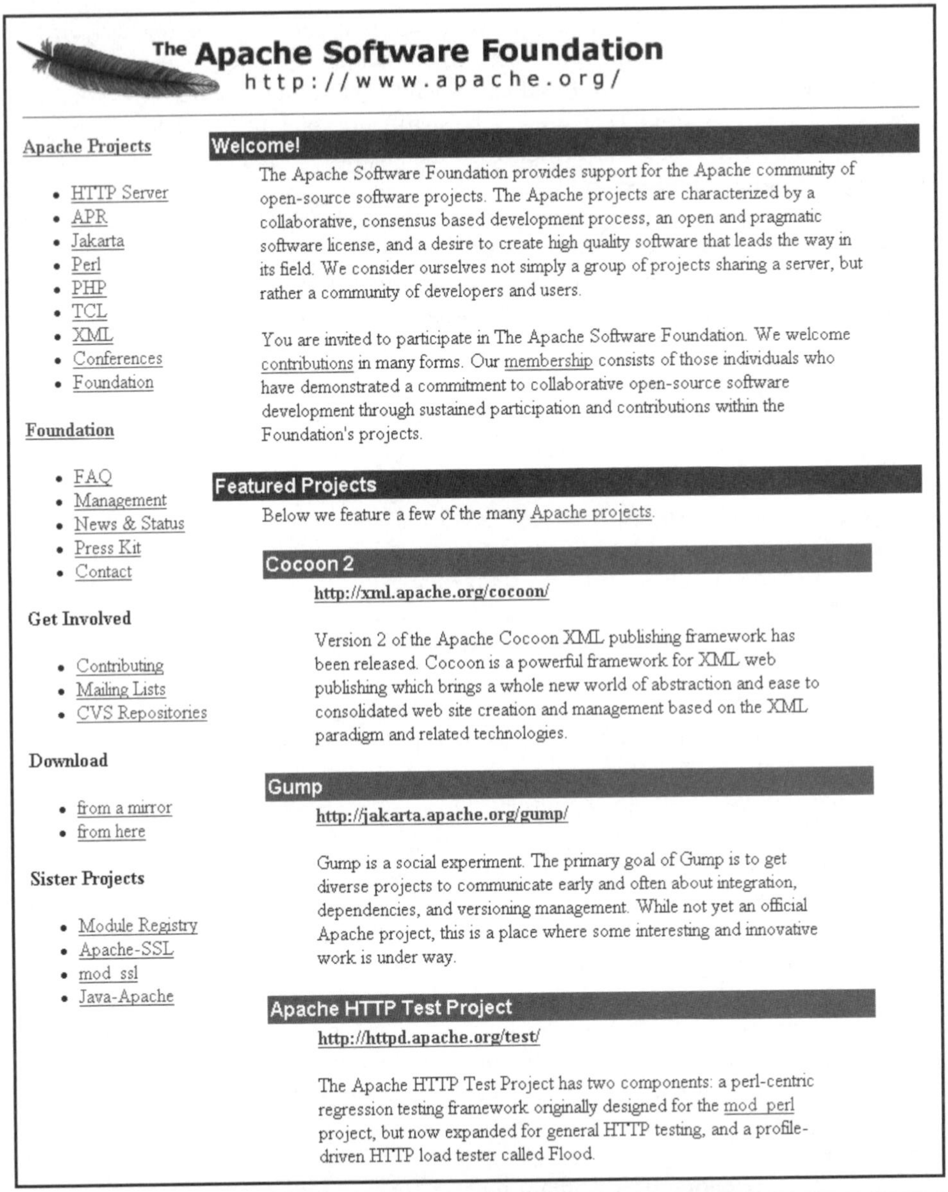

The Apache Software Foundation

http://www.apache.org/

Apache Projects

- HTTP Server
- APR
- Jakarta
- Perl
- PHP
- TCL
- XML
- Conferences
- Foundation

Foundation

- FAQ
- Management
- News & Status
- Press Kit
- Contact

Get Involved

- Contributing
- Mailing Lists
- CVS Repositories

Download

- from a mirror
- from here

Sister Projects

- Module Registry
- Apache-SSL
- mod_ssl
- Java-Apache

Welcome!

The Apache Software Foundation provides support for the Apache community of open-source software projects. The Apache projects are characterized by a collaborative, consensus based development process, an open and pragmatic software license, and a desire to create high quality software that leads the way in its field. We consider ourselves not simply a group of projects sharing a server, but rather a community of developers and users.

You are invited to participate in The Apache Software Foundation. We welcome contributions in many forms. Our membership consists of those individuals who have demonstrated a commitment to collaborative open-source software development through sustained participation and contributions within the Foundation's projects.

Featured Projects

Below we feature a few of the many Apache projects.

Cocoon 2

http://xml.apache.org/cocoon/

Version 2 of the Apache Cocoon XML publishing framework has been released. Cocoon is a powerful framework for XML web publishing which brings a whole new world of abstraction and ease to consolidated web site creation and management based on the XML paradigm and related technologies.

Gump

http://jakarta.apache.org/gump/

Gump is a social experiment. The primary goal of Gump is to get diverse projects to communicate early and often about integration, dependencies, and versioning management. While not yet an official Apache project, this is a place where some interesting and innovative work is under way.

Apache HTTP Test Project

http://httpd.apache.org/test/

The Apache HTTP Test Project has two components: a perl-centric regression testing framework originally designed for the mod_perl project, but now expanded for general HTTP testing, and a profile-driven HTTP load tester called Flood.

Figure 8-8   *The Apache home page*

Microsoft Internet Information Server

Microsoft Internet Information Server (IIS) comes bundled with Microsoft's Windows 2000 Server operating systems. IIS serves equally well as an intranet Web server or as a public Web server program, and thus it is popular for both public sites and corporate intranet sites. A robust and capable Web server program, IIS is suitable for any size site. Small sites running personal Web pages use IIS, as do some of the largest electronic commerce sites on the Web.

IIS, as a Microsoft product, was designed to run only on the Windows NT and 2000 operating systems. IIS includes an integrated search engine that allows users to create customized search forms with a variety of tools, including ASP, ActiveX Data Objects, and SQL database queries. IIS also includes Microsoft's FrontPage HTML development tool and other reporting tools. IIS supports FTP, allowing users to download files and data from the IIS server site using the FTP protocol.

IIS creates log files in a standard format. Like most other Web server products, IIS supports automatic cycling or archiving of log files. The Microsoft Management Console (MMC), which is included in IIS, provides central server management from any server on the network. IIS also permits administration from a remote browser. Because Windows NT lets you associate additional IP addresses with a single network interface card (NIC), IIS permits each virtual server to have its own IP address, known as multiple virtual hosts.

IIS's inclusion of ASPs provides an application environment in which HTML pages, ActiveX components, and scripts can be combined to produce dynamic Web pages. IIS's database support includes ODBC and Microsoft SQL. Figure 8-9 shows a Microsoft Internet Information Server information page complete with links to technical documents.

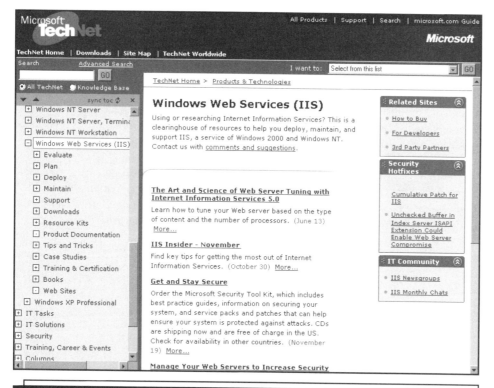

Figure 8-9   *Microsoft's Internet Information Server*

iPlanet Enterprise Server

A descendant of the original NCSA Web server program, Netscape Enterprise Server is now sold under the name iPlanet Enterprise Server. When AOL (now AOL Time Warner) purchased Netscape in 1999, the company formed a partnership with Sun Microsystems to support and continue to develop Netscape's server products. This partnership was named iPlanet and was a limited-term partnership that was scheduled to expire in 2002. Industry analysts believe that when the partnership expires, iPlanet will become a part of Sun, because the Web server and electronic commerce software that iPlanet now sells is more closely related to Sun's businesses than to AOL Time Warner's businesses.

Anyone developing a sophisticated, enterprise-strength Web site will appreciate iPlanet's extensive server features. Although the iPlanet server is not free, its $1500-per-CPU licensing fee is reasonable, and it allows a free 60-day trial download. The iPlanet software runs on many operating systems, including AIX, Digital UNIX, HP-UX, Solaris, and Windows. Some of the busiest and best-known sites on the Internet, including BMW, Dilbert, E*Trade, Excite, Lycos, and Schwab, run (or have run) some version of iPlanet Web Server. Although iPlanet appears to have a small share of the Web server market in the Netcraft survey results shown in Figure 8-7, these survey results can be somewhat misleading. The Netcraft numbers include all Web sites, including many very small sites or sites that are not engaged in commercial activity. Independent reports from consulting firms such as the GartnerGroup show that iPlanet Web server software is in use at more than 40 percent of all public Web sites and at more than 60 percent of the top 100 enterprise Web sites.

The iPlanet Web server provides a powerful development environment that supports development of Web-based applications that can be run on the Internet, an intranet, or an extranet.

The iPlanet Web server's management tools allow administrators to manage users and monitor server activity interactively. The software provides cluster management, which is a way for an administrator to manage multiple remote servers as a single group. This allows the administrator to update configuration files remotely or to start and stop a group of servers.

Like most other server programs, iPlanet supports dynamic application development, including CGI and the Java Servlet API for server-side applications. Its ODBC conformance means that iPlanet Enterprise Server provides connectivity to a number of database products as well. Figure 8-10 shows the iPlanet Enterprise Server page.

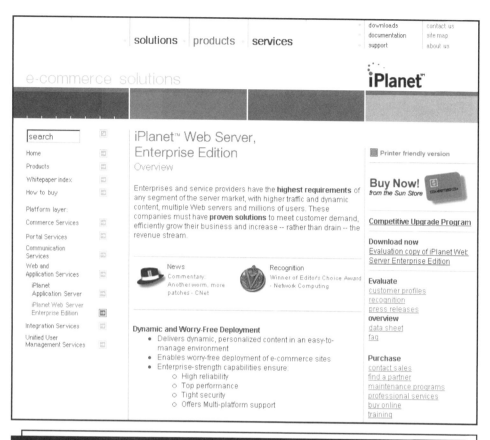

Figure 8-10 *iPlanet Enterprise Server*

Determining Web Server Hardware and Software Information

You can determine the type of hardware and software most Web sites are running by visiting **Netcraft**. On Netcraft's home page is a link named "What's that site running?" that opens a page with a Hostname text box. Type any Web address (for example, www.pepsi.com) in the text box and click the Examine button. Netcraft software examines the designated Web site and returns both Web server hardware and software information (see Figure 8-11). Occasionally, the information is not readily available, but Netcraft is almost always successful. The information provided in the figure for Pepsi shows that its site is running Microsoft IIS on a Windows operating system. At the bottom of the figure, a table containing the software history of the site is shown. You can see that Pepsi has been running the Microsoft products only since June 2001. Before that, Pepsi was running a Netscape Enterprise Web server (the predecessor to the iPlanet server) on the Solaris operating system. Netcraft can provide this historical information because the site continually checks Web sites for this information and stores it in a database.

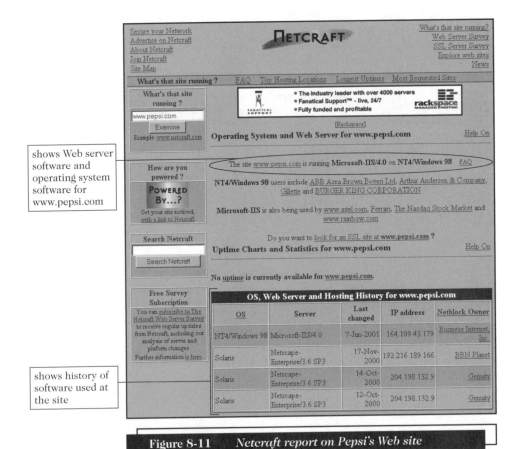

shows Web server software and operating system software for www.pepsi.com

shows history of software used at the site

Figure 8-11   *Netcraft report on Pepsi's Web site*

SERVER ARCHITECTURES AND WEB SERVER UTILITIES

Companies that operate more than one Web server must decide how to configure their servers to provide site visitors with the best service possible. The different ways that servers can be connected to each other and to related hardware such as routers and switches is called **server architecture**. There are several design options for Web server architectures and a number of utilities that companies can use to build effective electronic commerce sites. Load balancers, search engines, and intelligent software agents are three important utilities that can work with Web servers to make them more effective.

Web Server Architectures

Large electronic commerce Web sites must deliver millions of individual Web pages every day. They must also process thousands of customer and vendor transactions each day. Large Web sites must plan carefully to configure their server computers,

318

which can number in the hundreds or even thousands, to handle their daily Web traffic efficiently. These large collections of servers are called **server farms** because the servers are often lined up in large rooms, row after row, like crops in a field. One approach, sometimes called a **centralized architecture**, is to use a few very large and fast computers. A second approach is to use a large number of less-powerful computers and divide the workload among them. This is sometimes called a **distributed architecture** or, more commonly, a **decentralized architecture**.

Each approach has both benefits and drawbacks. The centralized approach requires expensive computers and is more sensitive to the effects of technical problems. If one of the few servers becomes inoperable, a large portion of the site's capability is lost. Thus, Web sites with centralized architectures must be very careful to have adequate backup plans. Any server problem, no matter how small, can threaten the operation of the site. The decentralized architecture spreads that risk over a large number of servers. If one server becomes inoperable, the site will continue to operate without much degradation in capability. The smaller servers used in the decentralized architecture are less expensive than the large servers used in the centralized approach. That is, the total cost of 100 small servers is usually less than the cost of one large server with the same capacity as the 100 small servers. However, the decentralized architecture does require additional hubs or switches to connect the servers to each other and to the Internet. Most large decentralized sites use load-balancing systems, which do cost additional money, to assign the workload efficiently.

Load-Balancing Systems

A **load-balancing switch** is a piece of network hardware that monitors the workloads of servers attached to it and assigns incoming Web traffic to the server that has the most available capacity at that instant in time. In a simple load-balancing system, the traffic that enters the site from the Internet through the site's router encounters the load-balancing switch, which then directs the traffic to the Web server best able to handle the traffic. Figure 8-12 shows a basic load-balancing system.

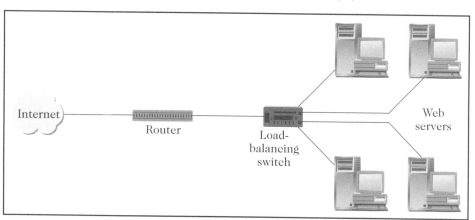

| **Figure 8-12** | *A load-balancing system in a decentralized architecture* |

In more complex load-balancing systems, the incoming Web traffic, which may enter from two or more routers in a larger Web site, is directed to groups of Web servers that have been dedicated to specific tasks. In the sample complex load-balancing system that appears in Figure 8-13, the Web servers have been gathered into groups of servers that handle delivery of static HTML pages, servers that coordinate queries of an information database, servers that generate dynamic Web pages, and servers that handle transactions. Load-balancing switches and the software that helps them do their work cost roughly between $10,000 and $50,000, and include products such as **E-Load**, **Loadrunner**, **ServerIron**, and **Silkperformer**.

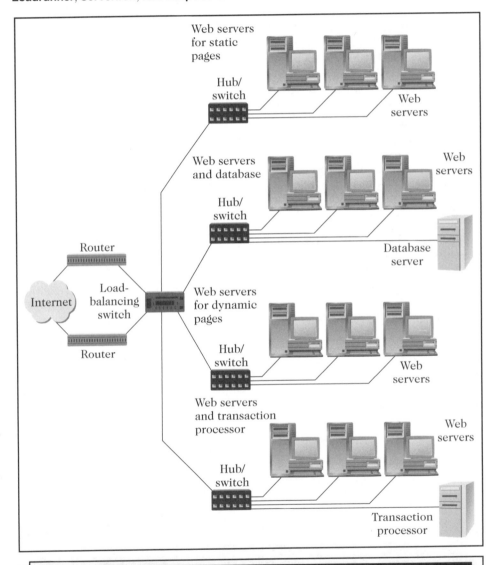

Figure 8-13 *Complex load balancing*

WEB SERVERS AT EBAY

Online auction site **eBay** is very popular, as you have learned in earlier chapters. Indeed, it is so popular that its Web servers deliver more than 150 million pages per day. These pages are a combination of static HTML pages and dynamically generated Web pages. The dynamic pages are created from queries run against eBay's Oracle database, in which it keeps all of the information about all auctions that are underway or closed within the most recent 30 days. With millions of auctions underway at any moment, this database is extremely large.

The combination of a large database and high transaction volume makes eBay's Web server operation an important part of the company's success and a potential contributor to its failure. The servers at eBay have failed more than 15 times during the company's life. The worst series of failures occurred during May and June of 2000, when the site went down more than four times within a two-month period. One of these failures kept the site offline for more than a day—a failure that cost eBay an estimated $5 million. The company's stock fell 20 percent in the days following that failure.

At that point, eBay decided it needed to make major changes in its approach to Web server configuration. Many of eBay's original technology staff had backgrounds at Oracle, a company that has a tradition of selling large databases that run on equally large servers. Further, the nature of eBay's business— any visitor might want to view information about any auction at any time—led eBay to implement a centralized architecture with one large database residing on a few large database server computers. It made sense to also use similar hardware to serve the Web pages generated from that database.

In mid-2000, following the worst site failure in its history, eBay decided to move to a decentralized architecture. This was a tremendous challenge, because it meant that the single large auction information database had to be replicated across groups, or clusters, of Web and database servers. However, eBay realized that using just a few large servers had made it too vulnerable to the failure of those machines. Once eBay had completed the move to decentralization, it found that adding more capacity was easier. Instead of installing and configuring a large server that might have represented 15 percent or more of the site's total capacity, clusters of six or seven smaller machines could be added that represented less than one percent of the site's capacity. Routine periodic maintenance on the servers also became easier to schedule.

The lesson from eBay's Web server troubles is that the architecture should be carefully chosen to meet the needs of the site. Web server architecture choices can have a significant effect on the stability, reliability, and, ultimately, the profitability of an electronic commerce Web site.

321

Search Engines

You learned about search engines and how they work in earlier chapters. Many Web sites include a search engine that allows site visitors to search the site, a group of sites, or even the entire Web. The companies that provide search engine technology for the major search engine sites and for Web portal sites also sell that technology to businesses that want to have a search function on their own Web sites. Some search engine developers, such as Google, operate their own search engine sites. Others such as Inktomi and Verity, create and sell the technology, but do not operate their own public search engine sites.

Intelligent Software Agents

Software agents have been around for a number of years. The Internet's rapid growth has sparked interest in Web agents. As you learned in Chapter 6, an intelligent software agent is a program that performs functions such as information gathering, information filtering, or mediation running in the background on behalf of a person or entity.

Dr. Pattie Maes, the founder of the MIT Software Agents Group, has led many major research efforts in the area of software agents. She and her team built the first truly successful agents for personalized information filtering. Her **Software Agents Group** investigates how to put computer systems to work on behalf of people by delegating tasks to software. The group describes agents as different from conventional software because "…they are long-lived, semi-autonomous, proactive, and adaptive."

Research reveals that software agents will become extremely important in the electronic commerce field sooner rather than later. For example, a purchasing manager could send a Web agent out onto the Internet to search for the best price and availability of 500 personal computers. Such an agent, armed with critical specifications supplied by a user, would first locate electronic commerce sites selling the desired products. Then, the agent would collect information about the price and characteristics of the equipment that the site sells from comparable selling agents that work on behalf of the commerce site. Armed with similar information from other Web sites selling computer equipment, the agent then performs its next critical task: determining from which of several computer-selling agents the company should purchase its equipment. Once the software agent identifies the best vendor, the agent can then negotiate any remaining terms of the transaction. Finally, the buying and selling software agents agree upon purchasing and delivery details.

Because software agents are always running in the background, they can help reduce the workload that people normally take on in locating, thinking about, negotiating, and purchasing goods and services on the Internet. Of course, the same is true for selling behaviors also. Currently, several agent systems have been implemented on the Web at sites such as **Best Web Buys** and **mySimon**, which you learned about in Chapter 6.

A simple example of how an intelligent agent can save a lot of drudgery and time is in procurement, which involves the systematic process of deciding what, when, and how much to purchase. Broader than just purchasing, procurement also includes ensuring that what you receive is the correct quantity and that you receive it on time. In a procurement application, agents notice the need for more materials, go out on the Internet to known suppliers, locate stock, negotiate the best price, and then arrange for delivery.

To simplify the example, imagine that you are a procurement officer and that one of your jobs is to ensure that the fabricating department has a continuous supply of rolled steel, enough so that the department can produce its expected weekly output of 55-gallon galvanized steel drums. If the department runs out of rolled steel midweek, 75 people and dozens of machines sit idle while they wait for raw materials. Rather than having to always be on the lookout for the best prices and delivery schedules for the raw materials, it would be much more efficient to use a software agent that monitors your company's inventory and then automatically orders new stock in just the right amount for delivery exactly when needed. Meanwhile, you are attending to less mundane tasks and handling procurement matters that cannot be given to agents.

Another example of agents at work is a stock alert. Customers of an online brokerage firm, for example, can establish criteria for purchasing or selling stocks when selected conditions occur. An intelligent agent monitors the stock and sends an alert when the specified conditions occur. For instance, an investor might want to purchase 100 shares of Microsoft if the stock price drops below $61 per share or sell Intel stock when the price-to-earnings ratio exceeds 3. Agents can perform either task. As researchers are able to make more progress in artificial intelligence and learning, you will see software agents enhanced with advances such as self-modifying behaviors and altered strategies that adapt to changing conditions and results.

Summary

Any company setting up a Web site must consider whether to purchase its own Web server hardware and software or to use a Web hosting company. Whether to self-host or pay someone else to host a site usually depends on the type of site being built. The most important consideration after deciding to self-host a site is to select a hardware and software combination that will allow the Web site to grow—to be scalable—as business grows. Running benchmark tests that simulate both heavy and light traffic to observe the Web server's performance is a good idea. The operating system, connection speed, user capacity, and the type of pages that the site will be serving also affect server performance. Each Web server software package offers slightly different capabilities.

Web site development tools provide features for creating and managing Web sites. Most Web server programs provide or work well with utility programs that include indexing, searching, and data analysis functions. Site management tools include link checkers and remote server administration. The most popular Web server programs are Apache HTTP Server, Microsoft Internet Information Server, and iPlanet Enterprise Server.

Web server administrators must decide whether to use a centralized or decentralized architecture. Both structures have advantages and disadvantages that must be carefully considered.

A search engine is a program that can be included on a site that lets visitors scan content for key words or topics. Intelligent software agents provide buyers with an automatic and persistent way to filter and collect information, locate goods and services, and negotiate price and delivery agreements.

Key Terms

Active server pages (ASP)
Application program interface (API)
Application service provider (ASP)
Archiving
Benchmarking
Centralized architecture
Co-location (collocation or colocation)
Commerce service provider (CSP)
Common gateway interface (CGI)
Cycling
Dead link
Decentralized architecture
Dedicated hosting
Distributed architecture
Dynamic content
Dynamic page
High availability server
High reliability server
Java servlet
Java server pages (JSP)
Link checker

Load-balancing switch
Log file
Managed service provider (MSP)
National Center for Supercomputing Applications (NCSA)
Orphan file
PHP: Hypertext Preprocessor (PHP)
Remote server administration
Response time
Reverse link checker
Scalable
Self-hosting
Server architecture
Server farm
Server-side scripting [server-side include (SSI) or server-side technologies]
Shared hosting
Static page
Throughput
Virtual host
Virtual server

Review Questions

1. List and briefly describe the advantages and disadvantages of hosting your own Web site instead of contracting with a service provider to perform this function.

2. Describe and briefly discuss two important measures of a Web site's performance.

3. Beginning with the links provided in the Online Companion, locate more information about two of the three Web servers mentioned in the chapter: Apache, Microsoft Internet Information Server, and iPlanet Enterprise Server. Write approximately 250 words about each of the two servers you choose. Include descriptions of six features for each Web server and indicate the computer platforms and operating systems on which each runs.

4. In 100 words or less, explain the differences between centralized and decentralized Web server architectures. Describe the characteristics of a Web site that could successfully use a centralized architecture.

5. Examine several of the research projects described on the **MIT Media Lab Software Agents Group** Web site. Choose one of these projects and write a 100-word analysis of the ultimate commercial potential for that project.

Exercises

1. Your friend Faye Borthick wants to set up a small Web site devoted to gardening. She believes her many years of experience in gardening give her an understanding of the kinds of gardening tools, fertilizers, soil amendment products, herbicides, pesticides, and plants that will appeal to the serious gardener. Right now Faye doesn't want to sell anything. She merely wants to display pages of plant photography, write and store short how-to papers for novice gardeners, and provide links to other gardening tips and traps on the Web. She wants your advice on whether to self-host the Web site or use an ISP to start her endeavor. Use **The List** or the **TopHosts** sites to locate information on the cost of using an Internet service provider to host a Web site. Then, estimate what a small Web site might cost in terms of the minimal configuration of hardware and software. Estimate the design and development costs and the annual maintenance costs. Then, select one of the Web server programs. Estimate the cost of a Web connection. Write a 400-word summary of everything you think Faye will need to know to use either of the two options (she builds it or she uses an ISP) for creating her site.

2. Your boss, Felicity Freedman, has sent you on a scavenger hunt. For three prominent electronic commerce sites and three university sites, Felicity would like you to determine which hardware platform they are running and which Web server software they are using. The six sites are Yahoo!, Apple Computer, Oracle Corporation, New York University, the University of Queensland (Australia), and Mount San Antonio College.

3. You have created a Web site for International Paper Products and Pulp complete with links to other pages on your site and to pages on the Internet. Bob Pardee, your supervisor, wants you to check periodically that the links on the corporate site are still valid. Instead of purchasing and installing a link-checking program, you decide to investigate online link checkers (Web sites that allow you to enter a Web site's root or home address and then check all the links that emanate from that site).

Use **Link Check** or **Elsop LinkScan** to check the links on any site of your choice. Print a few pages of the report and be prepared to turn them in to your instructor. Be patient. The program takes a little time to complete its work—especially on a Web page that has a large number of links.

For Further Study and Research

Abualsamid, A. 2001. "Dishing Up Dynamic Content," *Network Computing*, 12(8), April 16, 90–92.

Alwang, G. 1998. "Internet Web Servers," *PC Magazine*, 17(9), May 5, 184–208.

Ante, S. 2001. "Big Blue's Big Bet on Free Software," *Business Week*, December 10, 78–79.

Bannan, K. 2001. "Open the Floodgates: Web Accelerators Relieve Pressure on Your Servers," *PC Magazine*, 20(9), May 8, 143–152.

Baran, N. 2001. "Load Testing Web Sites," *Dr. Dobb's Journal: Software Tools for the Professional Programmer*, 26(3), March, 112–116.

Clyman, J. 2001. "Apache 2.0," *PC Magazine*, 20(13), July 3, 160.

Cohen, A. 1999. "Shopping Bots," *PC Magazine*, 18(13), July, 35.

Dragan, R. and F. Derfler. 2000. "Sun Microsystems Solaris 8.0 with iPlanet Web Server Enterprise Edition 4.1," *PC Magazine*, 19(10), May 23, 42.

Drucker, D. 2000. "Going Once, Going Twice, It's Sun," *InternetWeek*, May 22, 18.

Dustin, E. J. Rashka, and D. McDiarmid. 2002. *Quality Web Systems*, Boston: Addison-Wesley.

Dyck, T. and J. Rapoza. 2001. "IIS: Stay or Switch?" *eWeek*, October 29, 61–64.

The Economist. 2001. "Stealing Each Other's Clothes: Sun's Battle with IBM Raises Questions About Its Long Term Strategy," October 13, 61–63.

Harbaugh, L. 2000. "Balancing Act," *InternetWeek*, January 24, 26–30.

Kuo, J. 2001. "Work-Ready Linux," *InternetWeek*, September 10, 23–25.

Lee, Y. 2000. "Low-Cost Dedicated Servers," *Web Techniques*, 5(7), July, 88–89.

Lipschutz, R. 1999. "Internet in a Box," *PC Magazine*, 18(12), June 22, 189–202.

MacVittie, L. 2001. "IPlanet Goes Where No Server's Gone Before," *Network Computing*, 12(13), June 25, 96–99.

Maes, P., R. Guttman, and A. Moukas. 1999. "Agents that Buy and Sell," *Communications of the ACM*, 42(3), March, 81.

Machrone, W. 2000. "Picking the Right Server Is Key," *PC Magazine*, 19(10), May 23, 52.

Montalbano, E. 2001. "Sun Sets on IPlanet Alliance," *Computer Reseller News*, August 27, 12.

Morgan, C. 2000. "Web Content Management," *Computerworld*, 34(17), April 24, 72.

PC Magazine. 2001. "Servers," 20(14), August 12, 118.

Petreley, N. 2001. "The Cost of Free IIS," *Computerworld*, 35(43), October 22, 49.

Rapoza, J. 2001. "Zeus 4.0 Enters Web Server Pantheon," *eWeek*, November 19, 51–59.

Roberts-Witt, R. 2001. "The Internet Business," *PC Magazine*, 20(5), March 6, 8–17.

Roberts-Witt, R. 2001. "Web Server Brawn," *PC Magazine*, 20(10), May 22, 144–151.

Schroeder, E. 1999. "The Internet Rises and Sets with Sun," *PC Week*, March 8.

Schuchart, S. 2001. "IBM's Small Change," *Network Computing*, 12(9), April 30, 51–58.

Schwartz, J. 2001. "Update: How the NYSE Crashed," *InternetWeek*, June 8. Available online at: (http://www.internetwk.com/story/INW20010608S0006).

Shankland, S., M. Kane, and R. Lemos. 2001. "How Linux Saved Amazon Millions," *CNET News.com*, October 30. Available online at: (http://news.cnet.com/news/0-1003-200-7720536.html).

Stauffer, T. 2000. "New Features and Fast Load Times Keep LandsEnd.com Humming," *Publish*, November. Available online at: (http://www.publish.com/features/0011/feature3.html).

Vijayan, J. 2001. "Sun Attempts to Woo Users Away from IIS," *Computerworld*, 35(42), October 15, 28.

Wagner, M. and T. Kemp. 2001. "What's Wrong with EBay?" *InternetWeek*, January 15, 1–2.

ELECTRONIC COMMERCE
SOFTWARE

INTRODUCTION

WebMethods produces software that helps businesses conduct electronic commerce with their suppliers. Headquartered in Fairfax, Virginia, **webMethods** was founded by Australian Phillip Merrick and has attracted a number of major customers, including FMC Corporation and Dun & Bradstreet Corporation. **Covisint**, the auto industry procurement portal, uses webMethods software to integrate its existing Oracle database system with its Commerce One procurement and auction software, its Supply Solution supply chain execution software, and a variety of other vendors' software products.

WebMethods' software solves a difficult problem that businesses on the Internet face when trying to exchange information. The webMethods software allows companies to exchange business information with each other—information such as invoices and inventory tracking information—using XML (which you learned about in Chapter 2). With XML and webMethods' software, a manufacturing company's order is translated into a Web page that the manufacturer's software and Web server can understand. One of webMethods' largest customers, Dun & Bradstreet, compiles financial and credit information. It uses webMethods software to translate data from proprietary systems into a common format that any Dun & Bradstreet customer's computer can understand. WebMethods' software has saved Dun & Bradstreet and its customers money because its customers no longer need a customized program to interpret Dun & Bradstreet's data. Dun & Bradstreet no longer has to worry about supporting its regional data centers' many different financial data and credit information formats. WebMethods software takes care of translating the different formats into a single form.

As you will learn in this chapter, companies that undertake electronic commerce initiatives often combine software and tools from different vendors to accomplish their goals. Although small companies can sometimes use a single vendor to supply all of their electronic commerce software, most companies need to integrate a number of software products, each of which performs a particular task or process well.

LEARNING OBJECTIVES

In this chapter, you will learn about:

- Basic functions of electronic commerce software
- Characteristics to look for in an externally hosted electronic commerce solution
- Electronic store models
- Software for small electronic commerce sites
- Software for medium-sized to large electronic commerce sites
- Electronic commerce solutions for large organizations that have an existing infrastructure
- Supply chain management software
- Customer relationship management software
- Automated content management software

ELECTRONIC COMMERCE SOFTWARE BASICS

Just as the size and objectives of electronic commerce sites can be drastically different, the options for software and hardware solutions to set up an electronic business vary greatly. At the inexpensive end of the spectrum of electronic commerce solutions are choices such as externally hosted stores that provide software tools to build an online store on a host's site. At the other end of the range are sophisticated electronic commerce software suites that can handle high transaction volumes and include a broad assortment of features and tools.

The type of electronic commerce software an organization needs depends on several factors. One of the most important factors is the expected size of the enterprise and its projected traffic and sales. A high-traffic electronic commerce site with thousands of catalog inquiries each minute requires different software than a small online shop selling a dozen items. Another determining factor is budget. Creating an online store can be much less expensive than building traditional brick-and-mortar stores. The start-up costs of electronic commerce sites typically are a fraction of the costs of even the smallest conventional stores. A traditional store requires a physical location with leases, employees, utility payments, and maintenance. The cost of entering the electronic commerce market can be much lower.

Deciding on the target commerce audience will help merchants select the best electronic commerce software choices for their sites, or storefronts. If you are setting up a business-to-consumer (B2C) commerce site, your software choices will be different from those of a business-to-business (B2B) site. For example, B2C software accounts for and looks up sales tax rates for different states. On the other hand, B2B software is wholesale and can use existing extranet connections, does not include sales tax, and has a different set of rules than traditional B2C software. B2B electronic commerce can incorporate electronic data transfers between trading partners involving invoices, purchase orders, and other accounting information.

Another early decision is whether the company should use an external host (the ISPs, CSPs, and ASPs that you learned about in the previous chapter) or host the electronic commerce site in-house. The presence or absence of in-house information technology (IT) staff—programmers, Web designers, and network engineers—can be a critical factor in the decision. If an organization does not have or cannot easily hire people with the skills required to set up and maintain an electronic commerce site, it may want to outsource all or part of the job to an experienced service provider. If a company has an array of IT resources—including in-place hardware, servers, knowledgeable staff, and database systems—in-house commerce hosting may be neither difficult nor cost prohibitive.

Electronic Commerce Software Components

As you learned in Chapter 8, electronic commerce software must be hosted on a Web server or on a computer that is connected to a Web server. Once you have located or built a Web server, you can investigate and install electronic commerce software.

The specific duties that electronic commerce software will perform range from a few fundamental operations to a complete solution, from catalog display to fulfillment notification. Simply stated, all electronic commerce solutions must at least provide:

- A catalog display
- Shopping cart capabilities
- Transaction processing
- Tools to populate the store catalog and to maintain the catalog
- Tools to create and edit other site content

Larger and more complex electronic commerce sites will also use software that adds other features and capabilities to the basic set of commerce tools. These additional software packages can include:

- Supply chain management (SCM) software
- Customer relationship management (CRM) software

- Middleware that integrates the electronic commerce system with existing company information systems that handle inventory control, order processing, and accounting
- Automated content management software

Catalog Display

A small commerce site can have a very simple static catalog. A **catalog** is a listing of goods and services. A **static catalog** is a simple list written in HTML and displayed on a Web page or a series of Web pages. To add an item, delete an item, or change an item's listing, the company must edit the HTML of one or more pages. Larger commerce sites are more likely to use a dynamic catalog. A **dynamic catalog** stores the information about items in a database, usually on a separate computer that is accessible to the commerce server. A dynamic catalog can feature multiple photos of each item, detailed descriptions, and a search feature that allows customers to search for an item and determine its availability. Most of the Web stores you have seen in earlier chapters have been large, well-known sites. These sites include many features and have a professional look. Figure 9-1 shows the Web page of a small electronic commerce site, the **Women in Music Store**, that sells clothing and other logo products for the not-for-profit Women in Music National Network organization. This site uses simple, inexpensive electronic commerce software and has a clean look with few features beyond those necessary to make sales.

| Figure 9-1 | *Relatively small electronic commerce site* |

Small storefronts that sell fewer than 35 items, such as the Women in Music Store, need only a simple list of products or categories. Organization of the items is not particularly important. Companies that offer only a small number of items can provide a photo of each item on the Web page that is a link to more information about the product. Larger electronic commerce sites require more sophisticated navigation aids and better product organization. That is where a dynamic catalog becomes necessary.

A catalog organizes the goods and services being sold. To further organize its offerings, a retailer may break them down into departments. As in a physical store, merchandise in an online store can be grouped within logical departments to make locating an item, such as a camping stove, simpler. Web stores often use the same department names as their physical counterparts. In most physical stores, each product is kept in one place only. A Web store has the advantage of being able to include a single product in multiple categories. For example, running shoes can be listed as both footwear and athletic gear.

Good sites give buyers alternative ways to find products. Besides offering a well-organized catalog, large sites with many products should provide a search engine that allows customers to enter descriptive search terms, such as "men's shirts," so they can quickly find the Web page containing what they want to purchase. Remember, the most important rule of all commerce is: Never stand in the way of a customer who wants to purchase something.

Shopping Cart

In the early days of electronic commerce, shoppers selected items they wanted to purchase by filling out online forms. Using text box and list box form controls to indicate their choices, users entered the quantity of an item in the quantity text box, the SKU (stock-keeping unit) or product number in another text box, and the unit price in yet another text box. This system was awkward for ordering more than one or two items at a time. One problem with forms-based shopping was that shoppers usually had to write down product codes, unit prices, and other information about the product before going to the order form, which was inevitably on another page. Another problem was that customers sometimes forgot whether they had clicked the submit button to send in their orders. As a result, they either sent the same order twice (pressing the submit button when they had already done so) or thought they had submitted the order when they really had not (consequently failing to submit the order). The forms-based method of shopping was confusing and error-prone. Figure 9-2 illustrates the problems that shoppers faced with forms-based ordering systems. First, many customers found it difficult to remember the exact spelling of the coffee names. Second, customers were required to enter the coffee prices, which were located on a different Web page, in the text boxes. Thus, the customers needed to either write down or memorize the prices.

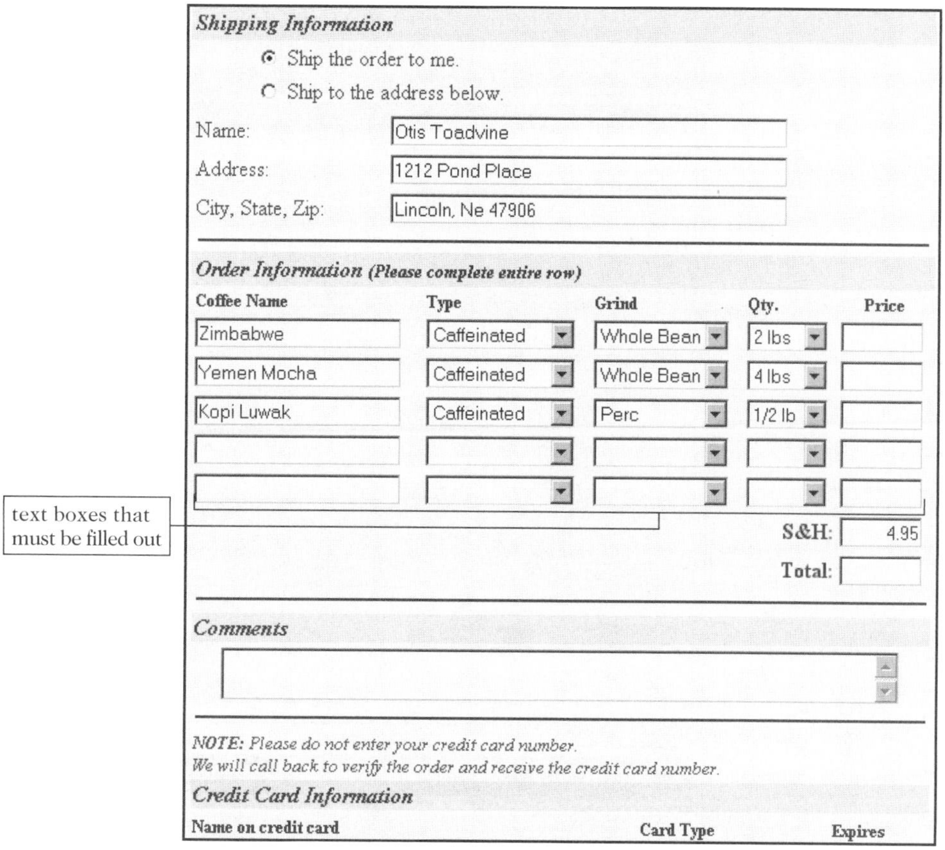

Figure 9-2   Using a form to enter an order

The forms-based method of ordering has given way to electronic shopping carts. Today, shopping carts are a standard of electronic commerce. A **shopping cart**, also sometimes called a shopping bag or shopping basket, keeps track of the items the customer has selected and allows customers to view the contents of their carts, add new items, or remove items. To order an item, the customer simply clicks that item. All of the details about the item, including its price, product number, and other identifying information, are stored automatically in the cart. If a customer later changes his or her mind about an item, he or she can view the cart's contents and remove the unwanted items. When the customer is ready to conclude the shopping session, the click of a button, usually labeled "Proceed to checkout" or something similar, commits the purchase transaction. Once committed, the customer cannot reverse the transaction. Figure 9-3 shows an example of a virtual shopping cart full of computer equipment.

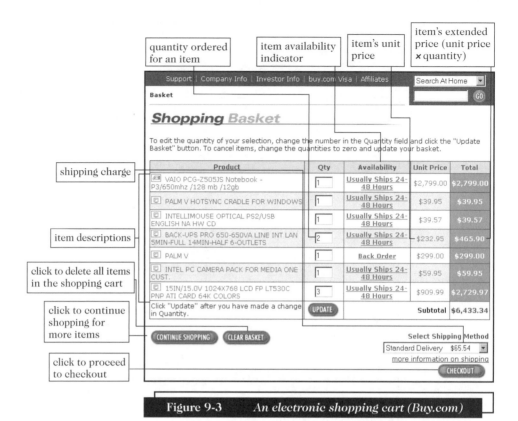

quantity ordered for an item

item availability indicator

item's unit price

item's extended price (unit price × quantity)

shipping charge

item descriptions

click to delete all items in the shopping cart

click to continue shopping for more items

click to proceed to checkout

Figure 9-3   *An electronic shopping cart (Buy.com)*

Clicking the "Checkout" button usually displays a screen that asks for billing and shipping information and that confirms the order. As you can see from the figure, the shopping cart software keeps a running total of each type of item. The shopping cart calculates a total as well as sales tax and shipping costs. Shopping cart software at some Web commerce sites allows the customer to fill a shopping cart with purchases, put the cart in virtual storage, and come back days later to confirm and pay for the purchases. A number of companies, including **BIZNET Internet Services**, **CartIt!**, **SalesCart**, and **WebGenie Software**, sell shopping cart software that sellers can add to their Web sites. These software packages range in price from a few hundred dollars to several thousand dollars, plus an ongoing monthly fee. The shopping cart software sold by SalesCart works with several different Web server software platforms, as shown in Figure 9-4.

shopping cart
software for
several different
Web server
platforms

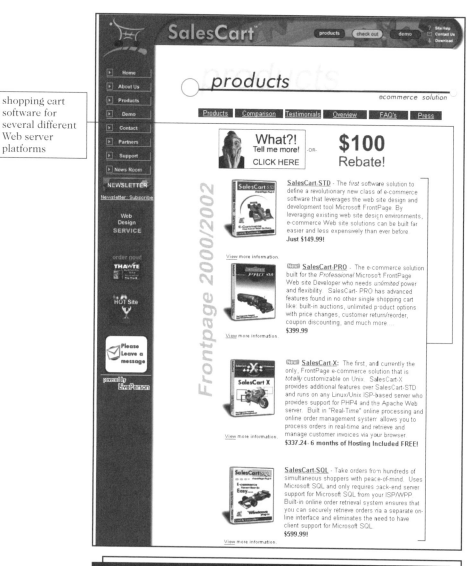

Figure 9-4 *SalesCart shopping cart software*

PDG SOFTWARE

PDG Software is a company based in Tucker, Georgia, that sells electronic commerce software to companies that operate small and midsized electronic commerce Web sites. PDG sells shopping cart software, auction software, shopping mall software and a number of other packages. Although it sells some of its software directly to the companies that use it, most of its sales are through resellers—firms that use PDG software as part of Web sites that they design, build, and deliver to customers as complete units.

In April 2001, a hacker discovered a vulnerability in the PDG software that allowed an intruder to enter the shopping cart and open the file that contained customer names, contact information, and credit card numbers. PDG developed a patch that would repair the software the same day it found out about the intrusions. PDG posted the patch on its Web site so that companies using the software could download and install the patch. Both PDG and the FBI issued press releases immediately to warn users of the problem with the shopping cart software and to encourage them to obtain the patch. Unfortunately, the users of the software that had purchased it as part of a complete electronic commerce Web site were, in many cases, unaware that their site included the PDG shopping cart software.

Because it took so long—several months, in some cases—to find and contact the companies using the software, online offenders had an excellent opportunity to exploit this vulnerability and collect thousands of credit card numbers. In most cases such as this, the difficulty of finding the sites that are running the vulnerable software helps slow down the attackers. Unfortunately, in this case, the intruder who discovered the opening also found that entering a specific word in a search engine's search expression would instantly return a list of the thousands of sites running the PDG software.

Most of the Web sites found out about the problem when their customers called them, suspicious because their credit card information had been compromised. The lesson from this failure is that companies that operate electronic commerce Web sites must know the source of the software used in creating and maintaining their sites and must monitor news about the security of that software.

Because the Web is a *stateless* system—unable to remember anything from one transmission or session to another—shopping cart information must be explicitly stored for the shopper to retrieve later. Furthermore, it must distinguish one shopper from another so that the purchases are not mixed up. One way to uniquely identify users and to store information about their choices is to create and store **cookies**, which, as you learned in earlier chapters, are bits of information stored on a client computer. When a customer returns to a site that issued a particular cookie, the shopping software reads either the cookie from the customer's computer or the database record from the merchant's server. If a shopper's browser does not allow storage of cookies, sites can use another way to preserve shopping cart information from one browser session to another. One popular electronic commerce software package, **ShopSite**, does this by automatically assigning a shopper a temporary number. The number is added to the end of the shopper's URL and persists as he or she navigates from one Web site to another. When the customer returns, the URL still contains the bits of information about his or her shopping cart. When the customer closes the browser, the temporary number is discarded and thus cannot be reused, even if the customer later reopens the browser and returns to the same Web site.

Transaction Processing

Transaction processing occurs when the shopper proceeds to the virtual checkout counter by clicking a checkout button. Then the electronic commerce software performs any necessary calculations, such as volume discounts, sales tax, and shipping costs. At checkout, the browser normally switches into a secure state of communication. Unless the user has disabled it, a dialog box appears indicating that the browser is entering or leaving a secure state. You will learn more about secure transmissions on the Internet in the next two chapters.

Transaction processing is the trickiest part of the electronic sale. Computing taxes and shipping costs are important parts of this process, and site administrators must continually check tax and shipping tables to make sure they are current. Some programs simplify shipping calculations by connecting directly to shipping companies to retrieve shipping costs. One site, **SmartShip**, determines which shipper offers the lowest price to deliver a package.

Transaction processing software must also handle additional details, such as tax-free sales. Many B2B transactions involve tax-free sales when items are purchased for resale or under other conditions. As you learned in Chapter 7, some B2C sales are also tax-free in some states. Other calculation complications include provisions for coupons, special promotions, and time-sensitive offers; for example, "purchase a round-trip ticket before the end of the month and receive a 50 percent discount". Some electronic commerce software provides connections to accounting software so that Web sales can be entered simultaneously in the company's internal accounting system. In most cases, however, the connections between the electronic commerce software and the accounting system are handled by a separate type of software called **middleware**. Some large companies that have sufficient IT staff write their own middleware; however, most companies purchase middleware that is customized for their business by the middleware vendor or by a consulting firm. Thus, most of the cost of middleware is not the software itself, but the consulting fees needed to make the software work in a given company. Making a company's information systems work together is called **interoperability** and is an important goal of companies when they install middleware. The total cost of a middleware implementation can range from $50,000 to several million dollars, depending on the complexity of the company's underlying operations and existing information systems. Major middleware vendors include **BEA Systems**, **Broadvision**, **Digital River**, and **IBM Tivoli Systems**. The BEA Systems Application Integration Web page appears in Figure 9-5.

337

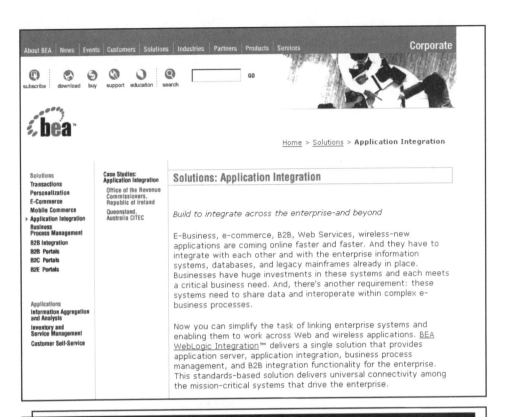

Figure 9-5   *BEA Systems Application Integration page*

Electronic Commerce Tools

Solutions for setting up shop online vary from inexpensive and simple to very expensive and complicated. The software options for setting up an online business fall into three categories: basic, midrange, and large. Each type of solution is described in detail in the remaining sections of this chapter; however, a brief overview of each is provided here.

At the low end of the electronic commerce software range are the very inexpensive storefronts that are offered by established portals, such as Yahoo!. Targeted at smaller stores, these systems offer hosting services and software that let merchants quickly create storefronts. Inexpensive, hosted electronic commerce stores provide the tools to quickly create a storefront. These Web stores include Web hosting for the site as well as tools for catalog creation, shopping carts, and transaction processing. Fees vary, with some sites providing online business capabilities for $100 per month or less. Other sites charge a monthly fee that is based on the number of items for sale. Some sites charge a setup fee of up to a few hundred dollars.

Midrange electronic commerce software is used by larger companies. These B2C and B2B systems have vastly different software and hardware requirements, requiring more transaction-handling power and page-presentation flexibility than the small storefront sites can offer.

Major online stores, such as Amazon.com, require electronic commerce software suites that run on large, dedicated computers and interact with database systems to display catalogs and process orders. These systems are expensive to create and operate and require a dedicated staff to oversee and maintain them.

Application Servers

A group of electronic software packages known as application servers give these companies the power and flexibility they need. An **application server** is software that takes the request messages received by the Web server and performs some kind of action based on the contents of these messages. The computer that runs application server software is often called an application server, also. The actions that the application server software performs are determined by the rules used in the business. These rules are called **business logic**. An example of a business rule is: When a customer logs in, check the password entered against the password file in the database.

Application server software is usually grouped into two types: page-based and component-based. A **page-based application server** returns pages generated by scripts that include the rules for presenting data on the Web page with the business logic. Common page-based server systems include Macromedia's ColdFusion, Java Server Pages (JSP), Microsoft's Active Server Pages (ASP), and PHP. For simpler systems, these page-based systems work quite well. Because they combine the page presentation logic with the business logic, however, they can be difficult to revise and update. Larger businesses often prefer to use a **component-based application server** approach that separates the presentation logic from the business logic. Each component of logic is created in its own module. This makes updating and changing elements of the system much easier—especially in large electronic commerce sites that are built and maintained by teams of programmers. The most common component-based systems in use today are **Enterprise JavaBeans** (EJBs), **Microsoft's Component Object Model** (COM), and the Object Management Group's **Common Object Request Broker Architecture** (CORBA).

339

Application server software licenses range in cost from $2000 to $35,000 per server. A number of companies sell products in this category, including Borland's AppServer, Hewlett-Packard's Total eServer, iPlanet's Application Server, IBM's WebSphere Application Server, Oracle's Oracle Application Server, and Sybase's EAServer. Microsoft's IIS Web server (which you learned about in Chapter 8) includes the ability to handle page-based application server duties. Many midsized companies that do not need the features of a component-based application server use Microsoft IIS to save the additional cost of separate application server software.

Application servers usually obtain the business logic information they use to build Web pages from databases. A **database manager** is software that stores information in a highly structured way. The structure of the database makes it easy for the database manager software to retrieve the information stored in the database. Smaller electronic commerce sites can use low-cost databases such as Microsoft Access. Larger sites need the power of more expensive database management software such as IBM DB2, Microsoft SQL Server, or Oracle. These database management software packages can be quite expensive. Typical installations cost between $10,000 and $200,000. Companies with very large databases that have operations in many locations must make their data available to users in those locations. Large information systems that store the same data in many different physical locations are called **distributed information systems** and the databases within those systems are called **distributed database systems**. The complexity of these systems leads to their high cost. Most companies that can afford it do use name brand database products; however, a number of companies are beginning to use mySQL, which was developed and is maintained by a community of programmers on the Web. Similar to the Linux operating system you learned about in earlier chapters, mySQL is open source software that can be downloaded and used at no cost. The term **open source** is used to describe such software because the "source code" of the software is freely available, or "open."

Larger software packages are geared toward firms conducting B2B electronic commerce over the Internet. These firms have extranets and intranets requiring tools and capabilities different from those needed to implement most B2C Web sites. In the case of B2B, both the buyer and seller systems are complex. Thus, commerce software to support B2B must similarly be more sophisticated than that for B2C. B2B systems, for example, usually require security tools not standard in business-to-consumer systems, such as encryption and authentication, as well as signed receipt notices.

Many B2B commerce systems must be able to connect to existing information systems such as enterprise resource planning software. **Enterprise resource planning (ERP)** software packages are business systems that integrate all facets of a business, including accounting, logistics, manufacturing, marketing, planning, project management, and treasury functions. The major ERP vendors include **Baan**, **J.D. Edwards**, **Oracle**, **PeopleSoft**, and **SAP**. A typical installation of ERP software costs between $2 million and $25 million, thus companies that are already running these systems usually want their electronic commerce sites to integrate with them. Figure 9-6 shows a typical architecture for a B2B application server that connects the existing information systems of several trading partners, including EDI and ERP systems.

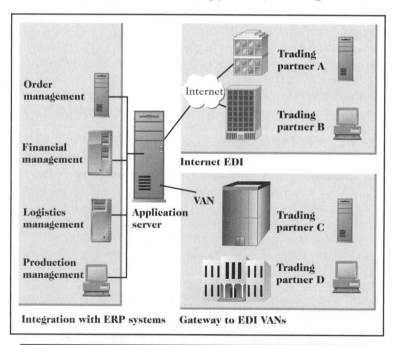

Figure 9-6   *Business-to-business commerce server topology*

With the exception of the very smallest stores, companies must carefully consider the features of commerce software packages before choosing one to test and use to run an online business. An important feature that most firms will need to consider is database support. Most Web storefronts operate with a database that stores product information, including size, color, type, and price details for the many products that midsized and larger sites sell. Usually, the database that serves an online store is the same one that is used by the existing corporate clients. It is better to have one database serving two separate entities because it eliminates parallel but distinct databases—something companies should avoid if possible. If no one at the company has database expertise, then the personnel to handle that part of the Web business must be hired, or Web products that do not use database backends must be used. If a company has existing inventory and product databases, then Web commerce software that does not support these systems should not be considered.

Content Management Tools

Most commerce software comes with wizards and other automated helpers that create template-driven pages, such as the home page, about pages, and contact pages. But most businesses want to customize Web pages with company and product pictures and text. Because someone at the company will have to perform Web site maintenance—for example, adding new products to the catalog—the software should be tested before committing to it to make sure that adding new categories of products and new items to an existing product is straightforward. For instance, creating sale item specials and displaying end-of-quarter sale items should not be difficult or time-consuming tasks.

A number of companies offer content management tools. Some of these tools are extensions of the Web site creation tools you learned about earlier in this book, such as **Macromedia Spectra**, which works with Macromedia's Web site creation tools Dreamweaver and Dreamweaver UltraDev.

COMMERCE HOSTING SERVICES

As you learned in Chapter 8, some large companies maintain their Web servers in-house. These companies are also likely to maintain their application servers in-house. For small and midsized companies, an attractive alternative is a commerce hosting service. Recall from Chapter 8 that commerce service providers (CSPs), which are also called application service providers (ASPs), provide a connection to the Internet just as ISPs do. In addition, CSPs (or ASPs) provide application server software, database management software, and electronic commerce expertise. Also recall that some people use "ISP" as a generic term to refer to all service providers, including CSPs (or ASPs).

CSPs have the same advantages that ISP hosting services offer, including spreading the cost of a large Web site over several "renters" hosted by the service. The biggest single advantage—low cost—occurs because the host provider has already purchased the server and configured it. The host provider has to worry about keeping it working through lightning storms and power outages. In all cases, look for Web hosts that offer reliability, good security, system simplicity, and widespread visibility.

ValueWeb, operating since 1996, is an example of a CSP. ValueWeb offers businesses comprehensive electronic commerce hosting services including shared hosting, dedicated hosting, and co-location services. Hosting more than 65,000 Web sites for customers in over 136 countries, ValueWeb has become one of the largest Web hosting companies in the world (see Figure 9-7).

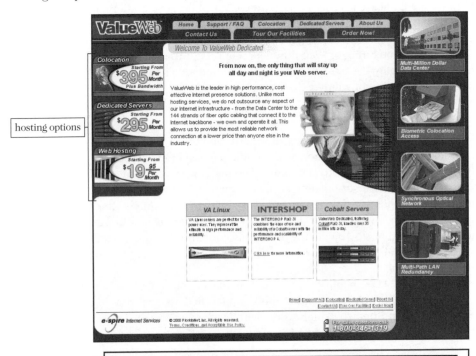

hosting options

Figure 9-7   *ValueWeb offers several Web hosting packages*

One of these general hosting arrangement types is a good fit for most types of online organizations. Cost, easy access, and levels of service determine the type of hosting a business will choose. Some of these are described further in this section. **TopHosts.com** (see Figure 9-8) features a comprehensive presentation about CSPs and hosting issues. This site contains hundreds of links and much good information.

Figure 9-8   *TopHosts.com home page*

BASIC ELECTRONIC COMMERCE SOFTWARE PACKAGES

Basic packages are free or low-cost electronic commerce software supplied by the Web host for building electronic commerce sites that will be kept on the host's server. Services in this category usually cost less than a few hundred dollars per month, and the software is available on the host site, allowing companies to immediately begin building and storing a storefront on the host's server. Many good packages fall into this basic packages group. This section first examines the low-cost host sites offering very fundamental features. Next, full-service mall-style host facilities are covered. Finally, operating expenses are reviewed.

Basic Host Services

Several Web hosting services offer free or low-cost commerce services designed for small online businesses selling only a small number of items (fewer than 50) and having relatively low transaction rates (fewer than 20 per day). Hosting services in this class offer adequate space for a Web store and forms-based shopping, but do not provide shopping carts. Often, they do not include transaction processing. In most cases, the hosting service will determine the domain name that can be used by the merchant. The free sites in this group can offer a great value for Web store first-timers unsure about investing in electronic commerce. The host makes money from advertising banners that are placed on the storefront's Web pages.

Typical examples of these types of hosts are **BizLand.com** and **HyperMart**. HyperMart's site contains useful tutorials and information about application server software, databases, and other electronic commerce site management topics. It allows use of any HTML editor and offers a turnkey, browser-based Web creation package to make creating store pages straightforward. BizLand.com offers an impressive list of tools for the small Web entrepreneur, including support for Microsoft FrontPage, a unique domain name, analysis reports providing statistics about user site visits, and registration with the major search engines. Figure 9-9 shows the HyperMart home page.

The host sites in this category all provide templates for constructing a commerce site. Everything is template driven, and most users can put together a site in 30 minutes or less. One major drawback of host sites in this group is that customer purchase transactions are handled by e-mail. When a customer orders items from a store, the hosting service groups and e-mails the orders to the merchant, who must then fill the orders and process the payment, including dealing with credit cards, checks, or money orders.

Another drawback is the banner advertising that appears on merchant sites. Because banner advertisements defray the costs of merchants' sites, they cannot be removed. However, for a monthly fee, storefronts can be banner-free.

The set of tools that is provided by basic hosting sites is the final drawback of such hosting arrangements. The basic tools are more difficult to use and lack the variety of styles and looks available with more expensive Web hosting services.

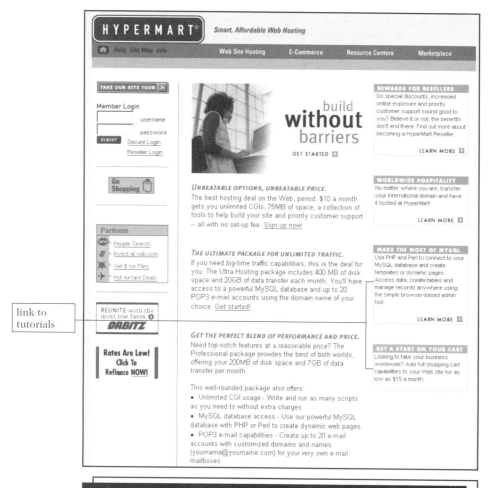

link to tutorials

Figure 9-9   *HyperMart home page*

Shared Mall-Style Hosting

Shared hosting sites provide online stores with good service, good Web creation tools, and little or no banner advertising clutter. Web hosts in this group charge a monthly fee, which is often higher than that of lower-end providers, and may also charge one-time setup fees as well as a percentage of or fixed amount for each customer transaction. These Web hosts also provide high-quality tools, storefront templates, an easy-to-use interface, and quick Web page generation capabilities and page maintenance.

Several significant differences exist between free Web host sites, described in the previous section, and these host sites. Mall-style hosts provide shopping cart software or the ability to use another vendor's shopping cart software. They also furnish comprehensive customer transaction processing through one of a few merchant

345

services. Such a service allows customers to choose to purchase their goods and services with a credit card, electronic cash, or other form of payment. The Web host processes the acceptance and authorization of credit cards on behalf of the merchant. Another benefit is that because they are paying a monthly fee to the Web hosting service, sites do not have to display any Web banners, which can sometimes be unattractive and distracting. The fourth benefit of the mall-style Web hosts is that they provide much higher quality Web store building and maintenance tools than do the free sites.

Quality Web hosting services in this category include **Bigstep**, **Yahoo! Store**, **FedEx eCommerce Builder**, and **eBay Stores**. These four are representative of most of the mall-style Web hosts and are quite popular. Some list a merchant's site in a directory listing on the host Web's opening page. The Yahoo! directory, for example, categorizes all the storefronts it hosts into fewer than a dozen major groups (Gifts and Collectibles, for example) and then divides each group into three to five subgroups, such as Arts and Crafts.

You can learn how capable these Web hosting services are by trying them. Yahoo!, for instance, allows a free 10-day trial of its hosting service and store-builder software. Bigstep has a 30-day trial and the FedEx eCommerce Builder service has a 60-day trial period. The eBay Stores service does not offer a free trial, but the charges are minimal and are based on the number of products listed and graphic images used. Creating an eBay test store with only a few items would cost less than $20. The Yahoo! and Bigstep hosts are described in the following sections.

Yahoo! Store

Yahoo! Store is the most expensive of this group of Web hosts. It serves as the business Web host not only for hundreds of small storefronts, but also for big names, such as the Kennedy Space Center Space Shop, The Sharper Image, PalmPilotGear, *Rolling Stone* magazine, and Cirque du Soleil.

Merchants can create, change, and maintain their Yahoo! storefronts through a Web browser. Yahoo! holds all the stores' pages in a proprietary format on its own site. Yahoo! processes merchant transactions on a secure server. When a merchant signs up for a Yahoo!-hosted storefront, its URL is a subdomain of the Yahoo! URL. For example, if a business's username is BoldImpressions, then the URL that customers visit is *http://store.yahoo.com/BoldImpressions*, a subdomain of Yahoo!. A different URL takes the merchant to the "employee's entrance" when it is time to make changes to the store. That URL is *http://store.yahoo.com*. Once a merchant logs on with a store owner username and password, he or she can edit and manage the store. Merchants can use Manager, a Yahoo!-supplied software product, to examine statistical information, set up acceptable payment methods, select preferred customer shipping methods, and change global site settings.

Bigstep

Bigstep has received many industry awards, including *PC Magazine*'s Editor's Choice award in 1999. Bigstep provides a well-designed storefront package and charges low monthly fees. Bigstep lets merchants build a test store for 30 days before charging for the service.

Similar to Yahoo! Store, Bigstep enables merchants to create, change, and maintain a storefront through a Web browser. Bigstep provides a comprehensive set of page building tools and a series of wizards that provide assistance.

To create a store, merchants must register with Bigstep. The registration process identifies the user with an e-mail address and password. If the merchant wants to build more than one storefront, he or she must use different e-mail addresses. New merchants can select any store name that is not already taken. As with Yahoo!, the store is a subdomain of Bigstep. After logging on with a store owner e-mail address and password, the merchant can create and manage the electronic store.

During the store editing process, Bigstep displays seven tabs corresponding to the sections of the store that define various aspects of the store's catalog, reporting, and utility functions. The tabs are Home, Site Building, Communication, Catalog, Commerce, Marketing, and Reporting.

Bigstep's reports provide data mining capabilities that search through site data collected in log files. **Data mining**—looking for hidden patterns in data—can help businesses find customers with common interests and discover previously unknown relationships among the data. Reports can indicate problematic pages in a store's design where, for example, a large number of customers get stuck and then leave the Web site. Other facts that Bigstep reports can reveal include the number of pages an average customer must load and display before locating the merchandise he or she wants. If customers have to load too many pages (three to seven pages is usually considered excessive), they often become impatient and leave without making a purchase. In other words, available reports can answer the following questions:

- How many visitors are coming to the site?
- What is the average length of stay for each visitor on each page?
- Which pages lead to actual purchases?
- What advertisements or links have brought qualified visitors to the page or site?
- What is the average number of pages that each visitor views?
- Are repeat customers attracted to the site?

If an electronic commerce site complements a merchant's brick-and-mortar store, Bigstep's built-in map locator can display the location of the store. The map locator uses Mapquest technology to map the location in the Map and Directions page of the Site Builder section based on the address supplied by the merchant. Additional Bigstep store features include automatic calculation of taxes and shipping, collection of customer data, merchant e-mail notification of sales, and customer e-mail confirmation when products ship.

Estimated Operating Expenses

The following table breaks down electronic commerce costs that a small entrepreneur who wants to put a store on the Web and offer, for example, 50 items for sale can expect to incur during the first year of business. The total omits transaction charges, which might average 50 cents per transaction and 2.0 percent of each sale's total.

Operating Costs	Cost Estimate
Initial site setup fee	$200
Annual maintenance fee (12 × $100)	1200
Domain name registration	70
Scanner for photo conversion or digital camera	500
Photo editing software	100
Occasional HTML and design help	400
Merchant credit card setup fee	200
Total first-year cost	**$2670**

The preceding costs are typical, but they can vary because different Web hosting sites charge varying fees for various services. Use these numbers as a guideline. Additional transaction processing fees can run into hundreds and thousands of dollars, but those fees occur only when a site makes sales. A good guideline for processing fees is to multiply expected annual gross sales by 3 percent. So, if a site's annual gross sales are $50,000, then the transaction fees should be approximately $1500. That value will adequately cover both the per-transaction fixed costs and the percentage of total sales costs charged by most merchant credit card processing agencies.

Contrast the preceding costs with comparable estimated costs for self-hosting a Web site. Setup and Web site maintenance costs include equipment, communications, physical location, and staff. Equipment—a server and networking gear—has a one-time cost ranging from $5000 to $15,000. A T1 connection or fraction thereof (see Chapter 2) costs from $8000 to $12,000 per year. A server must be housed in a room that is both secure and convenient to communications access. The cost to secure a room, properly air-condition it, and install a chemical fire extinguishing system can be nearly $5000 a year. A self-hosted system requires a staff of experts in a variety of Web-related languages, electronic commerce packages, and database management systems. Other technicians will likely be required. Staff costs are minimally between $50,000 and $100,000 annually. In total, annual operating costs for self-hosting approach $60,000 to $100,000 or more the first year. Costs for subsequent years will be about the same. Carefully compare self-host cost estimates with the costs charged by various hosting services.

Next, we will examine midrange electronic commerce packages. Midrange packages are suitable for running larger businesses and are more robust (and more expensive) than the Web hosting, template-driven packages described previously.

MIDRANGE PACKAGES

The line between basic electronic commerce software and midrange electronic commerce software can be blurry; however, midrange electronic commerce packages distinguish themselves in several clear ways. Midrange packages allow the merchant to have explicit control over merchandising choices, site layout, internal architecture, and remote and local management options. In addition, the midrange and basic electronic commerce packages differ on price, capability, database connectivity, software portability, software customization tools, and computer expertise required of the merchant.

Buying and using midrange electronic commerce software is significantly more expensive than using a hosted site, with prices ranging from $2000 to $50,000. Although not cheap, midrange software is far from expensive compared to high-end, enterprise-class systems that can cost millions of dollars. The midrange commerce packages provide more options and operate much more efficiently and capably than their low-end cousins. Midrange software traditionally has connectivity with sophisticated database systems and store catalog information. Having the catalog stored in a database simplifies product maintenance. Several of the midrange systems provide connections, sometimes called "hooks," into existing inventory and ERP systems. This can yield savings because there is no need to run duplicate inventory systems, and the cost of the existing systems is spread across several software systems.

All the midrange systems, including those described in this section, are hosted on the merchant's own computer. As with Web server software, not all electronic commerce software runs on every computer and operating system. Commerce systems, Web servers, and computer hardware must be matched carefully. Interestingly, some of these midrange packages can be turned into Web hosting packages that can serve many small storefronts, allowing a site to recapture some of its investment costs. Midrange systems are all highly customizable, allowing many elements to be altered, from the number of templates available to the type of credit cards and shipping combinations permitted. Perhaps the biggest difference between a typical midrange package and a basic electronic commerce system is that a midrange package usually requires part-time or full-time programming talent. In addition, expert advice may be needed to extend the package beyond its standard settings and capabilities.

Three midrange electronic commerce systems are described in this section. They are representative of the whole group, yet are different from one another in important ways. The systems are Intershop enfinity by Intershop Communications Inc., WebSphere Commerce Suite by IBM, and Commerce Server 2000 by Microsoft.

Intershop Enfinity

Intershop Enfinity, produced by Communications Inc., provides search and catalog capabilities, electronic shopping carts, online credit card transaction processing, and the ability to connect to existing backend business systems and databases. CSPs can provide Intershop's business hosting, storefront building, and management tools in much the same way as Yahoo! Store provides tools for its electronic commerce mall customers.

Intershop Enfinity has setup wizards and good catalog and data management tools. It provides many storefront templates. Management and editing of a storefront are done through a Web browser—either locally at the server or remotely through any Internet connection. Intershop's inventory manager tracks inventory levels and allows merchants to view the quantity of items available, create a list of inventory transactions, and enter new products into the inventory. Discount rules are also easy to enter. Merchants define the business rules for a discount and dates during which special discounts apply. Bundled with the software is a database management system. Alternatively, Intershop's access to DB2 (IBM's relational database) or Oracle can be used. If a merchant owns several online stores, then each store requires a separate database. Moving product information from the existing corporate databases into Intershop database files is straightforward using Intershop's Data Import wizard. Intershop contains all the standard features of commerce software, including a shopping cart, the automatic calculation of shipping charges, and local and state sales taxes. Customers receive e-mail to verify their orders. Intershop supports secure transactions. A wide variety of site and customer reports are available to track Web page visits and customer activities. Figure 9-10 shows an Intershop enfinity diagram illustrating how the product connects to other parts of an enterprise's systems and serves customers.

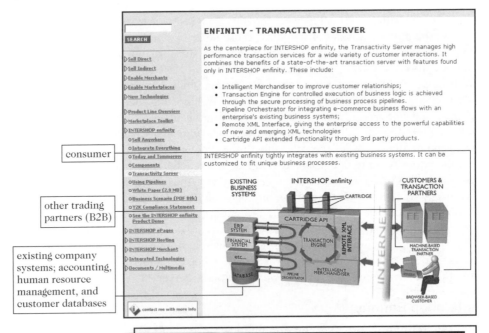

Figure 9-10 *Intershop Enfinity diagram*

WebSphere Commerce Suite

IBM produces the **WebSphere Commerce Suite**, which is a family of electronic commerce packages. IBM WebSphere Commerce Suite is a set of software components that provides software suitable for medium-sized to large businesses to sell goods and services on the Internet. It includes catalog templates, setup wizards, and advanced catalog tools to help companies create attractive and efficient electronic commerce sites. WebSphere Commerce Suite can be used both for business-to-business and business-to-commerce applications and provides a smooth connection to existing corporate systems such as inventory databases and procurement systems.

WebSphere Commerce Suite, Start Edition is available for Microsoft Windows NT and 2000 server operating systems. Merchants can begin with a small store and then move up to a bigger, more capable store as necessary. A wizard leads the merchant through the process of creating a starter store. Once that is up and working, more functionality can be added by executing commands and writing code. With the basic pages built, the catalog can be populated with products, prices, and product pictures. The WebSphere Commerce Suite also accommodates electronic download products, such as audio tracks or software.

For a more complete storefront, WebSphere offers a very large collection of functions, utility programs, and commands that allow further site customization. However, JavaScript, Java, or C++ expertise is required. Typical of commerce programs in this class, WebSphere can connect to existing databases and other legacy systems through DB2 or Oracle databases. A single store or several different stores can be administered from the same browser-based interface.

The system has all the standard electronic commerce features, including tools for a shopping cart, e-mail notifications upon sale completion, secure transaction support, promotions and discounting, shipment tracking, links to legacy accounting systems, and browser-based local and remote administration.

IBM makes a scaled-up version of the program for enterprise systems called WebSphere Commerce Suite, Pro Edition. WebSphere Commerce Suite, Pro Edition is available for the IBM S/390 platform, replacing IBM Net.Commerce and running on a variety of operating systems for larger systems, including IBM AIX, IBM AS/400, Windows NT, and Sun Solaris. Figure 9-11 shows the **WebSphere Commerce Suite, Start Edition** home page.

351

Figure 9-11   *WebSphere Commerce Suite, Start Edition home page*

Commerce Server 2000

Microsoft's **Commerce Server 2000** allows businesses to sell products or services on the Web using tools such as user profiling and management, transaction processing, product and service management, and target audience marketing. Commerce Server 2000 is not an out-of-the-box solution. Wizards help users build a site in several steps, but customized code is required to fit specific user needs. The Microsoft Visual InterDev tools, bundled with Commerce Server 2000, allow merchants to customize the sites they build.

Like other electronic commerce systems, Commerce Server 2000 has tools for the commerce cycles of engaging the customer (through marketing and advertising), transacting an order, and analyzing the sales information after the sale. Commerce Server 2000 also has powerful tools for advertising, promotions, cross-selling, and customer targeting and personalization.

Commerce Server 2000 provides many predefined reports for analyzing site activities and product sales data. Commerce Server 2000 can grow with increasing business demands. The system provides the standard electronic commerce capabilities, including several storefront templates, wizards for setting up and initializing a store, and database connections. In addition, Commerce Server 2000 provides a shopping cart, verifies completed sales transactions by e-mail, and supports secure transactions. It can connect to existing accounting systems, and the administrator can oversee the site through a Web browser. Commerce Server 2000's home page is shown in Figure 9-12.

Figure 9-12   *Microsoft Commerce Server 2000 home page*

SOLUTIONS FOR LARGE FIRMS

The distinction between midrange and large-scale electronic commerce software is much clearer than the one between basic systems and midrange systems. The telltale sign is price. Other elements, such as extensive support for business-to-business commerce, also indicate that the software is in this category. Commerce software in this class is sometimes called **e-business** (for electronic business) **software** or enterprise-class software. The term "enterprise" is used in information systems to describe a system that serves multiple locations or divisions of one company and encompasses all areas of the business or "enterprise." E-business software provides tools for both B2B and B2C commerce. In addition, e-business software interacts with a wide variety of existing systems, including database, accounting, and ERP systems. As electronic commerce has become more sophisticated, large companies have demanded that their Web sites and supporting information infrastructure do more things. The cost of these enterprise systems for large companies ranges from $200,000 for more basic systems to $10 million and more for encompassing solutions.

As you learned in Chapter 4, companies are storing data about site visitors in large databases and analyzing it to improve their relationships with those customers. Often called **clickstream**, such data tracks the path a visitor takes through a Web site, including which pages were viewed, the amount of time spent on each page, and the sequence in which pages were viewed. Thus, large electronic commerce sites must include customer relationship management software. In Chapter 5, you learned how companies are using the Web to integrate their supply chains. As a result, enterprise-class commerce Web sites must include or work with supply chain management software. In Chapter 6, you learned about companies that were building business portal sites to engage their customers and suppliers. A significant part of that strategy is providing useful, fresh content to attract site visitors to the portal. This need has given rise to software that automatically manages and rotates content on Web sites. An enterprise-class Web site will often include several of these types of software packages in its design.

E-business software running large online organizations requires one or more dedicated computers—in addition to the customary Web server front-end system and any necessary firewalls. An enterprise-scale solution also requires a Domain Name Server (DNS), an SMTP system to handle electronic mail, an HTTP server (Web front-end), an FTP server for upload and download capability, and a database server. Examples of e-business systems capable of running a large online company with high transaction rates include IBM's **WebSphere Commerce Suite, Pro Edition**, **Oracle E-Business Suite**, **Broadvision One-To-One Enterprise**, and Openmarket's **Transact**.

E-business software typically provides good tools for linking to and supporting supply and purchasing activities. A large part of B2B commerce is ordering supplies from trading or business partners and issuing the appropriate documents, such as purchase orders. For a selling business, e-business software provides standard electronic commerce activities, such as secure transaction processing and fulfillment, but it can also do more. For instance, it can interact with the firm's inventory system and make the proper adjustments to stock, issue purchase orders for needed supplies when they reach a critically low point, and generate other accounting entries in ERP, legacy accounting, or file systems. In contrast, both basic and midrange electronic commerce packages usually require an administrator to manually check inventory and explicitly place orders for items that need to be replenished.

In B2C situations, customers use their Web browsers to locate and browse a company's catalog. For electronic goods (software, research papers, music tracks, and so on), customers can download the items directly from the site or they can choose to complete order forms and have the hard copy versions of the products shipped to them. The Web server is linked to backend systems, including a database management system, a merchant server, and an application server. The database usually contains millions of rows of information about products, prices, inventory, user profiles, and user purchasing history. The history provides a way to recommend to a user on a return visit related items that he or she might wish to purchase. A merchant server houses the e-business system and key back office components. It processes payments, computes shipping and taxes, and sends a message to the fulfillment department when it must ship hard goods to a purchaser. Figure 9-13 shows a typical e-business system architecture.

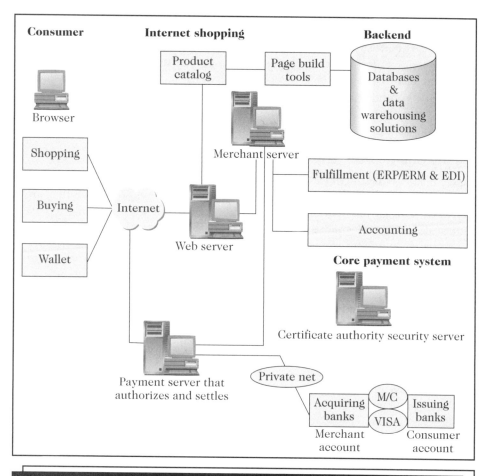

Consumer

Browser

Shopping

Buying

Wallet

Internet

Internet shopping

Product catalog

Page build tools

Merchant server

Web server

Backend

Databases & data warehousing solutions

Fulfillment (ERP/ERM & EDI)

Accounting

Core payment system

Certificate authority security server

Payment server that authorizes and settles

Private net

Acquiring banks

M/C

VISA

Issuing banks

Merchant account

Consumer account

Figure 9-13 *Typical, large e-business system architecture*

Customer Relationship Management Software

You learned about customer relationship management (CRM) in Chapter 4. The goal of CRM is to understand each customer's specific needs and then customize a product or service so that it meets those needs. The idea is that a customer whose needs are being met exactly is willing to pay more for the goods or services that are meeting those needs.

CRM software must obtain data from operations software that conducts activities such as sales automation, customer service center operations, and marketing campaigns. The software must also gather data about customer activities on the company's Web site and any other points of contact the company has with its customers and potential customers. CRM software uses this data to help managers conduct analytical activities such as gathering business intelligence, planning marketing strategies, customer behavior modeling, and customizing the products and services to meet the needs of specific customers or categories of customers.

Some companies create their own CRM software using outside consultants and their own IT staff. More companies are likely to buy a CRM software package. Siebel Systems' **E-Business 7** is one of the leading comprehensive CRM software packages available today. Siebel was the first company to specialize in CRM software and it has a large share of the market. Other major software firms have created products that compete with Siebel's including **Oracle Global CRM**, **PeopleSoft CRM**, and **MySAP CRM**. Prices for these systems start around $200,000 (on average, about $5000 per user).

Supply Chain Management Software

Supply chain management (SCM) software helps companies to coordinate planning and operations with their partners in the industry supply chains of which they are members. SCM software performs two general types of functions, planning and execution. Most companies that sell SCM software offer products that include both components, but the functions are quite different. SCM planning software helps companies develop coordinated demand forecasts using information from each participant in the supply chain. SCM execution software helps with tasks such as warehouse and transportation management. The two major firms offering SCM software are **i2 Technologies** and **Manugistics**.

The i2 Technologies product, RHYTHM, includes components that manage demand planning, supply planning, and demand fulfillment. The demand planning module includes proprietary algorithms customized for specific industry markets that examine customers' buying patterns and generate continuously updated forecasts. The supply planning module coordinates distribution logistics, inventory level forecasting, collaborative procurement, and supply allocations. The demand fulfillment module handles the execution elements, including order management, customer verification, backlog control, and order fulfillment.

The Manugistics SCM product includes a constraint-based master planning module that controls the other elements of the system. These other elements include modules for transportation management, replenishment management, manufacturing planning, scheduling, purchase planning, and materials control.

The cost of SCM software implementations varies tremendously depending on how many locations (retail stores, wholesale warehouses, distribution centers, and manufacturing plants) are in the supply chain. For example, a retailer with 500 stores might pay between $4 million and $10 million for an SCM package that includes both planning and execution functions, but a wholesaler with only three or four distribution centers might be able to install a good SCM product for $1 million.

Content Management Software

More large companies are finding new ways to use the Web to share information among their employees, customers, suppliers, and partners. Content management software helps companies control the large amounts of text, graphics, and media files that have become a key part of doing business. With the rise of wireless devices such as mobile phones, handheld computers, and personal digital assistants (PDAs), content management has become even more important.

Companies that need many different ways to access corporate information—for example, product specifications, drawings, photographs, or lab test results—often choose to manage the information and access to that information using content management software. The three leading companies that provide these tools are Documentum, Vignette, and WebMethods. Content management software costs between $200,000 and $500,000, but it can cost three or four times that much to customize, configure, and implement. Figure 9-14 shows the Documentum site.

Figure 9-14 *Documentum content management Web site*

357

Summary

In this chapter, you learned about electronic commerce software for small, medium-sized, and large businesses and the functions provided by each software type. The electronic commerce software a company chooses depends on its size, objectives, and budget, and requires making some major decisions. A company must first choose between paying a service provider to host the site or self host.

Small enterprises that are not sure whether electronic commerce is a good idea should consider using a service provider. Basic hosting services for small businesses provide a range of standard features, including tools for quickly creating storefronts, catalogs, and transaction processing. These packages are usually wizard- and template-driven.

If a company already has computing equipment and staff in place, purchasing a midrange package will provide more control over the site than basic hosting services. Midrange packages run on the merchant's server and interact with databases, allowing for dynamic catalogs, shopping carts, and order processing. These components should be tested for compatibility with a company's existing software.

Large enterprises that have high transaction rates, B2B partnerships, or a large investment in ERP and other backend systems will need to invest in larger, customizable systems that can provide security features. These packages can incorporate customer relationship management, supply chain management, and content management capabilities.

Key Terms

Application server

Business logic

Catalog

Clickstream

Component-based application server

Cookies

Data mining

Database manager

Distributed database systems

Distributed information systems

Dynamic catalog

E-business software

Enterprise resource planning (ERP)

Interoperability

Middleware

Open source

Page-based application server

Shopping cart

Static catalog

Transaction processing

Review Questions

1. List and describe the basic tasks that an electronic commerce site must accomplish.

2. How can the Internet, a stateless system, keep track of a person's preferences between browsing sessions? Are there any security restrictions that a browser could impose that would force a change in the way information about users is recorded? Explain one way to record information about a user that does not require accessing the user's computer.

3. List two disadvantages of hosting your commerce site on a host that is free or available at a very low cost. What is missing from such a host's services that would make your job as an online entrepreneur more difficult?

4. Describe in a paragraph the differences between basic electronic commerce software and midrange electronic commerce software. Discuss at least four differences and give examples of each type of software.

5. What are the characteristics of large firms conducting both business-to-business and business-to-consumer transactions that require more robust and capable electronic commerce systems? Consider the volume and types of transactions and store maintenance activities that differ between a small storefront operation and, for example, an Amazon.com-caliber store.

Exercises

1. FedEx eCommerce Builder allows you to create and save a storefront free for 60 days. You are to create a small store using **FedEx eCommerce Builder** as your Web host. When you have completed building your store, visit the store as a customer and print several of your store's Web pages. Hand in to your instructor your printed storefront Web pages. Your store should include at least eight products in at least three categories.

2. Annette Jackson owns a small crafts store in central Missouri. She wants to expand her store's reach outside the region to increase her profits and simultaneously reduce her inventory. Annette has been watching her teenage daughter, Kelly, use the Internet to order music CDs and books. After learning from Kelly how simple it is to order from online stores, Annette decided that she needs to create an online store. She has asked you to do a little research on how much it might cost in the first year to create a simple store selling 100 items.

   Annette wants you to investigate three CSPs and report back to her what you have found. Because her store is small, limit your research to fundamental host services and full-service hosting, and ignore larger electronic commerce software packages. You may want to begin your research with sites such as **Freemerchant.com** or **Bigstep**. Here's what Annette wants to know about the three host systems you examine:

   - What are the costs: initial setup fee, monthly fee, and transaction fees?
   - How much disk space does her 100-item store receive?
   - Does the CSP provide a search engine?
   - What promotion and marketing opportunities does it provide?
   - How are customers notified following a transaction?
   - Does the CSP provide a shopping cart? If not, how do customers enter orders?
   - Are storefront-building wizards available to help create a new store?
   - Are the transactions conducted in a secure environment?
   - Does a store get its own domain name? Is there an extra charge for a domain name and its registration?
   - Can you upload product names, descriptions, images, and costs from files or databases, or must you enter each item manually?
   - Does the CSP provide some sort of online user manual for the merchant?

   Produce a report of no more than 1000 words describing what you have learned about the CSPs. Clearly label each section, grouping your information by service provider.

3. Investigate and write a 400-word report summarizing the costs and features of any enterprise-class commerce package for large businesses. Pick 10 or fewer characteristics about the commerce package and describe them in some detail. Begin your research using the links provided in the Online Companion. You may want to use a good search engine such as **AltaVista**, **Google**, **Hotbot**, or others listed under Exercise 3 in the Online Companion.

For Further Study and Research

Apicello, M. 2000. "Internet-Based Procurement Ups Profits," *InfoWorld*, 22(46), 80.

Atanasov, M. 2001, "The ASP Trap," *Ziff Davis Smart Business*, 14(7), July, 58–63.

Buss, D. 2000. "The Big 'Vortal' Payoff," *Internet World*, April 15, 35.

Callaghan, D. 2000. "Keeping E-Carts Moving," *eWeek*, 17(20), May 15, 44.

Callaghan, D. 2000. "Web Tools Track Site Users' Actions," *eWeek*, 17(22), May 29, 43.

Caulfield, B., S. Finch, M. Maier, D. Orenstein, R. Tate, and O. Thomas. 2002. "The Who's Who of E-Business," *Business 2.0*, January. Available online at: (http://www.business2.com/articles/mag/0,1640,35691|2,FF.html).

Derfler, F. 2000. "Finding a Hassle-Free Host," *PC Magazine*, 19(19), November 7, 135–154.

Desai, G., E. Sanchez, and J. Fenner. 2001. "Web Application Servers Come of Age," *Network Computing*, 12(15), July 23, 63–71.

Doherty, S. 2001. "IBM WebSphere 4.0.1 Covers All the Bases for Development and Administration," *Network Computing*, 12(21), October 15, 32–33.

Dragan, R. 2001. "E-Store and More for Small Businesses," *PC Magazine*, 20(19), November 13, 47.

Duvall, M. 2001. "Bigstep Faces Big Rivals," *Interactive Week*, August 29. Available online at: (http://www.zdnet.com/intweek/stories/news/0,4164,2808573,00.html).

Dyck, T. 2001. "Web Server Brains," *PC Magazine*, 20(10), May 22, 124–132.

Fingar, P. 2002. "Web Services Among Peers," *Internet World*, January, 21.

Gambhir, S. and M. Muchmore, 2001. "Secrets," *PC Magazine*, 20(19), 130–131.

Ginsburg, L. and J. Pusedu. 2000. "Setting Up E-Shop," *WebTools*, November 22. Available online at: (http://www.webtools.com/story/ecommerce/TLS20001122S0001).

Gladwin, L. 2001. "Covisint Focuses on Tech Integration," *Computerworld*, 35(27), July 2, 10.

Graven, M. 2000. "e-Store Solutions," *PC Magazine*, 19(18), October 17, 146–167.

Herel, H. 2000. "E-Tailing: More than a Catalog," *PC Magazine*, 19(13), July, 147.

Information Security. 2001. "Crackers Steal Credit Card Data from Shopping Carts," May, 34.

King, J. 2000. "Companies Aren't Rushing to Conduct Business Online," *eWeek*, 17(13), March 27, 20.

King, J. 2000. "Filling Orders a Hot e-Business," *Computerworld*, 34(24), June 12, 3.

King, N. 1999. "To Host or Be Hosted?," *Internet World*, 10(10), October 15, 64–66.

Kumar, K. 2001. "Technology for Supporting Supply Chain Management," *Communications of the ACM*, 44(6), June, 6–9.

Lorek, L. 2001. "Open for Business," *Interactive Week*, 8(33), August 27, 29–30.

Maamar, Z., E. Dorion, and C. Daigle. 2001. "Toward Virtual Marketplaces for E-Commerce Support," *Communications of the ACM*, 44(12), December, 35–38.

McCright, J. and R. Ferguson. 2001. "Deals Will Add Hosted Apps to i2 Lineup," *eWeek*, 18(19), May 14, 22.

Munro, J. and D. Lidsky. 1999. "Web Storefront Software," *PC Magazine*, 18(1), January 5, 155–194.

Nielsen, J. 2001. "The End of Homemade Websites," *Alertbox*, October. Available online at: (http://www.useit.com/alertbox/20011014.html).

Pallato, J. 2002. "Power Play: Oracle9i Turns Up the Heat on BEA WebLogic, IBM WebSphere," *Internet World*, January, 58.

Roberts-Witt, S. 2001. "The Internet Business," *PC Magazine*, 20(5), March 6, 8–18.

Rubenking, N. 2001. "Hidden Messages," *PC Magazine*, 20(10), May 22, 86–88.

Rush, L. 2000. "Data Mining For Dummies," *Internet World*, June 19. Available online at: (http://ecommerce.internet.com/solutions/ectips/article/0,1467,6311_397341,00.html).

Schuff, D. and R. Saint Louis. 2001. "Centralization vs. Decentralization of Application Software," *Communications of the ACM*, 44(6), June, 88–94.

360

Schultz, K. 2001. " SCM Turned Inside Out: Vendors Incorporate Collaboration into the Supply Chain," *InternetWeek*, June 25, 25–30.

Schwartz, M. 2000. "Constructive Web Critics," *Computerworld*, 34(21), May 22, 48–52.

Seminerio, M. 2000. "Click-and-Mortar Brigade Born," *eWeek*, 17(24), June 12, 58.

Smetannikov, M. 2001. "MSPs Face Hard Questions About Software," *Interactive Week*, 8(34), September 3, 35.

Ulfelder, S. 2001. "The Web's Last Gap," *Computerworld*, 35(25), June 18, 48–49.

SECURITY THREATS TO ELECTRONIC COMMERCE

INTRODUCTION

On November 3, 1988, thousands of computer systems operators and systems administrators across the United States came to work and found their computer systems inoperable. Catatonic, the computers would not respond no matter what the administrators tried.

This disastrous event was eventually traced to a 23-year-old Cornell University graduate student named Robert Morris Jr., who had unleashed the **Internet Worm** on the Internet community—arguably the most infamous Internet attack ever. This worm (as you will learn in this chapter, a worm is one of several types of damage-causing programs that can be maliciously distributed through the Internet) caused thousands of computers around the country to slow down or cease production altogether. The program traveled across the Internet in a matter of minutes by exploiting a flaw in a popular UNIX e-mail program called sendmail. It managed to invade and infect over 6200 computers (10 percent of the computers on the Internet at the time) and cause a widespread slowdown. As news of the incident spread, many sites that were not yet infected disconnected their computers from the Internet. Several sources estimated that lost computing time was worth $24 million and that direct costs to eradicate the virus and bring computers back onto the Internet totaled $40 million. Other industry experts put the total cost of the worm closer to $100 million.

Industry experts who analyzed the worm discovered that it contained no destructive code. The damage was caused solely by its uncontrolled, cancer-like growth inside each computer. The worm eventually absorbed all computing resources until each infected machine failed. The Internet Worm was a wake-up call to the computer community. Since the Internet Worm attack, computer security has become an important consideration in all computer hardware and software acquisitions and management plans.

LEARNING OBJECTIVES

In this chapter, you will learn about:

- Computer and electronic commerce security terms
- The reasons that secrecy, integrity, and necessity are three parts of any security program
- Threats and countermeasures to eliminate or reduce threats
- Specific threats to Web client computers and Web servers
- Methods to enhance security in application servers and database servers
- How using certain Internet protocols can help increase security
- The roles that encryption and certificates play in assurance and secrecy

SECURITY OVERVIEW

In the early days of the Internet, electronic mail was one of its most popular uses. Despite e-mail's popularity, people have often worried that a business rival might intercept e-mail messages for competitive gain. Another fear was that employees' nonbusiness correspondence might be read by their supervisors, with negative repercussions. These were significant and realistic concerns.

Today, the stakes are much higher. The consequences of a competitor having unauthorized access to messages and digital intelligence are now far more serious than in the past. Electronic commerce, in particular, makes security a concern for all users. A typical worry of Web shoppers is that their credit card numbers will be exposed to millions of people as they travel across the Internet. A 2001 survey found that more than 90 percent of all Internet users have at least "some concern" about the security of their credit card numbers in electronic commerce transactions. This echoes the fear shoppers have expressed for many years about credit card purchases over the phone.

Today, consumers are more comfortable giving their credit card numbers and other information over the phone, but many of those same people fear providing the same information on a Web site. As you learned in Chapter 7, people are concerned about personal information they provide to companies over the Internet. Increasingly, people doubt that these companies have the willingness and the ability to keep customers' personal information confidential. This chapter examines the broad topic of computer security in the context of electronic commerce, presenting an overview of the important security issues and their current solutions.

Computer security is the protection of assets from unauthorized access, use, alteration, or destruction. There are two general types of security: physical and logical.

Physical security includes tangible protection devices, such as alarms, guards, fire-proof doors, security fences, safes or vaults, and bombproof buildings. Protection of assets using nonphysical means is called **logical security**. Any act or object that poses a danger to computer assets is known as a **threat**.

Countermeasure is the general name for a procedure, either physical or logical, that recognizes, reduces, or eliminates a threat. The extent and expense of counter-measures can vary, depending on the importance of the asset at risk. Threats that are deemed low risk and unlikely to occur can be ignored when the cost to protect against the threat exceeds the value of the protected asset. For example, it would make sense to protect from tornadoes a computer network in Oklahoma City, where there is a lot of tornado activity, but not to protect one in Los Angeles, where torna-does are rare. The risk management model shown in Figure 10-1 illustrates four gen-eral actions that an organization could take, depending on the impact (cost) and the probability of the physical threat. In this model, a tornado in Oklahoma would be in quadrant II, whereas a tornado in Southern California would be in quadrant III or IV.

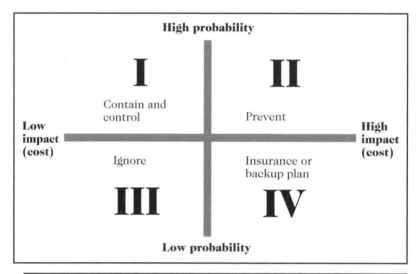

Figure 10-1   Risk management model

The same sort of risk management model applies to protecting Internet and electronic commerce assets from both physical and electronic threats. Examples of the latter include impostors, eavesdroppers, and thieves. An **eavesdropper**, in this context, is a person or device that can listen in on and copy Internet transmissions. People who write programs or manipulate technologies to obtain unauthorized access to computers and networks are called **crackers** or **hackers**.

A cracker is a technologically skilled person who uses those skills to obtain unauthorized entry into computers or network systems—usually with the intent of stealing information or damaging the information, the system's software, or even the system's hardware. Originally, the term hacker was used to describe a dedicated pro-grammer who enjoyed writing complex code that tested the limits of technology. Although the term hacker is still used in a positive way—even as a compliment—by computer professionals (who make a strong distinction between the terms hacker and cracker), the media and the general public usually use the term to describe those hackers who use their skills for ill purposes.

To implement a good security scheme, organizations must identify risks, determine how they will protect threatened assets, and calculate how much they can spend to protect those assets. In this chapter, the primary focus in risk management protection is on the central issues of identifying the threats and determining the ways to protect assets from those threats, rather than on the protection costs or value of assets.

Computer Security Classification

Computer security is generally classified into three categories: secrecy, integrity, and necessity (also known as denial of service). **Secrecy** refers to protecting against unauthorized data disclosure and ensuring the authenticity of the data's source. **Integrity** refers to preventing unauthorized data modification. **Necessity** refers to preventing data delays or denials (removal). Secrecy is the best known of the computer security categories. Every month, newspapers report on break-ins to government computers, or theft and use of stolen credit card numbers which are used to order goods and services. Integrity threats are reported less frequently and, thus, may be less familiar to the public. For example, an integrity violation occurs when an Internet e-mail message's contents are changed—possibly negating the message's original meaning. Necessity violations take several forms, and they occur relatively frequently. Delaying a message or completely destroying it can have huge consequences. Suppose that you send an e-mail message at 10:00 a.m. to E*Trade, an online stock trading company, with an order to purchase 1000 shares of IBM at market. Then, imagine the stockbroker does not receive the message (because an enemy delays it) until 2:30 p.m.—after the stock's price has increased 15 percent in the interim. The delay has cost you 15 percent of the value of the trade.

Security Policy and Integrated Security

Any organization concerned about protecting its electronic commerce assets should have a security policy in place. A **security policy** is a written statement describing which assets to protect and why they are being protected, who is responsible for that protection, and which behaviors are acceptable and which are not. The policy primarily addresses physical security, network security, access authorizations, virus protection, and disaster recovery. The policy develops over time, and is a living document that the company and security officer must review and update at regular intervals.

The first step an organization must take in creating a security policy is to determine what assets to protect and from whom (for example, credit cards are an asset to be protected from eavesdroppers). Then, the organization must determine who should have access to various parts of the system and who should not. Next, the organization determines what resources are available to protect the assets it has identified. Using the information it has acquired, the organization develops a written security policy. Finally, the organization commits resources to building or buying software, hardware, and physical barriers that implement the security policy. For example, if a security policy disallows any unauthorized access to customer information, including credit card numbers and credit history, then the organization must either create or purchase software that guarantees end-to-end secrecy for electronic commerce customers.

The **Center for Security Policy (CSP)** is a nonprofit organization that stimulates national and international debate about security policy—particularly security policies that address defense and technology interests of the United States. The CSP can provide guidance about the specifications a security policy should contain. Figure 10-2 displays the CSP "About Us" page.

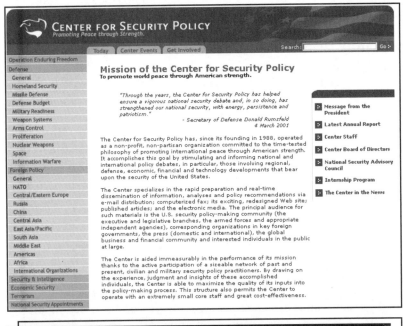

Figure 10-2   *Center for Security Policy About Us page*

Although absolute security is difficult to achieve, organizations can create enough barriers to deter most intentional violators. With good planning, organizations can also reduce the impact of natural disasters or terrorist acts.

Integrated security means having all security measures working together to prevent unauthorized disclosure, destruction, or modification of assets. A security policy covers many security concerns that must be addressed by a comprehensive and integrated security plan. Specific elements of a security policy address the following points:

- Authentication: Who is trying to access the electronic commerce site?
- Access control: Who is allowed to log on to and access the electronic commerce site?
- Secrecy: Who is permitted to view selected information?
- Data integrity: Who is allowed to change data, and who is not?
- Audit: Who or what causes selected events to occur, and when?

In this chapter, you will explore these security policy issues with a focus on how they apply to electronic commerce in particular. Next you will learn about threats to digital information, beginning with intellectual property threats.

INTELLECTUAL PROPERTY THREATS

Intellectual property threats are a larger problem than they were prior to the widespread use of the Internet. As you learned in Chapter 7, it is relatively easy to use existing material found on the Internet without the owner's permission. Actual monetary damage resulting from a copyright violation is more difficult to measure than damage from secrecy, integrity, or necessity computer security violations. However, the damage can be just as significant.

The Internet presents a particularly tempting target for two reasons. First, it is very easy to reproduce an exact copy of anything you find on the Internet, regardless of whether it is subject to copyright restrictions. Second, many people are simply unaware of the copyright restrictions that protect intellectual property. Instances of both unwitting and willful copyright infringements occur daily on the Internet. Most experts agree that copyright infringements on the Web occur because users are ignorant of what they can and cannot legally copy. Most people do not maliciously copy a protected work and post it on the Web.

Although copyright laws were enacted before the creation of the Internet, the Internet itself has complicated publishers' enforcement of copyrights. Recognizing unauthorized reprinting of written text is relatively easy; recognizing that a photograph has been borrowed, cropped, or illegally used on a Web page is a more difficult task.

Domain Names

Considerable controversy has arisen recently about intellectual property rights and Internet domain names. **Cybersquatting** is the practice of registering a domain name that is the trademark of another person or company in the hopes that the owner will pay huge amounts of money to acquire the URL. In addition, successful cybersquatters can attract many site visitors and, consequently, charge high advertising rates. A related problem, called **name changing**, occurs when someone registers purposely misspelled variations of well-known domain names. These variants sometimes lure consumers who make typographical errors in entering a URL. **Name stealing** occurs when someone posing as a site's administrator changes the ownership of the site's assigned domain name to another site and owner.

Cybersquatting

On November 29, 1999, the **U.S. Anticybersquatting Consumer Protection Act (ACPA)** was signed into law. Also known as the Trademark Cyberpiracy Prevention Act, it protects the trademarked names owned by corporations from being registered as domain names by parties that do not own the trademarked names. Under U.S. law, parties found guilty of cybersquatting can be found liable for damages of up to $100,000 per trademark. If the registration of the domain name is found to be "willful," damages can be as much as $300,000 (see the Null reference listed in the "For Further Study and Research" section at the end of this chapter). Recent U.S. cases that were finally settled out of court illustrate the problem. Three cybersquatters made headlines a few years ago when they tried to sell the URL barrydiller.com for $10 million. Barry Diller, the CEO of USA Networks, sued the trio and won.

Registering a generic name such as Wine.com is very different from registering a trademarked name in bad faith—cybersquatting. The former act is merely prospecting and is legal, whereas the latter is clearly extortion. If someone has already registered a domain name that would be an appropriate URL for another company, then that other business desiring the name might have to pay a high price to obtain it. Disputes that arise when one person has registered a domain name that is an existing trademark or company name are settled by the World Intellectual Property Association (WIPO). The WIPO home page appears in Figure 10-3.

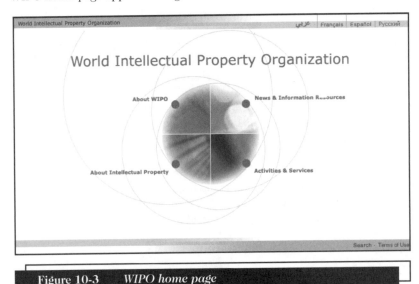

Figure 10-3   *WIPO home page*

One common type of dispute arises when a business has a trademark that is a common term. If a person obtains the domain name containing that common term, the owner of the trademark must seek resolution at the WIPO. Gordon Sumner, who has performed music for more than 20 years as Sting, filed a complaint with the WIPO because a Georgia man had obtained the domain name www.sting.com, and had reportedly offered to sell it to him for $25,000. In more than 80 percent of its cases, the WIPO has held for the trademark name owner; however, in this case, the WIPO noted that the word "sting" was in common and general use, and refused to award the domain to the performer (after the WIPO decision, the two parties came to undisclosed terms and the Sting official site is now at www.sting.com).

Name Changing

After obtaining their domain name, companies still face the possibility that someone will steal unsuspecting customers away from them by registering a domain name that is a slight variation or even a misspelling of a company's well-known domain name. A simple typo in a Web address could lead a Web surfer to LLBaen.com instead of LLBean.com. The Anticybersquatting Consumer Protection Act now helps define which cases are true cybersquatting and which are cases of permissible competition. Most businesses agree that the practice of name changing is annoying to affected online businesses and confusing to their customers. A company's best defense is to register as many variations in product and company spellings as possible. Unfortunately, there is no complete solution to this problem; as new high-level domains such as .biz become available, the name changing problem has recurred.

Name Stealing

Perhaps the most flagrant example of domain name abuse is name stealing. Name stealing occurs when someone other than a domain name's owner changes the ownership of the domain name. A **domain name ownership change** occurs when owner information maintained by a public domain registrar is changed in the registrar's database to reflect the new owner's name and business address. This usually happens only when safeguards are not in place. Once domain name ownership is changed, the name stealer can manipulate the site, post graffiti on it, or redirect online customers to other sites selling substandard goods. The main purpose of name stealing is to harass the site owner. The temporary loss of its domain name can cut off a business from its Web site for several days.

The ownership change can occur without notice because it is automated and because some domain name registrars' security procedures can be faulty. In one well-publicized case, six domain names disappeared quietly from the registry at domain name registrar **VeriSign Network Solutions**. One of the domain names was unavailable to its owner, an electronic commerce site, for five days. Such an interruption can add up to tremendous losses in sales.

Name stealing is fairly uncomplicated. A perpetrator can type in a registered administrator's address and change Web site ownership. Then, the perpetrator can register the stolen domain name with the perpetrator's IP address at another registrar's site. The name stealing does not go unnoticed for long; however, during the time it takes to return the IP address assignment to the correct domain name, customers attempting to visit the site will find it unavailable.

This chapter's remaining sections are organized around a three-element theme: protecting client computers, protecting the transmission of information on the Internet, and protecting the electronic commerce server.

THREATS TO THE SECURITY OF CLIENT COMPUTERS

You can study electronic commerce security requirements by examining the overall process, beginning with the consumer and ending with the commerce server. Each logical link in the process includes assets that must be protected to ensure secure electronic commerce: client computers, the messages traveling on the communication channel, and the Web servers—including any hardware connected to the Web servers. In this section, you will learn about the threats to client computers.

Until the debut of executable Web content, Web pages were static. Coded in HTML, the Web's standard page description language, static pages could do little more than display content and provide links to related pages with additional information. The widespread use of active content has changed that.

Active Content

Active content refers to programs that are embedded transparently in Web pages and that cause action to occur. Active content can display moving graphics, download and play audio, or implement Web-based spreadsheet programs. Active content is used in electronic commerce to place items you wish to purchase into a shopping cart and compute a total invoice amount, including sales tax, handling, and shipping costs. Developers use active content because it extends the functionality of HTML and adds excitement to Web pages. It also offloads some data processing chores from the busy server machine to the user's mostly idle client computer.

Active content is provided in several forms. The best-known active content forms are Java applets, ActiveX controls, JavaScript, and VBScript. (See the Online Companion links **JavaDomain.com** and the **Sun Java Security FAQ** for Web information about Java applets and Java security questions.) JavaScript and VBScript are known as scripting languages; they provide scripts, or commands that are executed. An **applet** is a small application program. Applets typically run within the Web browser. See **Cool Applets** for many examples of eye-catching Web applets. Figure 10-4 shows an example of one of the applets you can launch from the Cool Applets Web page.

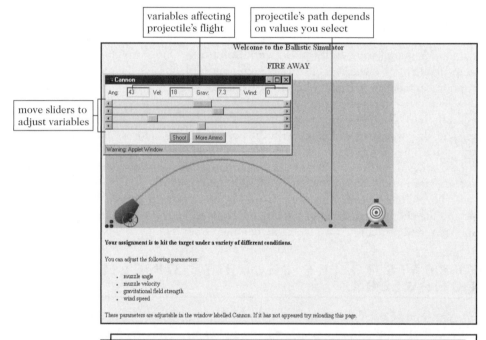

Figure 10-4 *Java applet example*

Other ways to provide Web active content include graphics and Web browser plug-ins. Graphics files can contain embedded invisible instructions that execute on the client computer when they are downloaded. Programs or interpreters that execute the instructions found in graphics programs and a variety of other formats can cause malicious instructions—hidden within legitimate graphics instructions—to be executed. **Plug-ins** are programs that interpret or execute instructions embedded in downloaded graphics, sounds, and other objects. Active content, including all forms, enables Web pages to take action. Buttons on a form, for instance, can activate embedded programs to calculate and display information or to send data from the client computer to a Web server. Active content gives life to static Web pages.

Active content is launched in a Web browser automatically when that browser loads a Web page containing active content. The applet automatically downloads along with the page, and begins running. Depending on how the browser's security settings are configured, the browser might open a warning dialog box, such as the one shown in Figure 10-5, announcing the active content and asking the user for permission to open that content.

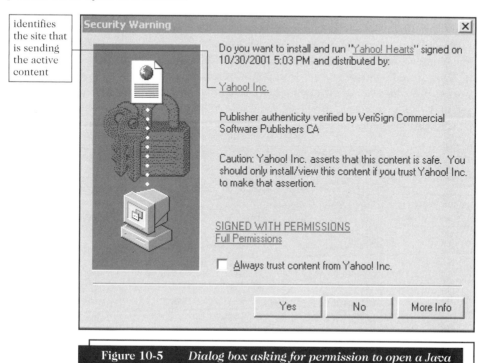

identifies the site that is sending the active content

Figure 10-5 *Dialog box asking for permission to open a Java applet from Yahoo! Games*

Because active content modules are embedded in Web pages, they can be completely transparent to anyone browsing a page containing them. Crackers intent on doing mischief to client computers can embed malicious active content in seemingly innocuous Web pages. This delivery technique, called a Trojan horse, immediately begins executing and taking actions that cause harm. A **Trojan horse** is a program hidden inside another program or Web page that masks its true purpose. The Trojan horse could snoop around a client computer and send back, to a cooperating Web server, information that is private—a secrecy violation. Worse yet, the program could alter or erase information on a client computer—an integrity violation. Zombies are equally threatening. A **zombie** is a program that secretly takes over another computer for the purpose of launching attacks on other computers. Zombie attacks cannot be traced to their creators.

Adding active content to Web pages involved in electronic commerce introduces several security risks. Malicious programs delivered through Web pages could reveal credit card numbers, usernames, and passwords that are frequently stored in special files called cookies, which you learned about earlier in this book. **Cookie Central** is a Web site devoted to Internet cookies. Malicious active content delivered by means of cookies can reveal the contents of files or even destroy files. A number of computer viruses can examine users' e-mail address books and mail malicious content to the addresses contained therein. Many of these viruses gain entry through e-mail accessed through a Web browser.

Although cookies are not inherently bad, some people dislike the idea of storing cookies on their computers. Web users can accumulate large numbers of cookies as they browse the Internet, and some cookies could contain sensitive, personal information. Current versions of all major Web browsers allow some control over cookies, and a variety of free and shareware programs are available to help identify, manage, display, and eliminate cookies. Examples are **Cookie Crusher**, which controls cookies *before* they are stored on a user's hard drive, and **Cookie Pal**. Figure 10-6 shows Cookie Pal's display of cookies stored on a computer. Notice that each cookie's origin (server), expiration date, and name are displayed. Cookies can be removed from a client computer by selecting the undesired cookie(s) and then clicking the Delete button.

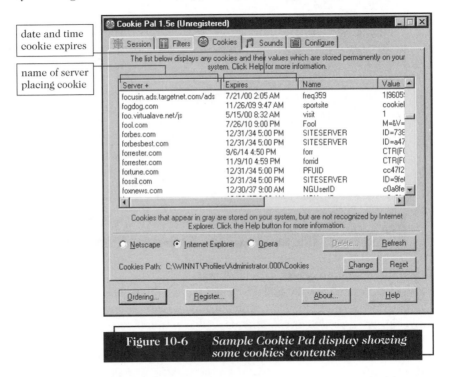

Figure 10-6 *Sample Cookie Pal display showing some cookies' contents*

Java, Java Applets, and JavaScript

Java is a high-level programming language developed by Sun Microsystems. Originally called OAK, Java was invented for embedded systems. Java's most popular use is in Web pages where thousands of applets implement a wide variety of client-side applications. The applets are downloaded along with the requested Web pages, and run on the client computer as long as the Web browser is Java-compatible. Both Netscape Navigator and Microsoft Internet Explorer are Java-compatible. Figure 10-7 shows the **Sun Java applets** page, which contains links to several applets you can purchase or download for free.

THE SOURCE FOR JAVA™ TECHNOLOGY
java.sun.com

APPLETS

An applet is a program written in the Java™ programming language that can be included in an HTML page, much in the same way an image is included. When you use a Java technology-enabled browser to view a page that contains an applet, the applet's code is transferred to your system and executed by the browser's Java Virtual Machine (JVM)

For information and examples on how to include an applet in an HTML page, refer to this description of the <APPLET> tag.

Products & APIs
Developer Connection
Docs & Training
Online Support
Community Discussion
Industry News
Solutions Marketplace
Case Studies

Applet Resources

▷ A list of sites where you can find applet resources.

 **Ten most recent highly rated submissions to JARS**

JDocHelper ★★★	J/Top ★★★★
Programming - Other	Utilities - Other
Designing-3DI ★★★	Color Wheel ★★★★
Multimedia - 3D Graphics	WWW - Other
Traffic 2000 ★★★	News Applet ★★★
Games - Arcade Games	WWW - Text Displays
ChatApplet v2 0 ★★★	JDK Commander V3 ★★★
Utilities - Chat	Programming - Other
Universal SQL Tools ★★★★	JCourier 2000 ★★★
Utilities - Databases	Utilities - Communication

Freebie Applets

▷ Here you will find free applets available for use on your web sites. All of the necessary java class files and example HTML markup are contained in a single zip file. The HTML markup necessary to use the applets is explained in detail on a separate page for each applet.

JDK™ Demo Applets

▷ Here you can find copies of demo applets included in Java™ Software Development Kits (JDK™ 1 0, JDK 1 1, and Java 2 SDK SE v1 2 releases)

▷ Applets from the *Java Tutorial* - These applets are part of the official Java tutorial.

Other Applets

AudioItem	Java Glossary	TumblingDuke
BouncingHeads	Hangman	UnderConstruction
Bubbles	ImageLoop	WordMatch
Bullets	ImageTest	XeoMenu
	ScrollingImages	

Applet Archive

Abacus	Escher	Pythagoras
Cannon	LED	Star Field
Crossword	Neon Sign	System Info
Dining Philosophers	Nuclear Plant	Voltage

Applets at Work

▷ To read about some fascinating examples of applets at work, see the following features on our Web site.

Applet Power!
Personal Applet Power
Applets Power the Client

Java™ Plug-in

▷ Java™ Plug-in software enables enterprise customers to direct applets or beans written in the Java programming language on their intranet web pages to run using Sun's Java Runtime Environment (JRE), instead of the browser's default. This enables an enterprise to deploy applets that take full advantage of the latest capabilites and features of the Java platform and be assured that they will run reliably and consistently.

links to Java applet downloads and information

Figure 10-7   Java applets page

373

Java can run outside the confines of a Web browser, too. Another reason Java is so popular is that it is platform-independent—it will run on many different computers. This "develop once, deploy everywhere" feature reduces development costs because only one source copy needs to be maintained for all machines.

Java adds functionality to business applications and can handle transactions and a wide variety of actions on the client computer. That relieves an otherwise busy server-side program from handling thousands of transactions simultaneously. Once downloaded, embedded Java code can run on a client's computer—which means that security violations can occur. To counter this possibility, a special security model, called the **Java sandbox**, has been developed. The Java sandbox confines Java applet actions to a set of rules defined by the security model. These rules apply to all untrusted Java applets. **Untrusted applets** are Java applets that have not been proven to be secure. When Java applets are run within the constraints of the security sandbox, they do not have access to security-compromising code in the system. For example, Java applets obeying the sandbox rules cannot perform file input, output, or delete operations. This prevents secrecy (disclosure) and integrity (deletion or modification) violations. See **Java Security White Paper** in the Online Companion for more information about the Java sandbox. Java application programs, unlike Java applets, run outside a Web browser and can perform *any* action on your computer—including malevolent acts.

Java applets that are loaded from a local file system are trusted. They do not run within the constraints of the Java applet sandbox. **Trusted applets** have full access to system resources on the client computer. They are "trusted" to not do any damage. **Signed Java applets** contain embedded digital signatures from a trusted third party, which are proof of the identity of the source of the applet. If the applet is signed, then it can be "let out of the sandbox" to use the full system resources. The notion is that if the user knows who produced the applet and trusts that party, then recourse is possible if the applet actually causes damage. Theoretically, malevolent applets are always produced anonymously. See **Security and Signed Applets** in the Online Companion for further details.

JavaScript is a scripting language developed by Netscape to enable Web page designers to build active content. Despite the similar-sounding names, JavaScript is completely unrelated to Sun's Java programming language. Supported by popular Web browsers, JavaScript shares many of the structures of the full Java language. When a user downloads a Web page with embedded JavaScript code, it executes on the user's (client) computer. Like other active content vehicles, JavaScript can invoke attacks by executing code that destroys the client's hard disk, discloses the e-mail stored in client mailboxes, or sends sensitive information to the attack perpetrator's Web server on the Internet. JavaScript code can also record the URLs of Web pages a user visits and capture information entered into CGI Web forms. For example, if a user enters credit card numbers while trying to reserve a rental car, a JavaScript program could copy the credit card number. JavaScript programs, unlike Java applets, do not operate under the restrictions of the Java sandbox security model.

In 1999, *The New York Times* revealed that RealNetworks, the maker of the popular RealPlayer multimedia Web browser plug-in, had been surreptitiously gathering information from its users. Easily downloaded and installed from the Internet, RealPlayer was recording user information such as the RealPlayer user's name, e-mail address, country, ZIP code, computer operating system, and other details. RealPlayer used the Internet connection to send back to RealNetworks the information that it had gathered. Soon after the discovery, and after considerable public embarrassment, RealNetworks issued a statement that a software patch was available for all current users. The patch would prevent the RealNetwork software from collecting and transmitting user information.

A JavaScript program cannot commence execution on its own, unlike Java programs or Java applets. To launch an ill-intentioned JavaScript program, a user must start the program. For example, a site with a retirement income calculator might require a visitor to click a button to see a retirement income projection. Once the user clicks the button, the JavaScript program launches and does its work—another example of a Trojan horse.

ActiveX Controls

ActiveX is an object, called a control, that contains programs and properties that Web designers place on Web pages to perform particular tasks. ActiveX components originate from many programming languages, such as C++ or Visual Basic. Unlike Java or JavaScript code, ActiveX controls run only on computers running Windows (95, 98, NT, Me, 2000, or XP) and only on browsers that support them. Once ActiveX code is completed, programmers put it inside an ActiveX envelope, compile the control, and place it on a Web page. When a Windows-based Web browser downloads a Web page containing an embedded ActiveX control, the control is executed on the client computer. Shockwave, an animation and entertainment control plug-in for Web browsers, is available as an ActiveX control. Other examples include Web-enabled calendar controls and many Web games. See **ActiveX Controls Download Page** in the Online Companion for a comprehensive list of ActiveX controls.

The security danger with ActiveX controls is that once they are downloaded, they execute like any other program on a client computer. They have full access to all system resources, including operating system code. This has very dangerous implications. An ill-intentioned ActiveX control could reformat a user's hard disk, send e-mail to all the people listed in his or her address book, or simply shut down the computer. Because ActiveX controls have full access to client computers, they can cause secrecy, integrity, or necessity violations. The actions of ActiveX controls cannot be halted once they begin execution. Most current versions of Web browser software can be configured to provide a notice when the user is about to download an ActiveX control. Figure 10-8 shows an example of the warning issued when Internet Explorer detects an ActiveX control.

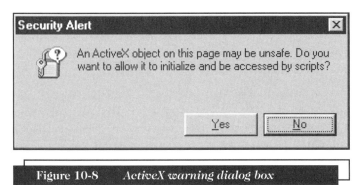

Figure 10-8   ActiveX warning dialog box

Graphics, Plug-Ins, and E-Mail Attachments

You learned earlier that graphics, browser plug-ins, and e-mail attachments can harbor executable content. Some graphics file formats have been specifically designed to contain instructions on how to render a graphic. That means that any Web page containing such a graphic could be a threat, because the code embedded in the graphic could cause harm to a client computer. Similarly, browser plug-ins, which are programs that enhance the capabilities of browsers, handle Web content that a browser cannot handle. Plug-ins are normally beneficial and perform tasks for a browser, such as playing audio clips, displaying movies, or animating graphics. Apple's QuickTime, for example, is a plug-in that downloads and plays movies that are stored in a special format.

Many plug-ins perform their duties by executing commands buried within the media they are manipulating. This opens the door to the possibility that someone intent on doing harm could embed commands within a seemingly innocuous video or audio clip. The ill-intentioned commands hidden within the object that the plug-in is interpreting could damage a client computer by erasing some (or all) of its files. Figure 10-9 shows the **Netscape Browser Plug-ins** page, which contains category listings of plug-in software it offers for free downloading.

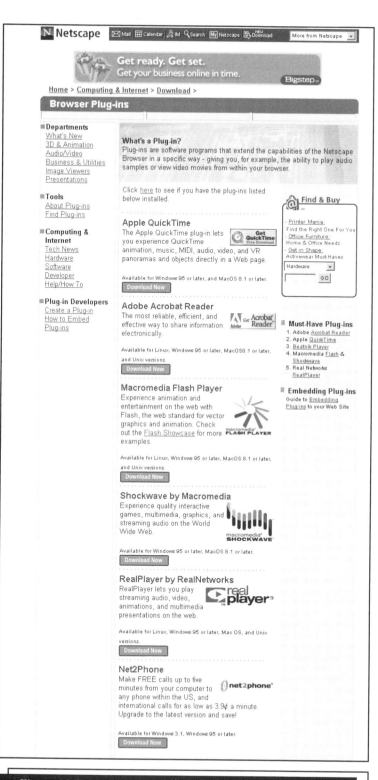

Figure 10-9 *Netscape Browser Plug-ins page*

The potential dangers lurking in e-mail attachments get a lot of news coverage and are the most familiar to the general population. E-mail attachments provide a convenient way to send nontext information over a text-only system—electronic mail. Attachments can contain word processing files, spreadsheets, databases, images, or virtually any other information you can imagine. Most programs, including the popular browser e-mail programs, display attachments by automatically executing an associated program; for example, the recipient's Excel program detaches an attached Excel workbook and displays it, or Word opens and displays a Word document. While this activity doesn't inherently cause damage, Word and Excel macro viruses *inside* the loaded documents and workbooks can damage a client computer and reveal otherwise confidential information. A **virus** is software that attaches itself to another program and can cause damage when the host program is activated. A **worm** is a type of virus, such as the Internet Worm described in this chapter's introduction, that replicates itself on the computers that it infects. Worms can spread quickly through the Internet. A **macro virus** is a type of virus that is coded as a small program, called a macro, and is embedded in a file. You have probably read about or experienced recent examples of e-mail attachment-borne virus attacks. **Symantec** and **McAfee**, among other companies, keep track of viruses and sell antivirus software.

E-mail attachments containing viruses and other malicious software are reported daily. Some of the most famous in recent years include the "ILOVEYOU" virus, also known as the "love bug," and its variants containing Visual Basic Script (VBScript) code. (**VBScript** is a programming language that runs on most microcomputers.) The "ILOVEYOU" virus has been attributed to a 23-year-old computer student who lived in the Philippines. The virus spread through the Internet with amazing speed as an e-mail message, infecting the computer of anyone who opened the e-mail attachment and clogging e-mail systems with thousands of copies of the useless e-mail message. The virus spread quickly because it automatically sent itself to as many as 300 addresses stored in a computer's Microsoft Outlook address book. Besides replicating itself explosively through e-mail, the virus caused other harm, destroying digital music and photo files stored on the target computers. The "ILOVEYOU" virus also searched for other users' passwords and forwarded that information to the original perpetrator. Within days, the virus spread to 40 million computers in more than 20 countries and caused an estimated $9 billion in damages—most of it in lost worker productivity.

In 2001, the incidences of virus and worm attacks increased. With more than 40,000 reported security violations occurring during the year, the parade of attacks included Code Red and Nimda virus-worm combinations, each affecting millions of computers and costing billions of dollars to clean up. Both Code Red and Nimda are examples of a **multi-vector virus**, so called because they can enter a computer system in several different ways (vectors).

Click **Virus Information** for a comprehensive Web site containing descriptions of thousands of viruses.

MICROSOFT INTERNET INFORMATION SERVER

As you learned in Chapter 8, Internet Information Server (IIS) is Microsoft's Web server software product. Microsoft supplies versions of the IIS software with its Windows server operating systems that are suitable for use in operating electronic commerce Web sites.

In August 2001, Microsoft faced an uncomfortable situation that many U.S. manufacturing companies had experienced over recalled defective products—Microsoft executives stood by at a news conference while a U.S. government official announced to gathered reporters that there was a serious flaw in a Microsoft product. The director of the FBI's National Infrastructure Protection Center was warning reporters that the Code Red worm, which was spreading through the Internet for the third time in as many weeks, was a serious threat to the continued operation of the Internet.

The Code Red worm exploits a vulnerability in Microsoft's IIS Web server software. When the worm was first identified, Microsoft rapidly made a patch available on its Web site. Microsoft also observed that Web server installations that had kept current with all of the updates and patches that Microsoft had issued would not be subject to attack by the worm.

Many of Microsoft's customers were outraged by these statements, noting that Microsoft had issued more than 40 software patches in the first half of 2001 and had issued 100 or more patches in each of several prior years. IIS users complained that keeping the software current was virtually impossible and called for Microsoft to deliver software that was more secure when first installed. Many IIS users began to consider switching to other Web server software. The GartnerGroup, a major IT consulting firm, recommended to its clients that they seriously consider alternatives to IIS for their critical Web server installations.

Many industry observers and software engineers agree that Microsoft was a victim of its own success. It had created a very popular and complex piece of software. It is extremely difficult to ensure that no bugs exist in complex software products and the popularity of the software made it an attractive target for crackers—one worm could bring down many of the servers operating on the Internet. These two factors, plus the likelihood that many IIS servers would not have all of the available security upgrades installed, combined to make it an irresistible target for a worm creator.

Microsoft has struggled to gain the confidence of large corporate IT departments. The company has worked hard in recent years to establish the reputation of its operating system software as reliable and trustworthy. The Code Red worm attack on its Web server software was a major setback in its reputation-building effort. You can review the **Microsoft Security Pages** through the link in the Online Companion.

The term **steganography** describes information (a command, for example) that is hidden within another piece of information. This information can be used for good or for bad purposes. Frequently, computer files contain redundant or insignificant information that can be replaced with other information. This other information resides in the background and is virtually undetectable. Steganography provides a way of hiding an encrypted file within another file so that a casual observer cannot detect that there is anything of importance in the containing file. Encrypting a file protects it from being read, and steganography makes it invisible.

Steganography is comparable to hiding an encrypted microdot containing trade secrets by gluing it to the pupil of a person's eye in a portrait. It is simultaneously hidden and secured. The casual viewer sees a person. Closer inspection reveals a microdot.

Many security analysts believe that the terrorist organization Al Qaeda used steganography to hide attack orders and other messages in images that it had its confederates post on Web sites. Messages hidden using steganography are extremely difficult to detect. This fact, combined with the fact that there are millions of images on the Web, makes the use of steganography by global terrorist organizations a deep concern of governments and security professionals. The Online Companion includes a link to a site with more information about **Steganography and Digital Watermarking**.

THREATS TO THE SECURITY OF COMMUNICATION CHANNELS

The Internet serves as the electronic connection between a consumer (client) and an electronic commerce resource (commerce server). Now that you understand security threats to client machines, the next asset to consider is the communication channel connecting clients to servers—namely, the Internet.

The Internet is not at all secure. Although the Internet has its roots in a military network, that network was built only to provide redundancy, not secure communications, in case one or more communications lines were cut. In other words, its original design goal was to provide several alternative paths on which to send critical military information. The military planned to send sensitive information in an encrypted form so that any messages traveling over the network would remain secret and tamperproof. However, the security of messages traversing the network was provided by software that converted the messages into unintelligible strings of characters called cipher text.

Today, the Internet remains unchanged from its original, insecure state. Messages on the Internet travel a random path from a source node to a destination node. A message passes through a number of intermediate computers on the network before reaching its final destination. The path can vary each time a message is sent between the same source and destination points. It is impossible to guarantee that every computer on the Internet through which messages pass is safe and secure. A message sent from a merchant in Manchester, England, to a supplier in Cairo, Egypt, may have passed through a competitor's computer located in Beirut, Lebanon. Because users cannot control the path and do not know where their message packets have been, it is quite possible that some intermediary can read the messages, alter them, or even completely eliminate messages from the Internet. That is, any e-mail message traveling on the Internet is subject to secrecy, integrity, and necessity violations. This section describes those problems in more detail. Chapter 11 describes several ways to deal with the security problems presented in this chapter.

This section discusses Internet channel security threats within the classifications of secrecy, integrity, and necessity. This organization provides a good structure for examining direct security threats to the Internet network itself.

Secrecy Threats

Secrecy is one of the highest profile security threats mentioned in articles and the popular media. Closely linked to secrecy is privacy, which also receives a great deal of attention. Secrecy and privacy, though similar, are different issues. Secrecy is the prevention of unauthorized information disclosure. **Privacy** is the protection of individual rights to nondisclosure. The **Privacy Council**, which helps businesses implement smart privacy and data practices, has created an extensive Web site surrounding privacy—covering both business and legal issues. Secrecy is a technical issue requiring sophisticated physical and logical mechanisms, whereas privacy protection is a legal matter. A classic example of the difference between secrecy and privacy is e-mail. A company may protect its e-mail messages against secrecy violations by using encryption (see Chapter 11). In encryption, a message is encoded into an unintelligible form that only the proper recipient can transcribe back into the original message. Secrecy countermeasures protect outgoing messages. E-mail privacy issues address whether company supervisors should be permitted to randomly read employees' messages. Disputes in this area center around who owns the e-mail messages: the company, or the employee who sent them. The focus in this section is on secrecy—keeping unauthorized persons from reading information they should not be reading.

You have already learned that a significant danger of conducting electronic commerce is theft of sensitive or personal information, including credit card numbers, names, addresses, and personal preferences. This kind of theft can occur any time anyone fills out a form or submits credit card information over the Internet, because it is not difficult for an ill-intentioned person to record information packets (a secrecy violation) from the Internet for later examination. The same problems can occur in e-mail transmissions. Special software applications called **sniffer programs** provide the means to tap into the Internet and record information that passes through a particular computer (router) while traveling from its source to its destination. Using a sniffer program is analogous to tapping a telephone line and recording a conversation. Sniffer programs can read e-mail messages as well as electronic commerce information.

Security experts periodically find electronic holes, called backdoors, in electronic commerce software. These can be left open accidentally by the software developer, or they can be left open intentionally—exposing clients to secrecy threats. A **backdoor** allows anyone with knowledge of the existence of a backdoor or a system password to cause damage by observing transactions, deleting data, or stealing data. In 2000, the Cart32 shopping cart software made by McMurtrey/Whitaker & Associates was found to have a backdoor through which credit card numbers could be obtained by anyone with a backdoor password. The company quickly supplied a patch to eliminate the backdoor. Although the backdoor resulted from a software programming error and not one created intentionally by a disgruntled employee, the consequences of a backdoor attack that releases customer credit card information can be disastrous, regardless of why it occurs.

Credit card number theft is an obvious problem, but proprietary corporate product information or prerelease data sheets mailed to corporate branches can be intercepted and passed along easily too. Often, corporate confidential information is even more valuable than a few credit cards, which usually have spending limits. Purloined corporate information can be worth millions of dollars.

Breaching secrecy on the Internet is not difficult. Here is an example of how you might inadvertently leak confidential information that an eavesdropper or another Web site server can retrieve afterwards. Suppose you log on to a Web site called www.storeone.com that contains a form with text boxes for your name, address, and e-mail address. When you fill out those text boxes and click the submit button, the information is sent to the Web server for processing. One way that a Web server can obtain and track that data is to collect your text box responses and place them at the end of the server's URL (which appears in the address box of your Web browser). This long URL (with your text box responses appended) is included in all HTTP request and response messages that travel between your browser and the storeone.com server. So far, no violations have occurred. Suppose, however, that you change your mind and decide not to wait for a response from the storeone.com server. Instead, you point your browser to another Web site, www.storetwo.com. The server at storetwo.com might be set up to collect Web demographics. If it is, it will log the URL from which you just came by capturing it from the HTTP request message that your browser sends. The site manager at storetwo.com can use this technique to determine how electronic commerce traffic has reached the site. However, by reading the part of the storeone.com URL that includes the information you entered into those text boxes on the storeone.com site, the storetwo.com server has obtained the confidential information you recently entered.

Web users are continually revealing information about themselves when they use the Web. This information includes IP addresses (Internet addresses) and the browser being used. This type of data exposure is also an example of a secrecy breach. Several Web sites offer an "anonymous browser" service that hides personal information from sites that you visit. One of these sites, **Anonymizer**, provides a measure of secrecy to Web surfers who use the site as a portal (the beginning site from which they visit other sites). Anonymizer places its address on the front end of any URLs that the user visits. This shield reveals only the Anonymizer Web site URL to other Web sites that the user visits. For example, if you visit Amazon.com through the Anonymizer site, Anonymizer would present this URL to the Amazon.com server: *http://www.anonymizer.com:8080/http://www.amazon.com*. Figure 10-10 shows Anonymizer's home page with a request to anonymously surf to Amazon.com's home page. This can make anonymous Web surfing possible, but tedious, because you must type each URL that you want to visit in the text box on the Anonymizer home page. To make the process easier, Anonymizer (and other companies) provide browser plug-in software (some of these plug-ins are free, others require purchase or a subscription) that users can download and install.

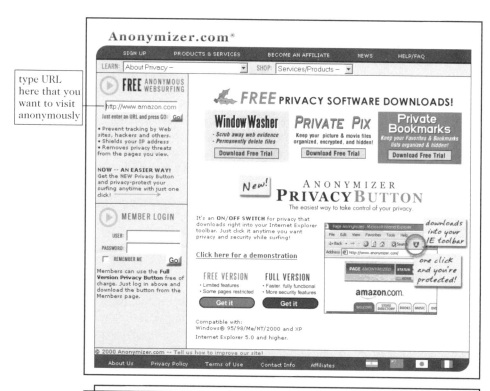

type URL here that you want to visit anonymously

Figure 10-10 *Using Anonymizer to visit Amazon.com anonymously*

Integrity Threats

An integrity threat, also known as **active wiretapping**, exists when an unauthorized party can alter a message stream of information. Unprotected banking transactions, such as deposit amounts transmitted over the Internet, are subject to integrity violations. Of course, an integrity violation implies a secrecy violation, because an intruder who alters information can read and interpret that information. Unlike secrecy threats, where a viewer simply sees information he or she should not, integrity threats can cause a change in the actions a person or corporation takes, because a mission-critical transmission has been altered.

Cyber vandalism is an example of an integrity violation. **Cyber vandalism** is the electronic defacing of an existing Web site's page. The electronic equivalent of destroying property or placing graffiti on objects, cyber vandalism occurs whenever someone replaces a Web site's regular content with his or her own. Recently, several cases of Web page defacing involved vandals replacing business content with pornographic material and other offensive content.

Masquerading or **spoofing**—pretending to be someone you are not or representing a Web site as an original when it really is a fake—is one means of disrupting Web sites. Using a security hole in a Domain Name Server (DNS), perpetrators can substitute the address of their Web site in place of the real one to spoof Web site visitors. For instance, a hacker could create a fictitious Web site masquerading as www.widgetsinternational.com

by exploiting a DNS security hole that substitutes his or her fake IP address for Widgets International's real IP address. All subsequent visits to Widgets International would be redirected to the fake Web site. There, the hacker could alter any orders to change the number of widgets ordered and redirect shipment of those products to another address. The integrity attack consists of altering an order and passing it to the real company's commerce server. The commerce server is unaware of the integrity attack and simply verifies the consumer's credit card number and passes the order on for fulfillment.

Integrity threats can alter vital financial, medical, or military information. You can imagine the impact if someone were to capture a message that credits a bank for $10 million dollars, and the credit is changed to a debit. Similarly, an integrity violation that alters an e-mailed résumé can affect the applicant's chances of being hired by the receiving company. Information alteration can have very serious consequences for businesses and people.

Necessity Threats

The purpose of a **necessity threat**, also known by other names such as a **delay, denial,** or **denial-of-service threat (DoS)**, is to disrupt normal computer processing or to deny processing entirely. A computer that has experienced a necessity threat slows processing to an intolerably slow speed. For example, if the processing speed of a single ATM transaction slows from one or two seconds to 30 seconds, users will abandon ATMs entirely. Similarly, slowing any Internet service will drive customers to competitors' Web or commerce sites—possibly discouraging them from ever returning to the original commerce site. In other words, slowing processing can render a service unusable or unattractive. For example, an online newspaper that reports three-day-old news is worth very little.

DoS attacks remove information altogether or delete information from a transmission or file. One documented denial attack caused selected PCs that have Quicken (an accounting program) installed to divert money to the perpetrator's bank account. The denial attack denied money from its rightful owners. In another famous DoS attack in early 2000, a rash of attacks occurred against high-profile electronic commerce sites such as Amazon.com and Yahoo!. The attackers used zombie computers to send a flood of data packets to overwhelm various electronic commerce servers and choke legitimate customers' access to them. Prior to the attack, perpetrators located vulnerable computers and loaded them with the software that attacked the commerce sites. The Internet Worm attack described in the opening case to this chapter is another example of a DoS attack.

THREATS TO THE SECURITY OF SERVER COMPUTERS

The server is the third link in the client-Internet-server trio embodying the electronic commerce path between the user and a commerce server. Servers have vulnerabilities that can be exploited by anyone determined to cause destruction or to acquire information illegally. One entry point is the Web server and its software. Other entry points are any backend programs containing data, such as a database and the server on which it runs. Perhaps the most dangerous entry points are

Common Gateway Interface (CGI) programs or utility programs residing on the server. While no system is completely safe, the commerce server administrator's job is to make sure that security policies are documented and considered in every part of the electronic commerce system.

Web Server Threats

Web server software, as you learned in Chapter 8, is designed to deliver Web pages by responding to HTTP requests. Although Web server software is not inherently high-risk software, it has been designed with Web service and convenience as the main design goals. The more complex the software, the higher the probability that it contains coding errors (bugs) and that it contains security weaknesses that provide openings through which unauthorized persons can obtain access to the computer on which the software runs.

Web servers running on most machines, including UNIX-based computers, can be set up to run at various privilege levels. The highest privilege level provides the most flexibility and allows programs, including Web servers, to execute all machine instructions and to have unlimited access to any part of the system, including highly sensitive and privileged areas. Correspondingly, the lowest privilege levels provide a logical fence around an executing program, preventing it from running whole classes of machine instructions and disallowing it access to all but the least sensitive areas of computer storage. Most security experts provide a program the lowest privilege level it needs to do its job. A system administrator who sets up accounts and passwords for users temporarily needs a very high privilege level—called *superuser* on computers that run UNIX operating systems—to modify sensitive and valuable areas of the system. Setting up a Web server to run in high-privilege status can lead to a Web server threat. Most of the time, a Web server provides ordinary services and mundane tasks that can be accomplished with a very low privilege level. If a Web server runs at a high privilege level, someone trying to exploit a Web server may be able to do so and subsequently execute instructions in privileged mode.

A Web server can compromise secrecy if it keeps the default setting of automatic directory listings selected. The secrecy violation occurs when the contents of a server's folder names are revealed to a Web browser. This frequently happens and is caused when a user enters a URL such as http://www.somecompany.com/FAQ/ and expects to see the default page in the FAQ directory. The default Web page that the server normally displays is named index.htm or index.html. If that file is not in the directory, the Web server frequently displays all the folder names in the directory. Then, visitors can click folder names at random and open folders that might otherwise be off limits. Figure 10-11 shows an example of displayed folder names.

385

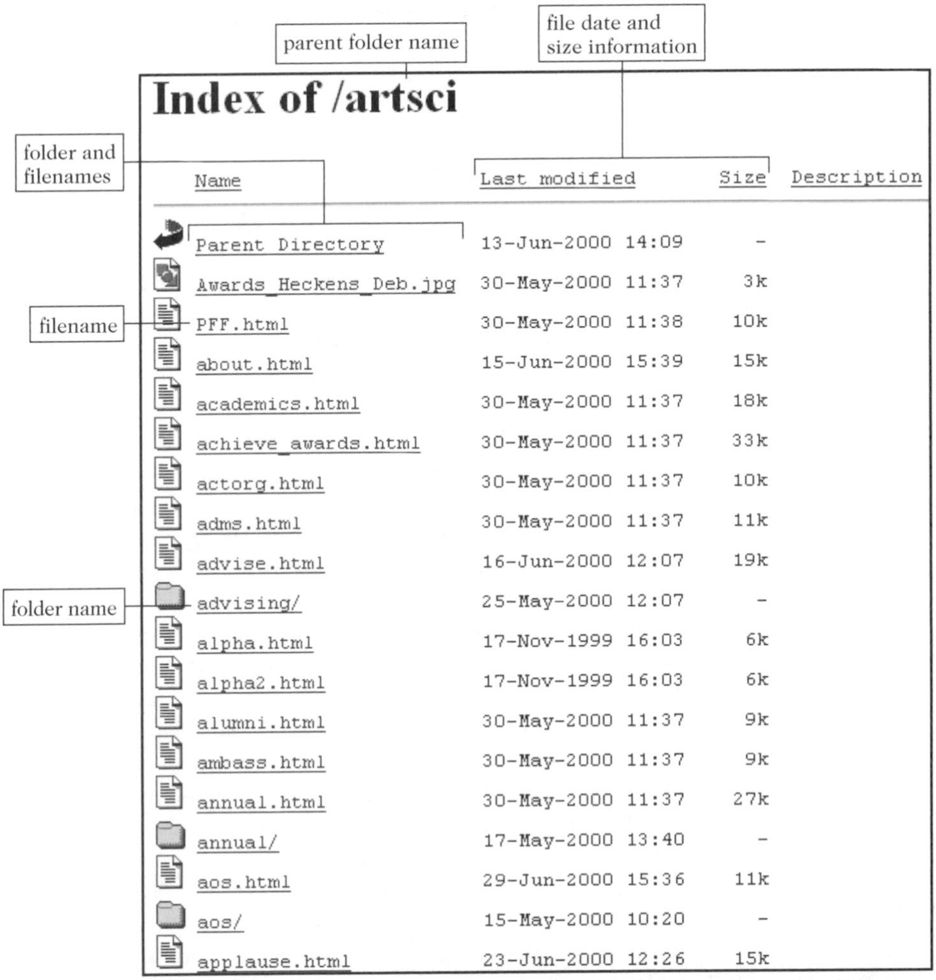
parent folder name
file date and size information
Index of /artsci
folder and filenames
Name
Last modified
Size
Description
Parent Directory 13-Jun-2000 14:09 -
Awards_Heckens_Deb.jpg 30-May-2000 11:37 3k
filename
PFF.html 30-May-2000 11:38 10k
about.html 15-Jun-2000 15:39 15k
academics.html 30-May-2000 11:37 18k
achieve_awards.html 30-May-2000 11:37 33k
actorg.html 30-May-2000 11:37 10k
adms.html 30-May-2000 11:37 11k
advise.html 16-Jun-2000 12:07 19k
folder name
advising/ 25-May-2000 12:07 -
alpha.html 17-Nov-1999 16:03 6k
alpha2.html 17-Nov-1999 16:03 6k
alumni.html 30-May-2000 11:37 9k
ambass.html 30-May-2000 11:37 9k
annual.html 30-May-2000 11:37 27k
annual/ 17-May-2000 13:40 -
aos.html 29-Jun-2000 15:36 11k
aos/ 15-May-2000 10:20 -
applause.html 23-Jun-2000 12:26 15k

Figure 10-11 *Displaying folder and filenames with a Web browser*

Careful site administrators turn off the folder name display feature. If a user attempts to browse a folder where protections prevent browsing, the Web server issues a warning message stating that the directory is not available.

Web servers can compromise security by requiring users to enter a username and password. The act of submitting a username to enter a particular part of the Web space is not in itself a secrecy violation. However, the confidential username and password can be subsequently revealed when the user visits multiple pages within the same Web server's protected area. The reason this can occur is that some servers require that users reestablish their usernames and passwords for each page they visit in the premium content area. This repeated information requirement is necessary because the Web is stateless—it cannot remember what happened during the last transaction. The most convenient way to remember a username and password is to store the user's confidential information in a cookie on his or her computer. That way, the Web server can request confirmation of the data by requesting that the computer send a cookie. Secrecy violations can occur because a cookie's information might be transmitted in an insecure way and copied by an eavesdropper. Although cookies are not inherently unsafe, a Web server should not ask a Web browser to transmit a cookie unprotected.

You learned in Chapter 8 that a Server Side Include (SSI), also called a servlet, is a small program embedded in a Web page that is executed by the server. Whenever a program is executed on a server and the program comes from an unknown or untrusted source, the SSI might request some unauthorized execution. The embedded SSI code could be an operating system directive that requests that the password file be displayed or sent back to a particular location. The **W3C Security FAQ** provides additional information about server security.

The File Transfer Protocol (FTP) program, described in Chapter 2, although not a Web server, is part of many Web server configurations. The FTP program can reveal threats to the Web server's integrity. One possibility for unauthorized information disclosure occurs when there are no protection mechanisms on the folders that an FTP user can browse. For instance, suppose a regular business client has an account on another business's computer and can periodically upload data to the business partner's computer. Using an FTP client program, a system administrator can log on to the business partner's computer, upload data, and then open and display the contents of other folders on the Web server computer. This is not difficult to do if protections are missing. With a Web client program, a user can double-click the parent directory folder to move up the folder hierarchy, double-click some other folder, such as the other company's privileged folder, and then download any information contained in that folder. This security breach is possible simply because the company has forgotten to restrict its business partner's browsing capabilities to a single folder.

One of the most sensitive files on a Web server is the file that holds Web server username and password pairs. If that file is compromised, an intruder can enter privileged areas masquerading as someone else. Such an intruder can obtain usernames and passwords if that information is readily available and not encrypted. Most Web servers provide secure storage of user authentication information. It is up to the Web server administrator to ensure that the Web server is instructed to always apply protection mechanisms to the data.

The passwords that users select can be a threat. Users sometimes select passwords that are easily guessed because they are their mother's maiden name, the name of one of their children, a telephone number, or some easily obtained identification number such as a social security number. **Dictionary attack programs** cycle through an electronic dictionary, trying every word in the book as a password. There are simple remedies for this attack, but users' passwords, once broken, may provide an opening for illegal entry into a server that can remain undetected for a long time.

Database Threats

Electronic commerce systems store user data and retrieve product information from databases connected to the Web server. Besides storing product information, databases connected to the Web contain valuable and private information that could irreparably damage a company if it were disclosed or altered. Most modern, large-scale database systems use extensive database security features that rely on usernames and passwords. Once a user is authenticated to a database, selected portions of the database are visible to that user. Security is enforced in databases through the use of privileges, which are stored in the database. However, some databases either store username/password pairs in a nonsecure way, or they fail to enforce security altogether and rely on the Web server to enforce security. If someone obtains user authentication information, then he or she can masquerade as a legitimate database user and reveal or download private and valuable information. Trojan horse programs hidden within the database system can also reveal information by downgrading the system (releasing sensitive information to a less protected area of the database that a broader population of users can peruse). When information is downgraded, all users have access—including potential intruders. Database security requires attention from a careful database administrator. Figure 10-12 shows a Web page describing Oracle's database security features.

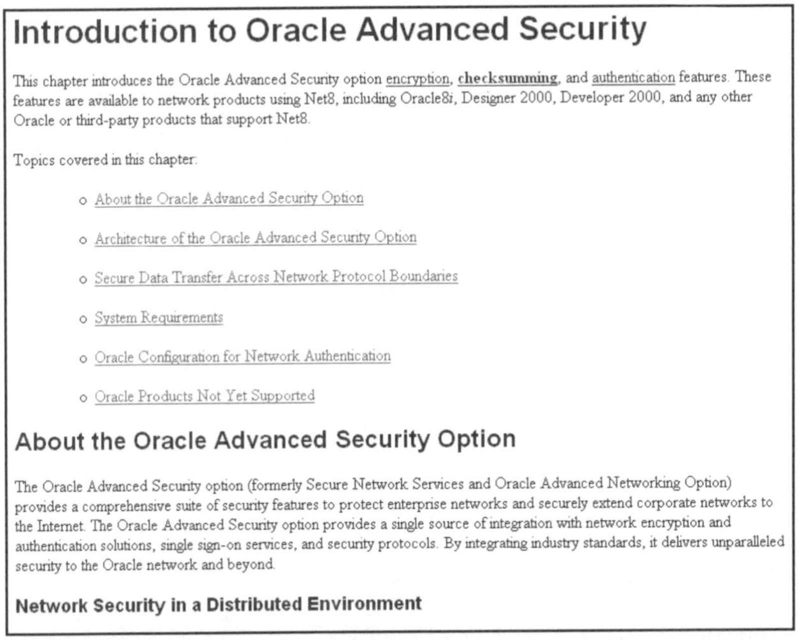

Introduction to Oracle Advanced Security

This chapter introduces the Oracle Advanced Security option encryption, checksumming, and authentication features. These features are available to network products using Net8, including Oracle8i, Designer 2000, Developer 2000, and any other Oracle or third-party products that support Net8.

Topics covered in this chapter:

- o About the Oracle Advanced Security Option

- o Architecture of the Oracle Advanced Security Option

- o Secure Data Transfer Across Network Protocol Boundaries

- o System Requirements

- o Oracle Configuration for Network Authentication

- o Oracle Products Not Yet Supported

About the Oracle Advanced Security Option

The Oracle Advanced Security option (formerly Secure Network Services and Oracle Advanced Networking Option) provides a comprehensive suite of security features to protect enterprise networks and securely extend corporate networks to the Internet. The Oracle Advanced Security option provides a single source of integration with network encryption and authentication solutions, single sign-on services, and security protocols. By integrating industry standards, it delivers unparalleled security to the Oracle network and beyond.

Network Security in a Distributed Environment

Figure 10-12   Oracle security features page

Common Gateway Interface Threats

Recall that a Common Gateway Interface (CGI) implements the transfer of information from a Web server to another program, such as a database program. CGIs and the programs to which they transfer data provide active content to Web pages. For example, a Web page might contain a list box asking you to fill in the name of your favorite professional sports team. Once you submit your choice, CGI programs process the information and look up the latest scores for the sports team you designated, place the scores into a Web page, and send the generated page back to your client browser.

Because CGIs are programs, they present a security threat if misused. Just like Web servers, CGI scripts can be set up to run with their privileges set to high—unconstrained. Defective or malicious CGIs with free access to system resources are capable of disabling the system, calling privileged (and dangerous) base system programs that delete files, or viewing confidential customer information, including usernames and passwords. When programmers discover inadequacies or errors in CGI programs, they rewrite and replace them. Older, retired CGIs that are not erased can provide openings into the system that have been long forgotten by the systems development staff. And, because CGI programs or scripts can reside almost anywhere on a Web server (that is, in any folder or directory), they are hard to track down and manage. However, anyone who is determined enough can track down replaced CGI scripts, examine them, ascertain their weaknesses, and exploit those weaknesses to gain access to a Web server and its resources. Unlike JavaScript, CGI scripts do not run inside a protective security perimeter or sandbox.

Other Programming Threats

Other Web server attacks can come from programs executed by the server. Java or C++ programs that are passed to Web servers by a client or that reside on a server frequently make use of a buffer. A **buffer** is an area of memory set aside to hold data read from a file or database. A buffer is necessary whenever any input or output operation takes place, because a computer can process file information much faster than the information can be read from input devices or written to output devices. A buffer serves as a holding area for incoming or outgoing data. Database information about to be processed, for example, is gathered in a buffer so that either the entire collection or a large quantity of it is in the computer's memory. Then, the data is made available to the processing unit for manipulation and analysis. Programs filling buffers can malfunction and overfill the buffer, spilling the excess data outside the designated buffer memory area. Usually, this occurs because the program contains an error or bug that causes the overflow. Sometimes, however, the mistake is intentional. In either case, buffer overflows can have moderate to very serious security consequences.

Anyone who has programmed has probably experienced the consequences of a buffer overflow or a runaway code segment that causes data or instructions to overwrite an out-of-bounds area in memory. The normal result of such a programming error is that the program halts with an exception—an error condition—and the process stops. Occasionally, the entire computer halts (crashes). Intentional crashes are deliberate denial attacks. The Internet Worm attack was such a program. It caused an overflow condition that eventually consumed all resources until the affected computer could no longer function.

A more insidious version of a buffer overflow attack writes *instructions* into critical memory locations so that when the intruder program has completed its work of overwriting buffers, the Web server resumes execution by loading internal registers with the address of the main attacking program's code. This type of attack can open the Web server to severe damage because the resumed program—which is now the attacker program—may regain control at a very high privilege or super user level. This opens up most files to disclosure and destruction by the attacking program.

Figure 10-13 shows a graphical representation of data being read from a file. The data flows into a memory buffer and then into a system area that is called the save area. A **save area** is where programs store critical information, such as the contents of all central processing unit registers and partial results of a program's computations just before control is passed to another program. When control returns to the original program, the save area contents are reloaded back into CPU registers, and control returns to the next instruction in the program. In an attack, however, control returns to the attacking program, not the benign program that gave up control. The **Red Hat Linux Buffer Overflow Attacks Page** link in the Online Companion describes the buffer vulnerabilities of Web servers that run on the Linux operating system.

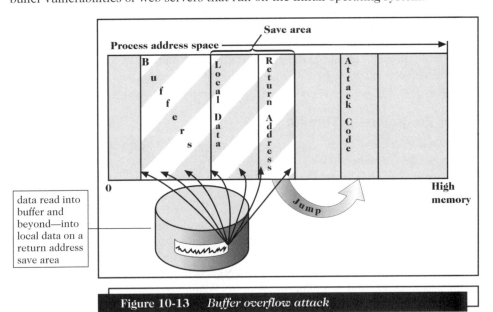

Figure 10-13   *Buffer overflow attack*

A similar attack, one in which excessive data is sent to a server, can occur on mail servers. Called a **mail bomb**, the attack occurs when hundreds or even thousands of people each send a message to a particular address. The attack might be launched by a large team of well-organized hackers, but more likely the attack is launched by one or a few hackers who have gained control over others' computers using a Trojan horse virus or some other method of turning those computers into zombies.

The accumulated mail received by the target of the mail bomb exceeds the allowed e-mail size limit and can cause e-mail systems to malfunction. While it is fairly easy to track the people responsible for the attack, it is debilitating nonetheless. Mail bombs may seem similar to spam, but they are just the opposite. Spamming occurs when one person or organization sends a *single* message to thousands of people, and is more of a nuisance than a security threat.

390

CERT

In 1988, a group of researchers met to study the infamous Internet Worm attack soon after it occurred. They wanted to understand how worms worked and how to prevent damage from future attacks of this type. The National Computer Security Center, part of the National Security Agency, initiated a series of meetings to figure out how to respond to future security breaks that might affect thousands of people. Soon after that meeting of security experts in 1988, the U.S. government created the Computer Emergency Response Team and housed it at Carnegie Mellon University in Pittsburgh. The operation is now operated as part of the federally funded Software Engineering Institute at Carnegie Mellon and has changed its legal name from the Computer Emergency Response Team (which had been abbreviated to "CERT" by most people who wrote and talked about it) to **CERT**. CERT still maintains an effective and quick communications infrastructure among security experts so that security incidents can be avoided or quickly handled.

Today, CERT responds to thousands of security incidents each year and provides a wealth of information to help Internet users and companies become more knowledgeable about security risks. CERT posts alerts to inform the Internet community about security events, and it is regarded as a primary authoritative source for information about viruses, worms, and other types of attacks. Figure 10-14 shows the CERT Vulnerabilities, Incidents & Fixes page, which many Web site administrators consult immediately when they suspect their sites are under attack.

<section>
Figure 10-14   CERT Vulnerabilities, Incidents & Fixes page
</section>

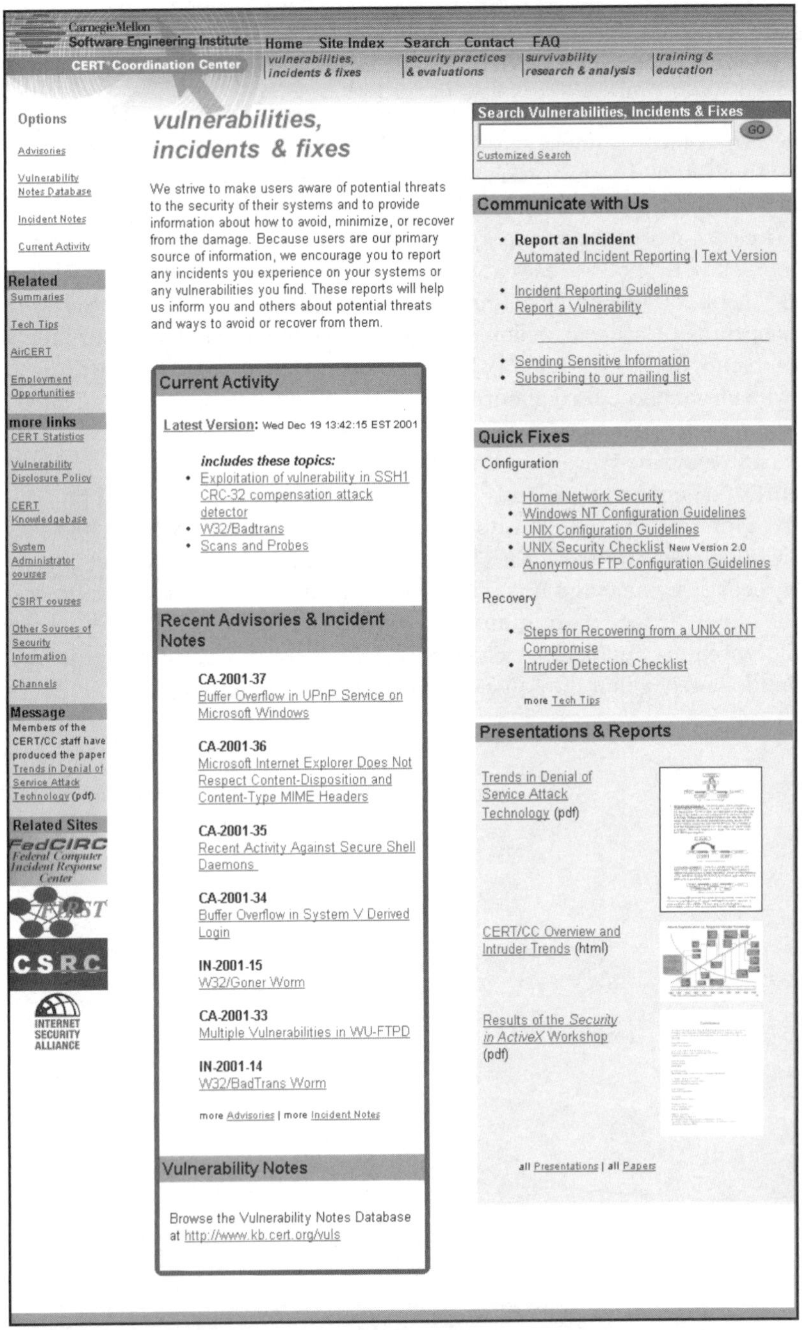

<section>
392
</section>

Summary

Electronic commerce is vulnerable to a wide range of security threats. Attacks against electronic commerce systems can disclose or manipulate proprietary information. Any commerce security policy must address secrecy, integrity, necessity, and intellectual property rights. Threats to commerce can occur anywhere in the commerce chain, beginning with a client computer and ending with the commerce and back office servers. News accounts of virus attacks have kept Web users aware of the security risks to client computers. More subtle threats are delivered as client-side applets. Java, JavaScript, and ActiveX controls run on client machines and have the potential to breach security.

Communication channels in general, and the Internet in particular, are especially vulnerable to attacks. The Internet is a vast network, and because no control exists over the nodes through which Internet traffic passes, information sent through the Internet is vulnerable to unauthorized disclosure. This can lead to disclosure of private information, alteration of critical business documents, and theft or loss of important business messages. The Internet Worm of 1988 is a classic example of a security threat that used the Internet as a vehicle to travel around the world and infect thousands of computers within minutes.

Commerce servers are susceptible to security breaches in many of the same ways as are client machines. Client security breaks can occur to any client that connects to the affected server. Common Gateway Interface (CGI) programs that run on servers have the potential to damage databases, abnormally terminate server software, and subtly change proprietary information. Attacks can come from within the server in the form of programs, or they can come from outside the server. One external attack occurs when a message overflows a server's internal storage region and overwrites crucial information. Overwritten information is replaced with either data or instructions that cause other programs on the server to execute.

The Computer Emergency Response Team, now known as CERT, was formed to address security outbreaks by linking knowledgeable security experts. When large security outbreaks occur, CERT members discuss methods to locate and eliminate the electronic attacker.

Key Terms

Active content

Active wiretapping

ActiveX

Applet

Backdoor

Buffer

Computer security

Countermeasure

Cracker

Cybersquatting

Cyber vandalism

Delay threat

Denial-of-service threat (DoS)

Denial threat

Dictionary attack program

Domain name ownership change

Eavesdropper

Hacker

Integrity

Java sandbox

JavaScript

Logical security

Macro virus

Mail bomb

Masquerading (spoofing)

Multi-vector virus

Name changing

Name stealing

Necessity

Necessity threat

Physical security

Plug-ins

Privacy

Save area

Secrecy   Trusted applet
Security policy   Untrusted applet
Signed Java applet   VBScript
Sniffer program   Virus
Steganography   Worm
Threat   Zombie
Trojan horse

Review Questions

1. Explain why Web sites use cookies. What problem do cookies solve? What does a cookie contain? How large are cookies? Where are they stored? Use the Online Companion to help with your research.

2. What is steganography? What does steganography have to do with security? Use the Online Companion to research the question. Write at least 100 words explaining steganography.

3. List some of the Web server security risks. See if you can find a Web site where you can list the names of the directories on the site and print them.

4. What security risks does the Internet pose? Are the risks mostly related to secrecy, or can a message's integrity be compromised also?

5. Why are programs such as CGI scripts and Java programs that run on client machines or on a Web server considered security threats? Explain how programs could breach security. Do Java script programs pose an equally serious security risk?

Exercises

1. Brought Back Bugs is a used Volkswagen dealer in Lincoln, Nebraska. The dealership has hired you to create a Web site for it. One of its requirements is that the site should display a few banner advertisements showing the week's specials. You decide that active content would be the thing to use to provide the rotating advertisement. Your immediate task is to investigate Java, JavaScript, Jscript, and Java applets—you need to learn everything there is to know about Java-related alternatives. Using the Online Companion and Web search engines, research Java applications and write a two-page paper about what you have discovered about Java, JavaScript, and the other Java namesakes. You might want to start by describing the history of Java. Who invented it? What are the advantages and disadvantages of Java over ActiveX controls? Who publishes JavaScript and Jscript? Are electronic commerce customers restricted to a particular browser to run Java-enhanced Web pages? Be sure to list the URLs for any Web sites that you use in your research.

2. You suspect that your electronic commerce site's Web server may be under attack from at least one source. Describe in a general way what types of threats are possible on a Web server. Consult outside sources on the Internet to help you answer this question. What are the three categories of security threats and how might they apply to your organization's Web server? Answer these two separate questions in 100–200 words.

3. Write a 300-word paper in which you evaluate the **CERT** organization. Include information about when it was founded, what groups or people are members, and where it is headquartered. Include in your discussion at least three current security alerts, specifying the name of the virus or attack program, the date the alert was posted, and two sentences about each reported security alert. Use the Online Companion, any of several Internet search engines, and the CERT home page to help you locate information. Use at least one of the sources in the references at the end of this chapter and include a citation to that reference.

For Further Study and Research

Ahuja, V. 1997. *Secure Commerce on the Internet.* Boston: AP Professional.

Alexander, S. 2000. "Viruses, Worms, Trojan Horses and Zombies," *Computerworld*, 34(18), May 1, 74.

Anderson, L. 2001. "They've Got Your Number," *The Industry Standard*, May 28. Available online at: (http://www.thestandard.com/article/0, 1902,24687,00.html).

Applegate, L., et. al. 1996. "Electronic Commerce: Building Blocks of New Business Opportunity," *Journal of Organizational Computing and Electronic Commerce*, 6(1), June, 1–10.

Briney, A. 2001. "Industry Survey 2001," *Information Security*, 4(10), October, 34–46.

Butler, R. and A. Goldstein. 2001. "Keeping the Hackers at Bay," *Time*, 158(23), November 26, 87.

Cohen, A. 2001. "When Terror Hides Online," *Time*, November 12, 65.

Colkin, E., A. Gilbert, G. Hulme, M. McGee, and J. Rendleman. 2001. "IT Security and the Law," *Information Week*, November 26, 22–24.

Connolly, P. 2001. "Low Cost Defense," *InfoWorld*, 23(70), September 24, 64.

Cope, N. 2000. "A Hit for Jethro Tull in Domain Name Dispute," *The Independent*, July 31, 15.

Dacey, R. 2001. *Information Security: IRS Electronic Filing Systems (GAO-01-306).* Washington, D.C.: United States General Accounting Office.

Delio, M. 2001. "Is This World Cyber War I?" *Wired News*, May 1. Available online at: (http://www.wired.com/news/politics/ 0,1283,43443,00.html).

Dyck, T. 2001. "How to Shore Up IT Resources to Prevent Attacks in the First Place," *eWeek*, 18(44), November 12, 65–66.

Ferguson, P. and D. Seine. 1998. "Network Ingress Filtering: Defeating Denial of Service Attacks Which Employ IP Source Address Spoofing," *Network Working Group, RFC 2267*, January.

Fisher, D. 2001. "Cracking Down on Hackers," *eWeek*, 18(44), November 12, 1–2.

Gardner, D. 1998. "E-Mail Bug Stirs Up a Scare," *Infoworld*, August 3, 20.

Gurley, J. 2001. "From Wired To Wiretapped," *Fortune*, 144(7), October 15, 214–215.

Hancock, B. 2001. "Terrorism and Steganography: Shaken, Not Stirred," *Computers & Security*, 20(2) 110–111.

Hardesty, L. 2001. "Stemming the Flood," *Technology Review*, 104(7), September, 28.

Harrison, A. 2000. "Denial-of-Service Victims Share Lessons Learned," *Computerworld*, 34(25), June 19, 8.

Hutheesing, N. 2001. "Master of Your Domain," *Forbes*, 167(5), Spring, 60.

Isenberg, D. 2000. "Many Trademarks, But Just One Domain Name," *Internet World*, July 1, 86.

Kennedy, S. 2001. "Security Technology and Other Issues," *Information Today*, 18(11), December, 34–35.

Kushner, D. 1999. "The Domain Name Game," *PC Magazine*, November 12. Available online at: (http://www.zdnet.com/pcmag/stories/reviews/ 0,6755,2385324,00.html).

Lawson, N. and J. Garris. 1999. "Plug Your Company's Common Security Holes," *PC Magazine*, 18(10), May 25, 217–218.

Lim, F. 1998. *ActiveX and the Internet.* El Granada: Scott/Jones, Inc.

Longstaff, T., J. Ellis, S. Hernan, H. Lipson, R. McMillan, L. Pesante, and D. Simmel. 1997. "Security of the Internet," *Froelich/Kent Encyclopedia of Telecommunications,* 231–255. New York: Marcel Dekker.

Machrone, B. 2000. "Always More Threats," *PC Magazine,* 19(13), July, 101.

Maney, K. 2001. "Osama's Messages Could Be Hiding in Plain Sight," *USA Today,* December 19, 6B.

McCullagh, D. 2001. "'Secure' U.S. Site Wasn't Very," *Wired News,* July 6. Available online at: (http://www.wired.com/news/privacy/0,1848,45031,00.html).

McGraw, G. and E. Felton. 1999. *Securing Java: Getting Down to Business with Mobile Code.* New York: John Wiley & Sons.

Merkow, M., J. Breithaupt, and K. Wheeler. 1998. *Building SET Applications for Secure Transactions.* New York: John Wiley & Sons.

Neuman, B. and T. Tso. 1994. "An Authentication Service for Computer Networks," *IEEE Communications,* 32(9), September, 33–38.

Null, C. 2000. "Name Grab," *PC Computing,* 13(4), April, 40–42.

O'Harrow, R. 2001. "Prozac Maker Reveals Patient E-Mail Addresses," *The Washington Post,* July 4, E1.

Oppliger, R. 1997. "Internet Security: Firewalls and Beyond," *Communications of the ACM,* 40(5), May, 92–102.

Radcliff, D. 2000. "Domain Name Game," *Computerworld,* 34(24), June 12, 71.

Rothstein, P. 2001. "Disaster Recovery: September 11 Changes Everything," *Information Security,* 4(11), 48–49.

Saita, A. 2001. "Deep Digital Cover," *Information Security,* 4(10), October, 22.

Scalet, S. 2001. "See You in Court: Information Security," *CIO,* November 1, 62–70.

Schwartz, E., B. Fonseca, D. Neel, S. Lee, and J. McCarthy. 2001. "Security Concerns Top Agenda," *InfoWorld,* 23(46), 17–18.

Shipley, G. 2001. "Growing Up with a Little Help from the Worm," *Network Computing,* 12(20), October 1, 39.

Spafford, E. 1988. "The Internet Worm Program: An Analysis," *Purdue University Computer Science Department Technical Report CSD-TR-823.*

Stein, L. 2000. "Napster: Asking for Trouble. Getting It." *Web Techniques,* 5(5), May, 12–15.

Sterling, B. 2001. "Steganography Goes Digital," *The New York Times,* December 9, 102.

Tippett, P. 2001. "The Crypto Myth," *Information Security,* 4(5), May, 38–40.

Trombly, M. 2001. "Wall Street Firms Look Externally for Web Security," *Securities Industry News,* 13(33), August 20, 3–4.

Verton, D. 2001. "Microsoft in Hot Seat After Code Red," *Computerworld,* 35(32), August 6, 1–2.

Verton, D. 2001. "Record-Breaking Year for Security Incidents Expected," *Computerworld,* 35(48), November 26, 12.

Vijayan, J. 2000. "Possible S & P Security Holes Reveal Risks of E-Commerce," *Computerworld,* 34(22), May 29, 6.

Vijayan, J. 2000. "Analysts: Better to Be Safe Than Sorry With Viruses," *Computerworld,* 34(26), June 26, 20.

Vijayan, J. 2000. "Joke Virus Spreads Fast, Clogs Servers," *Computerworld,* 34(26), June 26, 20.

Yeh, W-H. and J-J. Hwang. 2001. "Hiding Digital Information Using a Novel System Scheme," *Computers & Security,* 20(6), 533–538.

Zetter, K. 2001. "Holey Software!" *PC World,* 19(11), November, 135–140.

396

IMPLEMENTING ELECTRONIC COMMERCE SECURITY

INTRODUCTION

Jim Lockhart, IT help desk manager at CSD, Inc., first became aware of the problem late Tuesday evening.

Engineers who had been working on several different projects called him in quick succession complaining

that their computers had "died" and that files had disappeared. When Jim questioned each engineer,

several of them indicated that they kept electronic mail client programs open on their desktops and read

their e-mail occasionally. All the engineers had the same story: They were reading their e-mail, and when

they opened an e-mail attachment, their computers began acting strangely. The first symptoms were that

the programs (other than their e-mail client programs) reported file errors. When they checked the files,

they found that the files they had been working on had somehow been reduced to zero bytes in length.

The ExploreZip virus had struck CSD. ExploreZip is a Trojan horse virus that launches a worm

virus when it executes. It arrives by e-mail as an attachment. When the attachment (which is disguised

as a zipped file) is opened, the Trojan horse program opens and begins deleting files. The ExploreZip

worm program also becomes active. The worm program searches for available shared folders on any

network to which the infected computer is connected. It also launches an e-mail attack. When a computer infected with ExploreZip receives an e-mail message, the worm causes the e-mail program on that computer to reply to the message with a message that contains the ExploreZip virus as an attached file. In the message body carrying the attachment is the seemingly innocent text: "Hi (Recipient Name)! I received your email and I shall send you a reply ASAP. Until then, take a look at the attached zipped docs. Bye." When the unwary recipient opens the attachment, ExploreZip repeats its damage on the recipient's computer.

Protecting the electronic assets of electronic commerce systems is necessary if commerce is to thrive. The electronic world will always have to deal with viruses, worms, Trojan horses, eavesdroppers, and destructive programs whose goals are to disrupt, delay, or deny communications and the information flows between participants in industry value chains. Security protection must continually be developed to provide consumers with confidence in the online systems with which they interact, and through which they conduct business. This chapter describes security measures that protect client computers, the Internet over which commerce information flows, and the commerce server.

LEARNING OBJECTIVES

In this chapter, you will learn about:
- Security measures that can reduce or eliminate intellectual property theft
- How to secure client computers from attack by viruses and by ill-intentioned programs and scripts downloaded in Web pages
- How to authenticate users to servers and how to authenticate servers
- The protection mechanisms that are available to secure information sent between a client and server so that the information remains private
- How to secure message integrity, preventing another program from altering information as it travels on the Internet
- How firewalls can protect intranets and corporate servers against being attacked through the Internet
- What roles the Secure Sockets Layer, Secure HyperText Transfer Protocol, and secure electronic transaction protocols play in protecting electronic commerce

PROTECTING ELECTRONIC COMMERCE ASSETS

Regardless of whether companies are doing business over the Internet or face to face, security is a serious issue. Customers engaging in electronic commerce need to feel confident that their transactions are secure from prying eyes and safe from alteration. The volume of business sales conducted online is significant and will continue to grow over the coming years.

When businesses began using computers 50 years ago, security meant physical security: alarmed doors and windows, guards, security badges to admit people to sensitive areas, surveillance cameras, and so on. Back then, interactions between people and computers were limited to terminals (which had no processing capabilities of their own) connected directly to large mainframe computers. There were no other connections to computers. Computer security meant dealing with the few people who had access to terminals. Anyone wishing to run programs did so by submitting programs in the form of decks of punched cards fed into card readers. People reclaimed their card decks and the output results—usually piles of fan-folded, green-bar paper—from the input/output clerk. Security was pretty simple.

Both the population of computer users and the methods to access computing resources have increased tremendously since those early years of computing. Millions of people now have access to computing power over both private and public networks that connect millions of computers. It is not a simple matter to determine who is using a computing resource, because the user could be located in South Africa, but using a computer in California. A whole new series of security tools and methods have evolved and are employed today to protect electronic assets. The transmission of valuable information such as electronic receipts, purchase orders, credit card numbers, and order confirmations has drastically changed the way security is viewed, and has introduced new electronic and automatic methods to deal with security threats.

Data security measures date back to the time of the Roman Empire, when Julius Caesar coded information to prevent enemies from reading secret war and defense plans. Modern electronic security methods also trace their roots to the defense sector. The U.S. Department of Defense was the main driving force behind both early security requirements and more recent advances. In the late 1970s, the Defense Department formed a committee to develop computer security guidelines for handling classified information on computers. The result of that committee's work was *Trusted Computer System Evaluation Criteria*, known in defense circles simply as the "Orange Book," because its cover was orange. It spelled out rules for mandatory access control—the separation of confidential, secret, and top secret information—and established criteria for certification levels for computers ranging from D (not trusted to handle multiple levels of classified documents at once) to A1 (the most trustworthy level).

Although that work was groundbreaking in defining security terms, conditions, and tests for security, it did not address how to handle electronic commerce computer security. Nonetheless, that early security work has been beneficial in developing electronic commerce computer security because it spawned commerce security research, which resulted in commercially applicable and practical security solutions. The early work also provided a much-needed formal approach to security. For example, security experts have learned that an organization cannot hope to produce secure commerce systems unless it has a written security policy in place. As you learned in Chapter 10, a security policy must spell out the assets to protect, what is needed to protect those assets, an analysis of the likelihood of the threats, and the rules to enforce to protect those assets. The security policy must be regularly reviewed and revised as threat conditions change. Without a written policy, it is difficult to implement any security at all.

Both defense and commercial security guidelines state that organizations must protect assets from unauthorized disclosure, modification, or destruction. However, military security policy differs from commercial policy because military applications stress separation of levels of security. Corporate information is usually classified as either "public" or "company confidential." The typical security policy concerning confidential company information is straightforward: Do not reveal company confidential information to anyone outside the company.

As mentioned in Chapter 10, a security policy should protect a system's privacy, integrity, and availability (necessity) and authenticate users. When you recast these goals for electronic commerce, they become those shown in Figure 11-1. You can find some example security policies and guidelines for writing security policies at the **SANS Security Policy Project** Web page. The SANS (System Administration, Networking, and Security) Institute is a research and education organization that gives systems administrators, computer security professionals, and network administrators a place to share information about the challenges they face in their daily work.

Requirement	Meaning
Secrecy	Prevent unauthorized persons from reading messages and business plans, obtaining credit card numbers, or deriving other confidential information.
Integrity	Enclose information in a digital envelope so that the computer can automatically detect messages that have been altered in transit.
Availability	Provide delivery assurance for each message segment so that messages or message segments cannot be lost undetectably.
Key management	Provide secure distribution and management of keys needed to provide secure communications.
Nonrepudiation	Provide undeniable, end-to-end proof of each message's origin and recipient.
Authentication	Securely identify clients and servers with digital signatures and certificates.

Figure 11-1   *Minimum requirements for secure electronic commerce*

This chapter examines security by looking at how to protect assets, beginning with the client. First, you will learn about protections available for intellectual property and privacy online. Then, you will learn about protections for client computers, for the transmission media (the Internet itself), and, finally, for Web servers.

PROTECTING INTELLECTUAL PROPERTY AND PRIVACY ONLINE

Protecting digital intellectual property poses problems that are different from traditional intellectual property security. You learned in Chapter 7 that traditional intellectual properties, such as written works, art, and music, are protected by national and, in some cases, international laws. Digital intellectual properties, including art, logos, and music posted on Web sites, are also protected by laws. Although those laws act as a deterrent, they do not prevent violations from occurring, and they do not provide a means to reliably trace the path taken by the violators to acquire the intellectual property.

The real dilemma for individuals presenting digital property is how to display and make available intellectual property on the Web, while simultaneously protecting those copyrighted works. Although absolute intellectual property protection so far has proven elusive, there are measures available that provide some level of protection and accountability for copyrights held on digital works.

Protecting Intellectual Property

In the United States, Congress has been trying to deal legislatively with digital copyright issues. The U.S. Department of Justice maintains the **Cybercrime** site to provide information and updates on hacking, software piracy, and the latest security information, as well as the latest information on cyber crime prosecutions. Part of that site is devoted to protecting intellectual property, and provides valuable information about both intellectual property attacks and countermeasures that companies and individuals can employ to protect intellectual assets (see Figure 11-2).

The **Information Technology Association of America (ITAA)**, which is a trade organization representing U.S. information technology, has proposed some solutions to the current problems in digital copyright protection, including the following:

- Host name blocking
- Packet filtering
- Proxy servers

401

Computer Crime and
Intellectual Property Section (CCIPS)

Protecting Intellectual Property Rights: Copyrights, Trademarks and Trade Secrets

A. Intellectual Property Policy and Programs
B. Intellectual Property Cases
C. Prosecuting Intellectual Property Crimes Guidance
D. Criminal Intellectual Property Laws
E. Economic Espionage Act
F. Intellectual Property Documents

I. Intellectual Property Policy and Programs

The Department of Justice and other agencies are continually working to improve protections for intellectual property rights and the enforcement of intellectual property laws. You can find information on DOJ initiatives, summits, and speeches in this section. This section also contains information on U.S. interagency efforts, such as NIPLECC, as well as international efforts to protect intellectual property rights.

A. Joint Anti-Piracy Initiative Launched on July 23, 1999
B. DOJ Speaks out on Intellectual Property Rights
C. The Audio Home Recording Act and Napster

II. Intellectual Property Cases

Here you can find information about many of the criminal intellectual property cases that the Department of Justice has prosecuted. Press releases regarding these cases may be found here as well.

A. Recently Prosecuted Intellectual Property Rights Cases
B. Operation "Counter Copy"

III. Prosecuting Intellectual Property Crimes Guidance

Figure 11-2   *The Cybercrime Web site*

All three approaches illustrate how an Internet service provider might try to block access to an entire offending site. However, none of these approaches is really effective in preventing theft or providing identification of property obtained without the copyright holder's permission.

Several methods show promise in the battle to protect digital works, but they only provide partial protection. New and improved methods are continually being developed. One promising technique employs steganography to create a **digital watermark**. The watermark is a digital code or stream embedded undetectably in a digital image or audio file. It can be encrypted to protect its contents, or simply hidden among the bits—digital information—comprising the image or recording. **Verance Corporation** is a company that provides, among other products, digital audio watermarking systems to protect audio files on the Internet. Its systems identify, authenticate, and protect intellectual property. Verance's ARIS MusiCode system enables recording artists to monitor, identify, and control the use of their digital recordings.

The audio watermarks do not alter the audio fidelity of the recordings in which they are embedded. The Verance SoniCode product provides verification and authentication tools. SoniCode was originally developed by ARIS Technologies, which is now owned by Verance Corporation. SoniCode can ensure that telephonic conversations have not been altered. The same is true for audiovisual transcripts and depositions. **Blue Spike** produces a watermarking system called Giovanni. Like the SoniCode system, the Giovanni watermark authenticates the copyright and provides copy control. **Copy control** is an electronic mechanism for limiting the number of copies that one can make of a digital work.

A group of more than 180 companies and organizations devoted to providing protection for intellectual property—digital music in this case—is the **Secure Digital Music Initiative (SDMI)** organization. Its members include information technology and consumer electronics companies, security technology firms, Internet service providers, and the music recording industry. SDMI's charter is to develop open, public technology specifications that protect the playing, storing, and distributing of digital music.

Digimarc Corporation is another company providing watermark protection systems and software. Its products embed a watermark that allows any works protected by its Digimarc system to be tracked across the Web. In addition, the watermark can link viewers to commerce sites and databases. It can also control software and playback devices. Finally, the imperceptible watermark contains copyright information and links to the image's creator. That enables nonrepudiation of a work's authorship and facilitates electronic purchase and licensing of the work. Figure 11-3 shows Digimarc's home page.

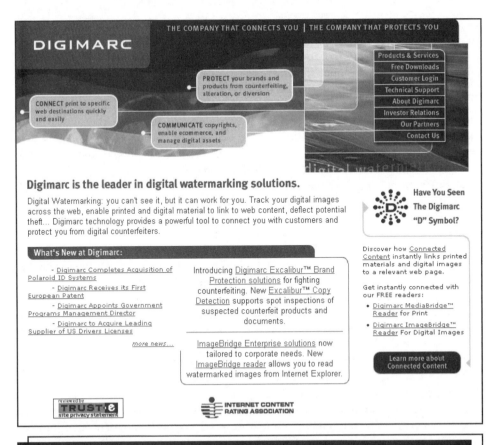

Figure 11-3 *Digimarc Corporation home page*

Protecting Privacy

You have learned about cookies in earlier chapters of this book. These small pieces of text are stored on client computers and can contain private information. Although servers that write cookies usually write them in an encrypted form, nothing prevents a Web server from writing an unencrypted cookie to a client computer. Anyone with access to the computer (physical access or access through the Internet) could read and interpret an unencrypted cookie. The information stored in a cookie could include credit card data, passwords, and login information. Because most cookies are like tickets that allow admission to Web sites that the client computer has previously visited, a cookie can provide entrance to anyone who has the cookie in his or her possession. Although cookies do not harm client machines directly, they can lead to security violations.

Cookies are designed to solve a problem inherent with Web servers: saving information about a Web user from one session to another. There are two kinds of cookies: **session cookies**, which exist until you shut down your browser, and **persistent cookies**, which can exist indefinitely. An electronic commerce site may use both kinds of cookies. For example, a session cookie might contain information about a particular

404

shopping session, whereas a persistent cookie contains information to help the Web site recognize a visitor the next time he or she shops there. Each time a browser moves to a different part of a merchant's Web site, the merchant's Web server asks the visitor's computer to send back the cookie that the Web server previously stored on the visitor's computer.

The privacy problem exists because *any* computer that supplies any part of a Web page can send a cookie to the visitor's computer for storage and subsequent retrieval. As you learned in earlier chapters, advertisers can learn a great deal about visitors' browsing habits from retrieving cookies stored on their computers.

Sometimes an image from an advertiser is not visible on the Web page. A **Web bug** is a tiny graphic on a Web page. A Web bug's only purpose is to provide a way for a Web site to place cookies on a visitor's computer. The Internet advertising community often refers to Web bugs as "clear GIFs" or "1-by-1 GIFs" because they can have a color value of "transparent" and can be as small as one pixel by one pixel.

WebSideStory provides software that analyzes Internet traffic data and provides reports to Web sites about who visits their site and what sites the visitors came from. WebSideStory's HitBox software technology collects and warehouses data from Web site visitors remotely, securely, and anonymously. Figure 11-4 shows an example of the information stored in a cookie collected by a Web site—information specific to the computer that browsed the Web site.

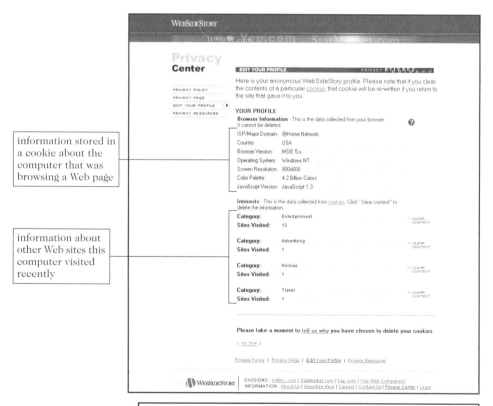

information stored in a cookie about the computer that was browsing a Web page

information about other Web sites this computer visited recently

405

Figure 11-4   *Private information stored in a cookie on a client computer*

The most complete way for Web site visitors to protect themselves from revealing private information or being tracked by cookies is to disable cookies entirely. Web surfers using Netscape Navigator can select Edit, Preferences, and then select the Advanced item in the left column. Then, they should click the Disable Cookies option button. Microsoft Internet Explorer users must select Tools, Internet Options, and then click the Privacy tab. Next, users must click the Advanced button and select Block for both First-party and Third-party Cookies. The problem with this approach—denying all cookies—is that useful cookies are blocked along with the others, requiring visitors to enter more information each time they revisit a Web site. The full resources of some sites are not available to visitors unless their browsers are set to allow cookies.

A better approach is to use one of the many third-party programs, called **cookie blockers**, that prevent cookie storage selectively. Some of these programs, such as **WebWasher**, plug into a browser and allow users to block cookies from the Web servers that load advertising banners into Web pages. Other cookie blockers allow cookies to be filtered by Internet (IP) address, allowing in the "good" cookies and denying storage to all others. The Additional Resources section of the Online Companion for this chapter includes links to a number of cookie blockers.

PROTECTING CLIENT COMPUTERS

Client computers, usually PCs, must be protected from threats that originate in software and data that are downloaded to the client computer from the Internet. Chapter 10 provides a detailed account of client computer threats, but here is a quick recap: Ordinary Web pages that are delivered to a host computer in response to a browser's request are static displays of information and are completely harmless to the client computer. Active content, delivered over the Internet in dynamic Web pages, is not harmless. Such Web pages can be one of the most serious threats to client computers.

Recall that active content consists of programs that are embedded in Web pages. Most of the time, the programs perform only their appointed duties; they are not dangerous. Occasionally, however, threats masquerade as harmless active content but cause damage when they are executed on the client computer. The other popular active content tools are ActiveX controls. Besides threats from programs inside Web pages, downloaded graphics, browser plug-ins, and e-mail attachments can harbor threats that could harm client computers when the hidden programs are activated.

Another threat to client computers is a malevolent server site masquerading as a legitimate Web site. Users and their client computers can be duped into revealing information to malicious Web sites. This section discusses protection mechanisms that can prevent or greatly reduce the threats to client computers.

Monitoring Active Content

The Netscape Navigator and Microsoft Internet Explorer browsers are equipped to recognize when they are about to download Web pages containing active content. When a browser downloads Web pages and runs programs embedded in them, it gives the user a chance to confirm that the programs are from a known and trusted

406

source. The way that both of these popular browsers ensure security varies slightly, so they are presented separately in this section. But first, you will learn about digital certificates, which are essential in providing assurance to clients and servers that the participant is authenticated.

Digital Certificates

A **digital certificate**, also known as a **digital ID**, is an attachment to an e-mail message or a program embedded in a Web page that verifies that a user or Web site is who it claims to be. In addition, the digital certificate contains a means to send an encrypted message—encoded so others cannot read it—to the entity that sent the original Web page or e-mail message. In the case of a downloaded program containing a digital certificate, the encrypted message identifies the software publisher (ensuring that the identity of the software publisher matches the certificate) and indicates whether the certificate has expired or not (is still valid). The digital certificate is a **signed** message or code. Signed code or messages serve the same function as a photo on a driver's license or passport. They provide proof that the holder is the person identified by the certificate. Just like a passport, a certificate does not imply anything about either the usefulness or quality of the downloaded program. The certificate only supplies a level of assurance that the software is genuine. The idea behind certificates is that if the user trusts the software developer, signed software can be trusted because, as proven by the certificate, it came from that trusted developer.

Digital certificates are used for many different types of online transactions, including electronic commerce, electronic mail, and electronic funds transfers. A digital ID verifies a Web site to a shopper and, optionally, identifies a shopper to a Web site. Web browsers or e-mail programs automatically exchange digital certificates invisibly when they are requested to validate the identity of each party involved in a transaction.

Digital signatures cannot be forged easily. Figure 11-5 shows the general structure of a digital certificate. You will learn in the sections that follow how and when certificates are exchanged to provide assurance between clients and servers. Figure 11-6 displays the digital certificate owned by Amazon.com. Whenever a browser indicates that it has established secure communications with a Web site—when a lock appears in the browser's status line—the user can double-click the lock to display the Web site's digital certificate.

407

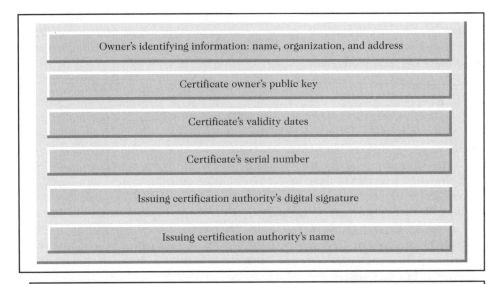

Figure 11-5   *Structure of a digital certificate*

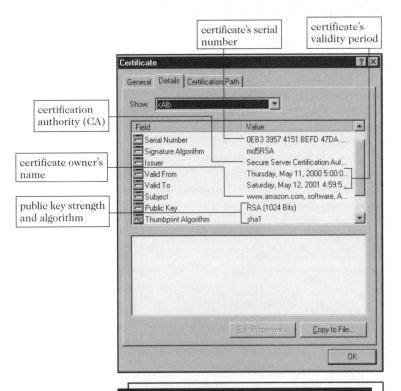

Figure 11-6   *Amazon.com's digital certificate*

408

The software publisher is not the same as the entity that signs the certificate. The certificate is only an approval of the code and does not indicate who authored it. Companies that sign the software first obtain a software publisher certificate from a primary or secondary certification authority. A **certification authority (CA)** issues a digital certificate to an organization or individual. A CA requires entities applying for digital certificates to supply appropriate proof of identity. Once the CA is satisfied, it issues a certificate. Then, the CA signs the certificate—its stamp of approval is affixed—in the form of a public encryption key, which "unlocks" the certificate for anyone who receives the certificate attached to the publisher's code.

A **key** is simply a number—usually a long binary number—that is used with the encryption algorithm to "lock" the characters of the message being protected so that they are undecipherable without the key. Longer keys usually provide significantly better protection than shorter keys. In effect, the CA is guaranteeing that the individual or organization that presents the certificate is who it claims to be. There are only a small number of CAs. One of the oldest and best known is **VeriSign**. The certificates it issues are as trustworthy as VeriSign itself. Other CAs are listed in the Online Companion. Figure 11-7 shows VeriSign's products and services page, which includes links to a number of its digital certificate products.

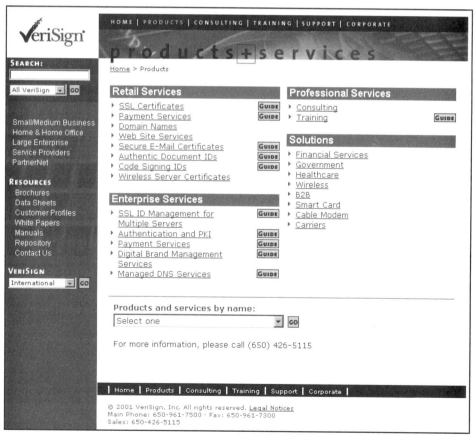

Figure 11-7   VeriSign's products and services page

Identification requirements vary from one CA to another. One CA may require a driver's license for individuals' certificates, while others may require a notarized form or fingerprints. CAs usually publish their identification requirements so that any Web user or site accepting certificates from each CA understands how stringent the CA's validation procedures are. Certificates are classified as low, medium, or high assurance, based largely on the identification requirements imposed on certificate seekers. The fees charged by CAs vary with the level of assurance provided; higher levels are more expensive.

VeriSign provides certificate issuing and revocation services and offers several classes of certificates—from Class 1 through Class 4—that are differentiated by their assurance level, which is the confidence level one can assume based on the process the CA uses to verify the owner's identity. Class 1 certificates are the lowest level and bind e-mail addresses and associated public keys. Class 4 certificates apply to servers and the server organizations. Requirements for Class 4 certificates are significantly greater than those for Class 1. VeriSign's Class 4 certificate, for example, offers assurance of the individual's identity and of that person's relationship to the specified company or organization.

Digital certificates expire after a period of time (often, one year). This built-in limit provides protection for both users and businesses. Limited-duration certificates guarantee that businesses and individuals must submit their credentials for reevaluation periodically. The expiration date appears in the certificate itself and in the dialog boxes that browsers display when a Web page or applet that has a digital certificate is about to be opened. Besides becoming invalid when their time is expired, certificates can be revoked. If the CA determines that a corporation has a history of delivering malicious code, it can refuse to issue new certificates and revoke all existing certificates.

The next sections describe the security features built into the two most popular Web browsers, Microsoft Internet Explorer and Netscape Navigator.

Microsoft Internet Explorer

Microsoft Internet Explorer provides client-side protection inside the browser. Internet Explorer also reacts to ActiveX and Java-based active content. Internet Explorer uses Microsoft Authenticode technology to verify the identity of downloaded active contents, which are programs. Authenticode can check for two important items from a downloaded ActiveX control: who has signed the code, and whether the code has been modified since it was signed. Authenticode technology verifies that the program has a valid certificate. However, it does not prevent a malicious program from being downloaded and run on a client computer. In other words, Authenticode technology can only verify that XYZ Corporation, which the user trusts, has signed the code. If a publisher has not attached a certificate to the active content, users can set up Internet Explorer so that the Web page's code is not downloaded at all. Unfortunately, Authenticode cannot guarantee that XYZ Corporation's Java or ActiveX control will perform flawlessly. Therefore, users must decide whether they trust active content from individual companies, or so-called "zones."

Here's how Authenticode works: When you download a page containing a certificate and active content, Authenticode detaches the certificate (sometimes called the signature block), verifies the identity of the certification authority, verifies that the content is from the publisher, and ensures that the program is unaltered from its original state. A list of trusted CAs is built into Internet Explorer along with their public keys, and Authenticode scans the list to locate a match with the CA supplying the certificate. If the public key in the list matches the public key in the certificate, the CA is known to be a genuine one. The CA's public key is used to unlock the certificate, and within the unlocked certificate is the software publisher's signed digest—a summary of the certificate itself. If the signed digest proves that the software publisher signed the downloaded code, the certificate is displayed. That display assures you that the supplier is valid.

Figure 11-8 shows a security warning and certification validation dialog box. Authenticode determines that the active content has a signed, valid certificate. In this case, Beatnik, Inc., is the publisher, VeriSign is the CA, and the software being downloaded is the Beatnik Player Web browser plug-in. If a user were to download a Web page containing active content that was not signed, the dialog box would indicate that there is no valid certificate. Whether Internet Explorer displays a security warning depends on how security is configured on the browser.

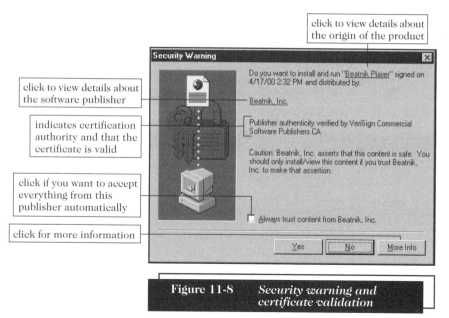

Figure 11-8 *Security warning and certificate validation*

Users can specify different security settings that determine how Internet Explorer handles the programs and files it downloads, depending on the source of the files being downloaded. Microsoft Internet Explorer divides the Internet into zones (categories). That way, users can classify particular Web sites into one of the zones and then assign security levels appropriate for each zone, or group of Web sites. The four zones are: *Internet, local intranet, trusted sites*, and *restricted sites*. The Internet zone is anything that is not on the client computer, not on an intranet, or not assigned to any other zone. The local intranet zone typically contains Web sites that do not require a proxy server (any sites on the Connections tab [see Figure 11-9], for example), the internal corporate network to which the client computer is attached, and other local intranet sites. The trusted sites zone contains sites the user trusts. The restricted sites zone contains Web sites the user does not trust. These might not be sites the user has identified as dangerous—they might just be sites with which he or she is not yet familiar. Figure 11-9 shows that security level options are Low, Medium-Low, Medium, and High. Users can click the Custom Level... button to fine-tune those designations. Figure 11-9 also shows the four Internet Explorer zones and the dialog box that allows users to customize security for a level.

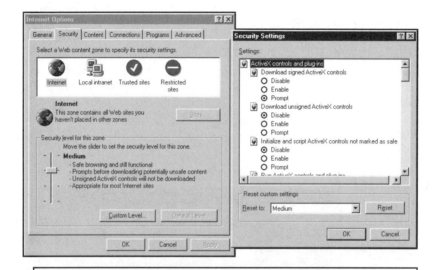

Figure 11-9   *Internet Explorer zones and their security levels*

Figure 11-10 summarizes the default behavior of the four security settings.

Security level	Default security protection provided
High	Provides the safest way to browse, but less functional. Disables less secure features. Disables cookies.
Medium	Safe browsing is functional. Prompts before downloading potentially unsafe content. Does not download unsigned ActiveX components.
Medium-Low	Downloads everything without prompts. Runs most content without prompts. Does not download unsigned ActiveX components.
Low	Supplies minimal safeguards and warnings. Downloads and runs most content without prompts. Can run all active content.

Figure 11-10   *Internet Explorer security zone default settings*

Authenticode technology boils down to a yes/no decision on who and what the user trusts. While security settings can be fine-tuned, the protections are still a choice between running or not running active code. Nothing in Authenticode provides ongoing monitoring of code *during* its execution. So, seemingly safe code that Authenticode permits into a computer can still malfunction—either because of a programming mistake or an intentional act. In other words, once a user passes judgment on the trustworthiness of a site, zone, or vendor, security breaches are still possible when permitting downloaded content.

Netscape Navigator

This section includes examples from Netscape Navigator version 4.79. Although versions in the 6.x series are available as this book goes to press, most people who use Netscape are still using the 4.7x series versions. The options described here are all available in the 6.x series browsers, but the screens look different.

Netscape Navigator allows users to control whether active content is downloaded to their computers. If a user decides to allow Netscape Navigator to download active content, he or she can view the signature attached to Java and JavaScript controls. Since ActiveX controls are a proprietary Microsoft technology, they do not execute with Netscape Navigator. Security is set in the Preferences dialog box. To display this dialog box, a user must select Edit, Preferences. Then, he or she must click Advanced in the left panel of the Preferences dialog box. The right panel displays the security settings, as shown in Figure 11-11. Java and JavaScript can be enabled or disabled. The Preferences dialog box also provides three option buttons that allow the user to specify how to handle cookies. All cookies can be unconditionally accepted, select cookies can be sent back to the server, or cookies can be disallowed completely. A separate check box allows users to receive a warning before accepting cookies.

413

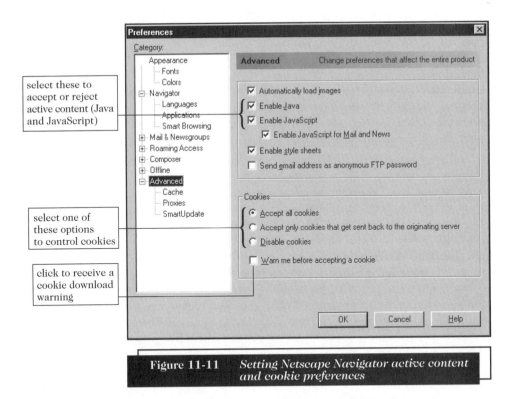

select these to
accept or reject
active content (Java
and JavaScript)

select one of
these options
to control cookies

click to receive a
cookie download
warning

Figure 11-11 *Setting Netscape Navigator active content and cookie preferences*

Users who choose to allow Java or JavaScript active content will always receive an alert from Netscape Navigator. The alert indicates whether the active content is signed and allows the user to view the attached certificate (if available) to determine whether to grant or deny permission to download the active content. Figure 11-12 shows a Netscape Navigator alert from attempting to download a plug-in from Headspace, Inc. Notice that Netscape Navigator judges the risk to be high. Clicking the Details button on the security alert displays more information about the current download request. Clicking the Grant button allows the download process to proceed. Clicking the Deny button denies access, and the Java applet or JavaScript is not downloaded. The vendor's certificate attached to the active content can be viewed by clicking the Certificate button (see Figure 11-12). Figure 11-13 shows the Headspace certificate that is attached to the plug-in installation program.

414

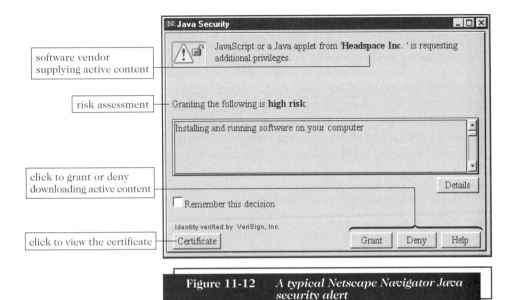

software vendor
supplying active content

risk assessment

click to grant or deny
downloading active content

click to view the certificate

Figure 11-12 *A typical Netscape Navigator Java security alert*

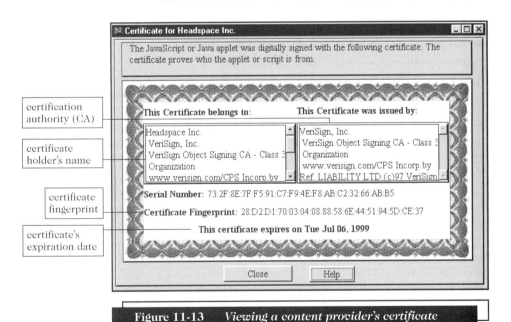

certification
authority (CA)

certificate
holder's name

certificate
fingerprint

certificate's
expiration date

Figure 11-13 *Viewing a content provider's certificate*

415

Notice that the content provider's certificate has a serial number unique to the certificate and a signature (the list of numbers and letters following the Certificate Fingerprint label). The certificate has a limited lifetime. Certificates are often renewed annually, though some certificates last more than a year.

Using Antivirus Software

Antivirus software is a key element in defending client computers. Although this book does not provide a detailed examination of antivirus software, it is important that this valuable type of protection be included in any security plan. Antivirus software only protects computers from viruses that are already downloaded to them.

One of the most likely places to find viruses is in electronic mail attachments. Some e-mail systems, such as Yahoo! Mail, lets users scan attachments using antivirus software before downloading e-mail. Many of the most damaging viruses arrive as electronic mail attachments. When e-mail recipients open the attachments, the viruses launch their attacks. A good antivirus program will scan e-mail attachments and warn of the embedded viruses.

Antivirus software is only effective if the antivirus data files are kept current. The data files contain virus-identifying information that is used to detect viruses on a client computer. Because people generate new viruses by the hundreds every month, users must be vigilant and update their antivirus data files regularly so that the newest viruses are recognized and eliminated. The Online Companion contains links to the leading antivirus software companies.

Computer Forensics and Ethical Hacking

A small group of firms, endorsed by corporations and security organizations, have the unlikely job of breaking into client computers. Called **computer forensics experts** or **ethical hackers**, these computer sleuths are hired to probe PCs and locate information that can be used in legal proceedings. The field of **computer forensics** is responsible for the collection, preservation, and analysis of computer-related evidence. Ethical hackers are often hired by companies to test their computer security safeguards. Links to the Web sites of several companies that offer computer forensics and ethical hacking services are included in the Additional Resources section of the Online Companion for this chapter. The home page of one of these firms, Berryhill Computer Forensics, appears in Figure 11-14.

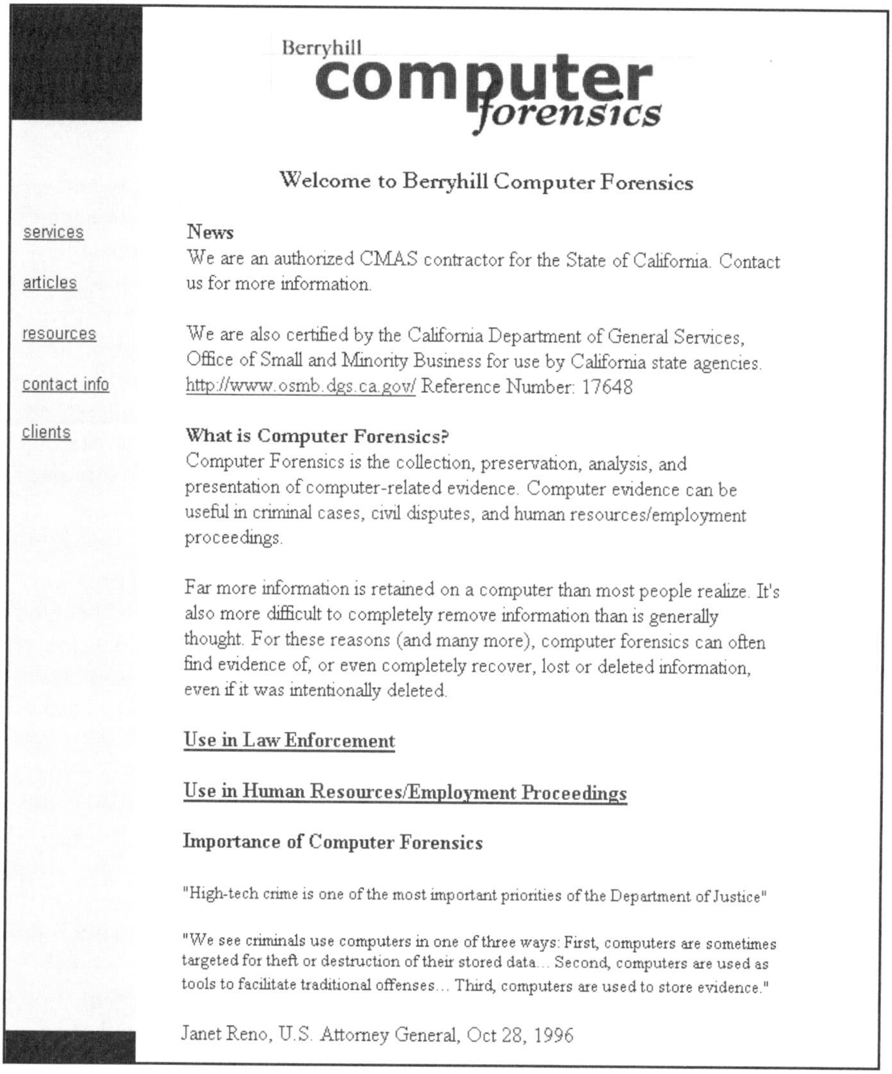

Berryhill
computer
forensics

Welcome to Berryhill Computer Forensics

services

articles

resources

contact info

clients

News

We are an authorized CMAS contractor for the State of California. Contact us for more information.

We are also certified by the California Department of General Services, Office of Small and Minority Business for use by California state agencies. http://www.osmb.dgs.ca.gov/ Reference Number: 17648

What is Computer Forensics?

Computer Forensics is the collection, preservation, analysis, and presentation of computer-related evidence. Computer evidence can be useful in criminal cases, civil disputes, and human resources/employment proceedings.

Far more information is retained on a computer than most people realize. It's also more difficult to completely remove information than is generally thought. For these reasons (and many more), computer forensics can often find evidence of, or even completely recover, lost or deleted information, even if it was intentionally deleted.

Use in Law Enforcement

Use in Human Resources/Employment Proceedings

Importance of Computer Forensics

"High-tech crime is one of the most important priorities of the Department of Justice"

"We see criminals use computers in one of three ways: First, computers are sometimes targeted for theft or destruction of their stored data... Second, computers are used as tools to facilitate traditional offenses... Third, computers are used to store evidence."

Janet Reno, U.S. Attorney General, Oct 28, 1996

Figure 11-14   *Berryhill Computer Forensics home page*

PROTECTING ELECTRONIC COMMERCE COMMUNICATION CHANNELS

Protecting electronic commerce communication channels is by far the most visible segment of computer security. Hardly a week passes without a newspaper or magazine article detailing attacks on the Internet or descriptions of attackers gaining entrance to a computer system by way of an insecure communications channel, such as intranets, extranets, or the Internet. Consequently, a great deal of attention has been given to protecting assets while they are in transit between client computers and remote servers.

Providing commerce channel security means providing channel secrecy, guaranteeing message integrity, and ensuring channel availability. In the first section below, you will learn how authentication is part of the protocols that provide security services.

Providing Transaction Privacy

Since eavesdroppers cannot be prevented from snooping on the Internet, businesses must use techniques that prevent eavesdroppers from reading Internet messages that they intercept. Sending a message over the Internet is like sending a postcard through the mail; it will probably reach its destination, but everyone involved with delivering it can read the message. The only way to prevent snoopers from copying your credit card number, for example, is to encrypt it before you send it over the Internet. Encrypting electronic mail or Internet commerce transactions is like writing a message on the postcard in a language only you and the recipient understand. No one else understands that language, so even though other people might intercept the message, it will make no sense to them unless they are the intended recipient.

Encryption

Encryption is the coding of information by using a mathematically based program and a secret key to produce a string of characters that is unintelligible. The science that studies encryption is called **cryptography**, which comes from a combination of the two Greek words *krupto* and *grafh*, which mean "secret" and "writing," respectively. That is, cryptography is the science of creating messages that only the sender and receiver can read.

Cryptography is different from steganography, which makes text invisible to the naked eye. Cryptography does not hide text; it converts it to other text that is visible but that does not appear to have any meaning. What an unauthorized reader sees is a string of random text characters, numbers, and punctuation.

The program that transforms normal text, called **clear text**, into **cipher text** (the random assemblage of bits) is called an **encryption program**. Messages are encrypted just before they are sent over a network or the Internet. Upon arrival, each message is decoded, or **decrypted**, using a **decryption program**—a type of encryption-reversing procedure. Encryption programs, and the logic behind them, called **encryption algorithms**, are considered so vitally important to preserving security within the United States that the National Security Agency has control over their dissemination. Some encryption algorithms are considered so important that the U.S. government has banned publication of details about them. Currently, it is illegal for U.S. companies to export some of these encryption algorithms. Web pages containing software whose distribution is restricted contain warnings about U.S. export laws. The Freedom Forum Online contains a number of articles on lawsuits and legislation surrounding encryption export laws. Critics consider publication restrictions a freedom of speech issue. If you are interested in reading more about the latest arguments in the ongoing debates over freedom of speech and export law, search the **Freedom Forum** using the keyword "encryption" as the search term.

One property of encryption programs, or algorithms, is that someone can know the details of the encryption program and still not be able to decipher the encrypted message without the key used in the process of encoding the message. The resistance of an encrypted message to attack attempts is directly dependent on the size (bits) of the key used in the encryption procedure. A 40-bit key is considered minimal, whereas longer keys, such as 128-bit keys, provide much more secure encryption. A sufficiently long key can make the security of messages unbreakable.

The type of key and associated encryption program used to "lock" a message or otherwise manipulate a message subdivides encryption into three functions:

- Hash coding
- Asymmetric encryption
- Symmetric encryption

Hash coding is a process that uses a **hash algorithm** to calculate a number, called a **hash value**, from a message of any length. It is a fingerprint for the message, because it is almost certain to be unique for each message. Due to the design of good quality hash algorithms, the probability of two different messages resulting in the same hash value, which would create a **collision**, is extremely small. Hash coding is a particularly convenient way to tell whether a message has been altered in transit, because its original hash value and the hash value computed by the receiver will not match after a message is altered.

Asymmetric encryption, or **public-key encryption**, encodes messages by using two mathematically related numeric keys. In 1977, Ronald Rivest, Adi Shamir, and Leonard Adleman invented the RSA Public Key Cryptosystem while they were professors at MIT. Their invention revolutionized the way sensitive information is exchanged. In their system, one key of the pair, called a **public key**, is freely distributed to the public at large—to anyone interested in communicating securely with the holder of both keys. The public key is used to encrypt messages. The second key—called a **private key**—belongs to the key owner, who carefully keeps the key secret. The owner uses the private key to decrypt messages sent to him or her.

Here is an overview of how the encryption system works: If Herb wants to send a message to Allison, then he obtains Allison's public key from any of several well-known public places. Then, he encrypts his message to Allison using her public key. Once the message is encrypted, only Allison can read the message by decrypting it with her private key. Because the keys are unique, only one secret key can open the message encrypted with a corresponding public key, and vice versa. Reversing the process, Allison can send a private message to Herb using Herb's public key to encrypt the message. When he receives Allison's message, Herb uses his private key to decrypt the message and then read it. If they are sending e-mail to one another, the message is secret only while in transit. Once a message is downloaded from the mail server and decoded, it is stored in plain text on the recipient's machine for all to view.

Symmetric encryption, also known as **private-key encryption**, encodes a message by using a single numeric key, such as 456839420783, to encode and decode data. Because the same key is used, both the message sender and the message receiver must know the key. Encoding and decoding messages using symmetric encryption is very fast and efficient. However, the key must be guarded. If the key is made public, then all messages previously sent using that key are vulnerable, and both the sender and receiver must use new keys for future communications. It is difficult to securely distribute new keys to authorized parties. The catch is that to transmit *anything* privately, it must be encrypted. This includes the new, secret key. Another significant problem with private keys is that they do not scale well in large environments such as the Internet. Each pair of users on the Internet who want to share information privately must have their own private key. That is a huge number of key-pair combinations, and is similar to a telephone system of private lines without switching stations. Enabling 12 people to have a private key pair between all pairs (or private telephone lines between each pair), would require 66 private keys. In general, N individual Internet clients require $(N(N-1))/2$ private key pairs.

In secure environments such as the defense sector, using private-key encryption is simpler, and it is the prevalent method to encode sensitive data. Distribution of classified information and encryption keys is straightforward in the defense sector. It requires guards (two-person control) and secret transportation plans. The **Data Encryption Standard (DES)** is an encryption standard adopted by the U.S. government for encrypting sensitive or commercial information. It is the most widely used private-key encryption system. However, the DES private key size is increased periodically, because individuals are using increasingly fast computers to break messages encoded with shorter keys. Not long ago, for example, the Electronic Frontier Foundation's Deep Crack key breaker used 100,000 PCs on the Internet to break a DES-encrypted test message in under 23 hours. (See **Cracking the 56-bit DES system** in the Online Companion.)

Today, the U.S. government uses a stronger version of the Data Encryption Standard, called **Triple Data Encryption Standard (3DES)**. Triple DES offers good protection—it cannot be cracked even with today's supercomputers—and will do so for several years to come. However, the U.S. government's **National Institute of Standards and Technology (NIST)** has developed a new encryption standard designed to keep government information secure. The new standard is called the Advanced Encryption Standard (AES). In February 2001, the NIST announced that the four-year development process had been successful and that two cryptography researchers from Belgium had created the algorithm chosen for AES. The algorithm's name is Rijndael (pronounced "rain doll") and you can learn more about the development process and the algorithm at the **AES** Web site.

Public-key systems provide several advantages over private-key encryption methods. First, the combination of keys required to provide private messages between enormous numbers of people is small. If N people want to share secret information with one another, then only N unique public key pairs are required—far fewer than an equivalent private-key system. Second, key distribution is not a problem. Each person's public key can be posted anywhere and does not require any special handling to distribute. Third, public-key systems make implementation of digital signatures possible. This means that an electronic document can be signed and sent to any recipient with nonrepudiation. That is, with public-key techniques, it is not possible for anyone other than the signer to have electronically produced the signature; in addition, the signer cannot later deny electronically signing the electronic document.

Public-key systems have disadvantages. One disadvantage is that public-key encryption and decryption are significantly slower than private-key systems. This extra time can add up quickly as individuals and organizations conduct commerce on the Internet. Public-key systems do not replace private-key systems, but serve as a complement to them. Public-key systems are used to transmit private keys to Internet participants so that additional, more efficient communications can occur in a secure Internet session. Figure 11-15 shows a graphical representation of the hashing, private-key, and public-key encryption methods: Figure 11-15a shows hash coding, Figure 11-15b depicts private-key encryption, and Figure 11-15c illustrates public-key encryption.

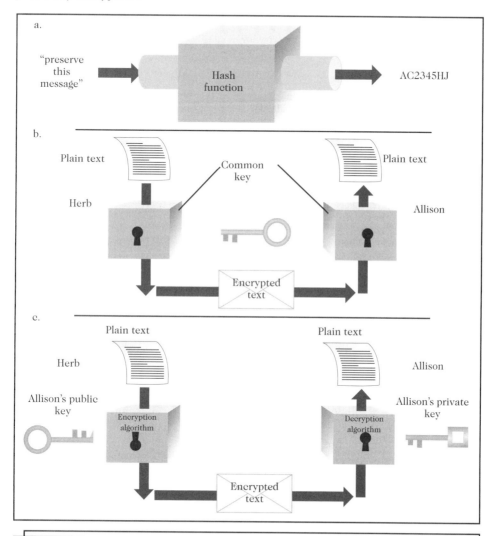

Figure 11-15 *(a) Hash coding, (b) private-key, and (c) public-key encryption*

Encryption Algorithms and Standards

Several encryption, or cipher, algorithms exist that can be used with secure commerce servers. The U.S. government approves the use of several of these inside the

United States, whereas others—typically weaker algorithms—are approved for use outside the United States. Secure commerce servers usually can accommodate most, if not all, of these different algorithms, because they must be able to communicate with browsers. To accommodate various browsers and different versions of those browsers, servers must offer a small array of ciphers. Some of the more visible and important algorithms are briefly described in Figure 11-16.

Algorithm	Type	Comments
AES	Private key	Just a standard; promises to be stronger than 3DES
Blowfish	Private key	Block cipher; developed by Bruce Schneier
DES	Private key	Block cipher; developed in the 1970s
ECC	Public key	
IDEA	Private key	Block (considered the best algorithm available)
LUC	Public key	
MD2	Digest (Hash)	Considered dead; developed by RSA Security, Inc.
MD4	Digest (Hash)	Considered insecure; 128-bit digest; developed by RSA Security, Inc.
MD5	Digest (Hash)	Fair to good security; 128-bit digest; developed by RSA Security, Inc.
RC2	Private key	Block cipher; developed by RSA Security, Inc.
RC4	Private key	Stream cipher
RC5	Private key	Block cipher
RC6	Private key	Block cipher
RSA	Public key	Block cipher; developed by RSA Security, Inc.
SHA1	Digest (Hash)	Replaces SHA algorithm; 160-bit hash value
Skipjack	Private key	Block cipher
Triple DES	Private key	Encrypt-decrypt-encrypt sequence with three keys

Figure 11-16   Significant encryption algorithms and standards

The encryption names in the left column of Figure 11-16 are arranged in alphabetical order, not in order of importance. If you are interested in learning more details about any particular algorithm, you can use a search engine and locate complete specifications on any of the algorithms shown in Figure 11-16. They are listed here so you can become familiar with their names, and a few of them are mentioned later in this chapter. Any secure server or browser uses one or more of these algorithms when it encodes information.

Algorithms in Figure 11-16 are of three different types, indicated in the Type column. You have already learned two of the types: private-key and public-key. Several different algorithms, or methods, are available for public-key encryption, and Figure 11-16 lists several different private-key algorithms. Each of these algorithms has a different level of strength, and some of these algorithms are older and have proven to be inadequate for modern uses and high-speed central processing units.

The series of algorithms called MD2, MD4, and MD5 in Figure 11-16 are message digest (thus the abbreviation MD) algorithms. MD2 was once considered a good function. Today, it is not used because better digest functions are available (MD5, for example).

Secure Sockets Layer Protocol

The **Secure Sockets Layer** (SSL) system from Netscape Communications and the Secure HyperText Transfer Protocol (S-HTTP) from CommerceNet are two protocols that provide secure information transfer through the Internet. SSL and S-HTTP allow both the client and server computers to manage encryption and decryption activities between each other during a secure Web session.

SSL and S-HTTP have very different goals. Whereas SSL secures connections between two computers, S-HTTP sends *individual* messages securely. Encryption of outgoing messages and decryption of incoming messages happen automatically and transparently with both SSL and S-HTTP.

SSL provides a security "handshake" in which the client and server computers exchange a brief burst of messages. In those messages, they agree upon the level of security they will use to exchange digital certificates and perform other tasks. Each computer unfailingly identifies the other. It is not a problem if the client does not have a certificate, because the client is the one who is sending sensitive information. On the other hand, the server with whom the client is doing business ought to have a valid certificate. Otherwise, the client cannot be certain the commerce site is actually who it says it is. After identification, the SSL encrypts and decrypts information flowing between the two computers. This means that information in both the HTTP request and any HTTP responses are encrypted. Encrypted information includes the URL the client is requesting, any forms containing information the user has completed (which might include a credit card number), and HTTP access authorization data such as usernames and passwords. In short, *all* communication between SSL-enabled clients and servers is encoded. When SSL encodes everything flowing between the client and server, an eavesdropper will receive only unintelligible information.

SSL can secure many different types of communications between computers in addition to HTTP. For example, SSL can secure FTP sessions, enabling private downloading and uploading of sensitive documents, spreadsheets, and other electronic data. SSL can secure Telnet sessions in which remote computer users can log on to corporate host machines and send their passwords and usernames. The protocol that implements SSL is HTTPS. By preceding the URL with the protocol name HTTPS, the client is signifying that it would like to establish a secure connection with the remote server. For example, if a user were to enter the protocol and URL https://www.amazon.com, a secure link with Amazon.com would be immediately established. The locked padlock in the browser status bar verifies a secure connection.

Secure Sockets Layer comes in two strengths: 40-bit and 128-bit. The designations indicate the length of the private session key generated by every encrypted transaction. A **session key** is a key used by an encryption algorithm to create the cipher text from plain text during a single secure session. The longer the key, the more resistant the encryption is to attack. A browser that has entered into an SSL session shows the lock in the browser status bars as closed (locked). Otherwise, the lock appears to be open. This is true for both Microsoft Internet Explorer and Netscape Navigator. Once the session is ended, the session key is discarded permanently and not reused for subsequent secure sessions.

Here is how SSL works with an exchange between a client and an electronic commerce server site: Remember that SSL has to authenticate the commerce site (at least) and encrypt any transmissions between the two computers. When a client browser sends a request message to a server's secure Web site, the server sends a hello request to the browser (client). The browser responds with a client hello. The exchange of these greetings, or the handshake, allows the two computers to determine the compression and encryption standards that they both support.

Next, the browser asks the server for a digital certificate—proof of identity. In response, the server sends to the browser a certificate signed by a recognized certification authority. The browser checks the serial number and certificate fingerprint on the server certificate (see Figure 11-13) against the public key of the CA stored within the browser. Once the CA's public key is verified, the endorsement is verified. That action authenticates the commerce server.

Both the client and server agree that their exchanges should be kept secure because they involve transmitting credit card numbers, invoice numbers, and verification codes over the Internet. To implement secrecy, SSL uses public-key (asymmetric) encryption and private-key (symmetric) encryption. Although public-key encryption is handy, it is slow compared to private-key encryption. That is why SSL uses private-key encryption for nearly all its secure communications. Since it uses private-key encryption, SSL must have a way to get the key to both the client and server without exposing it to an eavesdropper. SSL accomplishes this by having the browser generate a private key for both to share. Then the browser encrypts the private key it has generated using the server's public key. The server's public key is stored in the digital certificate that the server sent to the browser during the authentication step. Once the key is encrypted, the browser sends it to the server. The server, in turn, decrypts the message with its private key and exposes the shared private key.

From this point on, public-key encryption is no longer used. Instead, only private-key encryption is used. All messages sent between the client and the server are encrypted with the shared private key, also known as the session key. When the session ends, the session key is discarded. A new connection between a client and a secure server starts the entire process all over again, beginning with the "hello browser," "hello server" exchange. The client and server can agree to use 40-bit encryption or a 128-bit encryption. The algorithm may be DES, triple DES, or the RAS encryption algorithm. Whichever combination is used, both client and server have agreed beforehand on the encryption "language" they will use. Figure 11-17 illustrates the SSL handshake that occurs before a client and server exchange private-key encoded business information for the remainder of the secure session.

424

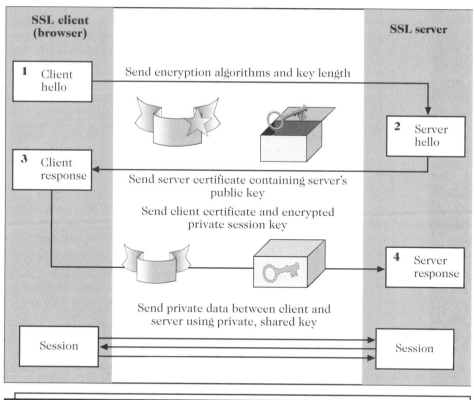

SSL client (browser)

1 Client hello

Send encryption algorithms and key length

2 Server hello

3 Client response

Send server certificate containing server's public key

Send client certificate and encrypted private session key

4 Server response

Send private data between client and server using private, shared key

Session

Session

SSL server

Figure 11-17   *Establishing an SSL session*

You can determine details about an SSL site by visiting Netcraft's Web Server Survey site. Netcraft's surveys were first mentioned in Chapter 8, where you learned that if you enter a URL at Netcraft's site, Netcraft will return details about the Web server software. The same is true for SSL-enabled servers, which run different Web server software than sites that are not SSL-enabled.

To discover what software and encryption algorithms a commerce site supports, click the Online Companion link **What's that SSL site running?** to start the query process. Then, enter a URL and click the Examine button to return information about the commerce. Frequently SSL queries take longer to process than the Netcraft Web server query for sites not running SSL—as much as 45 seconds or so. Figure 11-18 shows an example of the output returned, which is information about Sun Microsystems' SSL-enabled Web server. At the top of the query results page is a link called "What does it all mean?" You can click that link for an explanation of each field of the returned SSL query.

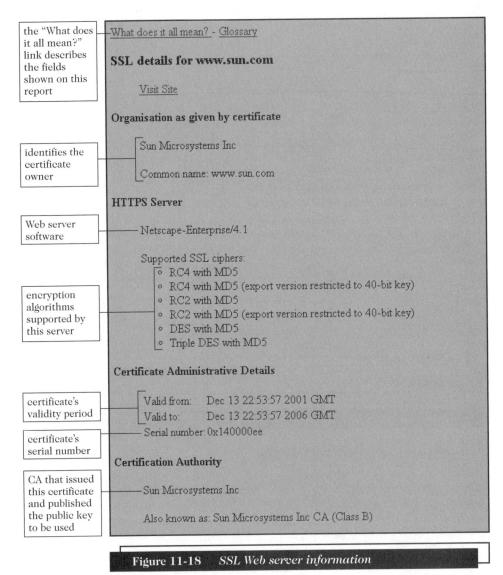

the "What does it all mean?" link describes the fields shown on this report

What does it all mean? - Glossary

SSL details for www.sun.com

Visit Site

Organisation as given by certificate

identifies the certificate owner

Sun Microsystems Inc

Common name: www.sun.com

HTTPS Server

Web server software

Netscape-Enterprise/4.1

Supported SSL ciphers:
- RC4 with MD5
- RC4 with MD5 (export version restricted to 40-bit key)
- RC2 with MD5
- RC2 with MD5 (export version restricted to 40-bit key)
- DES with MD5
- Triple DES with MD5

encryption algorithms supported by this server

Certificate Administrative Details

certificate's validity period

Valid from: Dec 13 22:53:57 2001 GMT
Valid to: Dec 13 22:53:57 2006 GMT

certificate's serial number

Serial number: 0x140000ee

Certification Authority

CA that issued this certificate and published the public key to be used

Sun Microsystems Inc

Also known as: Sun Microsystems Inc CA (Class B)

Figure 11-18   *SSL Web server information*

Secure HTTP (S-HTTP) Protocol

Secure HTTP (S-HTTP) is an extension to HTTP that provides a number of security features, including client and server authentication, spontaneous encryption, and request/response nonrepudiation. The protocol was developed by **CommerceNet**, a consortium of organizations interested in promoting electronic commerce. The S-HTTP protocol provides symmetric encryption for maintaining secret communications, public-key encryption (from RSA Security, Inc.) to establish client/server authentication, and **message digests** (summaries of messages as small integer numbers) for data integrity. (Data integrity is described in the next section, "Ensuring Transaction Integrity.") Interestingly, the client or the server can use S-HTTP techniques separately. That is, a client browser may require security through the use of a private (symmetric) key, whereas the server may require client authentication by using public-key techniques.

The details of S-HTTP security are conducted during the initial negotiation session between the client and server. Either the client or the server can specify that a particular security feature be required, optional, or refused. When one party stipulates that a particular security feature be required, the client or server will continue the connection only if the other party (client or server) agrees to enforce the specified security. Otherwise, no secure connection is established. Suppose the client browser specifies that encryption is required to render all communications secret. In such a situation, the transactions of a high-fashion clothing designer purchasing silk from a Far East textile house will remain confidential. Eavesdropping competitors cannot learn which fabrics are featured next season. On the other hand, the textile mill may insist that integrity be enforced so that quantities and prices quoted to the purchaser remain intact. In addition, the textile mill may want assurances that the purchaser is who he or she claims to be, and not an imposter. A form of nonrepudiation, this security property provides positive confirmation of an offer by a client and makes it impossible for the client to deny ever having made the offer. It is, in effect, a secure digital signature (defined and described in the next section).

S-HTTP differs from SSL in the way it establishes a secure session. Whereas SSL carries out a client/server handshake exchange to set up a secure communication, S-HTTP sets up security details with special packet headers that are exchanged in S-HTTP. The headers define the type of security techniques, including the use of private-key encryption, server authentication, client authentication, and message integrity. Header exchanges also stipulate which specific algorithms each side supports, whether the client or the server (or both) supports the algorithm, and whether the security technique (for example, secrecy) is required, optional, or refused. Once the client and server have agreed to security implementations enforced between them, all subsequent messages between them during that session are wrapped in a secure container, sometimes called an envelope. A **secure envelope** encapsulates a message and provides secrecy, integrity, and client/server authentication. In other words, it is a complete package. With it, all messages traveling on the network or Internet are encrypted so that they cannot be read. Messages cannot be *undetectably* altered because integrity mechanisms provide a detection code that signals that a message has been altered. Clients and servers are authenticated with digital certificates issued by a recognized certification authority. The secure envelope includes all of these security features.

You have learned how encryption provides message secrecy and confidentiality, and you have learned how digital certificates serve to authenticate a server to a client, and vice versa. However, implementing message integrity may be a mystery to you still. The methods that allow you to ensure that an interloper does not change a message in transit appear in the next section.

Ensuring Transaction Integrity

Electronic commerce ultimately involves a client browser sending payment information, order information, and payment instructions to the commerce server, and the commerce server responding with an electronic confirmation of the order details. If an Internet interloper alters any of the order information in transit, harmful consequences can result. For instance, the perpetrator could alter the shipment address or quantity so that he or she receives the merchandise instead of the original customer. This is an example of an **integrity violation**, which occurs whenever a message is altered while in transit between the sender and receiver.

Although it is difficult and expensive to *prevent* a perpetrator from altering a message, there are security techniques that allow the receiver to *detect* when a message has been altered. When the receiver—a commerce server, for example—receives a damaged message, the receiver simply asks the sender to retransmit the message. Apart from being annoying, a damaged message harms no one as long as both parties are aware of the alteration. Damage occurs when unauthorized message changes go undetected by the message's sender and receiver.

Hash Functions

A combination of techniques creates messages that are both tamperproof and authenticated. Additionally, those techniques provide the property of nonrepudiation—making it impossible for the message's creator to claim that the message was not his or hers and that he or she did not send it. To eliminate fraud and abuse caused by commerce messages being altered, two separate algorithms are applied to a message. First, a hash algorithm (see Figure 11-16) is applied to the message. Hash algorithms are **one-way functions**, meaning that there is no way to transform the hash value back to the original message. This approach is acceptable, because a hash value is only compared with another hash value to see if there is a match—the original, pre-hash values are never compared with one another.

MD5 is an example of a hash algorithm used extensively in electronic commerce. A hash algorithm has these characteristics: It uses no secret key; the message digest it produces cannot be inverted to produce the original information; the algorithm and information about how it works are publicly available; and hash collisions are nearly impossible.

Once the hash function computes a message's hash value, that value is appended to the message. Suppose the message is a purchase order containing the customer's address and payment information. When the merchant receives the purchase order and attached message digest, he or she calculates a message digest for the message (exclusive of the original attached message digest). If the message digest value that the merchant calculates matches the message digest attached to the message, the merchant then knows the message is unaltered—that is, no interloper altered the amount or the shipping address information. Had someone altered the information, then the merchant's software would compute a message digest different from the message digest that the client calculated and sent along with the purchase order.

Digital Signatures

Hash functions are not a complete solution. Because the hash algorithm is public and (by design) widely known, anyone could intercept a purchase order, alter the shipping address and quantity ordered, re-create the message digest, and send the message and new message digest on to the merchant. Upon receipt, the merchant would calculate the message digest and confirm that the two message digests match. The merchant is fooled into concluding that the message is unadulterated and genuine. To prevent this type of fraud, the sender encrypts the message digest using his or her private key.

An encrypted message digest (message hash value) is called a **digital signature**. A purchase order accompanied by a digital signature provides the merchant positive identification of the sender and assures the merchant that the message was not altered.

Since the message digest is encrypted using public-key techniques, only the owner of the public/private key pair could have encrypted the message digest. Thus, when the merchant decrypts the message with the user's public key and subsequently calculates a matching message digest value, the result is proof that the sender is authentic. Furthermore, matching hash values prove that only the sender could have authored the message (nonrepudiation), because only his or her private key would yield an encrypted message that could be successfully decrypted by an associated public key. This solves the spoofing problem.

If necessary, both parties can agree to provide transaction secrecy in addition to the integrity, nonrepudiation, and authentication that the digital signature provides. Simply encrypting the entire string—digital signature and message—guarantees message secrecy. Used together, public-key encryption, message digests, and digital signatures provide quality security for Internet transactions. Figure 11-19 illustrates how a digital signature and a signed message are created and sent.

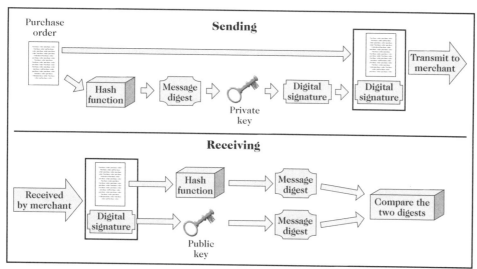

Figure 11-19   *Sending and receiving a signed message*

429

During the summer of 2000, U.S. President Bill Clinton signed a bill giving digital signatures the same legal status as traditional signatures. Clinton first signed the paper version of the new digital signature legislation with a pen. Then, he signed the electronic version of the bill with a smart card (see Chapter 12) containing his digital signature. After doing so, the name "Bill Clinton" appeared on the screen under the text of the new law entitled *Electronic Signatures in Global and National Commerce Act.* People can now electronically sign all sorts of legal documents, such as online car lease agreements, electronic loan papers, and purchase orders.

Guaranteeing Transaction Delivery

As you learned in Chapter 10, denial or delay of service attacks remove or absorb resources. In such an attack, Java programs could download with a Web page and then proceed to tie up a client computer until the mouse freezes and the computer no longer reacts to keyboard keystrokes. The same sort of attack could occur on the commerce channel—a network or the Internet. One way to deny service is to flood the Internet with a large number of packets to crash the server or to slow it down to levels of service that are unacceptable to everyone attempting to do business. Some attacks have caused a crash of the operating system itself. In other cases, the Web server has crashed and temporarily made the electronic commerce function of the site unavailable to customers. If this happens frequently to a particular commerce site, shoppers might avoid that site.

Neither encryption nor digital signatures protect information packets from theft or slowdown. However, the Transmission Control Protocol (TCP) half of the TCP/IP pair is responsible for end-to-end control of packets. When it reassembles packets at the destination in the correct order, it handles all the details when packets do not appear. Among TCP's duties are to request that the client computer resend data when packets seem to be missing. That is, no special computer security protocol beyond TCP/IP is required as a countermeasure against denial attacks. The TCP/IP protocol builds checks into the data so that it can tell when data packets are altered, inadvertently or otherwise. For more details about the TCP/IP protocol, you should consult one of the several good textbooks available on the subject.

PROTECTING THE WEB SERVER

The security protections described in previous sections centered around protecting the client computer and protecting electronic commerce transactions on the Internet or commerce channel. Although much discussion in the security world focuses on protecting the Internet from hackers and intruders, there is very little discussion about protecting the electronic commerce Web server and associated servers connected to it.

Some companies rely on their Internet service provider to help them with Web server security. Major ISPs that offer managed services, such as **Genuity**, **PSINet**, and **Verio**, often include Web server security as an add-on service. Other companies have chosen to hire smaller, specialized security service providers to handle security (see

Learning From Failures—Pilot Network Services to learn more about one counterexample to this approach). Having a service provider handle security usually adds an additional $1000 to $3000 per month to the bandwidth charges. The specialized security firms often charge two to three times more than that for their services. This section presents an overview of several security solutions to the server threats that you learned about in Chapter 10.

LEARNING FROM FAILURES

PILOT NETWORK SERVICES

Pilot Network Services began operations in 1993, at the dawn of commercial use of the Internet. Its goal was to build a network that would be secure for electronic commerce activities. It built a network that included its own carefully monitored connections to the Internet and a database of attack signatures. Attack signatures are descriptions of the Internet traffic characteristics that indicate a hacker attack on a Web server. Pilot, as a firm specializing in security services, built an excellent collection of attack signatures and kept it updated much better than other firms that were not security specialists.

Pilot maintained the Web servers for many of its clients, and it used special versions of the operating systems and Web server software that it had customized to be especially resistant to attacks. Pilot's engineers meticulously applied patches for all known points of access to the software and worked to identify new, as yet unknown, points of vulnerability—for which they immediately created and applied protective patches. For its customers hosting their own servers, Pilot provided the Internet connection through its own secure network. The router between the client's network and Pilot's network—and the operating system running the Pilot network—were customized to eliminate any known security loopholes.

Pilot had 24/7 monitoring of its network by computer security experts in addition to the network technicians that any other Web hosting company would provide as part of a managed services offering. Because it offered high quality services, its fees were considerably higher than the security service charges imposed by other service providers. Typical charges were $6000 per month for the basic connection plus $4000 per month for each Web server.

Even at these high prices, Pilot had many fans among the Fortune 500. Pilot never had more than 300 customers, but it monitored more than 70,000 individual networks for a customer list that included General Electric, PeopleSoft, Sovereign Bancorp, *The Washington Post*, and many other major accounts. By 1999, Pilot appeared to be doing well. Its revenue had increased more than 80 percent over 1998. News releases were issued regularly announcing new customers.

In late 2000, Pilot's stock price began to fall, along with the stock prices of many companies in Internet-related businesses. Although Pilot's sales were

(continued)

growing, its costs were escalating at an even more rapid rate. The company had never reported a profit, and its annual losses had increased to $21.7 million in 2000. Pilot executives assured its customers that the company was financially sound, but the ability of companies in Internet-related businesses to survive on the promise of future earnings had disappeared. Pilot's ability to raise the cash it needed to continue operating had vanished.

In early 2001, some of Pilot's customers noticed that the service was failing. Phone calls and e-mails were not being returned quickly. On the afternoon of April 25, 2001, Pilot's employees received four e-mails. The first explained that telephones would be disconnected that evening. The second asked all employees to turn in their mobile phones and pagers. The third announced that the chief financial officer had resigned. The final e-mail stated that all employees were out of a job as of 4:30 p.m.

Pilot's clients, many of whom found out about the collapse from the Pilot employees who had been servicing their accounts, were in serious trouble. Their connections to the Internet had vanished with no warning. The companies that had used Pilot to host their entire Web operations were in an even worse situation. A group of Pilot customers convinced AT&T (the provider of Pilot's Internet connections) to continue to carry traffic from Pilot, even though Pilot had not paid AT&T. Providian Financial, a major bank holding company and credit card processor, sent its own employees into Pilot's operations centers to keep Providian's Web servers operating. Other Pilot customers that were Providian's competitors protested loudly. Most Pilot customers were concerned that their Web servers were suddenly open and vulnerable to attack.

Several of Pilot's competitors tried to raise funding to take over the business, but all of those attempts failed, and, on May 9, 2001—two weeks after the collapse—AT&T cut Internet service and Pilot was liquidated. Pilot's former customers were scrambling to hire security staff, find alternative hosting firms, or join forces with other companies to keep their electronic commerce sites operating. The lesson from this failure is that security is a critical part of an electronic commerce operation. It should be handled with the same care that a company would use to protect any physical asset. If any part of the security function is handed over to another company, that company's condition becomes an important concern and must be carefully monitored.

Access Control and Authentication

Access control and authentication refers to controlling who and what has access to the commerce server. Recall that authentication is verification of the identity of the entity wanting access to the computer—principally through digital certificates. Just as users can authenticate servers with whom they are interacting, servers can authenticate individual users. When a server requires positive identification of a client computer and its user, it requests that the client send a certificate.

The server can authenticate a user in several ways. First, the certificate represents the user's admittance voucher. If the server cannot decrypt the user's digital signature contained in the certificate using the user's public key, then the certificate did not come from the true owner. Otherwise, the server is certain that the certificate came from the owner. This procedure prevents fraudulent certificates of "admission" to a secure server. Second, the server checks the timestamp on the certificate to ensure that the certificate has not expired (see Figure 11-18). A server will reject an expired certificate and provide no further service.

Third, a server can use a callback system in which the user's client computer name and address are checked against a list of usernames and assigned client computer addresses. Such a system works especially well in the confines of an intranet where usernames and client computers are closely controlled and systematically assigned. On the Internet, a callback system is more difficult to manage, particularly if client users are mobile and work from different locations. It is easy to see how certificates issued by trusted CAs play a central role in authenticating client computers and their users. Certificates provide attribution—irrefutable evidence of identity—if a security breach occurs.

Usernames and passwords have been used for decades to provide some element of protection for servers through attribution and quasi-secure identification. You use passwords every day for access to your e-mail server, to your university or company network, and to log on to subscription services on the Internet. To authenticate users using passwords and usernames, the server must acquire and store a database containing rightful users' passwords and usernames. The system always allows users to be added or deleted, and it usually provides a facility to change passwords. Most modern systems provide a hint style of memory refresher in case the user forgets his or her password. Usually, a user can retrieve a forgotten password by requesting that the server mail the user the password, though this requires an e-mail account outside the system to which access is being requested.

Many Web server systems store usernames and passwords in a file. Large electronic commerce sites usually keep username/password combinations in a database with built-in security features. It makes sense to use that same database to store user preferences, as well as usernames and passwords. To do otherwise would cause large delays, with thousands of users attempting to log on simultaneously.

Regardless of where login information is stored, the most popular and the safest way to store passwords—a method used by UNIX systems—is to store usernames in clear text and encrypt passwords. When you create a new username and password combination (or when a server computer does it for you, as in some systems), the password is encrypted using a one-way encryption algorithm. With the clear text username and encrypted password stored, the system can validate users when they log on by checking the username they enter against the list of usernames stored in the database. The password that a user enters when he or she logs on to a system is encrypted. Then the resulting encrypted password from the user is checked against the encrypted password stored in the database. If the two encrypted versions of the password match for the given user, the login is accepted. That is why even a system administrator cannot tell you what your forgotten password is on some systems. Instead, the administrator must assign a new temporary password that the user can change to another password.

Passwords are not immune to discsovery, and a person truly intent on stealing a password can often figure out a way to do so. Figure 11-20 shows an example of a login box for accessing subscription-only areas of *The Wall Street Journal*'s Web site. Notice that the site visitor can save his or her username and password as a cookie on the client computer, which allows access to subscription areas of the site without entering the username and password on subsequent site visits. The trouble with that system of cookies is that the information might be stored on the client computer in clear text. If the cookie contains login and password information, then that information is visible to anyone who has access to the user's computer.

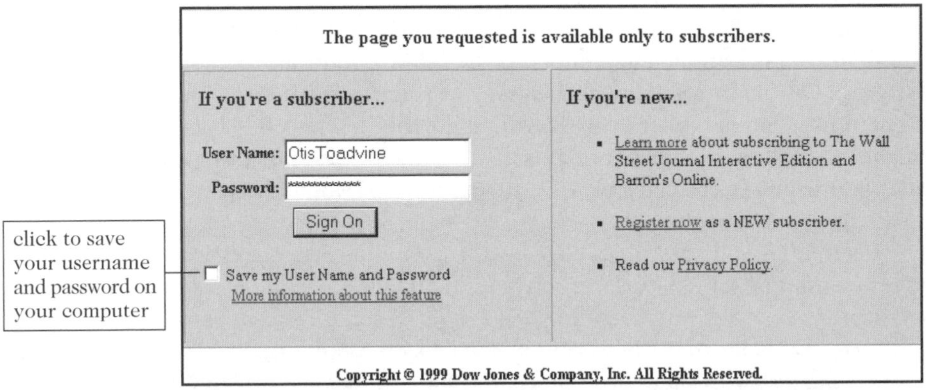

The page you requested is available only to subscribers.

If you're a subscriber...

User Name: OtisToadvine
Password: ************

Sign On

click to save your username and password on your computer

☐ Save my User Name and Password
More information about this feature

If you're new...

- Learn more about subscribing to The Wall Street Journal Interactive Edition and Barron's Online.
- Register now as a NEW subscriber.
- Read our Privacy Policy.

Copyright © 1999 Dow Jones & Company, Inc. All Rights Reserved.

Figure 11-20   *Logging on with a username and password*

Web servers often provide access control list security to restrict file access to selected users. An **access control list** (**ACL**) is a list or database of files and other resources and the usernames of people who can access the files and other resources. Each file has its own access control list. When a client computer requests Web server access to a file or document that has been configured to require an access check, the Web server checks the resource's ACL file to determine if the user is allowed to access that file. This system is especially convenient to restrict access of files on an intranet server so that individuals can only access selected files on a need-to-know basis. The Web server can exercise fine control over resources by further subdividing file access into the subactivities of read, write, or execute. For example, some users may be permitted to read the corporate employee handbook, but not allowed to update or write to the file. Only the human resources (HR) manager would have write access to the employee handbook, and that access privilege is stored along with the HR manager's ID and password in an ACL.

Operating System Controls

Most computer operating systems (even including most of today's personal computers) have a username and password user authentication system in place. That system provides the security substructure for Web servers residing on the host computer on which the operating system runs. The UNIX operating system (and its variants) is the operating system that runs the majority of the Web server platforms today. UNIX contains several native protection mechanisms that prevent unauthorized disclosure and that enforce integrity at a file level. UNIX has many different implementations, including AIX, Irix, Linux, HP-UX, SCO, Solaris, SunOS, and

Ultrix. Each one is a particular vendor's implementation of the original UNIX created by AT&T's Bell Labs (now called AT&T Labs) in 1969. Access control lists and user-name/password protections are probably the best known of the UNIX security features. For further details about operating system (UNIX, Windows 2000, and so on) security features, you should consult some of the literature that is available on this topic. The Online Companion includes links to some of this information.

Firewalls

A **firewall** is a computer and software combination that is installed at the Internet entry point of a networked system. The firewall provides a defense—sometimes the first line of defense—between a network to be protected and the Internet or other network that could pose a threat. All corporate communication to and from the Internet flows through firewalls. The protected network and computers are inside the firewall, and any other network is outside. Firewalls are computers that have the following characteristics:

- All traffic from inside to outside and from outside to inside the network must pass through it.
- Only authorized traffic, as defined by the local security policy, is allowed to pass through it.
- The firewall itself is immune to penetration.

Those networks inside the firewall are often called **trusted**, whereas networks outside the firewall are called **untrusted**. Acting as a filter, firewalls permit selected messages to flow into and out of the protected network. For example, one security policy a firewall might enforce is to allow all HTTP (Web) traffic to pass back and forth, but to disallow FTP or Telnet requests either into or out of the protected network. Ideally, firewall protection should prevent access to networks inside the firewall by unauthorized users, and thus prevent access to sensitive information. Simultaneously, a firewall should not obstruct legitimate users. Authorized employees outside the firewall ought to have access to firewall-protected networks and data files. Firewalls can separate corporate networks from one another and prevent personnel in one division from accessing information from another division of the same company. Using firewalls to segment a corporate network into secure zones serves as a coarse, need-to-know filter.

Large companies that have multiple sites and many locations must install a firewall at each location that has an external connection to the Internet. Such a system ensures an unbroken security perimeter that is effective for the entire corporation. In addition, each firewall in the corporation must follow the same security policy. Otherwise, one firewall might permit one type of transaction to flow into the corporate network that another excludes. The result is an unwanted access that is permitted throughout the corporation because one firewall left a small security door open to the entire network.

Firewalls should be stripped of any unnecessary software. For example, a newly purchased UNIX-based computer comes with many software programs that provide a complete computing environment. All software shipped with the computer must be examined and removed if it does not serve a purpose other than support of the operating system. Because the firewall computer is used only as a firewall and not as a

435

general-purpose computing machine, only essential operating system software and firewall-specific protection software should remain on the computer. Having fewer software programs on the system should reduce the chances for malevolent software security breaches. Access to a firewall should be restricted to a console physically connected directly to the firewall machine. Otherwise, remote administration of the firewall must be provided, which opens up the possibility of a break in the firewall by an imposter remotely accessing the firewall along the same path that an administrator would use.

Firewalls are classified into the following categories: packet filter, gateway server, and proxy server. **Packet-filter firewalls** examine all data flowing back and forth between the trusted network (within the firewall) and the Internet. Packet filtering examines the source and destination addresses and ports of incoming packets, and denies or permits entrance to the packets based on a preprogrammed set of rules.

Gateway servers are firewalls that filter traffic based on the application they request. Gateway servers limit access to specific applications such as Telnet, FTP, and HTTP. Application gateways arbitrate traffic between the inside network and the outside network. In contrast to a packet filter technique, an application-level firewall filters requests and logs them at the application level, rather than at the lower IP level. A gateway firewall provides a central point where all requests can be classified, logged, and later analyzed. An example is a gateway-level policy that permits incoming FTP requests, but blocks outgoing FTP requests. That policy prevents employees inside a firewall from downloading potentially dangerous programs from the outside.

Proxy servers are firewalls that communicate with the Internet on the private network's behalf. When a browser is configured to use a proxy server firewall, the firewall passes the browser request to the Internet. When the Internet sends back a response, the proxy server relays it back to the browser. Proxy servers are also used to serve as a huge cache for Web pages.

Summary

This chapter presents solutions to the security threats described in Chapter 10. Several techniques are available and are being developed to protect intellectual property. The three general assets to protect are client computers, electronic commerce channels, and the Web server. Key security provisions in each of these parts of the customer-Internet-commerce server linkage are secrecy, integrity, and available service.

Encryption provides secrecy, and several forms of encryption are available. They include private-key and public-key techniques. Although public-key encryption eliminates the problem of sharing a secret key, it is much slower than private-key encryption. Private-key encryption is used during most commerce sessions because it is fast and efficient. Integrity protections ensure that transactions—messages and electronic commerce transactions—are not altered. Digital certificates provide both integrity controls and user authentication. A trusted third party or its trusted subordinate organizations, known as certification authorities, provide digital certificates to users and organizations. Several Internet protocols, including Secure Sockets Layer and Secure HTTP, provide secure Internet transmission capabilities.

The Web server must be protected. Protections for the server include access control and authentication provided by username and password login procedures and client certificates. Firewalls can be used to separate trusted inside computer networks and clients from untrusted outside networks, including other divisions of a company's enterprise network system and the Internet.

Key Terms

Access control list (ACL)

Asymmetric encryption

Certification authority (CA)

Cipher text

Clear text

Collision

Computer forensics

Computer forensics expert

Cookie blocker

Copy control

Cryptography

Data Encryption Standard (DES)

Decrypted

Decryption program

Digital certificate (digital ID)

Digital signature

Digital watermark

Encryption

Encryption algorithm

Encryption program

Ethical hacker

Firewall

Gateway server

Hash algorithm

Hash coding

Hash value

Integrity violation

Key

Message digest

One-way function

Packet-filter firewall

Persistent cookie

Private key

Private-key encryption

Proxy server

Public key

Public-key encryption

Secure envelope

Secure Sockets Layer

Session cookie

Session key

Signed (message or code)

Symmetric encryption

Triple Data Encryption Standard (3DES)

Trusted (network)

Untrusted (network)

Web bug

437

Review Questions

1. Prepare a 200-word overview on the subject of watermarking. Explain how it works and what it protects.

2. Using your favorite search engine or Web directory site, research and briefly describe the contents of a digital certificate. Write 100 words and include a diagram, if necessary.

3. Write a 200-word essay in which you outline the field of computer forensics and ethical hacking. In your essay, describe at least one real situation in which computer forensic experts or ethical hackers used their talents to help the legal system.

4. In about 200 words, explain the differences between public-key encryption and private-key encryption. List advantages and disadvantages of each encryption method. Be sure to mention any restrictions on the use of certain strong algorithms.

5. Write 250 words about how hash algorithms work and how they differ from encryption methods (either public-key or private-key encryption). Using the Online Companion and your favorite search engine or Web directory sites, list the names of three hash algorithms and describe their differences in a few words. For example, mention how long the hash value is, and who proposed or invented the particular algorithm.

Exercises

1. Historians believe that Julius Caesar and his military commanders used secret codes, which are known today as Caesar ciphers. There are 26 such ciphers in the English alphabet. In the simplest, A is replaced with B, B is replaced with C, and so on, up to Z, which is replaced with A. Thus, "Caesar" becomes "dbftbs." This type of code is called a "rotate-one Caesar cipher" because it rotates the alphabet one place. A rotate-two cipher replaces A with C, B with D, and so on until Z, which is replaced with B. The following line is encrypted using a simple Caesar rotation cipher.

   ```
   Mjqqt Hfjxfw. Mtb nx dtzw
   hnvmjw? Xyfd fbfd kwtr ymj
   Xjsfyj ytifd.
   ```

   In the preceding cipher text, the letters have been changed, but spaces and punctuation marks have not. Capitalization has been preserved. See if you can decipher the preceding message using a simple rotate-n cipher where the value of n is less than 10. For your assignment, write the cipher text first, and then below it, write the plain text decryption of the message. Good luck! This problem is not very difficult.

2. Obtain your own temporary certificate from VeriSign. Click the VeriSign link in the Online Companion and locate the link on VeriSign's Web pages that asks you to try out a 30-day certificate. Once you receive the certificate, see if you can print out your public key. (Make sure you *do not* print out your private key.) Alternatively, you can obtain a free *permanent* certificate from Thawte rather than VeriSign to complete this exercise. Click the **Thawte** link under Exercise 2 in the Online Companion.

3. Your colleague, Andrea Flemington, is responsible for security of your company's electronic commerce Web site. Her manager has sent her, rather unexpectedly, to one of the company's divisions in Sydney. Before leaving, she left a note on your desk asking you to help her with a brief analysis of other electronic commerce sites. In her note, she

requested that you conduct a survey of 10 Web sites that employ the Secure Sockets Layer (SSL) protocol to protect their transactions. Andrea indicated that you could select six sites yourself, but she wanted you to be sure to include these sites in the survey: www.amazon.com, www.ebay.com, www.fedex.com, and www.gateway.com. Print out the query results from each of the 10 sites for reference as you write a report about your research. Create a table containing 11 rows and five columns. In the leftmost column of the table, list the name (or abbreviation) of each company. Place the following labels in the first row of the table: Company Name, HTTPS Server, SSL Ciphers, Validity Period, and CA's Name. Fill in each row with the appropriate information. Here is an example of the table:

Company Name	HTTPS Server	SSL Ciphers	Validity Period	CA's Name
Sun Microsystems	Netscape Enterprise 3.0	RC4 with MD5 RC2 with MD5 DES with MD5 Triple DES	January 1, 2001 to January 1, 2002	VeriSign

For Further Study and Research

Anderson, R. and F. Petitcolas. 1998. "On the Limits of Steganography," *IEEE Journal of Selected Areas in Communications*, 16(4), May, 474–481.

Aucsmith, D. (ed.) 1998. *Proceedings of the Second International Workshop on Information Hiding.* Portland, OR, April 14–17, Springer Verlang, 219–239.

Baltazar, H. 2000. "Hacker Attacks Welcomed," *eWeek*, 17(26), June 26, 30–34.

Betts, M. 2000. "Digital Signatures Law to Speed Online B-to-B Deals," *Computerworld*, 34(26), June 26, 8.

Busby, M. 1999. *Demystifying TCP/IP*. Plano, TX: Wordware Publishing.

Curtin, M. 2000. "On Guard: Fortifying Your Site Against Attack," *Web Techniques*, April, 46–50.

Delio, M. 2001. "SirCam: Devious, But Not Sinister," *Wired News*, July 25. Available online at: (http://www.wired.com/news/technology/ 0,1282,45506-2,00.html).

DeMaria, M. 2001. "Symantec Firewall/VPN Devices Secure the Small Office at the Right Price," *Network Computing*, 12(24), November 26, 30–31.

Derfler, F. and L. Freed. 2000. "Crash Proof Your Web Site," *PC Magazine*, 19(7), April 4, 134–136.

DoD Directive 5215.1 CSC-STD-001-83. 1983. *Department of Defense Trusted Computer System Evaluation Criteria* (the "Orange Book"), Washington, D.C.

Fonseca, B. 2000. "RSA Releases Patent Early," *InfoWorld*, 22(37), September 11, 27.

Fratto, M. 2001. "Firewalls at your Service," *Network Computing*, 12(23), November 12, 71–77.

Fratto, M. 2001. "Buyers Guide: Enterprise Firewalls," *Network Computing*, 12(25), December 10, 90–92.

Gardner, E. 2000. "ZixIt Corp.," *Internet World*, July 1, 28.

439

Garfinkel, S. and G. Spafford. 1996. *Practical UNIX & Internet Security*. Cambridge, MA: O'Reilly & Associates.

Glass, B. 2000. "Keeping Your Private Information Private," *PC Magazine*, 19(11), June 6, 118–130.

Gurley, W. 2001. "From Wired to Wiretapped," *Fortune*, October 15, 214–215.

Harrison, A. 2000. "Advanced Encryption Standard," *Computerworld*, 34(22), May 29, 57.

Harrison, A. 2000. "Basically Uncrackable," *Computerworld*, 34(25), June 19, 82.

Information Technology Association of America (ITAA). 2000. *Intellectual Property Protection in Cyberspace*. Arlington, VA: ITAA.

Isenberg, D. 2000. "Enforcing E-Contracts," *Internet World*, August 1, 62.

Katzenheisser, S. and F. Petitcolas (eds.). 1999. *Information Hiding Techniques for Steganography and Digital Watermarking*, Norwood. MA: Artech House.

Keen, P. 2000. "Designing Privacy for Your E-Business, *PC Magazine*, 19(11), June 6, 132.

Kennedy, S. 2001. "Security Technology and Other Issues," *Information Today*, 18(11), December, 34–35.

Kutter, M. 1998. "Watermarking Resistance to Translation, Rotation, and Scaling," *Proceedings of SPIE: Multimedia Systems and Applications*, Vol. 3528, Boston, November 1–6, 423–431.

Lehman, D. 2000. "Privacy Policies Missing On 77% of Web Sites," *Computerworld*, 34(16), April 17, 103.

Machrone, B. 2001. "Fashionably Paranoid," *PC Magazine*, 20(10), May 22, 79.

Manes, S. 2001. "Security, Microsoft Style: No Safety Net?" 19(11), November, 210.

Merkow, M., J. Breithaupt, and K. Wheeler. 1998. *Building SET Applications for Secure Transactions*. New York: John Wiley & Sons.

Nash, K. 2000. "Fine-Tuned Security," *Computerworld*, 34(20), May 15, 76.

Palmer, C. 2001. "Ethical Hacking," *IBM Systems Journal*, 40(3), 769–780.

Petreley, N. 2001. "The Cost of Free IIS," *Computerworld*, 35(43), October 22, 49.

Pleas, K. 1999. "Certificates, Keys, and Security," *PC Magazine*, 18(8), April 20, 227–230.

Popovich, K. 2000. "Biometrics Gets Thumbs Up," *eWeek*, 17(24), June 12, 37.

Radcliff, D. 2000. "Digital Signatures," *Computerworld*, 34(15), April 10, 64.

Rivest, R. 1992. *The MD5 Message-Digest Algorithm*, IETF RFC 1321.

Rubin, A. 2000. "None of Your E-Business: Protecting Your Privacy in Cyberspace," *Web Techniques*, April, 55–58.

Rutrell, Y. 2001. "So Many Patches, So Little Time," *InternetWeek*, October 8, 1–2.

Seltzer, L. 2001. "Internet Professional: Solutions for Web Designers and Builders," *PC Magazine*, 20(19), November 13, IP01–IP04.

Smetannikov, M. 2000. "Pilot Shields Networks," *Interactive Week*, 7(19), May 15, 64.

U.S. National Institute of Standards and Technology. 1993. *Data Encryption Standard (DES): Federal Information Processing Standards Publication 46–2*. Gaithersburg, MD: U.S. Computer Systems Laboratory.

Vijayan, J. 2001. "Corporations Left Hanging as Security Outsourcer Shuts Doors," *Computerworld*, 35(18), April 30, 13.

Weber, T. 2001. "Why You Need to Install A Firewall—And Why You Shouldn't Have To," *The Wall Street Journal*, July 30, B1.

Wilson, T. 2001. "VPNs Don't Fly Outside Firewalls," *InternetWeek*, May 28, 1–2.

Zelnick, N. 2000. "Digital Signatures Won't Get Such a Quick Sign-on," *Internet World*, June 15, 20.

440

PAYMENT SYSTEMS FOR ELECTRONIC COMMERCE

INTRODUCTION

In 1991, a teenager named Max Levchin immigrated from the Ukraine to the United States. Settling in

Chicago, Levchin had a burning interest in cryptography. Growing up in a Soviet police state convinced

him that the ability to send coded messages that could not be read or intercepted was both important

and useful. He majored in computer science at the University of Illinois and spent many hours at the

school's **Center for Supercomputing**, pursuing his passion for making and breaking codes. When

he graduated in 1998, he wanted to follow the American dream of turning his knowledge into money,

so he headed for the heart of the computer industry in Palo Alto, California. Levchin's plan to build the

ultimate transmission encryption scheme has not yet panned out, but he has managed to turn his

knowledge into a successful business. As cofounder and chief technical officer of **PayPal**, an online

payment processing company that you will learn about in this chapter, Levchin has used his expertise

in cryptography and computer security to protect the firm from losses that could destroy it.

PayPal, founded in 1999, operates a service that lets people exchange money over the Internet.

It has become the most-used payment system for clearing auction transactions on eBay. People can

also use PayPal to send money to anyone who has an e-mail address, and to receive money. PayPal

charges very small fees to business users and no fees at all to individuals, so its profit margins are small. However, it has grown so rapidly that its thin profit margins are realized on a very large number of users. The company has been able to raise $211 million in funding (almost half of that amount after the dot-com bubble had burst) and its prospects look good. A single, well-organized, large-scale fraud attack on PayPal, however, could put the company out of business quickly. Levchin's current contribution to the company's success is his development of payment surveillance software that continually monitors PayPal transactions. The software searches millions of transactions as they occur every day and looks for patterns that might indicate fraud. The software notifies PayPal managers immediately when it finds something suspicious.

The software appears to be working very well. Companies that process credit card transactions have experienced much larger fraud occurrence rates on the Web (about 1.13 percent) than in physical stores (about .70 percent). PayPal claims to have kept its fraud rate below .50 percent. As long as PayPal can keep its fraud rate low, it can continue to charge lower transaction fees than its competitors and still make a profit. Some industry observers believe that PayPal's ability to avoid high fraud rates could make it a serious competitor to banks in other areas of financial transaction handling, such as credit card processing.

LEARNING OBJECTIVES

In this chapter, you will learn about:

- How companies operating online collect payments from customers
- Credit and debit card processing for electronic commerce transactions
- The history and future of electronic cash
- The implementation of electronic cash systems
- How electronic wallets work
- The use of stored-value cards in electronic commerce
- Protocols used to protect credit card transactions

ELECTRONIC PAYMENT SYSTEMS

One major function of electronic commerce sites is the handling of payments over the Internet. Electronic commerce involves the exchange of some form of money for goods and services.

Implementation of electronic payment systems is still evolving. As a result, a number of proposals and implementations of electronic payment systems compete for dominance. However, electronic payments are far cheaper than the dead-tree method of mailing out paper invoices and then processing payments received. Electronic payments can be convenient for customers and can save companies a lot of money. Estimates of the cost of billing one person vary between $1 and $1.50. Sending bills and receiving payments over the Internet promises to drop the transaction cost to an average of 50 cents per bill. The total savings is huge when the unit cost is multiplied by the number of customers who could use electronic payment. For example, a telephone company in a major metropolitan area might have 5 million customers, each of whom receives a bill every month. A savings of 50 cents on each of those 60 million bills would add up to about $30 million in a year. The environmental impact is also significant. Those 60 million paper bills weigh about 1.7 million pounds. It takes 2200 trees to make that much paper—along with the energy consumed and the wastes generated in the paper-making process.

Today, three basic ways to pay for purchases dominate both traditional and electronic business-to-consumer commerce. Cash, checks, and credit cards account for more than 90 percent of all consumer payments in the United States. Figure 12-1 shows the proportion of payments made during 2000 in the United States for all types of consumer commerce.

Type	Number of Transactions	Dollar Value of Transactions
Cash	42%	17%
Checks	32%	52%
Credit Cards	18%	21%
Electronic	1%	3%
Other	7%	7%

Figure 12-1 *Payment methods for U.S. consumer transactions, 2000*

Credit cards are by far the most popular form of consumer electronic payments online. Recent surveys have found that over 80 percent of Internet purchases are paid for with credit cards. Electronic cash distribution and payment can be handled by wallets, smart cards, or proprietary, limited-use scrip. **Scrip** is digital cash minted by a company instead of by a government. Scrip can be converted into cash, but usually it is not. It must be exchanged for goods or services by the company that issued the scrip. Scrip is like a gift certificate that is good at more than one store. Several companies, such as **eCash Technologies**, sell software that enables Web merchants to offer a variety of payment systems. A page from the eCash Technologies Web site that lists the software products it offers appears in Figure 12-2.

Figure 12-2 *Software offerings of eCash Technologies*

FLOOZ AND BEENZ

Flooz and Beenz were two pioneers in the business of issuing scrip for use on the Web. The scrip created by these two companies could be bought, traded, and exchanged for merchandise, or discounts on merchandise, at a number of Web retailers.

In 1998, Beenz began offering its scrip for sale on its Web site. The scrip was called beenz, and the company's logo included a small kidney-bean shape. A number of merchants agreed to accept the beenz scrip and by mid-2000, Beenz had more than a million customers who were accumulating and using beenz to buy merchandise on the Web. Beenz formed a partnership with Columbus Bank and Trust that allowed beenz holders to transfer their beenz value to a debit card that they could use in the physical world.

Flooz began selling its scrip product, flooz, in late 1999. Flooz had overwhelming support from major partners such as NextCard, and quickly signed an agreement with BarnesandNoble.com, in which the bookseller would accept flooz scrip for purchases on its Web site. Flooz undertook major promotional activities, including an $8 million advertising campaign featuring Whoopi Goldberg.

By August 2001, both companies had ceased operations. Although the idea of using scrip was novel, it did not solve any major problems for consumers on the Web. The use of beenz and flooz scrip also required consumers to learn a new and different technique for making Web payments. Neither product meshed very well with existing payment systems.

The lesson from the Flooz and Beenz failures is that any Web product or service must meet a real need of consumers, and it must not require those consumers to learn a new way to do something that they are already comfortable doing. The new product or service must also integrate well with existing systems and practices.

Merchants should offer their customers payment options that are safe, convenient, and widely accepted. The key is to determine which choices work the best for the company and its customers. In this chapter, you will learn more about how to make those decisions.

Four technologies are discussed in this chapter. They are electronic cash, software wallets, smart cards, and credit and debit cards. Each has unique properties, costs, advantages, and disadvantages. Some are methods that are already popular and widely accepted; others are only beginning to catch on and have an unclear future. Regardless of how popular a particular payment solution is today, you will learn what has worked and what appears not to be working.

All of these electronic payment methods can work well for B2C Web commerce sites. B2B transactions often use different ways to invoice and pay for materials and services, both online and offline. Recent industry estimates are that 16 billion business invoices and billing statements are generated in the United States each year, and 20 percent of those bills are paid electronically. Of these B2B electronic payments, fewer than 10 percent are made using credit cards.

DEBIT CARDS, CREDIT CARDS, AND CHARGE CARDS

Credit cards are by far the most popular form of online payments for consumers. A **credit card**, such as a **Visa** or a **MasterCard**, has a spending limit based on the user's credit history; a user can pay off the entire credit card balance or pay a minimum amount each billing period. Credit card issuers charge interest on any unpaid balance. Many consumers already have credit cards, or are at least familiar with how they work. Credit cards are widely accepted by merchants around the world and provide assurances for both the consumer and the merchant. A consumer is protected by an automatic 30-day period in which he or she can dispute an online credit card purchase. Paying for online purchases with a credit card is just as easy as in a brick-and-mortar store. Merchants who already accept credit cards in an offline store can accept them immediately for online payment because they already have a merchant credit card account. Online purchases require an extra degree of security not required in offline purchases, because a card holder is not present and cannot provide proof of identity as easily as he or she can when standing at the cash register.

A debit card looks very much like a credit card, but it works quite differently. Instead of charging purchases against a credit line, a **debit card** removes the amount of the charge from the cardholder's bank account and transfers it to the seller's bank account. Debit cards are issued by the cardholder's bank and usually carry the name of a major credit card issuer, such as Visa or MasterCard, by agreement between the issuing bank and the credit card issuer.

A **charge card**, such as one from **American Express** or **Diner's Club**, carries no spending limit, and the entire amount charged to the card is due at the end of the billing period. Charge cards do not involve lines of credit and do not accumulate interest charges. In the United States, many retailers, such as department stores and the oil companies that own gas stations, issue their own charge cards. In the rest of this chapter, the term *payment card* will refer to credit cards, debit cards, and charge cards.

Advantages and Disadvantages of Payment Cards

Payment cards have several features that make them an attractive and popular choice, both with consumers and merchants, in online and offline transactions. For merchants, payment cards provide fraud protection. When a merchant accepts payment cards for online payment or for orders placed over the telephone—called **card not present** transactions because the merchant's location and the purchaser's location are different—the merchant can authenticate and authorize purchases using a payment card processing network. For consumers, payment cards are advantageous because the Consumer Credit Protection Act limits the cardholder's liability to $50 if the card is used fraudulently. Once the cardholder notifies the card's issuer of the card theft, the cardholder's liability ends. Frequently, the payment card's issuer waives the $50 consumer liability when a stolen card is used to purchase goods.

Perhaps the greatest advantage of using payment cards is their worldwide acceptance. Payment cards can be used anywhere in the world, and the currency conversion, if needed, is handled by the card issuer. For online transactions, payment cards are particularly advantageous. When a consumer reaches the electronic checkout, he or she enters the payment card's number and his or her shipping and billing information in the appropriate fields to complete the transaction. The consumer does not need any special hardware or software to complete the transaction.

Payment cards have very few disadvantages, but they do have one when compared to cash. Payment card service companies charge merchants per-transaction fees and monthly processing fees. These fees can add up, but merchants view them as a cost of doing business. Any merchant who does not accept payment cards for purchases risks losing a significant portion of sales to other merchants who do accept payment cards. The consumer pays no direct transaction-based fees for using payment cards, but the prices of goods and services are slightly higher than they would be in an environment free of payment cards. Most consumers also pay an annual fee for credit cards and charge cards. This annual fee is much less common on debit cards.

Payment cards provide built-in security for merchants because merchants have a higher assurance that they will be paid through the companies that issue payment cards than through the sometimes-slow direct invoicing process. To process payment card transactions, a merchant must first set up a merchant account. The series of steps in a payment card transaction are usually transparent to the consumer. Several groups and individuals are involved: the merchant, the merchant's bank, the customer, the customer's bank, and the company that issued the customer's payment card. All of these entities must work together for customer charges to be credited to merchant accounts (and vice versa when a customer receives a payment card credit for returned goods).

Payment Acceptance and Processing

Most people are familiar with the use of payment cards: When a purchase is made, the clerk runs the card through the online payment card terminal and the card account is charged immediately. The process is slightly different on the Internet, although the purchase and charge processes follow the same rules. Payment card processing has been made easier over the past two decades because Visa and Mastercard, along with Mastercard's European affiliate, **Europay**, have implemented a single standard for the handling of payment card transactions called the **EMV standard** (EMV is derived from the names of the companies: Europay, Mastercard, and Visa).

In a brick-and-mortar store, customers walk out of the store with purchases in their possession, so charging and shipment occur nearly simultaneously. Online stores and mail order stores in the United States must ship merchandise within 30 days of charging a payment card. Since the penalties for violating this law can be significant, most online and mail order merchants do not charge payment card accounts until they ship merchandise.

Payment card transactions follow these general steps once the merchant receives a consumer's payment card information, which is usually sent using SSL or SHHP:

1. The merchant authenticates the payment card to ensure it is valid and not stolen.

2. The merchant checks with the payment card issuer to ensure that credit or funds are available and puts a hold on the credit line or the funds needed to cover the charge.
3. Settlement occurs, usually a few days after the purchase, which means that funds travel between banks through the automated clearinghouse system into the merchant's account.

Open and Closed Loop Systems

In some payment card systems, the card issuer pays the merchants that accept the card directly and does not use an intermediary such as a bank or clearinghouse system. These types of arrangements are called **closed loop systems** because no other institution is involved in the transaction. American Express and Discover Card are examples of closed loop systems.

Open loop systems involve three or more parties. Suppose an Internet shopper uses his or her Visa card issued by the First Bank of Woodland to purchase an item from Web Wonders, whose bank account is at the Hackensack Commerce Bank. The banking system includes one or more intermediary banks that coordinate the transfer of funds from the First Bank of Woodland to the Hackensack Commerce Bank. Whenever a third party, such as the intermediary banks in this example, processes a transaction, the system is called an **open loop system**. Systems using Visa or MasterCard are the most visible examples of open loop systems. Many banks issue both cards. Unlike American Express or Discover, neither Visa nor MasterCard issue cards directly to consumers. Member banks are responsible for establishing customer credit limits.

Setting up a Merchant Account

A **merchant bank** or **acquiring bank** is a bank that does business with sellers (both Internet and non-Internet) who want to accept payment cards. In other words, to process payment cards for Internet transactions, an online merchant must set up a **merchant account**. When the merchant's bank collects credit card receipts on behalf of the merchant from the payment card issuer, it credits their value to the merchant's account.

A merchant must provide business information before the bank will provide an account through which the merchant can process payment card transactions. Typically, a new merchant must supply a business plan, details about existing bank accounts, and a business and personal credit history. The merchant bank wants to have confidence that the merchant has a good prospect of staying in business and wants to minimize its risk. An online merchant that appears disorganized is less attractive to a merchant bank than a well-organized online merchant.

The type of business also influences the bank's likelihood of granting the account. In some industries, merchant banks will be reluctant to offer a merchant account because of the type of business; some businesses have a higher likelihood of customers repudiating payment card charges than others. For example, a business that sells a guaranteed weight loss scheme—a business in which many customers will want their money back—will find many merchant banks unwilling to provide an account. The bank will assess the level of risk in the business based on the type of business and the credit information that is provided. Merchant banks must estimate what percentage of

447

sales are likely to be contested by cardholders. When a cardholder successfully contests a charge, the merchant bank must retrieve the money it placed in the merchant account in a process called a **chargeback**. To ensure that sufficient funds are available to cover chargebacks, a merchant bank might require a company to maintain funds on deposit in the merchant account. For example, a new and risky business that plans to make $100,000 in sales each month might be required to keep $50,000 or more on deposit in its merchant account.

Several third-party Internet and Web-based services are available to handle all the details of processing payment card transactions. The next section discusses payment card processing options for Internet stores.

Processing Payment Cards Online

Software packaged with your electronic commerce software can handle payment card processing automatically, or you can contract with a third party to handle all your payment card processing.

Several companies provide payment-processing services. **InternetSecure**, for example, allows merchants to concentrate on business while it provides secure payment card payment services. InternetSecure supports payments with Visa and MasterCard for Canadian and United States accounts. The company provides risk management and fraud detection, and handles transactions from online merchants using existing bank-approved payment card processing infrastructure, secure links, and firewall technology to ensure complete security. InternetSecure notifies the merchant of all approved orders and supplies authorization codes to customers who purchase soft goods that are downloaded upon payment card approval. InternetSecure is responsible for ensuring that all transactions that it processes for payment card companies are credited to the merchant's account.

First Data provides merchant payment card processing services with the **ICVERIFY**, **PCAuthorize**, and **WebAuthorize** programs. ICVERIFY is intended for Windows systems. PCAuthorize is intended for smaller commerce service providers, and WebAuthorize is for large, enterprise-class merchant sites. PCAuthorize automatically accepts merchant forms containing consumers' payment card information, captures the data, and deposits the amount in the merchant's payment card account. Once the merchant receives payment from a customer through a Web page form, the merchant passes the Web page with customer payment card information on to the processing bank.

Services such as PCAuthorize and WebAuthorize both connect directly to a network of banks called the Automated Clearing House (ACH) and to credit card authorization companies. You can learn more about the ACH by following the Online Companion link to **NACHA - The Electronic Payments Association**. Banks connect to the ACH through highly secure private leased telephone lines. The merchant sends the card information to a payment card authorization company, which reviews the customer account and, if it approves the transaction, sends the credit authorization to the issuing bank. Then the issuing bank deposits the money in the merchant's bank account through the ACH. The merchant's Web site receives confirmation of the acceptance of the consumer transaction. After receiving notification of acceptance or rejection of the transaction, the merchant Web site confirms the sale to the customer over the Internet. In addition, the merchant site usually sends an e-mail confirmation of the sale to the consumer with details about the purchase price and shipping information. Figure 12-3 is a graphic representation of the process.

Other payment card processing companies include VeriSign's **PayFlow** system, **iAuthorizer**, and **Authorize.Net**. PayFlow is the online payment system developed by CyberCash that is now operated by VeriSign. iAuthorizer is a payment service that links a Web site directly to the payment portal. It provides order forms, CGI scripts, shopping cart features, and transaction processing. Transactions are processed on a secure host at the iAuthorizer center, and iAuthorizer merchants access their reporting and administration information through a secure system that requires a merchant to type a username and password for access. Both merchants and the merchants' customers receive notification of transaction approvals by e-mail. Authorize.Net is an online, real-time payment card processing service that allows merchants to link their sites to the Authorize.Net system by simply inserting a small block of HTML code into their transaction page. With Authorize.Net, a customer's order is encrypted and transferred to the Authorize.Net server. The server, in turn, relays the transaction to a bank network through a private, leased line. Merchants must have an Authorize.Net account to use the service. Check the Online Companion links for more details about these services.

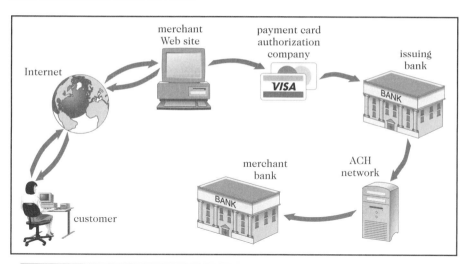

| Figure 12-3 | *Processing a payment card* |

ELECTRONIC CASH

Although credit cards dominate online payments today, electronic cash shows promise for the future. Consulting firm GartnerGroup estimates that electronic cash will be used in more than 60 percent of all online transactions by 2009. **Electronic cash** (also called e-cash or digital cash) is a general term that describes the attempts of several companies to create a value storage and exchange system that operates online in much the same way as government-issued currency operates in the physical world.

As you learned in the previous section, banks that issue credit cards make money by charging merchants a processing fee on each transaction. This fee ranges from 1 percent to 4 percent of the value of the transaction. Often, banks impose a minimum fee of 20 cents or more per transaction. Many banks charge electronic commerce sites more than similar brick-and-mortar stores—up to $1 more per credit card transaction. The cost of an online transaction can be 50 percent higher than the cost to process the same transaction for a brick-and-mortar retailer.

You might have observed that stores accepting credit cards often require a minimum purchase amount of $10 or $15. Merchants impose a minimum purchase amount because the bank fees for small purchase amounts would be greater than their profits on those transactions. The same is true for Internet purchases. Small purchases are not profitable for merchants who accept only credit cards for payment. There is a market for small purchases on the Internet—purchases below $10. This is one potentially significant market for electronic cash. With very low fixed costs, electronic cash provides the promise of allowing users to spend, for example, 50 cents for an online newspaper, or 80 cents to send an electronic greeting card.

Most of the world's population does not have credit cards. Many adults cannot obtain credit cards due to minimum income requirements or past debt problems. Children and teens—eager purchasers representing a significant percentage of online buyers—are ineligible, simply because they are too young. People living in many countries other than the United States hold fewer credit cards because they have traditionally made their purchases in cash. For all of these people, electronic cash provides the solution to paying for online purchases.

Even though there have been many failures in the last few years in electronic cash introductions, electronic cash just refuses to die. Compaq and IBM are among several companies that think electronic cash schemes are in their infancy, and these companies envision a profitable future for such methods. Electronic cash is attractive in two arenas: in the sale of goods and services of less than $10—the lower threshold for credit card payments—and in the sale of higher-priced goods and services to those without credit cards.

Internet payments for items costing from a few cents to approximately a dollar are called **micropayments**. Micropayment champions see many applications for such small transactions, such as paying 5 cents for an article reprint or 25 cents for a complicated literature search. However, micropayments have not been implemented very well on the Web yet. The payments that are between $1 and $10 do not have a generally accepted name (some industry observers use the term micropayment to describe any payment of less than $10); in this book, the term **small payments** will be used to include all payments of less than $10.

All electronic payment schemes have issues that must be satisfactorily resolved to allay consumers' fears and give them confidence in the technology. Concerns about electronic payment methods include privacy and security, independence, portability, and convenience.

Privacy and security questions are probably the most important issues that have to be addressed with any consumer. Consumers want to know whether transactions are vulnerable, and whether the electronic currency can be copied, reused, or forged.

Electronic cash brings with it unique security problems. Electronic cash should have two important characteristics in common with physical currency. First, it must be possible to spend electronic cash only once, just as with traditional currency. Second, electronic cash ought to be anonymous, just as hard currency is. That is, security procedures should be in place to guarantee that the entire electronic cash transaction only occurs between two parties, and that the recipient knows that the electronic currency being received is not counterfeit or being used in two different transactions. Ideally, consumers should be able to use electronic cash without revealing their identities—this prevents sellers from collecting information about individual or group spending habits. Figure 12-4 shows the home page of **eCharge**, one company that provides electronic cash services. Other companies in the electronic cash business include **iPIN** and **InternetCash**.

Figure 12-4 *ECharge home page*

Electronic cash has the advantages of being independent and portable. When electronic cash is independent, it is unrelated to any network or storage device. That is, electronic cash is really not free-floating currency if its existence depends on a particular proprietary storage mechanism that is specially designed to hold one type of electronic cash. Electronic cash should ideally be able to pass transparently across international borders and be automatically converted to the recipient country's currency. Electronic cash portability means that it must be freely transferable between any two parties in all forms of peer-to-peer transactions. Credit and debit cards do not possess this property of portability or transferability between every combination of two parties. In a credit card transaction, the credit card payment recipient must already have a merchant account established with a bank—a condition that is not required with electronic cash.

Perhaps the most important characteristic of cash is convenience. If electronic cash requires special hardware or software, it will not be convenient for people to use. Chances are good that people will not support an electronic cash system that is difficult to use, and that the system will not survive.

Holding Electronic Cash: Online and Offline Cash

Two widely accepted approaches to holding cash exist today: online storage and offline storage. Online cash storage means that the consumer does not personally possess electronic cash. Instead, a trusted third party—an online bank—is involved in all transfers of electronic cash and holds the consumers' cash accounts. Online systems work by requiring merchants to contact the consumer's bank to receive payment for a consumer purchase, which helps prevent fraud by confirming that the consumer's cash is valid. This resembles the process of checking with a consumer's bank to ensure that a credit card is still valid and that the consumer's name matches the name on the credit card.

Offline cash storage is the virtual equivalent of money kept in a wallet. The customer holds it, and no third party is involved in the transaction. Protection against fraud is still a concern (you will learn why later), so either hardware or software safeguards must be used to prevent double or fraudulent spending. Smart cards, discussed later in this chapter, are the hardware solution to storing electronic money. Encrypted—and thus tamperproof—methods are the software solution (described later) that prevents double spending. **Double spending** is spending a particular piece of electronic cash twice by submitting the same electronic currency to two different vendors. By the time the same electronic currency clears the bank for a second time, it is too late to prevent the fraudulent act.

The Advantages and Disadvantages of Electronic Cash

Billing for goods and services that customers have purchased is part of any business. Traditional billing methods in the brick-and-mortar paradigm are costly and involve generating invoices, stuffing envelopes, buying and affixing postage to the envelopes, and sending the invoices to the customers. Meanwhile, the accounts payable department must keep track of incoming payments, post accounts in the database, and ensure that customer data is current.

Online stores have many of the same payment collection inefficiencies as their brick-and-mortar cousins. Most online customers use credit cards to pay for their purchases. Online auction customers also use conventional payment methods, including checks and money orders. Electronic cash systems, though less popular than other payment methods, provide advantages and disadvantages that are unique to electronic cash.

For the most part, electronic cash transactions are more efficient (and therefore less costly) than other methods, and that efficiency should foster more business, which eventually means lower prices for consumers. Transferring electronic cash on the Internet costs less than processing credit card transactions. Conventional money exchange systems require banks, bank branches, clerks, automated teller machines, and an electronic transaction system to manage, transfer, and dispense cash. Operating this conventional money exchange system is expensive.

Electronic cash transfers occur on an existing infrastructure—the Internet—and through existing computer systems. Thus, the additional costs that users of electronic cash must incur are nearly zero. Because the Internet spans the globe, the distance that an electronic transaction must travel does not affect cost. When considering moving physical cash and checks, distance and cost are proportional—the greater the distance that the currency has to go, the more it costs to move it. However, moving electronic currency from Los Angeles to San Francisco costs the same as moving it from Los Angeles to Hong Kong. Merchants can pay other merchants in a business-to-business relationship, and consumers can pay each other. Electronic cash does not require that one party obtain an authorization, as is required with credit card transactions.

Electronic cash has disadvantages, and they are significant. Using electronic cash provides no audit trail. In other words, electronic cash is just like real cash in that it cannot be easily traced. Because true electronic cash is not traceable, another problem arises: money laundering. Money laundering can occur through purchases of goods and services with electronic cash. Ill-gotten electronic cash is spent anonymously for like-valued goods. The goods are then sold for real cash on the open market.

Just as physical currency can be counterfeited, electronic cash is susceptible to forgery. It is possible, though increasingly difficult, to create and spend forged electronic cash. Aside from the risk of forgery, there are several other potentially damaging digital economic factors that might result from the use of electronic cash. These factors have to do with the expansion of the money supply when banks loan electronic cash on consumer and merchant accounts in traditional bank accounts. You can learn more about these economic factors by following the links to **Understanding the Digital Economy** and **The Economic and Social Impacts of Electronic Commerce** in the Online Companion.

Electronic cash has not been a commercial success thus far. Merchants worldwide have been slow to adopt electronic cash as an acceptable payment system. Making electronic cash a popular alternative payment system requires wide acceptance and a solution to the problems of multiple electronic cash standards. Customers do not want to have to carry a dozen different brands of electronic cash to be able to purchase goods from a majority of the merchants that accept electronic cash. Establishing electronic cash as a popular payment method requires that a standard be developed for electronic cash disbursement and acceptance—a standard that individual vendors then implement for their individual electronic cash systems. Electronic cash from different vendors must be easily interchangeable so that customers can exchange one cash type for another when needed.

How Electronic Cash Works

To establish electronic cash, a consumer opens an account with a bank and presents physical identity documentation. The consumer can then withdraw electronic cash by accessing the bank through the Internet and presenting proof of identity— usually a digital certificate issued by a certification authority. After the bank verifies the consumer's identity, it issues the consumer a specific amount of electronic cash and deducts the same amount from the consumer's account. In addition, the bank may charge a small processing fee. The consumer stores the electronic cash in an electronic wallet (described later in this chapter) on his or her computer's hard disk, or on a smart card (also described later in this chapter).

Consumers can spend their electronic cash at electronic commerce sites that accept electronic cash for payment. Briefly, the consumer sends electronic cash (the logistics of this are described later in the chapter) to the merchant for the specified total cost of the goods or services. Then, the merchant validates the electronic cash; that is, the merchant determines that it is not forged and it belongs to the consumer. Only when the goods or services are shipped to the consumer can the merchant present the electronic cash to the issuing bank for deposit. The bank then credits the merchant's account for the transaction amount, minus a small service charge.

Throughout this process, the electronic cash must be protected from both theft and alteration. Additionally, the bank and merchant must be able to verify that the electronic cash belongs to the consumer who spent it.

Providing Security for Electronic Cash

You have already learned about one significant problem with electronic cash: its potential for double spending. The main deterrent to double spending is the threat of prosecution. Complex cryptographic algorithms are the keys to creating tamper-proof electronic cash that can be traced back to its origins. A two-part lock provides anonymous security that also signals when someone is attempting to double spend cash. When a second transaction occurs for the same electronic cash, a complicated process comes into play that reveals the identity of the original electronic cash holder. Otherwise, electronic cash that is used correctly maintains a user's anonymity. This double-lock procedure protects the anonymity of electronic cash users and simultaneously provides built-in safeguards to prevent double spending. Figure 12-5 shows a graphic representation of this double spending detection process using a double-lock system.

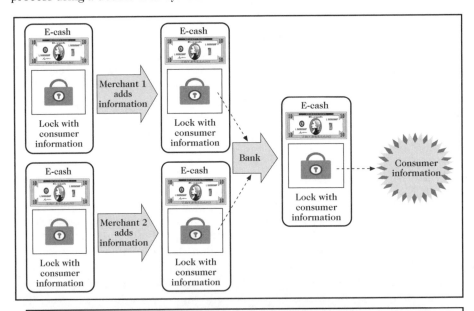

Figure 12-5   *Detecting double spending of electronic cash*

Double spending can neither be detected nor prevented with truly anonymous electronic cash. **Anonymous electronic cash** is electronic cash that, like bills and coins, cannot be traced back to the person who spent it. One way to be able to trace electronic cash is to attach a serial number to each electronic cash transaction. That way, cash can be positively associated with a particular consumer. That does not solve the double spending problem, however. Although a single issuing bank could detect if two deposits of the same electronic cash are about to occur, it is impossible to ascertain who is at fault in such a situation—the consumer or the merchant. Of course, electronic cash that contains serial numbers is no longer anonymous, and anonymity is one reason to acquire electronic cash in the first place. Electronic cash containing serial numbers also raises a number of privacy issues, because merchants could use the serial numbers to track spending habits of consumers.

Creating truly anonymous electronic cash requires a bank to issue electronic cash with embedded serial numbers such that the bank can digitally sign the electronic cash while removing any association of the cash with a particular customer. The process begins when a consumer creates a random serial number that he or she sends to the bank issuing the electronic cash. The bank uses the consumer's random serial number along with the bank's digital signature and sends the random number, electronic cash, and digital signature as one package back to the user. When the user receives the electronic cash bundle, the user extracts the original random serial number and keeps the bank's digital signature. The consumer can now spend the electronic cash, which is digitally signed by the bank. When the consumer spends the electronic cash and the merchant passes it along to the issuing bank, the bank validates the electronic cash because it contains the bank's digital signature. However, the bank cannot determine who the spender is. It only knows that the electronic cash is genuine.

Electronic Cash Systems

Electronic cash has not been nearly as successful in the United States as it has been in Europe and Japan. In the United States, most consumers have credit cards, debit cards, charge cards, and checking accounts. These payment alternatives work well for U.S. consumers in both online and offline transactions. In most other countries of the world, consumers overwhelmingly prefer to use cash. Since cash does not work well for online transactions, electronic cash fills an important need in those countries as they begin conducting B2C electronic commerce. This type of need does not exist in the United States because consumers there already use payment cards for traditional commerce and these payment cards work well for electronic commerce.

Compaq Computer's electronic cash technology allows people to use its NetCoin electronic cash to pay 45 cents for an hour's worth of playing a video game or 4 cents to download a piece of clip art. KDD Communications (KCOM) is the Internet subsidiary of Kokusai Denshin Denwa, which is Japan's largest global phone company. KCOM offers its own NetCoin electronic cash system and offers electronic cash through its NetCoin Center. Shoppers can go to the NetCoin Center and fill their electronic wallets (see the next section on "Electronic Wallets") with electronic cash. Then, they can shop online for recipes or travel directories (10 cents each), or download MP3 music for less than a dollar per song. Other content providers, such as Japanese newspapers, provide access to their newspaper archives and charge a small fee to retrieve articles. There are even plans in Japan to open a donation site where browsers can donate electronic coins to charitable organizations.

Specific reasons for past failures of electronic cash systems in the United States are not completely clear. Some industry observers blame the failure on the way that many electronic cash systems were implemented. Most of these systems required the user to download and install complicated client-side software that ran in conjunction with the browser. Also, there were a number of competing technologies and therefore no standards were ever developed for the entire electronic cash system. The absence of electronic cash standards means that consumers are faced with choosing from an array of proprietary electronic cash alternatives—none of which are interoperable. **Interoperable software** runs transparently on a variety of hardware configurations and on different software systems.

Despite their rough start, not all U.S.-based electronic cash ventures have failed. Next, you will learn about some of the Internet companies that are currently offering electronic cash services and bill presentment and payment systems.

CheckFree

CheckFree, the largest online bill processor in the world, provides online payment processing services to both large corporations and individual Internet users. CheckFree provides infrastructure and software that permits users to pay all their bills with online electronic checks. Figure 12-6 shows a demonstration of a facsimile check in the process of being constructed and paid by an online consumer. To view the demonstration, click the **CheckFree** link in the Online Companion and then click the "Demo" link to see a page of CheckFree product demonstrations. One of the demonstrations walks you through the process of paying a bill online. CheckFree provides part of the technology that Web portal Yahoo! uses to provide its **Yahoo! Bill Pay** service (see Figure 12-7).

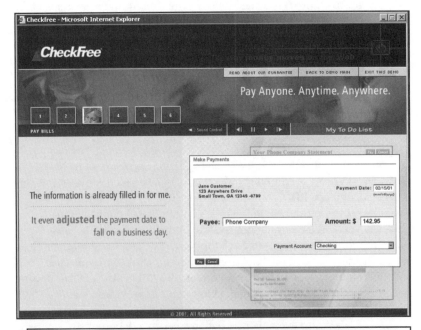

Figure 12-6   *CheckFree online bill-payment demonstration*

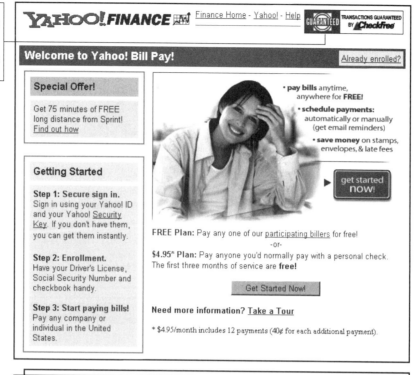

Yahoo! Bill Pay service uses CheckFree transaction processing

Figure 12-7 *Yahoo! Bill Pay service*

Clickshare

Clickshare is an electronic cash system aimed at magazine and newspaper publishers. Clickshare's technology has occasionally been called a micropayment-only system; however, the ability to make micropayments is only one of Clickshare's features. Users with an ISP that supports Clickshare are automatically registered with Clickshare. When users click links leading to other sites that are registered with Clickshare, they can make purchases on those sites without having to register with Clickshare again. Clickshare keeps track of transactions and bills the user's ISP. The ISP, which already has an account relationship with the user, then bills the user for his or her purchases.

Another feature of Clickshare is that it tracks where a user travels on the Internet. This feature has significant value to advertisers and marketers who want to measure audience preferences; however, it does defeat anonymity, and anonymity is one reason that consumers might want to use Clickshare. The micropayment capability is, according to the company, a by-product of the core functionality of tracking identified users. Clickshare tracks users with the standard HTTP Web protocol and does not require cookies or software wallets. Clickshare claims to be the only company that can do this. (Click **Clickshare Example Scenario** for a diagram and explanation of how users are billed for the hyperlinks that they click.) Figure 12-8 shows Clickshare's home page.

457

clickshare
clickshare service corporation

Increase the value of your customers, create new revenue streams, discover new ways to sell content and services on wired and wireless devices

New Clients

Buy articles, books, music, images, classes -- anything digital -- with one click, one account one bill... privately and securely!

Consumers

Access up-to-date business, marketing and technical information

Client Support

Clickshare
makes content sales work.
Sell the article...
sell the photo...
sell the series...
sell the track.

Home | Solutions | News | Consumer | Support | About us | Contact us

copyright 2001, clickshare service corp. all rights reserved

| **Figure 12-8** | *Clickshare electronic cash system* |

ECoin.Net

ECoins are electronic tokens issued by **eCoin**. Consumers can use the tokens to pay for online goods. ECoins provide a way to make micropayments online. The electronic cash is stored in an eCoin wallet on the consumer's computer. As with similar micropayment systems, people can purchase and download a single newspaper article for a few cents, browse a pay-per-view Web site, or download a single music track for 75 cents. Each merchant sets its own fees. Of course, a merchant's Web site must be set up to handle eCoin electronic cash in order for the consumer to use this currency. A consumer can use eCoins by first installing the wallet software as a plug-in to his or her Web browser. Merchants that accept eCoins do not have to install software. Instead, an eCoin-compatible commerce site creates special invoice tags in its HTML pages to enable the consumer's eCoin manager to work with the merchant's site.

The eCoin system uses a three-link chain consisting of a consumer, a merchant, and the eCoin server. The eCoin server operates as a broker that maintains and updates consumer and merchant accounts, accepts requests for payment from the consumer's software, and computes invoices for the merchant site. The eCoin server runs on eCoin's own site. The eCoin system employs security features that prevent double spending. The eCoin structure makes consumers anonymous to merchants but not to the eCoin server. This is a conscious decision by eCoin.net to make all eCoin transactions traceable during the early phase of eCoin's operation. Figure 12-9 shows the eCoin home page.

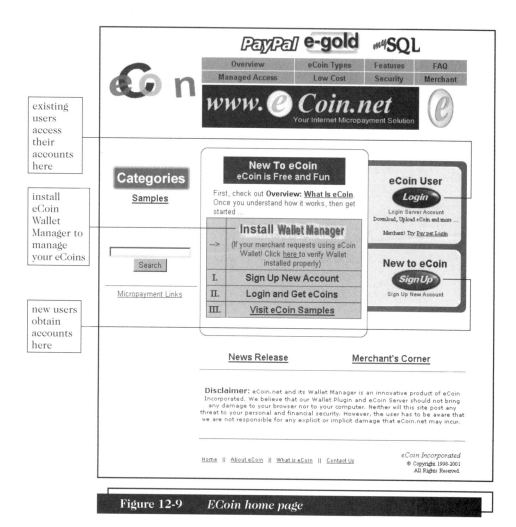

existing users access their accounts here

install eCoin Wallet Manager to manage your eCoins

new users obtain accounts here

Figure 12-9   ECoin home page

InternetCash

Merchants who want to provide their online customers with the option of using cash rather than credit cards to pay for purchases will want to consider **InternetCash**. It provides electronic currency that is very similar to traditional cash. Customers must first purchase an InternetCash card from stores such as Circle K, Sunglass Hut, or any of almost 10,000 brick-and-mortar stores. Similar to prepaid phone cards, the InternetCash cards come in denominations of $10, $20, $50, and $100. After purchasing a card, customers go online and activate their cards by entering a 20-digit activation code found on the back of the card and creating a personal identification number. A **personal identification number**, or **PIN**, is a random series of digits chosen by the customer that serves as a password. Figure 12-10 shows the InternetCash activation display.

459

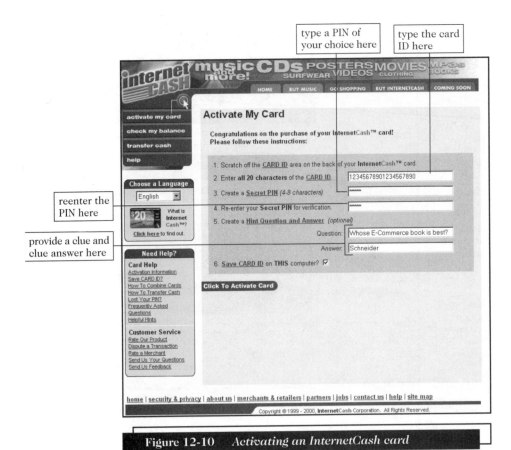

type a PIN of your choice here

type the card ID here

reenter the PIN here

provide a clue and clue answer here

Figure 12-10 *Activating an InternetCash card*

Once the card is activated, the customer can pay for purchases using the InternetCash card at any electronic merchant that accepts it. Customers choose to use InternetCash at checkout, where it appears as a payment choice, along with popular credit cards such as MasterCard and Visa. Selecting InternetCash as the payment choice opens a secure browser window. When the customer types his or her InternetCash PIN in the secure browser window and the card is accepted, InternetCash's Web site creates a digital signature for the purchase. After verifying the payment information, the merchant submits the payment request to InternetCash for processing. InternetCash, in turn, verifies that (by checking the PIN) the customer has the right to use that account. After verifying the customer's identity, InternetCash transfers the money to the merchant's account and sends the merchant a payment authorization number. The transaction's value is automatically deducted from the value of the card, which is maintained in InternetCash's centralized database. A customer's PIN automatically eliminates any customer nonrepudiation issues—the customer cannot claim that he or she did not order the merchandise. After the merchant receives confirmation that the customer's InternetCash payment was successful, the merchant then sends the customer an order confirmation number indicating that the transaction has been completed.

InternetCash is already part of the shopping cart packages of several commerce service providers, including Microsoft's Site Server, IBM's Net.Commerce, Intershop, and Mercantec's SoftCart. Unfortunately, InternetCash is not interchangeable with other electronic cash types. InternetCash makes money by subtracting a transaction fee before remitting customer payments to merchants. Customers do not pay a fee to use InternetCash, however. Because InternetCash values are stored on a secure InternetCash server, customers do not have to worry about which client computer they are using to make InternetCash purchases, and InternetCash provides a convenient online purchase solution for teens and others who do not have access to credit cards.

PayPal

PayPal is the popular electronic cash payment system that you learned about in the opening case of this chapter. PayPal provides payment processing services to businesses and to individuals. PayPal earns a profit on the **float**, which is money that is deposited in PayPal accounts and not used immediately. After two years in business, PayPal began charging a transaction fee to businesses that use the service to collect payments. Individuals who use PayPal to send money to other individuals do not pay a transaction fee. The free payment clearing service that PayPal provides to individuals is called a **peer-to-peer payment system** because the payments are from one type of entity to another of the same type (in the case of PayPal, the entities are individuals).

PayPal eliminates the need to pay for online purchases by writing and mailing checks or using payment cards. PayPal allows consumers to send money instantly and securely to anyone with an e-mail address, including an online merchant. PayPal is a convenient way for auction bidders to pay for their purchases, and sellers like it because it eliminates the risks of accepting other types of online payments. PayPal transactions clear instantly so that the sender's account is reduced and the receiver's account is credited when the transaction occurs. Anyone with a PayPal account—online merchants or eBay auction participants alike—can withdraw cash from their PayPal accounts at any time by requesting that PayPal send them a check or make a direct deposit to their checking accounts. Figure 12-11 shows PayPal's home page.

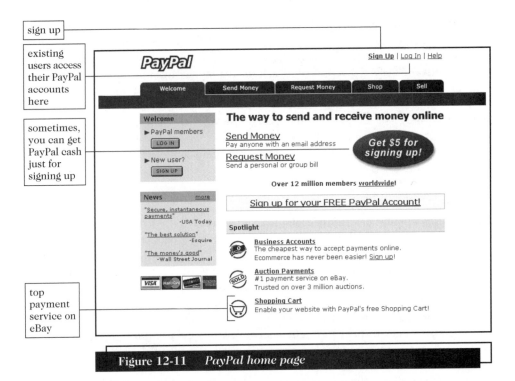

Figure 12-11 *PayPal home page*

To use PayPal, merchants and consumers first must register for a PayPal account. There is no minimum amount that a PayPal account must contain, and customers add money to their PayPal accounts by sending a check or using a credit card. Once members' payments are approved and deposited into their PayPal accounts, they can use their PayPal money to pay for purchases.

Merchants must have PayPal accounts to accept PayPal payments. Using PayPal to pay for auction purchases is very popular. Figure 12-12 shows a partial list of the thousands of items for which auction participants will accept PayPal cash (notice that these sellers elected to include the word *PayPal* in their auction descriptions). A consumer can use PayPal to pay a seller for purchases even if the seller does not have a PayPal account. PayPal sends the seller an e-mail message indicating that a payment is waiting at the PayPal Web site. To collect PayPal cash, the seller or merchant who received the e-mail message must register and provide PayPal with payment instructions. PayPal will then either send the merchant a check, or will deposit funds directly into the merchant's checking account.

Picture	Item Title	Price	Bids	Ends PST
	ALL FLASH QUARTERLY #1, *paypal* *Buy It Now*	$3,150.00	-	Dec-30 14:32
	Laubin Oboe Excellent Condition Paypal NR	$2,800.00	3	Dec-30 21:38
	NEW Russound CA6.4L LCD keypads PayPal *Buy It Now*	$1,799.00	-	Jan-03 16:36
	Russound Multi Room CA 6.4 system PayPal *Buy It Now*	$1,499.00	-	Jan-03 16:36
	ADVENTURE COMICS #58, Jan. 1941 *paypal* *Buy It Now*	$1,350.00	-	Dec-30 13:32
	NEW! Sony FX150 Laptop 128mb Paypal Verified *Buy It Now*	$1,250.00	-	Dec-31 23:18
	1995 JC Penny Treasure Hunt Set!!! Paypal *Buy It Now*	$1,025.00	2	Jan-02 18:39
	Chevrolet:Corsica 1994 CHEVROLET CORSICA PAYPAL	$1,025.00	5	Dec-30 17:20
	Yamaha:SPECIAL 650 1979 YAMAHA 650 SPECIAL PAYPAL OK	$830.00	23	Dec-30 17:33
	FLASH COMICS #89, Nov. 1947 *paypal* *Buy It Now*	$675.00	-	Dec-30 16:24
	Russound Multi Room Complete System PayPal *Buy It Now*	$569.00	-	Dec-30 22:09
	BATMAN #34, May 1946 *paypal* *Buy It Now*	$550.00	-	Dec-30 11:53
	FLASH COMICS #16, April 1941 *paypal* *Buy It Now*	$540.00	-	Dec-30 16:22
	BATMAN # 3, *paypal* *Buy It Now*	$525.00	-	Dec-30 10:52
	DETECTIVE COMICS #103, Sept 1945 *paypal* *Buy It Now*	$499.00	-	Dec-30 14:37
	ADVENTURE COMICS #80, *paypal* *Buy It Now*	$499.00	-	Dec-30 14:17
	HEAT PRESS 14 BRAND NEW w/1y Wrnty CC/PayPal *Buy It Now*	$479.00	-	Dec-30 20:36
	NEW YAMAHA RX-V1000 RECEIVER, PAYPAL, NR	$475.00	2	Dec-31 22:22
	SUPERMAN #12, Oct. 1941 *paypal* *Buy It Now*	$475.00	-	Dec-30 17:59
	ARTOGRAPH DB400 With table mount *paypal *Buy It Now*	$395.00	-	Dec-30 18:28

Figure 12-12 *Many eBay auctions accept PayPal*

Billpoint

PayPal grew rapidly by serving the needs of buyers and sellers on auction sites such as eBay, Yahoo! Auctions, and Amazon Auctions. This success and its potential for profits did not go unnoticed by the management team at eBay. In May 1999, eBay purchased a small electronic payments company and, one year later, sold a 35 percent stake in that company to Wells Fargo bank. This company, **Billpoint**, is now operated as a joint venture between eBay and Wells Fargo. Figure 12-13 shows Billpoint's home page.

eBay affiliation is prominently displayed

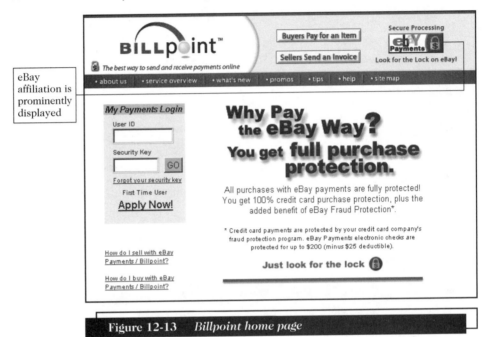

Figure 12-13 *Billpoint home page*

Unlike PayPal, Billpoint does not currently allow its members to maintain cash deposits or to transfer cash directly from their checking accounts. Billpoint requires all of its members to have a credit card and to use that credit card to pay for their Billpoint charges. The only exception is for members who elect to use Wells Fargo's electronic check service, which has a limit of $200 per check. Also unlike PayPal, Billpoint only handles payments for eBay auctions; it does not offer peer-to-peer payment services to individuals, nor does it provide payment services for other auction sites. Billpoint has grown rapidly, but PayPal has maintained its first-mover advantage and is still by far the most widely used payment processing system on eBay.

Many sellers on eBay accept both Billpoint and PayPal (along with other payment methods, such as credit cards, checks, and money orders). If you examine Figure 12-12 carefully, you will notice that two of the auction listings include a small padlock icon, which indicates that the seller accepts payment through Billpoint. Since eBay owns Billpoint, it gives the service prominent placement in the payment section of the auction pages of sellers that accept it. Figure 12-14 shows an information page within the eBay site that describes the Billpoint service as "eBay Payments." The only association with Billpoint is a small logo at the bottom of the page.

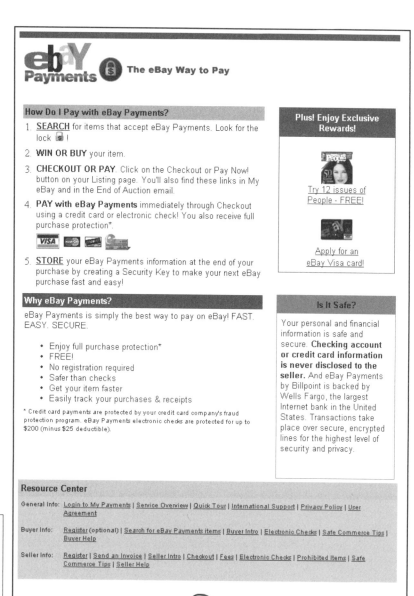

identification as Billpoint payment services is done with a small logo at the bottom of the page

Figure 12-14   Billpoint services described as "eBay Payments"

Other companies have entered this business as well. Western Union offers what it calls electronic money orders that customers can use to settle auction transactions through its **BidPay** site. Traditional banks have also created Internet payment sites, including Bank One's **eMoneyMail** and Citibank's **c2it** services.

ELECTRONIC WALLETS

As consumers are becoming more enthusiastic about online shopping, they have begun to tire of repeatedly entering detailed shipping and payment information each time that they make an online purchase. Research repeatedly has shown that filling out forms ranks high on online customers' lists of gripes about online shopping. That is one problem that electronic wallet technology intends to solve. The other problem electronic wallets solve is providing a secure storage place for credit card data and electronic cash. Thus, an **electronic wallet** (sometimes called an **e-wallet**), serving a function similar to a physical wallet, holds credit card numbers, electronic cash, owner identification, and owner contact information, and provides that information at an electronic commerce site's checkout counter. Some electronic wallets contain an address book too.

Electronic wallets make shopping more efficient. When consumers select items to purchase, they can then click their electronic wallet to order the items quickly. Although they do not do it yet, wallets could serve their owners by tracking purchases that they make and maintaining receipts for those purchases. Maintaining records of a consumer's purchasing habits is something that online giants such as Amazon.com have mastered, but an enhanced digital wallet could reverse that process and use a Web robot to suggest where the consumer might find a lower price on an item that he or she purchases regularly.

Electronic wallets fall into two categories based on where they are stored. A **server-side electronic wallet** stores a customer's information on a remote server belonging to a particular merchant or belonging to the wallet's publisher. The main weakness of server-side electronic wallets is that a client-server security breach could reveal thousands of users' personal information—including credit card numbers—to unauthorized parties. Typically, server-side electronic wallets employ strong security measures that minimize or eliminate the possibility of unauthorized disclosure.

A **client-side electronic wallet** stores a consumer's information on his or her own computer. Storing an electronic wallet on the user's computer shifts the responsibility for maintaining security to the user. Because no user information is stored on a central server, there's no chance that an attack on an electronic-wallet vendor will yield consumer information such as credit card numbers. Many of the early electronic wallets were client-side wallets that required users to download the wallet software. This need to download software onto every computer used to make purchases continues to be a chief disadvantage of client-side wallets.

Server-side wallets, on the other hand, remain on a server and thus require no download time or installation on a user's computer. A disadvantage of client-side wallets is that they are not portable. For example, a client-side wallet is not available when a purchase is made from a computer other than the computer on which the wallet resides.

For a wallet to be useful at many online sites, it should be able to populate the data fields in any merchants' forms at any site that the consumer visits. This accessibility means that the electronic wallet manufacturer and merchants from many sites must coordinate their efforts so that a wallet can recognize what consumer information goes into each of a given merchant's forms.

Electronic wallets store shipping and billing information, including a consumer's first and last names, street address, city, state, country, and ZIP or postal code. Most electronic wallets also can hold many credit card names and numbers, affording the consumer a choice of credit cards at the online checkout. Some electronic wallets also hold electronic cash from various providers, such as eCash.

Electronic wallets can save shoppers time; when shoppers have filled their shopping carts, they proceed to the electronic checkout counter to confirm their choices. At the checkout counter, they are confronted with a form (or series of forms) into which they must enter their name, address, credit card number, and other personal information. Figure 12-15 shows an example of one part of a multiscreen order form. The consumer must fill in all the information boxes to complete the checkout process.

Please enter your details here. You are now on a Secure Server so you may send your details with confidence. Please wait while this page loads, and then enter all requested details.

	Billing Details	
First Name		*
Last Name		*
Company:		
Address		*
City:		*
State		
Zip / Postcode		*
Country	United States	
Phone daytime		*
Email		*
Phone Evening		
Please enter a password		*

This password is useful for you to track your order, or use for your next order and save having to give us your details all over again. Enter any letters and number between 4 and 12 characters.

* = Required information If Billing address is different than shipping address, please contact your credit card company and inform them to note the account. Their toll free number is on the back of your credit card.

Figure 12-15   A typical electronic checkout counter form

Forms like the one in Figure 12-15 take some time to complete and can be a deterrent to customers completing their purchases. Research studies have concluded that Web consumers are less likely to abandon their shopping carts at the checkout counter when they use electronic wallets, which enter required information into checkout forms automatically. The difference between single-click systems such as Amazon's 1-Click technology and a digital wallet is that the single-click systems each work in a particular store. Electronic wallets, in theory, work at many merchants' stores—every merchant that accepts the system.

A number of companies entered the electronic wallet business, including major firms such as MasterCard. Most of these companies have abandoned their efforts because current versions of all major browsers now include a feature that remembers names, addresses, and other commonly requested information, and provides a one-click completion of fields on Web forms that request that information.

Microsoft .NET Passport

Microsoft .NET Passport (often referred to as Passport or Microsoft Passport) is an electronic wallet operated by Microsoft. Anyone who obtains a Hotmail account, which is Microsoft's free e-mail service, is automatically signed up for a Passport account. Passport functions in the same way as most other electronic wallets, by automatically completing order forms. All of the personal data entered into a Passport wallet is encrypted and password-protected.

Passport consists of four integrated services: Passport single sign-in service (SSI), Passport Wallet Service, Kids Passport service, and public profiles. The sign-in service allows a user to sign in at a participating Web site using his or her username and password. The Passport Wallet service provides standard electronic wallet functions, such as secure storage and form completion of credit card and address information. When requested by a participating merchant, a consumer's secure information is released to the merchant so that the consumer does not need to enter data into a form. The Kids Passport service helps parents protect and control their children's online privacy, and the public profiles service allows consumers to create a public page of information about themselves. Figure 12-16 shows the Passport home page.

Figure 12-16 *Microsoft.NET Passport home page*

W3C Micropayment Standards Development Activity

The World Wide Web Consortium (W3C) conducted an active standards development activity for micropayments in electronic commerce for several years. Although the activity has now been closed, the **W3C Electronic Commerce Interest Group (ECIG)** developed a set of standards called the **Common Markup for Micropayment Per-Fee-Links** before it ended its activities. The ECIG wanted to stimulate comment on its proposed draft micropayment standard. The proposed ECIG micropayment links standard sets guidelines for a system that provides an extensible and interoperable way to embed micropayment information in a Web page. An **extensible system** is one that developers can add to (or extend) without voiding any earlier work on the system. The ECIG draft proposal identifies existing system micropayment types of online connections, stored-value systems (such as the smart cards described in the next section), and combined online-offline systems.

Merchants who want the widest Internet consumer audience must be willing to offer support for several different payment systems. To do so, the merchants' Web pages must embed, within each Web page, payment information that is specific to each payment system that different consumers have adopted. This redundancy has motivated the W3C to develop a common Web page markup system that all micropayment systems can support. The ECIG draft standard proposes the following architecture: The client (the consumer's Web browser) initiates the micropayment activity by contacting the merchant server. The client browser consists of the browser itself, a module called Per Fee Link Handler (PFLH), and one or more electronic payment wallets. On the merchant side of the client-server architecture is an HTTP server. The W3C document proposes new HTML tags to carry the embedded micropayment information. Figure 12-17 shows a list of these proposed micropayment tags. It is now up to the individual electronic wallet and electronic cash vendors to develop and revise their software to conform to the new standard.

Field Name	Short Description	Format	Requirements
merchanturl	Identifies the merchant site.	URL	MUST be provided
merchantname	Specifies a merchant designation.	character string	MAY be provided
buyurl	Identifies what the client is buying.	relative URL	MUST be provided
textlink	Describes textually what the client is buying. The text source of the fee link.	character string	MUST be provided
imagelink	Describes graphically what the client is buying. The graphic source of the fee link. (textlink provides a textual equivalent of the image for accessibility.)	URL	MAY be provided
price	Specifies amount and currency.	character string	MUST be provided
duration	Indicates the time after purchase any URL can be retrieved with payment.	integer number	SHOULD be provided
longdesc	Describes in detail the content of the client's purchase.	character string	SHOULD be provided
requesturl	Identifies what the client is actually requesting.	relative URL	MAY be provided
expiration	Indicates a date until which the offer from the merchant is valid.	character string: YYYY-MM-DDThh:mm:ssTZD	MAY be provided
specific field	Provides information unique to each payment system.	URL and character string	MAY be provided

Figure 12-17   W3C proposed micropayment HTML tags

The ECML Standard

A consortium of several high-tech companies and credit card companies proposed another standards initiative that would replace the competing electronic wallet standards with a single standard. The consortium of companies, which includes America Online, Compaq, Dell, IBM, Microsoft, Visa U.S.A., and MasterCard, has agreed on a technology called **ECML**, or **electronic commerce modeling language**. Users can enter their credit card and address information once into an ECML-capable electronic wallet. Any existing wallet can be redesigned to follow the ECML standard, although none currently do. Users control access to their ECML electronic wallets, and the wallets will be accepted at all commerce sites if the consortium is successful in generating widespread acceptance. So far, several electronic wallet makers have accepted and implemented the standard. They include IBM, Microsoft, and Trintech.

The ECML standard will expedite online processing for customers by simplifying the form-filling procedure. Currently, the field names for various forms vary from one online merchant to another. For example, one merchant may refer to the telephone number field as "phone," while another merchant may call it "PhoneNumber." The ECML format will work by providing uniform field names upon which all merchants can build their input forms. An ECML-compliant Web site would allow a customer to enter billing, shipping, and payment information once. If the customer visits another ECML-compliant site, the same customer information is already filled out when the customer proceeds to the checkout counter. Figure 12-18 displays some of the ECML-standard fields and their characteristics.

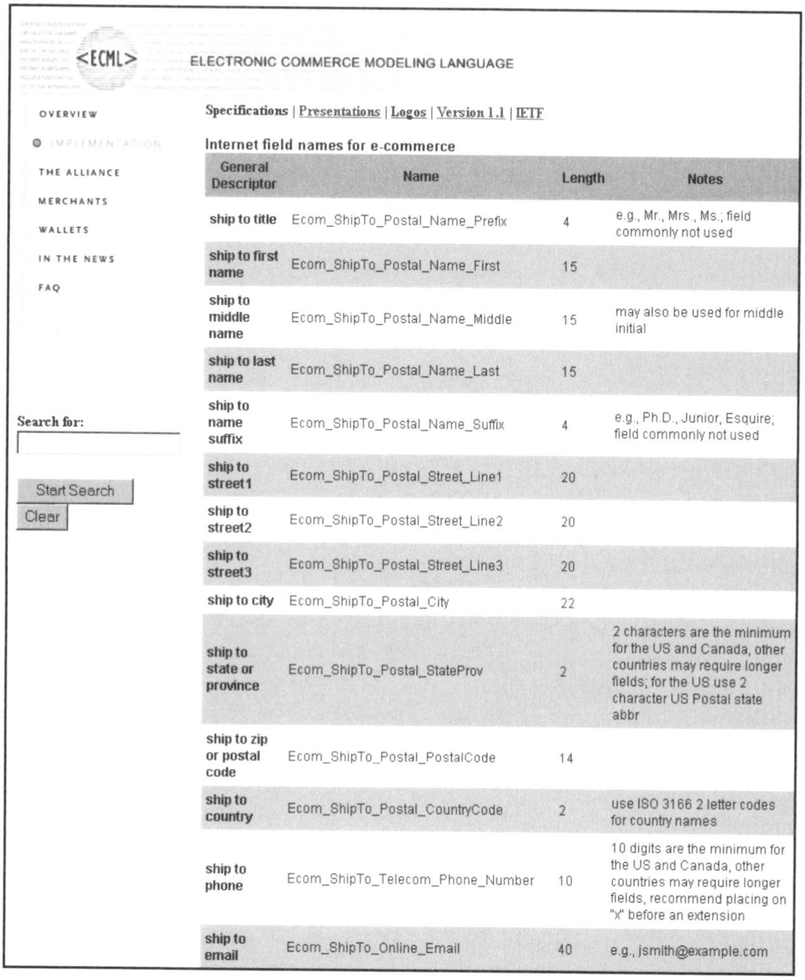

<ECML> ELECTRONIC COMMERCE MODELING LANGUAGE

OVERVIEW

● IMPLEMENTATION

THE ALLIANCE

MERCHANTS

WALLETS

IN THE NEWS

FAQ

Search for:

Start Search

Clear

Specifications | Presentations | Logos | Version 1.1 | IETF

Internet field names for e-commerce

General Descriptor	Name	Length	Notes
ship to title	Ecom_ShipTo_Postal_Name_Prefix	4	e.g., Mr., Mrs., Ms.; field commonly not used
ship to first name	Ecom_ShipTo_Postal_Name_First	15	
ship to middle name	Ecom_ShipTo_Postal_Name_Middle	15	may also be used for middle initial
ship to last name	Ecom_ShipTo_Postal_Name_Last	15	
ship to name suffix	Ecom_ShipTo_Postal_Name_Suffix	4	e.g., Ph.D., Junior, Esquire; field commonly not used
ship to street1	Ecom_ShipTo_Postal_Street_Line1	20	
ship to street2	Ecom_ShipTo_Postal_Street_Line2	20	
ship to street3	Ecom_ShipTo_Postal_Street_Line3	20	
ship to city	Ecom_ShipTo_Postal_City	22	
ship to state or province	Ecom_ShipTo_Postal_StateProv	2	2 characters are the minimum for the US and Canada, other countries may require longer fields; for the US use 2 character US Postal state abbr
ship to zip or postal code	Ecom_ShipTo_Postal_PostalCode	14	
ship to country	Ecom_ShipTo_Postal_CountryCode	2	use ISO 3166 2 letter codes for country names
ship to phone	Ecom_ShipTo_Telecom_Phone_Number	10	10 digits are the minimum for the US and Canada, other countries may require longer fields, recommend placing on "x" before an extension
ship to email	Ecom_ShipTo_Online_Email	40	e.g., jsmith@example.com

Figure 12-18   Some ECML-standard fields

ECML does not provide standardized names for fields such as age, shopping preferences, or birthday—information that merchants usually collect to provide customers with a more personalized shopping experience. The ECML standard might include more advanced features in the future—features such as purchase receipts and package tracking.

Assuming that an acceptable standard will evolve, the ultimate success of electronic wallets will depend on the confidence that Internet users have in the technology. As the NetBank story (see the Learning from Failures feature) illustrates, customer confidence is an important part of the success of any Internet technology, especially when that technology controls a person's financial welfare.

LEARNING FROM FAILURES

NETBANK

CompuBank and NetBank were two of the first Internet banks to open in the United States. They were both pure Internet banks; that is, neither was founded by an existing bank with a physical presence. After four years of operation, CompuBank had about 50,000 accounts, $64 million of deposits, and was losing more than $20 million per year. NetBank had done considerably better, with 160,000 accounts, $1 billion of deposits, and had been profitable for 10 consecutive quarters.

In early 2001, CompuBank decided to close its operations and it found NetBank to be a willing purchaser of its accounts. When a bank buys accounts from another bank, it performs a series of procedures called due diligence. These **due diligence** procedures include checking the new customers' credit histories and banking records. Due diligence is usually performed before the transaction is completed and before the closing bank's customers look to the buying bank as the institution that will handle their accounts.

For a number of reasons, not all of which are clear, the due diligence process was still underway on the date that the transfer of accounts was to take place. NetBank placed holds on many accounts and sent letters to many account holders explaining that they were not acceptable customers by NetBank's standards. For any bank, this would have been a difficult situation, but the nature of the two banks as Internet-only operations made things considerably worse for everyone.

Press accounts of the fiasco included stories of the problems that between 4000 and 8000 CompuBank depositors experienced. Some were small—online bill payments did not occur, debit and credit cards were rejected at stores and restaurants, ATMs would not yield cash—while others were much larger. One couple who had kept the money to cover closing costs on a house purchase in a CompuBank account found that NetBank had placed a hold on the money. Since they could not pay the closing costs, they were forced to find another mortgage lender. In the suit they filed against NetBank, the couple asserted that the increased rate on the mortgage loan would cost them tens of thousands of dollars. Other CompuBank customers were irritated that they lost access to their money for weeks. Some customers could not determine whether the bills they had set up to be automatically paid had, in fact, been paid.

NetBank admitted failures in customer service related to the incident. Many customers who called to complain or ask for explanations experienced waits on hold of 45 minutes and then were transferred to the bank's security department where a recording answered and asked callers to leave their social security number and wait to be called back. None of the customers reported being called back. The timing of NetBank's notification was problematic, too. Many customers reported receiving a letter from NetBank indicating that there were problems with their accounts. The letter, dated April 30, was received by the customers on or after May 14. The letter included a telephone number to call for assistance, but that number had been disconnected on May 12. Many of the unhappy customers found each other on Internet discussion boards and compared notes.

NetBank has not disclosed the number of customers it lost by its handling of this transition; indeed, it may not know. CompuBank's customers were largely experienced Internet users who had elected to be part of the leading edge in handling their financial affairs. Many of them, after this experience, have sworn that they will never again do business with a bank that does not have a physical presence. The lesson from NetBank's experience is that customer service and the ability to communicate with customers becomes extremely important for companies that process electronic payments or are responsible for their customer's finances.

STORED-VALUE CARDS

Today, most people carry a number of plastic cards—credit cards, debit cards, charge cards, driver's license, health insurance card, employee or student identification card, and others. One solution that could reduce all those cards to a single plastic card is called a stored-value card.

A **stored-value card** can be an elaborate smart card or a plastic card with a magnetic strip that records the currency balance. The main difference is that a smart card can store larger amounts of information and includes a processor chip on the card. Common stored-value cards include prepaid phone, copy, subway, and bus cards. Most magnetic strip cards can be recharged by inserting them into the appropriate machines, inserting currency into the machine, and withdrawing the card; the card's strip stores the increased cash value. Magnetic strip cards are passive; that is, they cannot send or receive information, nor can they increment or decrement the value of cash stored on the card. The processing must be done on a device into which the card is inserted. Although both magnetic strip cards and smart cards can store electronic cash, a smart card is better suited for Internet payment transactions because it has some processing capability.

473

Smart Cards

A **smart card** is a stored value card that is a plastic card with an embedded microchip that can store information. Credit, debit, and charge cards currently store limited information on a magnetic strip. A smart card can store about 100 times the amount of information that a magnetic strip plastic card can store. A smart card can hold private user information such as financial facts, encryption keys, account information, credit card numbers, health insurance information, medical records, and so on.

Smart cards are safer than conventional credit cards because the information stored on a smart card is encrypted. For example, conventional credit cards show your account number on the face of the card and your signature on the back. The card number and a forged signature are all that a thief needs to purchase items and charge them against your card. With a smart card, credit theft is much more difficult because the key to unlock the encrypted information is a PIN; there is no visible number on the card that a thief can identify, nor is there a physical signature on the card that a thief can see and use as an example for a forgery.

Smart cards have been in use for more than a decade. Popular in Europe and parts of Asia, smart cards so far have not been as successful in the United States. In Europe and Japan, smart cards are being used to pay for pay phones and television. The cards are very popular in Hong Kong, too, where practically every retail counter and restaurant cash register has a smart card reader on it. The city's transportation companies—subways, buses, railways, trams, and ferries—joined together and created a smart card called the Octopus that lets commuters use one card for all of their public transportation needs. The Octopus can be reloaded at any transportation location or at 7-Eleven stores throughout Hong Kong. The **Hong Kong Citybus** Web page with information about the Octopus card appears in Figure 12-19.

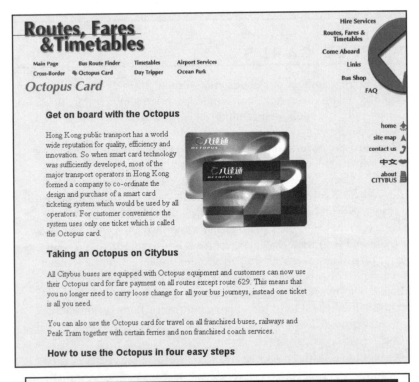

Figure 12-19   *Octopus smart card information on the Hong Kong Citybus site*

Smart cards are beginning to appear in the United States. In San Francisco, the Bay Area **Metropolitan Transportation Commission** created a smart card system patterned after the Octopus card. This system, TransLink, is the first integrated ticketing system for public transportation in the United States. The transportation smart card, implemented in late 2001, allows commuters to ride most modes of public transit available in the city, including trains, buses, cabs, and ferries, by simply waving a single card near a reader device in transit vehicles or in stations. TransLink users can reload their smart cards at several retail outlets or directly from their bank accounts.

Visa recently introduced its smart card, the **smart Visa Card**. One of the first promotions of the new Visa card is slated for late 2002 when retailer Target introduces its Target Visa smart card for use in Target stores and on the Target.com Web site. The Target Visa will include electronic wallet and automated login information for the Target.com Web site, but it will also function as a normal Visa card at other merchants. American Express has also released a smart card called **Blue**. Figure 12-20 shows a Web page leading to information about three versions of the American Express Blue smart card.

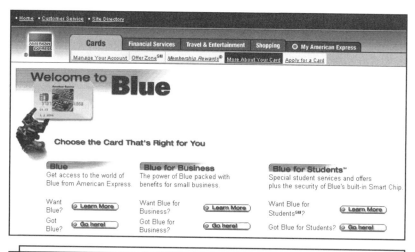

Figure 12-20   American Express Blue smart card home page

In the United States, the **Smart Card Alliance** promotes the benefits of smart cards. The organization promotes the widespread acceptance of multiple-application smart card technology. Its members include companies in banking, financial services, computer technology, health care, telecommunications, and a number of government agencies. The Alliance focuses on information exchange and member interaction. Every member of the Alliance recognizes that smart cards will succeed in the United States only if a critical mass of smart cards supports applications—both physical and Internet-based—of interest to consumers. The Alliance promotes compatibility among smart cards, card-reader devices, and applications.

Mondex is a smart card that holds and dispenses electronic cash. As it gains acceptance on the Internet and in the general marketplace, the Mondex smart card will allow other applications to reside on its microchip. The Mondex card was invented in 1990 and is now a part of MasterCard International. Mondex's Hong Kong pilot program took place in 1996 and was the main force behind the general acceptance of smart cards in Hong Kong. The Mondex card gave people in Hong Kong, who traditionally have used cash, a new and appealing way to make payments on the Internet and in the physical world. Trials of the Mondex card have not been successful in the United States and Canada because consumers there already have payment methods that work for them, such as credit cards, debit cards, and paper checks.

Mondex smart cards can accept electronic cash directly from a user's bank account. Cardholders can spend their electronic cash with any merchant who has a Mondex card reader. Two cardholders can even transfer cash between their cards over a telephone line. That is an advantage of Mondex—a single card will work both in the online world of the Internet and the offline world of ordinary merchant stores, while simultaneously being less susceptible than credit cards to threats of theft. The Mondex card provides electronic cash that is anonymous.

Another advantage of Mondex is that the cardholder always has the correct change for vending machines of various types. Coca Cola has reported that as much as 25 percent of its vending machine sales are lost because consumers do not have change. Mondex electronic cash supports micropayments as small as 3 cents.

Mondex has some disadvantages, too. The card carries real cash in electronic form, and the risk of theft of the card may deter users from loading it with very much money. Mondex does not allow the deferred payment you can obtain with a credit card—you can defer paying your charge or credit card bill for almost a month without incurring any interest charges. Mondex cards dispense their cash immediately. Transactions completed using a Mondex card do not provide receipts.

A Mondex transaction has several steps. These steps ensure that the transferred cash safely reaches the correct destination. The following are the steps in using a Mondex card to transfer electronic cash from buyer to seller:

1. The card user inserts the Mondex card into a reader. The merchant and the card user are both validated to ensure that both the user and the merchant are still authorized to make transactions.
2. The merchant's terminal requests payment while simultaneously transmitting the merchant's digital signature.
3. The customer's card checks the merchant's digital signature. If the signature is valid, then the transaction amount is deducted from the cardholder's card.
4. The merchant's terminal checks the customer's just-sent digital signature for authenticity. If the cardholder's signature is validated, the merchant's terminal acknowledges the cardholder's signature by again sending back to the cardholder's card the merchant's signature.
5. Once the electronic cash is deducted from the cardholder's card, the same amount is transferred into the merchant's electronic cash account. Waiting until after the transaction amount is deducted from the cardholder's card ensures that electronic cash is neither created nor lost. That is, serializing the deduction and crediting events before signifying a complete transaction eliminates the creation or loss of cash if the system malfunctions in the middle of the process.

Figure 12-21 shows an overview of this process, beginning with the cardholder filling his or her card with electronic cash through the purchase transaction, and ending with the merchant exchanging accumulated electronic cash for credit in his or her bank account.

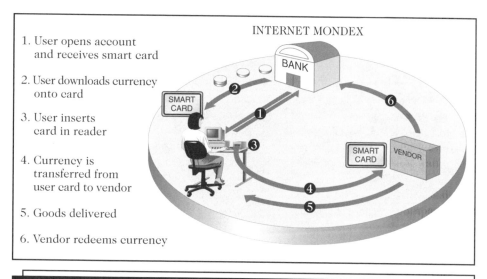

1. User opens account and receives smart card

2. User downloads currency onto card

3. User inserts card in reader

4. Currency is transferred from user card to vendor

5. Goods delivered

6. Vendor redeems currency

Figure 12-21 *Mondex smart card processing*

SECURE ELECTRONIC TRANSACTION (SET) PROTOCOL

Secure electronic transaction, or **SET**, is a secure protocol jointly designed by MasterCard and Visa with the backing of Microsoft, Netscape, IBM, GTE, SAIC, and other companies. The purpose of SET is to provide security for card payments as they traverse the Internet between merchant sites and processing banks. Although the Secure Socket Layers (SSL) protocol transmits payment data and other sensitive information securely between merchants and consumers, SSL does not verify that the consumer is the payment holder who owns the payment card.

SET was designed to improve the security and nonrepudiation of online payments. After four years of being promoted by its sponsors and not being adopted by anyone, SET has been abandoned. SET consumed a considerable amount of money and hard work, but no company found it to provide a compelling solution. The sponsoring organization still maintains a Web site with information about the protocol that you can visit by following the Online Companion link to **SETCo**.

477

Summary

Online stores can accept a variety of forms of payment. Electronic cash, one form of online payment, has been slow to catch on in the United States. A number of companies have faltered in recent years as they attempted to introduce electronic cash to the online world. Electronic cash is especially useful for making micropayments because the cost of processing payment cards for small transactions is greater than the profit on such transactions. Electronic cash shares several benefits with real cash: it is portable, anonymous, and usable for international transactions. Electronic cash can be stored online or offline. A third party, such as a bank, stores online electronic cash. The consumer holds offline cash in specially designed wallets.

Credit, debit, and charge cards (payment cards) are the most popular forms of payment on the Internet. They are ubiquitous, convenient, and easy to use.

Electronic wallets provide convenience to online shoppers because they hold payment card information, electronic cash, and personal consumer identification. Electronic wallets eliminate the need for consumers to reenter payment card and shipping information at a site's electronic checkout counter. Instead, the electronic wallet automatically fills in form information at sites that recognize the particular wallet software's technology. One persistent problem with electronic wallets is the lack of an internationally accepted standard. The W3C developed one such standard and the ECML standards group hopes that its standard will be universally adopted. With a single wallet standard, merchants would be more willing to install electronic wallet-friendly software on their commerce sites.

Smart cards are physical devices that contain an embedded microcomputer chip that stores financial information about the cardholder. Smart cards are intended to replace the collection of plastic cards people now carry, including payment cards, drivers' licenses, and insurance cards. Trials of smart cards in a few U.S. cities have proved disappointing; however, smart cards are popular in other parts of the world. Visa and American Express have introduced smart cards. Unlike electronic cash or payment cards, smart cards require merchants to install new hardware—smart card readers.

Key Terms

Acquiring bank
Anonymous electronic cash
Card not present
Chargeback
Charge card
Client-side electronic wallet
Closed loop system
Credit card
Debit card
Double spending
Due diligence
Electronic cash
Electronic wallet (e-wallet)
EMV standard
Extensible system

Float
Interoperable software
Merchant account
Merchant bank
Micropayments
Open loop system
Peer-to-peer payment system
Personal identification number (PIN)
Scrip
Secure electronic transaction (SET)
Server-side electronic wallet
Small payments
Smart card
Stored-value card

Review Questions

1. Write two paragraphs in which you explain the advantages and disadvantages of electronic cash.

2. In about 200 words, discuss why anyone with a credit card would want to use an electronic payment system such as PayPal or Billpoint for an Internet transaction.

3. In one paragraph, outline the problems that a company might encounter if it were to conduct international transactions using electronic cash.

4. Explain in about 100 words what electronic wallets are and why they are useful.

5. Explain the advantages and disadvantages of a smart card such as Mondex.

Exercises

1. Matt Remes has formed a small business and has just completed building an electronic commerce Web site that sells subscriptions to special interest newsletters. The titles range from *Apple Growers Digest and Newsletter* to *Wilderness Backpacking Newsletter*. Many organizations and individuals produce the newsletters, and Matt's role is to raise the visibility of these sometimes-obscure publications produced in out-of-the-way places. All the newsletters are published and available biweekly or monthly. Unlike traditional subscription services, Matt's business has an agreement from all newsletter publishers that he can sell subscriptions for single issues or subscriptions for periods of up to three years. He does not want to allow subscribers to use their payment cards to purchase a subscription that is less than two years in duration. But he finds that nearly 60 percent of the first-time customers on his site prefer to sample issues before committing to a subscription of a year or more. Discuss this case and present possible solutions to the problem. In about 200 words, describe existing systems that Matt could use to provide his subscribers with a simple system that does not depend on payment cards. Use **Google, HotBot**, or your favorite search engine to locate 10 newsletters that are published online.

2. Bonnie Carson has owned and managed her gift and card shop in the Central Shopping Mall for three years. Business has been good, but Bonnie wanted to expand her business. One year ago, she hired a Web designer and built a Web site hosted by a national Internet service provider. Part of the monthly ISP fee for her merchant site includes the software needed to process credit card purchases. She has obtained a merchant account with a national credit card processing company. Bonnie's Web-based business is beginning to pick up. She wants to provide more payment options to her customers. Write a report in which you advise Bonnie on the use of payment processing services such as **PayPal**. Identify at least three reasons that Bonnie should use such a service and at least three reasons why she should not. Be sure to evaluate any competitors to PayPal that you can find as part of your analysis.

3. Evan Moskowitz and you have formed an Internet company called Teach-U-Comp to market and sell computer courses online. The first courses you will offer online are all on computer programming languages, including Visual Basic, Java, and C++. Students can sign up for as many courses as they would like, and each course takes four weeks to

complete. Each course costs $55, and students receive continuing education units (CEUs) based on the duration of the course and its level of difficulty. Evan is busy creating the online content and installing the course delivery software, and he has asked you to investigate and report back to him about the feasibility of implementing electronic wallet payment systems, in addition to your site's existing credit card payment system. Your job is to investigate electronic wallet software and

write a 400-word report about what you found. Begin by looking at **Microsoft Passport**. Second, look at **Gator**. Unless restrictions prohibit you from doing this, download the Gator software and install it. Describe in another paragraph the process of downloading and installing the wallet. Finally, review the current status of the electronic commerce modeling language (**ECML**). Be sure to include in your report the URLs of any Web sites that you find to be useful.

For Further Study and Research

Andrews, W. 1999. "The Digital Wallet: A Concept Revolutionizing E-Commerce," *Internet World*, October 15, 35–45.

Bannan, K. 2001. "No Credit? No Problem! Digital Cash Made Easy," *PC World*, 19(2), February, 60–61.

Berst, J. 1999. "Ecommerce Breakthrough. Finally, an Easy Way to Pay Online," *PC Week*, 16(24), June 14.

Bills, S. 2001. "Microsoft Says Aggregation on Site Doesn't Make It a Foe," *American Banker*, 166(167), August 29, 1–2.

Bills, S. 2001. "Billpoint CEO to Banks: We Could Use Help," *American Banker*, 166(239), December 14, 18.

Boss, S., D. McGranahan, and A. Mehta. 2000. "Will the Banks Control Online Banking?" *The McKinsey Quarterly*, 3, 70–77.

Bryant, A. 2000. "Plastic Is Getting Smarter," *Newsweek*, October 16, 80.

Card Technology. 2000. "The Long Climb Ahead," March, 30–34.

DeCarmo, L. 2000. "Security Protocols and Performance," *Dr. Dobb's Journal*, 25(11), November, 40–48.

Essick, K., T. Busse, and R. Guth. 1998, "Ready, SET, Wait," *Info World*, 20(21), May 25, 1.

Gallbraith, J. 1995. *Money: Whence it Came, Where it Went*. London: Penguin Books.

Gilbert, J. 2001. "Target's Use of Technology Boosts Its Brand Image," *Business 2.0*, June. Available online at: (http://www.business2.com/marketing/2001/06/brand_technology.htm).

Grant, D. 2001. "Internet Banking Nightmare: Couple Sue After Access to Their Funds Was Cut Off for 10 Crucial Days," *EastSideJournal.com*, June 10. Available online at: (http://www.eastsidejournal.com/sited/story/html/56486).

Graven, M. 2000. "Statements and Payments," *PC Magazine*, 19(13), July, 155–156.

Grimes, B. 2001. "Forgot Your Site Password? Just Yodlee," *PC World*,19(2), December, 59.

Haskins, W. 2000. "Cashing In on E-Shopping," *PC Magazine,* 19(19), November 7, 85.

Hisey, P. 2001. "Credit Card Fraud Hurts E-Tailers," *Retail Merchandiser*, 41(9), September, 33–34.

Janal, D. 1998. *Risky Business: Protect Your Company from Being Stalked, Conned or Blackmailed on the Web*. New York: John Wiley & Sons.

Johnson, A. 2000. "Looking for Big Profits In Small Purchases," *Computerworld*, 34(20), May 15, 78.

Keizer, G. 2001. "Online Payment Options: PayPal or Plastic?" *Small Business*, May 25. Available online at: (http://www.zdnet.com/smallbusiness/stories/general/0,5821,2765483-5,00.html).

Kingston, J. 2001. "The Tech Scene: Don't Spend Your Last Flooz on Web Money." *American Banker*, 166(157), August 15, 1–2.

Kuykendall, L. 2001. "U.S. Smart Cards Seen Flourishing—Eventually," *American Banker*, 166(216), November 9, 7.

Lewis, H. 2001. "NetBank, CompuBank Merge, Customers Get Squashed," *Bankrate.com*, May 22. Available online at: (http://www.bankrate.com/bzrt/news/ob/20010521a.asp).

Magnusson, P. 2001. "Yes, They Certainly Will," *Business Week*, November 5, 90–91.

Marlin, S. 2000. "Wells, eBay Launch E-Payment Service," *Bank Systems & Technology*, 37(5), May, 14.

Mearian, L. 2001. "Visa Smart-Card Hardware Ready, But Software Isn't," *Computerworld*, 35(20), May 14, 10.

Merkow, M., J. Breithaupt, and K. Wheeler. 1998. *Building SET Applications for Secure Transactions*. New York: John Wiley & Sons.

Nerurkar, U. 2000. "Security Analysis & Design," *Dr. Dobb's Journal*, 25(11), November, 50–56.

Ptacek, M. 2001. "CompuBank's Demise May Signal a New Era," *American Banker*, 166(63), April 2, 16.

Roberts-Witt, S. 2001. "Show Me the Money," *PC Magazine*, 20(6), March 20, 13–15.

Robinson, B. 2001. "Is It Too Late for Smart Cards?" *Information Week*, March 19, 81–83.

Rosen, C. 2001. "RFID and Smart Cards Get Customers on Their Way," *Information Week*, August 6, 24.

Rosen, S. 2001. "The E-Buck Stops Here," *Information Week*, June 4, 55–57.

Roth, A. 2001. "ACH Volume Growth Slowest in Decade," *American Banker*, 166(77), April 23, 1–2.

Roth, A. 2001. "CompuBank Merge Nettles NetBank," *American Banker*, 166(119), June 21, 1–2.

Rubin, A. 2000. "Kerberos Versus the Leighton-Micali Protocol," *Dr. Dobb's Journal*, 25(11), November, 21–26.

Sapsford, J. 2000. "American Express Credit Cards to Offer Disposable Numbers for Web Shopping," *The Wall Street Journal*, September 8, B6.

Schwartz, E. 2001. "Digital Cash Payoff," *Technology Review*, 104(10), December, 62–68.

Scucka, D. 2001. "Charging Into Japan: eCharge Thinks Japan's Consumers Will Take to Its Net-Based System," *J@pan Inc*, 3(4), April, 70–72.

Stallings, W. 2000. "The SET Standard & E-Commerce," *Dr. Dobb's Journal*, 25(11), November, 30–36.

Stock, H. 2000. "Digicash Idea Finds New Life in More Flexible eCash," *American Banker*, 165(67), April 6, 9.

Tedeschi, B. 2001. "Online Billing, An Idea That Looked Like a Winner, Has Held Its Own," *New York Times*, June 4, C8.

Ulfelder, S. 2000. "Timing Is Everything To Industry Veteran," *Computerworld*, 34(21), May 22, 85.

Walker, L. 2001. "Shooting Back at Gator's 'Ambush Ads,'" *The Washington Post*, August 30, E1.

Walker, L. 2001. "The Slow Evolution of Online Bill Paying," *The Washington Post*, September 6, E1.

Wingfield, N. 1999. "Single Standard Is Set in Online Wallet," *The Wall Street Journal*, June 14, B10.

481

PLANNING FOR
ELECTRONIC BUSINESS

INTRODUCTION

AlliedSignal (now **Honeywell**) is a diversified manufacturing and technology business selling products in

the aerospace, automotive, chemicals, fibers, and plastics industries. In 1999, the company had more than

70,000 employees and annual sales exceeding $15 billion. Although some of AlliedSignal's products used

new technologies or helped other firms create new technologies, many of the products were commodity

items that were manufactured and sold just as they had been for decades. In early 1999, AlliedSignal's

CEO, Larry Bossidy, called together the heads of the company's business units for a one-day conference.

He invited Michael Dell, chairman and CEO of **Dell Computers**, and John Chambers, CEO of **Cisco**

Systems, to speak about their companies' electronic commerce implementation successes.

At the end of the day, Bossidy gave the business unit heads their marching orders. They were to

take what they had learned and create a strategy for implementing electronic commerce in their busi-

ness units—in two months. Bossidy told the room full of rather stunned managers that, although most

of their business units were at or near the top of their industries, the Internet would change everything.

He believed that the kinds of electronic commerce strategies that had worked so well for Dell and

Cisco in the computer industry would also work in AlliedSignal's businesses. He wanted to make sure

that AlliedSignal was the first to exploit those strategies and any other strategies that the business managers could devise. In two months, each manager reported back with a strategy that included multiple electronic commerce projects, such as Web sites for selling products, providing customer service, improving corporate infrastructure, managing supply chains, coordinating logistics, holding auctions, and creating virtual communities. These plans were evaluated in the company's annual strategic planning process, and the best ones were chosen for funding and immediate implementation. In a matter of months, one of the largest industrial enterprises in the world had drastically altered its course, setting sail for the uncharted waters of electronic commerce.

The ability of companies to plan, design, and implement cohesive electronic commerce strategies will make the difference between success and failure for the majority of them. The tremendous leverage that firms can gain by being the first to do business a new way on the Web has caught the attention of top executives in many industries. The keys to successful implementation of any information technology project are planning and execution. This chapter will provide some useful guidelines for those who will manage the planning, implementation, and continuing operations of electronic commerce initiatives.

LEARNING OBJECTIVES

In this chapter, you will learn about:

- Identifying the value of electronic commerce initiatives
- Aligning implementation plans with strategies
- Deciding which electronic commerce project elements to outsource
- Selecting Web hosting services
- Using incubators and fast venturing techniques to launch Internet business initiatives
- Using project and portfolio management techniques to plan and control electronic commerce activities
- Staffing electronic commerce activities

PLANNING THE ELECTRONIC COMMERCE INITIATIVE

A successful business plan for an electronic commerce initiative should include activities that will identify the initiative's specific objectives and link those objectives to business strategies (identified in Chapters 3, 4, 5, and 6).

In setting the objectives for an electronic commerce initiative, managers should consider the strategic role of the project, its intended scope, and the resources available for executing it. In this section, you will learn how to identify objectives and link those business objectives to business strategies. In later sections of this chapter, you will learn about Web site development strategies and how to manage the implementation of an electronic commerce initiative.

Identifying Objectives

Businesses undertake electronic commerce initiatives for a wide variety of reasons. Common objectives that a business might hope to accomplish through electronic commerce could include increasing sales in existing markets, opening new markets, serving existing customers better, identifying new vendors, coordinating more efficiently with existing vendors, or recruiting employees more effectively.

Resource decisions for electronic commerce initiatives should consider the expected benefits and expected costs of meeting the objectives. These decisions should also consider the risks inherent in the electronic commerce initiative and compare them to the risks of inaction—a failure to act could concede a strategic advantage to competitors.

Linking Objectives to Business Strategies

Businesses can use **downstream strategies**, which are tactics that improve the value that the business provides to its customers. Alternatively, businesses can pursue **upstream strategies** that focus on reducing costs or generating value by working with suppliers or inbound shipping and freight service providers.

You have already learned about many of the things that companies are doing on the Web. Although the Web is a tremendously attractive sales channel for many firms, companies can use electronic commerce in a variety of ways to do much more than selling. They can use the Web to complement their business strategies and improve their competitive positions. As you learned in earlier chapters of this book, electronic commerce opportunities can inspire businesses to undertake activities such as:

- Building brands
- Enhancing existing marketing programs
- Selling products and services
- Selling advertising
- Understanding customer needs better
- Improving after-sale service and support
- Purchasing products and services
- Managing supply chains
- Operating auctions
- Building virtual communities and Web portals

Although the success of each of these activities is measurable to some degree, many companies have undertaken these activities on the Web without setting specific, measurable goals. In the mid-1990s—the early days of electronic commerce—companies that had good ideas could start a business activity on the Web and not face competition. Successes and failures were measured in broad strokes. A company would either become the eBay of its industry or it would disappear—either slipping into bankruptcy or being acquired by another company.

As electronic commerce is now maturing, more companies are taking a closer look at the benefits and costs of their electronic commerce projects. Measuring both benefits and costs is becoming more important. A good implementation plan will set specific objectives for benefits to be achieved and costs to be incurred. In many cases, a company will create a pilot Web site to test an electronic commerce idea, and then release a production version of the site when it works well. Companies must specify clear goals for their pilot tests so that they know when the site is ready to go into full operation.

Measuring Benefit Objectives

Some benefits of electronic commerce initiatives are tangible and easy to measure. These include things such as increased sales or reduced costs. Other benefits are intangible and can be much more difficult to measure, such as increased customer satisfaction. When identifying benefit objectives, managers should try to set objectives that are measurable, even when those objectives are for intangible benefits. For example, success in achieving a goal of increased customer satisfaction might be measured by counting the number of first-time customers who return to the site and buy again.

Many companies create Web sites to build their brands or enhance existing marketing programs. These companies can set goals in terms of increased brand awareness, as measured by market research surveys and opinion polls. Companies that sell goods or services on their sites can measure sales volume in units or dollars. A complication that occurs in measuring either brand awareness or sales is that the increases can be caused by other things that the company is doing at the same time or by a general improvement in the economy. A good marketing staff or outside consulting firm can help a company sort out the effects of marketing and sales programs. Firms may need these groups to help set and evaluate these kinds of goals for electronic commerce initiatives.

Companies that want to use their Web sites to improve customer service or after-sale support might set goals of increased customer satisfaction or reduced costs of providing customer service or support. For example, **Philips Lighting** wanted to use the Web to provide an ordering system for its smaller customers that did not use EDI. The primary goal for this initiative was to reduce the cost of processing smaller orders. Philips had identified that over half the cost of processing smaller orders was the cost of responding to inventory availability and order status inquiries. Customers who placed small orders often called or sent faxes asking for this information.

Philips built a pilot Web site and invited a number of its smaller customers to try it. The company found that customer service phone calls from the test group of customers dropped by 80 percent. Based on that measurable increase in efficiency, Philips decided to invest in additional hardware and personnel to staff a version of the Web site that could handle virtually all of its smaller customers. The reduction in the cost of handling small orders justified the additional investment.

Companies can use a variety of similar measurements to assess the benefits of other electronic commerce initiatives. Supply chain managers can measure supply cost reductions, quality improvements, or faster deliveries of ordered goods. Auction sites can set goals for the number of auctions, the number of bidders and sellers, the dollar volume of items sold, the number of items sold, or the number of registered

485

participants. The ability to track such numbers is usually built into auction site software. Virtual communities and Web portals measure the number of visitors and try to measure the quality of their visitors' experiences.

Some sites use online surveys to gather this data; however, most settle for estimates based on length of time that each visitor remains on the site and how often visitors return. A summary of benefits and measurements that companies can make to assess the value of those benefits appears in Figure 13-1.

Electronic commerce initiatives	Common measurements of benefits provided
Build brands	Surveys or opinion polls that measure brand awareness
Enhance existing marketing programs	Change in per-unit sales volume
Improve customer service	Customer satisfaction surveys, the number of customer complaints
Reduce cost of after-sale support	Quantity and type (telephone, fax, e-mail) of support activities
Improve supply chain operation	Cost, quality, and on-time delivery of materials or services purchased
Hold auctions	Quantity of auctions, bidders, sellers, items sold, registered participants; dollar volume of items sold
Provide portals and virtual communities	Number of visitors, number of return visits per visitor, and duration of average visit

Figure 13-1 *Measuring the benefits of electronic commerce initiatives*

No matter how a company measures the benefits provided by its Web site, it usually tries to convert the raw activity measurements to dollars. Having the benefits measured in dollars lets the company compare benefits to costs and compare the net benefit (benefits minus costs) of a particular initiative to the net benefits provided by other projects. Although each activity provides some value to the company, it is often difficult to measure that value in dollars. Usually, even the best attempts to convert benefits to dollars yield only rough approximations.

486

Measuring Cost Objectives

At first glance, the task of identifying and estimating costs may seem much easier than the task of setting benefits objectives. However, many managers have found that information technology project costs can be as difficult to estimate and control as the benefits of those projects. Since Web development uses hardware and software technologies that change even more rapidly than those used in other information technology projects, managers often find that their experience does not help much when they are making estimates. Most changes in the cost of hardware are downward, but the increasing sophistication of software provides an ever-increasing demand for more of the newer, cheaper hardware. This often yields a net increase in overall hardware costs. The more sophisticated software, of course, usually costs more than the amount originally budgeted, too. Even though electronic commerce initiatives tend to be completed within a shorter time frame than many other information technology projects, the rapid changes in Web technology can quickly destroy a manager's best-laid plans.

In addition to hardware and software costs, the project budget must include the costs of hiring, training, and paying the personnel who will design the Web site, write or customize the software, create the content, and operate and maintain the site. As more companies build electronic commerce sites, people who have the skills necessary to do the work are commanding increasingly higher compensation.

For many companies, one of the largest and most significant costs associated with electronic commerce initiatives is the cost of *not* undertaking such an initiative. The foregone benefits that a company could have obtained from an electronic commerce initiative that they chose not to pursue are costs. Managers and accountants use the term **opportunity cost** to describe such lost benefits from an action not taken.

Based on data collected in separate recent surveys, International Data Corporation and the GartnerGroup both estimated that the cost for a large company to build and implement an adequate entry-level electronic commerce site was about $1 million. About 79 percent of this cost was labor-related; 10 percent was the cost of software and 11 percent was the cost of hardware. GartnerGroup concluded that it would take between $2 million and $5 million to build a site that would compare favorably to leading sites. International Data Corporation noted that 10 of the top 100 electronic commerce sites had spent over $10 million for development and implementation.

A 1999 *Advertising Age* survey of smaller companies showed that they were spending an average of $78,000 to build new electronic commerce sites, an increase of 75 percent over the previous year's average. More recent estimates of the cost to build small Web sites have continued to increase as more companies establish themselves on the Web, and as expensive features such as shopping carts and search engines become standard on even the most basic sites. Some industry analysts have pegged the minimum dollar amount needed to open an entry level electronic commerce Web site at $150,000. The GartnerGroup estimates that establishing a basic electronic commerce operation on the Web today will cost a company between $300,000 and $1 million, and that creating a site that is a true differentiator will cost a minimum of $15 million. Figure 13-2 summarizes industry estimates for the cost of creating a Web business at three different levels: a basic entry level, a level comparable to most existing Web competitors, and a level that makes the Web site stand out as truly different from competitors' sites.

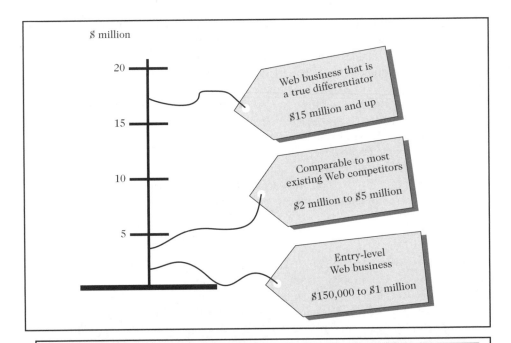

Figure 13-2   Starting a Web business: three price tags

The initial cost of building an electronic commerce site is not the whole story, unfortunately. Since Web technology continues to evolve at a rapid pace, most businesses will want to take advantage of what that technology offers to remain competitive. Most experts agree that the annual cost to maintain and improve a site once it is up and running—whether it is a small site or a large site—will be between 50 percent and 200 percent of its initial cost.

In a 2001 article, members of the management consulting firm McKinsey & Company reported a study that estimated startup and ongoing costs for magazine publishers' Web sites (see the Barsh, et al. reference in the "For Further Study and Research" section at the end of this chapter for a reference to the full report). The McKinsey study estimated costs for two types of magazine sites: a full portal site that would serve as a destination in itself, and a more limited magazine companion site that would complement a printed magazine. The full portal site cost estimate was $2.4 million to build and $4.3 million per year to maintain, with a staff of 35 people; the companion site cost estimate was $150,000 to build and $270,000 per year to maintain, with a staff of two people. Both of these estimates exclude the cost of developing content for the site and assume that the magazine publisher already has an existing IT infrastructure for a print publishing business serving a subscriber base of 300,000. Figure 13-3 shows the approximate breakdown of these costs.

As an increasing number of traditional businesses create Web versions of their physical stores, the cost to build an online business that is a true differentiator—a site that stands out and offers something new to customers—will continue to increase. Much of the cost in such a Web site is for elements that make a major difference in how well the site works, but are not readily apparent to a site visitor.

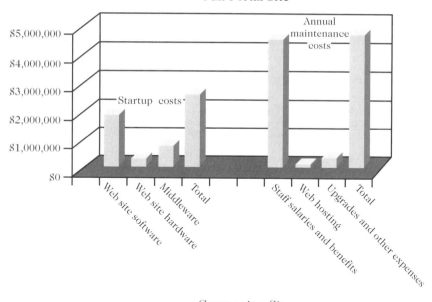

Full Portal Site

Startup costs

Annual maintenance costs

Web site software · Web site hardware · Middleware · Total · Staff salaries and benefits · Web hosting · Upgrades and other expenses · Total

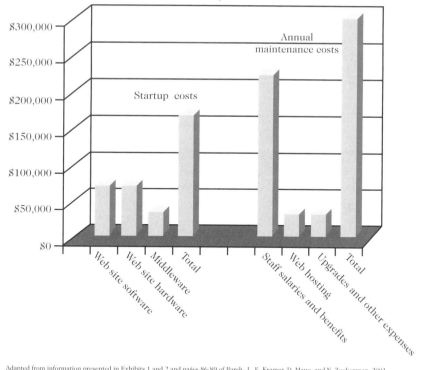

Companion Site

Startup costs

Annual maintenance costs

Web site software · Web site hardware · Middleware · Total · Staff salaries and benefits · Web hosting · Upgrades and other expenses · Total

Adapted from information presented in Exhibits 1 and 2 and pages 86-89 of Barsh, J., E. Kramer, D. Maue, and N. Zuckerman. 2001. "Magazine's Home Companion," *The McKinsey Quarterly*, 2, June, 83-91.

Figure 13-3   *Cost estimates for building and operating two types of magazine publisher Web sites*

Planning for Electronic Business

For example, Kmart's Web store, **BlueLight.com**, cost more than $140 million to create. The site's home page, shown in Figure 13-4, is certainly well-designed and highly functional, but the typical visitor would never guess how much this site cost to build. Much of the site's cost is hidden; the money was used to buy and customize middleware that connects the Web site to Kmart's vast inventory and logistics databases.

reference to brand that has been successful in physical stores

promotion designed to appeal to Web visitors

Figure 13-4 *Kmart's BlueLight.com home page*

Comparing Benefits to Costs

Most companies have procedures that call for an evaluation of any major expenditure of funds. These major investments in equipment, personnel, and other assets are called **capital projects** or **capital investments**. The techniques that companies use to evaluate proposed capital projects range from very simple calculations to complex computer simulation models. However, no matter how complex the technique, it always reduces to a comparison of benefits and costs. If the benefits exceed the cost of a project by a comfortable margin, the company invests in the project.

A key part of creating a business plan for electronic commerce initiatives is the process of identifying potential benefits (including intangibles such as employee satisfaction and company reputation), identifying the costs required to generate those benefits, and evaluating whether the benefits exceed the costs. Companies should evaluate each element of their electronic commerce strategies using this cost/benefit approach. A simplified representation of the cost/benefit approach appears in Figure 13-5.

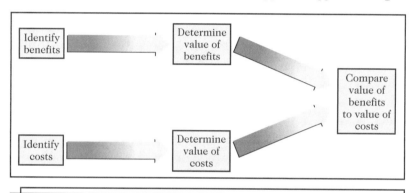

Figure 13-5 *Cost/benefit evaluation of electronic commerce strategy elements*

You might have learned techniques for capital project evaluation, such as the payback method or the net present value method, in your accounting or finance courses. These evaluation approaches provide a quantitative expression of a comfortable benefit-to-cost margin for a specific company. They can also mathematically adjust for the reduced value of benefits that the investment will return in future years (benefits received in future years are worth less than those received in the current year). Managers often use the term **return on investment (ROI)** to describe any capital investment evaluation technique, even though ROI is the name of only one of these techniques.

Although most companies evaluate the anticipated value of electronic commerce initiatives in some way before approving them, many companies see these projects as absolutely necessary investments. Thus, they might not subject them to the same close examination and rigid requirements as they do other capital projects. These companies fear being left behind as competitors stake their claims in the online marketspace. The value of early positioning in a new market is so great that many companies are willing to invest large amounts of money with few near-term profit prospects.

Newspaper Web sites are a good example of this desire to establish a foothold in the online marketspace. Profitable electronic commerce initiatives in the newspaper business, such as Gannet's **USA Today** and *The Wall Street Journal*'s **WSJ.com** sites, are few. Most newspaper sites continue to lose money. Despite the losses, most newspaper companies believe that they cannot afford to ignore the long-term potential of the Web and feel compelled to continue to support their online investments. These companies have estimated their opportunity costs of not being present on the Web (for example, the loss of future profits to be earned from the Web site, or the risk of losing market share to competitors) to be greater than the losses they are experiencing in the near term.

STRATEGIES FOR WEB SITE DEVELOPMENT

When companies began establishing their presences on the Web, the typical Web site was a static brochure that was not updated frequently with new information and seldom had any capabilities for helping the company's customers or vendors transact business. As Web sites have become the home not only of transaction processing but also of automated business processes of all kinds, these Web sites have become important parts of companies' information systems infrastructures. The evolution of Web site functions—from the static brochures of the early days of electronic commerce, to transaction processing tools, to today's automated homes for business processes of all kinds—appears in Figure 13-6.

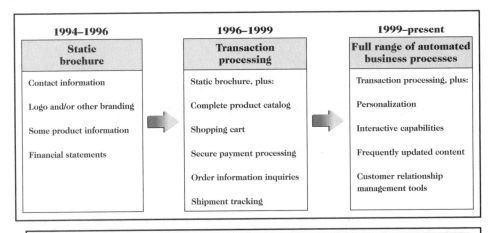

1994–1996	1996–1999	1999–present
Static brochure	**Transaction processing**	**Full range of automated business processes**
Contact information	Static brochure, plus:	Transaction processing, plus:
Logo and/or other branding	Complete product catalog	Personalization
Some product information	Shopping cart	Interactive capabilities
Financial statements	Secure payment processing	Frequently updated content
	Order information inquiries	Customer relationship management tools
	Shipment tracking	

Figure 13-6   *Increasing complexity of Web site functions*

492

This transformation occurred rapidly—taking only a year or two in most companies—and very few businesses have caught up with the changes in terms of how they develop and manage Web sites. Thus, the purposes and scope of Web sites have increased greatly, but few businesses today manage them as the dynamic business applications they have become. The tools that companies have developed over the years to manage software development projects are designed to help those companies meet the needs of their current customers and operate more effectively within existing value chains.

Many large and medium-sized companies have found it extremely difficult to develop new information systems and Web sites that work with such systems to create new markets or reconfigure their supply chains. In the past, companies that have had success in exploring new ways of working with their customers and suppliers by reconfiguring supply chains have had the luxury of time—in many cases, years—to complete those reconfigurations. However, the speed at which the Internet has changed markets and marketing channels throughout entire industry value chains precludes lengthy reconfigurations. Now, companies that want to successfully adapt to the changed business environment of the information age must explore alternatives to traditional systems development methods.

Internal Development vs. Outsourcing

Although many companies would like to think that they can avoid electronic commerce site development problems by outsourcing the entire project, savvy leaders realize that they cannot. No matter what kind of electronic commerce initiative a company is contemplating, the initiative's success depends on how well it is integrated into and supports the activities in which the business is already engaged. However, few companies are large enough or have sufficient in-house expertise to launch an electronic commerce project without some external help. Even Wal-Mart, with annual sales of more than $150 billion, did not undertake its 2000 Web site relaunch alone. The key to success is finding the right balance between outside and inside support for the project. Hiring another company to provide the outside support for all or part of the project is called **outsourcing**.

The Internal Team

The first step in determining which parts of an electronic commerce project to outsource is to create an internal team that is responsible for the project. This team should include people with enough knowledge about the Internet and its technologies to know what kinds of things are possible. Team members should be creative thinkers who are interested in taking the company beyond its current boundaries, and they should be people who have distinguished themselves in some way by doing something very well for the company. If they are not already recognized by their peers as successful individuals, the project may suffer from lack of credibility.

Some companies make the mistake of appointing as electronic commerce project leader a technical wizard who does not know much about the business and is not well-known throughout the company. Such a choice can greatly increase the likelihood of failure. Business knowledge, creativity, and the respect of the firm's operating function managers are all much more important than technical expertise in establishing successful electronic commerce.

Measuring the achievements of this internal team is very important. The measurements do not have to be monetary. Achievement can be expressed in whatever terms are appropriate to the objectives of the initiative. Customer satisfaction, number of sales leads generated, and reductions in order-processing time are examples of metrics that can provide a sense of the team's level of accomplishment. The measurements should show how the project is affecting the company's ability to provide value to the consumer. Many consultants advise companies to set aside between 5 percent and 10 percent of a project's budget for quantifying the project's value and measuring the achievement of that value.

Increasingly, companies are recognizing the value of the intellectual capital they have built up in the form of employees' knowledge about the business and its processes. In the past, many companies ignored the value of their human assets because they did not appear in the accounting records or financial statements.

Leif Edvinsson has pioneered the use of human capital measures at Skandia Group, a large financial services company in Sweden. In addition to acknowledging employees' competencies, Edvinsson's measures include the value of customer loyalty and business partnerships as part of a company's intellectual capital. This networking approach to evaluating intellectual capital shows promise as a tool for assessing and tracking the value of internal teams and their connections to external consultants. Although these measurements are just now being adapted for use in measuring systems development efforts, we have included references to books by Edvinsson and Max Boisot, another proponent of human capital measurement, in the "For Further Study and Research" section at the end of this chapter.

The internal team should hold ultimate and complete responsibility for the electronic commerce initiative, from the setting of objectives to the final implementation and operation of the site. The internal team will decide which parts of the project to outsource, to whom those parts will be outsourced, and what consultants or partners the company will need to hire for the project. Consultants, outsourcing providers, and partners can be extremely important early in the project because they often develop skills and expertise in new technologies before most information systems professionals do.

Early Outsourcing

In many electronic commerce projects, the company outsources the initial site design and development to launch the project quickly. The outsourcing team then trains the company's information systems professionals in the new technology before handing the operation of the site over to them. This approach is called **early outsourcing**. Since operating an electronic commerce site can rapidly become a source of competitive advantage for a company, it is best to have the company's own information systems people working closely with the outsourcing team and developing ideas for improvements as early as possible in the life of the project.

Late Outsourcing

In the more traditional approach to information systems outsourcing, the company's information systems professionals do the initial design and development work, implement the system, and operate the system until it becomes a stable part of the business operation. Once the company has gained all the competitive advantage provided by the system, the maintenance of the electronic commerce system can be outsourced so that the company's information systems professionals can turn their attention and talents to developing new technologies that will provide further competitive advantage. This approach is called **late outsourcing**. Although for years late outsourcing has been the standard for allocating scarce information systems talent to projects, electronic commerce initiatives lend themselves more to the early outsourcing approach.

Partial Outsourcing

In both the early outsourcing and late outsourcing approaches, a single group is responsible for the entire design, development, and operation of a project group—either inside or outside the company. This typical outsourcing pattern works well for many information systems projects. However, electronic commerce initiatives can benefit from a partial outsourcing approach, too. In **partial outsourcing**, which is also called **component outsourcing**, the company identifies specific portions of the project that can be completely designed, developed, implemented, and operated by another firm that specializes in a particular function.

Many smaller Web sites outsource their e-mail handling and response function. Customers expect rapid and accurate responses to any e-mail inquiry they make of a Web site with which they are doing business. Many companies send the customer an automatic order confirmation by e-mail as soon as the order or credit card payment is accepted. A number of companies provide e-mail auto-response functions on an outsourcing basis.

Another common example of partial outsourcing is an electronic payment system. Many vendors are willing to provide complete customer payment processing. These vendors provide a site that takes over when customers are ready to pay and returns the customers to the original site after processing the payment transaction.

One of the most common elements of electronic commerce initiatives that companies outsource using this approach is the Web hosting activity that you learned about in Chapter 9. Internet service providers (ISPs) offer Web hosting services to companies that want to operate electronic commerce sites, but that do not want to invest in the hardware and staff needed to create their own Web servers. ISPs are usually willing to accommodate requests for a variety of service levels. Small businesses can rent space on an existing server at the ISP's location. Larger companies can purchase the server hardware and have the ISP install and maintain it at the ISP's location. The ISP provides the continuous staffing and expertise needed to keep an electronic commerce site up and running 24 hours a day, seven days a week (this kind of service is often called **24/7 operation**). Most ISPs offer a wide range of services, including personal Web access for individuals. Some ISPs specialize in services to business. These larger ISPs cater to companies that want to operate electronic commerce sites. They usually offer wider bandwidth connections to the Internet than smaller ISPs and offer more reliable continuous service.

A number of ISPs and other firms offer services beyond basic Internet connectivity to companies that want to do business on the Web. Many of these services were described earlier as candidates for partial outsourcing strategies and include automated e-mail response, transaction processing, payment processing, security, customer service and support, order fulfillment, and product distribution. Recall from Chapter 11 that a company that offers these services may be called a commerce service provider (CSP) or, if the service uses specific application software (such as an automated e-mail response service), it may be called an application service provider (ASP).

NORDISK AVIATION

Nordisk Aviation is a subsidiary of the Norwegian Norsk Hydro Group. It designs, manufactures, and repairs air cargo containers for both freight and passenger baggage for major airlines throughout the world, and for freight carriers such as FedEx and UPS. It also designs and sells handling systems and pallets that work with the containers. The company has annual sales of more than $200 million and employs more than 500 people at its 22 locations around the world.

Nordisk was a strong believer in using the outsourcing approach for its IT projects—its IT department included only two people. These two IT staff members worked as the overseers of every IT design and implementation project for the company. They also managed the ongoing IT services provided to Nordisk by other companies.

In late 2000, Manfred Gollent, the president of Nordisk, decided it was time to upgrade the company's Web site—which had been operating as an information site for several years—to include portal features that would allow Nordisk customers to check order status and learn about current developments in container and container handling systems design. The logical approach for Nordisk was to find a company to which it could outsource the project.

The two members of Nordisk's IT staff went to work finding suitable Web developers. The previous Web developer had disappeared; they were unable to find any trace of the person who had created the existing Web site. The developer had created the Web site so that it used a number of programs to deliver dynamic pages. Unfortunately, the developer had given Nordisk only the executable code and not the actual programs. He also did not provide Nordisk with any documentation of the programs.

When the Web site was initially created, it was not an important strategic project for Nordisk. The IT staff members, who were busy with other important projects did not ensure that the application code and documentation were received at the time. Nordisk had to hire a company to rebuild the site completely to obtain the additional portal functions it wanted to add to the site. The lesson from the Nordisk case is that even when a company is outsourcing virtually all of its Web development, it must have procedures in place to ensure that the project is internally managed and documented.

496

Selecting a Hosting Service

The internal team should be responsible for selecting the ISP that will provide the site's hosting service. For smaller electronic commerce projects, teams can consult an ISP directory such as **The List**, which you learned about in Chapter 8. These sites provide a search engine that helps visitors choose an ISP, Web hosting service, or ASP that meets their needs from the sites' thousands of listings.

For larger Web site implementations, the team will want to obtain the advice of consultants or other firms that rate service providers (ISPs, ASPs, and CSPs), such as **HostCompare.com** and **Keynote Systems**. The most important factors to evaluate when selecting a hosting service include:

- Functionality
- Reliability
- Bandwidth and server scalability
- Security
- Backup and disaster recovery
- Cost

Companies that sell hosting services provide different features and different levels of service. The functionality offered by a service provider can include credit card processing and the ability to link to existing databases that store customer and product information. Almost all hosting services offer site visitor tracking, but as you learned in Chapter 9, the capabilities of different tracking software packages are not all equal. Some tracking software provides much more detailed information and easier-to-use report generators than other tracking software. You should determine the functionality offered by a hosting service and carefully evaluate whether that functionality will be sufficient to meet the needs of your Web site.

The service should offer a guarantee that limits possible downtime. Electronic commerce buyers expect hosting services to be up and running 24 hours a day, every day. Of course, no hosting service can promise it will never fail, but it can provide staffing and backup hardware that minimizes reliability problems. Coordination of this function with the service provider can be very important. Usually, a business must have some around-the-clock staff available or on-call to work with the service provider when an interruption occurs.

The bandwidth of the service's connection to the Internet must be sufficient to handle the peak transaction loads that its customers require. Sometimes a service provider will sign up new accounts faster than it can expand the bandwidth of its connections, resulting in access bottlenecks. A guarantee that specifies bandwidth availability or server response times is worth negotiating into a service provider contract. If a company expects its site's traffic to increase rapidly, it is important that its service provider can rapidly increase the server capacity and the bandwidth provided. In general, larger hosting services can scale up more easily than smaller hosting services. Again, it is worth negotiating some scalability into the service provider contract in such situations.

Since the company's information on customers, products, pricing, and other data will be placed in the hands of the service provider, the vendor's security policies and practices are very important. You learned about electronic commerce security issues and practices in Chapters 10 and 11. The service provider should specify the types of security it provides and how it implements security. No matter what security guarantees the service provider offers, the company should monitor the security of the electronic commerce operation through its own personnel or by hiring a security consulting firm. Security consultants can periodically test the system and can launch attacks on the security features used by the service provider to determine whether they are easily breached.

The hosting service should be able to guarantee close to 100 percent reliability by having a workable disaster recovery plan in place. In addition to having off-site data backup or mirroring, the hosting service should have a way to restore a company's site very quickly in the case of a natural disaster.

Service providers offer many different pricing plans for different levels of service. Knowing what types of server hardware and software your site will require, and having a good estimate of the range of transaction loads the site is likely to generate, can help in negotiating a price for the hosting service.

New Methods for Implementing Partial Outsourcing

In the past five years, new ways of implementing the partial outsourcing strategy have evolved specifically for Web businesses. The next two sections describe two of the more popular of these methods; incubators and fast venturing.

Incubators

An **incubator** is a company that offers start-up companies a physical location with offices, accounting and legal assistance, computers, and Internet connections at a very low monthly cost. Sometimes, the incubator will offer seed money, management advice, and marketing assistance. In exchange, the incubator receives an ownership interest in the company, typically between 10 percent and 50 percent.

When the company grows to the point that it can obtain venture capital financing or launch a public offering of its stock, the incubator sells all or part of its interest and reinvests the money in a new incubator candidate. One of the first Internet incubators was **idealab!**, which helped companies such as **CarsDirect.com**, **Overture**, and **Tickets.com** get their start.

Some companies have created internal incubators. A number of companies used internal incubators in the past to develop technologies that the companies planned to use in their main business operations. Most of these programs, such as the Kodak internal venturing program of the 1980s, were unsuccessful and, ultimately, were shut down. Employees in internal incubators found it difficult to maintain an entrepreneurial spirit when they knew that the technology they were developing would ultimately be taken away and controlled by the parent company.

More recently, companies such as Matsushita Electric's U.S. Panasonic division have started internal incubators to help launch new companies that will grow to become important strategic partners. The companies launched in the incubator will retain their individual management teams and the assets they develop. The prospects for these strategic partner incubators appear to be much brighter than those of the old-style technology development incubators.

Fast Venturing

Often, large companies struggle to emulate the entrepreneurial spirit of smaller companies as they launch their Internet business initiatives. Many of these companies are trying to expand the internal incubator model and create an effective support system for new business and technology ideas, such as electronic commerce initiatives. One approach that is becoming popular is called fast venturing.

In **fast venturing**, an existing company that wants to launch an electronic commerce initiative joins external equity partners and operational partners that can offer the experience and skills needed to develop and scale up the project very rapidly. Equity partners are usually banks or venture capitalists that sometimes offer money, but are more likely to offer experience gained from guiding other start-ups that they have funded. Operational partners are firms, such as systems integrators, consultants, and Web portals, that have experience in moving projects along and scaling up prototypes. The roles of each participant in fast venturing are described in Figure 13-7.

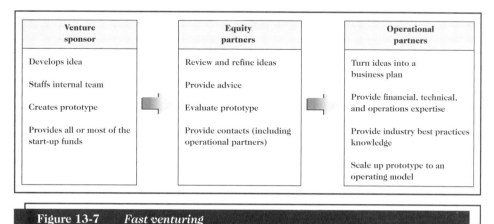

Venture sponsor	Equity partners	Operational partners
Develops idea	Review and refine ideas	Turn ideas into a business plan
Staffs internal team	Provide advice	Provide financial, technical, and operations expertise
Creates prototype	Evaluate prototype	
Provides all or most of the start-up funds	Provide contacts (including operational partners)	Provide industry best practices knowledge
		Scale up prototype to an operating model

Figure 13-7   *Fast venturing*

The venture sponsor is the existing company that wants to launch the electronic commerce initiative. The equity partners are entities that have provided start-up money to new ventures in the past and have developed knowledge about operating new ventures. The equity partners provide advice based on this knowledge to the venture sponsor, which typically has little experience in developing new ventures. The operational partners are people and companies that have built Web business sites before. Thus, they can provide expertise in the technologies and business practices needed to create a successful operating electronic commerce site.

MANAGING ELECTRONIC COMMERCE IMPLEMENTATIONS

The best way to manage any complex business software implementation is to use formal project management techniques. Project management was developed by the U.S. military and the defense contractors that worked with the military in the 1950s and the 1960s to develop weapons and other large systems. Not only was defense spending increasing in those years, but individual projects were becoming so large that it became impossible for managers to maintain control over them without some kind of assistance.

Project Management

Project management is a collection of formal techniques for planning and controlling the activities undertaken to achieve a specific goal. The project plan includes criteria for cost, schedule, and performance—it helps project managers make intelligent trade-off decisions regarding these three criteria. For example, if it becomes necessary for a project to be completed early, the project manager can compress the schedule by either increasing the project's cost or decreasing its performance.

Today, project managers use specific application software called **project management software** to help them manage projects. Project management software products, such as **Microsoft Project** and **Primavera Project Planner**, give managers an array of built-in tools for managing resources and schedules. The software can generate charts and tables that show, for example, which parts of the project are critical to its timely completion, which parts can be rescheduled or delayed without changing the project completion date, and where additional resources might be most effective in speeding up the project.

In addition to managing the people and tasks of the internal team, project management software can help the team manage the tasks assigned to consultants, technology partners, and outsourced service providers. By examining the costs and completion times of tasks as they are completed, project managers can learn how the project is progressing and continually revise the estimated costs and completion times of future tasks.

Information systems development projects have a well-deserved reputation for running out of control and ultimately failing. They are much more likely to fail than other types of projects, such as building construction projects. The main causes for information systems project failures are rapidly changing technologies, long development times, and changing customer expectations. Because of this vulnerability, many teams rely on project management software to help them achieve project goals.

Although electronic commerce certainly uses rapidly changing technologies, the development times for most electronic commerce projects are relatively short—often they are accomplished in under six months. This gives both the technologies and the expectations of users less time to change. Thus, electronic commerce initiatives are, in general, more successful than other types of information systems implementations.

You can learn more about project management by reading the references listed in the "For Further Study and Research" section at the end of this chapter, or by clicking the Online Companion link for the **Project Management Institute**, a not-for-profit organization devoted to the promotion of professional project management practices.

Project Portfolio Management

Larger organizations often have many IT implementation projects going on simultaneously—a number of which could be electronic commerce implementations or updates. Some chief information officers (CIOs) of larger companies now use a portfolio approach to managing these multiple projects. **Project portfolio management** is a technique in which each project is monitored as if it were an investment in a financial portfolio. The CIO records the projects in a list (usually using spreadsheet or database management software) and updates the list regularly with current information about each project's status.

Project management software performs a function similar to this for the tasks within a project, but most project management software packages are designed to handle individual projects and do not do a very good job of consolidating activities across multiple projects. Also, the information used in project portfolio management differs somewhat from the information used to manage specific projects. Project management software tracks the details of how each project is accomplishing its specific goals. In project portfolio management, the CIO assigns a ranking for each project based on its importance to the strategic goals of the business and its level of risk (probability of failure).

To develop these rankings, the CIO can use any of the methods that financial managers use to evaluate the risk of making investments in business assets. Indeed, using the tools of financial management helps the CIO to explain electronic commerce projects as investments in assets—using the language that financial managers (and often the CEO) understand. You can learn more about project portfolio management by reading the Berinato article cited in the "For Further Study and Research" section at the end of this chapter.

Staffing the Operation

Regardless of whether the internal team decides to outsource parts of the design and implementation activity, it must determine the staffing needs of the electronic commerce initiative. The general areas of staffing that are most important to the success of an electronic commerce initiative include:

- Business management
- Account managers
- Applications specialists
- Web graphics designers
- Content creators
- Content managers or editors
- Customer service staff
- Systems administration
- Network operations staff
- Database administration

The business management function should include internal staff. The business manager should be a member of the internal team that sets the objectives for the project. The **business manager** is responsible for implementing the elements of the business plan and reaching the objectives set by the internal team. If revisions to the plan are necessary as the project proceeds, the business manager develops specific proposals for plan modifications and additional funding and presents them to the internal team and top management for approval.

The business manager should have experience and knowledge related to the business activity that is being implemented in the electronic commerce site. For example, if business managers are assigned to a retail consumer site, they should have experience managing a retail sales operation. In the future, a company might want to hire an experienced electronic commerce business manager from another company; however, electronic commerce is too new for many managers to have obtained experience yet. Thus, most companies try to find internal candidates for this position.

In addition to including the business manager, the business management function in large electronic commerce initiatives may include other individuals who carry out specialized functions, such as project management or account management, that the business manager does not have time to handle personally. An **account manager** keeps track of multiple Web sites in use by a project or keeps track of the projects that will combine to create a larger Web site.

501

Most larger projects will have a test version, a demonstration version, and a production version of the Web site located on different servers. The test version is the "under construction" version of a Web site. Since most sites are frequently updated with new features and content, the test version gives the company a place to make sure that each new feature works before exposing it to customers. The demonstration version has features that have passed testing and must be demonstrated to an internal audience (for example, the marketing department) for approval. The production version is the full operating version of the site that is available to customers and other visitors. The account manager supervises the location of specific Web pages and related software installations as they are moved from test, to demonstration, to production.

As more vendors provide packaged software solutions for electronic commerce, such as those you learned about in Chapter 9, companies will need information systems staff that can install and maintain the software. Most large businesses have **applications specialists** who maintain accounting, human resources, and logistics software. Similarly, electronic commerce sites that buy software to handle catalogs, payment processing, and other features, need applications specialists to maintain the software. Although the installation of these software packages can be outsourced, most companies prefer to train their own staff to serve in this function when the site becomes operational.

Since the Web is a visual medium, the role of graphic elements on individual Web pages is important. A company must either retain the services of a graphics design firm, a Web design firm that includes graphics designers, or must hire employees with graphic design skills. A **Web graphics designer** is a person trained in art, layout, and composition, who also understands how Web pages are constructed. The Web graphics designer, or design team for larger sites, must ensure that the Web pages on the site are visually appealing, easy to use, and make consistent use of graphics elements from page to page.

Most larger sites and many smaller sites will have content created specifically for the Web site. Other sites will adapt content from existing sources within the company for use on the Web site or will purchase content to use on the site. These activities require that the company hire **content creators** to write original content and **content managers** or **content editors** to purchase existing material and adapt it for use on the site.

The Web offers businesses a unique opportunity to reach out to their customers. Thus, business-to-consumer and business-to-business sites that want to capitalize on that opportunity must include a customer relationship management function. **Customer service** personnel help design and implement customer relationship management in the electronic commerce operation. They can, for example, issue and administer passwords, design customer interface features, handle customer e-mail and telephone requests for service or follow-up action, and conduct telemarketing for the site. Companies strive to provide the best possible service to satisfy the demands of their customers. The increasing power of customers to organize and express their expectations on the Web is a natural extension of the increase in consumerism that has occurred over the past two decades.

Some companies outsource parts of their customer relationship management operation to independent call centers. A **call center** is a company that handles incoming customer telephone calls and e-mails for other companies. Using a call center often makes sense for smaller companies that do not have the volume of customer inquiries to justify creating their own internal call center operation. Some call centers work with a variety of businesses; others focus on one specialty area. For example, a specialized call center might contract with software manufacturers to provide installation help for their software products. Call center employees who are skilled in helping customers install one software package are often able to learn how to support other software packages very quickly.

A systems administrator who understands the server hardware and operating system is an essential part of a successful electronic commerce implementation. The **systems administrator** is responsible for the system's reliable and secure operation. If the site operation is outsourced to a service provider, the service provider will supply this function. If the site is hosted by the company, it will need to devote at least one person to this job. In addition, the internal system administrator needs sufficient staff to maintain full 24/7 operation and site security. These **network operations** staff functions include load estimation and load monitoring, resolving network problems as they arise, designing and implementing fault-resistance technologies, and managing any network operations that are outsourced to service providers or telephone companies.

Most electronic commerce sites will require some kind of **database administration** function to support activities such as transaction processing, order entry, inquiry management, or shipment logistics. These activities require either an existing database into which the site is being integrated, or a separate database established for the electronic commerce initiative. It is important to have a database administrator who can effectively manage the design and implementation of this function.

Post-Implementation Audits

After an electronic commerce site is successfully launched, most of the project's resources are devoted to maintaining and improving the site's operations. However, an increasing number of businesses are realizing the value of a post-implementation audit. A **post-implementation audit** is a formal review of a project after it is up and running.

The post-implementation audit gives managers a chance to examine the objectives, performance specifications, cost estimates, and scheduled delivery dates that were established for the project in its planning stage, and compare them to what actually happened. In the past, most project reviews focused on identifying individuals to blame for cost over-runs or missed delivery dates. Since many external forces in technology projects can overwhelm the best efforts of managers, this blame identification approach was generally unproductive, as well as uncomfortable, for the managers on the project.

A post-implementation audit allows the internal team, the business manager, and the project manager to raise questions about the project's objectives and provide their "in-the-trenches" feedback on strategies that were set in the project's initial design. By agreeing beforehand not to lay blame, the company obtains valuable information that it can use in planning future projects and gives the participants a meaningful learning experience.

503

Summary

This chapter provided an overview of key elements that are typically included in business plans for electronic commerce implementations. The first step is setting objectives. Specific objectives derive from the initiative's overall goals and include planned benefits and planned costs. The benefit and cost objectives should be stated in measurable terms, such as dollars or quantities. Before undertaking an electronic commerce project, most companies will evaluate its estimated costs and benefits.

Businesses use a number of evaluation techniques; however, most businesses calculate projects' return on investment to gauge their value. Many companies have undertaken electronic commerce projects without evaluating their costs and benefits in detail because they fear being left out of the Internet's marketspace.

Companies must decide how much, if any, of an electronic commerce project they will outsource. The first step in determining an outsourcing strategy is to form an internal team that includes knowledgeable individuals from within the company. The internal team develops the specific project objectives and is responsible for meeting those objectives. The internal team designs an outsourcing strategy, selects a hosting service (or decides to have the company host its own Web server), and supervises the staffing of the project.

Project management is a formal way to plan and control specific tasks and resources used in a project. It provides project managers with a tool they can use to make informed trade-offs among the project elements of schedule, cost, and performance. Large organizations are beginning to use project portfolio management techniques to track and make trade-offs among multiple ongoing projects. Electronic commerce initiatives are usually completed within a short time frame and thus are less likely to run out of control than other information systems development projects.

The company must staff the electronic commerce initiative regardless of whether portions of the project are outsourced. Critical staffing areas include business management, application specialists, customer service staff, systems administration, network operations staff, and database administration. A good way for all participants to learn from project experiences is to conduct a post-implementation audit that compares project objectives to the actual results.

Key Terms

24/7 operation	Fast venturing
Account manager	Incubator
Applications specialist	Late outsourcing
Business manager	Network operations
Call center	Opportunity cost
Capital investment	Outsourcing
Capital project	Partial outsourcing
Component outsourcing	Post-implementation audit
Content creator	Project management
Content editor	Project management software
Content manager	Project portfolio management
Customer service	Return on investment
Database administration	Systems administrator
Downstream strategies	Upstream strategies
Early outsourcing	Web graphics designer

504

Review Questions

1. Name three benefit objectives that a business might decide to measure in an electronic commerce business plan.

2. Why have some firms approved electronic commerce projects without taking a close look at their return on investment numbers?

3. Why is late outsourcing seldom used in electronic commerce projects?

4. In fewer than 200 words, name and briefly describe four factors that a company should evaluate when selecting an ISP, ASP, or CSP to provide Web hosting services.

5. Why should the head of the business management function of an electronic commerce initiative be an employee of the company implementing the project? Limit your answer to 250 words.

Exercises

1. The Grover Cams Company manufactures cams and other components for diesel engines. As Webmaster for Grover, you have created an attractive Web site that includes information about the company's history, its financial statements, and digitized depictions of the company's main products. You have been talking with your manager, Chief Information Officer Tom Buckles, for several months about adding electronic commerce features to the Web site that will allow your smaller customers to order directly from Grover instead of through their local distributors. Tom finally created a capital budget proposal for the Web site expansion and submitted it to Grover's board of directors. The board always calculates and evaluates a capital project's return on investment before approving it. Tom came back from the board meeting looking unhappy. The board told him that the project did not provide a high enough financial return to approve it. However, the board realized that electronic commerce initiatives could be important to Grover's future strategic position in the business; thus it would be willing to consider nonmonetary factors as a basis for approving the project. Tom would like to take the project back to the board next month, but he does not have a good sense of what nonmonetary factors might persuade the board to approve the project. He wants you to write a memo that outlines some of those factors and explains why they are important to Grover's future strategic position. In addition to considering the discussion in this chapter, you may want to use the Online Companion and draw on resources at **Business Week's e.biz**, CIO's **E-Business Research Center**, Internet.com's **Electronic Commerce Guide**, or ZDNet's **eBusiness Update** as you prepare your memo.

2. You are working for International Delicacies, which has become successful selling unusual food and other gift items through its mail-order catalog. Your manager, Jagdish Singh, is interested in exploring electronic commerce options for the company. He wants you to be a member of the internal team for the project and has asked you to identify some potential ISPs, ASPs, or CSPs to which the company could outsource the Web hosting portion of the project. Since the project is at a very early stage, the team has not specified the exact nature of the Web hosting function; however, the team would like to get a rough idea of what the setup costs and monthly charges might be. Use **The List**, **HostCompare.com**, or another directory of service providers to find three providers in your area. Follow the directory's links to your chosen service providers to learn more about them. Write a memo to Jagdish that summarizes the services offered by each Web hosting service provider and that

505

compares the costs charged by each for those services. Based on your research, evaluate each service provider and recommend whether the team should further consider each service provider as a potential outsourcing partner by the team.

3. As manager of networks and computing operations for Fashion Land, a retailer of women's clothing and accessories, you have seen the business grow from seven stores in Kansas City to over 100 stores located throughout the Midwest. Fashion Land's marketing research team has found that many members of its target customer group—females between the ages of 15 and 35—are becoming regular users of the Web. The researchers have found that these customers would not want to buy major clothing items on the Web; however,

they would like to buy accessories. Alone, or in a team assigned by your instructor, do the following:

- Outline a business strategy for Fashion Land's electronic commerce initiative. The outline should include a list of specific objectives and the costs and benefits of accomplishing each objective. The outline should also include recommendations regarding what to outsource, what Web hosting services will be needed, and what staff should be hired.

- Prepare a memo that lists and briefly describes the major hardware, software, security, payment processing, advertising, international, legal, and ethics issues that might arise in the development of this electronic commerce site.

For Further Study and Research

Abdel-Hamid, T. and S. Madnick. 1991. *Software Project Dynamics: An Integrated Approach.* Englewood Cliffs, NJ: Prentice Hall.

Abdel-Hamid, T., K. Sengopta, and C. Sweet. 1999. "The Impact of Goals on Software Project Management: An Experimental Investigation," *MIS Quarterly,* 23(4), December, 531–555.

Barker, R. 2000. "When to Outsource," *Business 2.0,* March 1. Available online at: (http://www.business2.com/content/magazine/indepth/2000/03/01/11125).

Barsh, J., E. Kramer, D. Maue, and N. Zuckerman. 2001. "Magazine's Home Companion," *The McKinsey Quarterly,* 2, June, 83–91.

Berinato, S. 2001. "Do the Math," *CIO,* 15(1), October 1, 53–60.

Berry, J. 2001. "Sometimes It's OK to Skip ROI Model," *InternetWeek,* October 22, 41.

Boisot, M. 1999. *Knowledge Assets: Securing Competitive Advantage in the Information Economy.* New York: Oxford University Press.

Borck, J. 2001. "A Balancing Act to ROI," *InfoWorld,* 23(30), July 23, 54.

Brache, A. and J. Webb. 2000. "The Eight Deadly Assumptions of E-Business," *Journal of Business Strategy,* 21(3), May–June, 13–17.

Brooks, F. 1995. *The Mythical Man-Month: Essays on Software Engineering, Anniversary Edition.* Reading, MA: Addison-Wesley.

Copeland, R. 2001. "ROI: The IT Department's Moving Target," *Information Week,* August 6, 45–47.

Edvinsson, L. and M. Malone. 1997. *Intellectual Capital: Realising Your Company's True Value by Finding its Hidden Brainpower.* New York: HarperCollins.

Eng, S. 2000. "Hatching Schemes," *The Industry Standard,* November 27, 174–175.

Fogarty, K. 2000. "Muddling Through to a Disaster," *Computerworld,* 34(16), April 17, 42.

Fryer, B. 2000. "Consulting's Next Big Thing," *Computerworld,* 34(1), January 3, 94–95.

Gido, J. and J. Clements. 1999. *Successful Project Management.* Cincinnati, OH: South-Western College Publishing.

Glass, R. 1997. *Software Runaways: Lessons Learned from Massive Software Project Failures*. Upper Saddle River, NJ: PTR Prentice Hall.

Goldratt, E. 1997. *Critical Chain*. Great Barrington, MA: North River Press.

Grackin, A. 2000. "Internet-Driven Business Models," *Manufacturing Systems*, 18(6), June, 56.

Griffin, J. 2000. "Rethinking Internet Valuation," *Business 2.0*, June 13, 239.

Hellweg, E. and S. Donahue. 2000. "The Smart Way to Start an Internet Company," *Business 2.0*, March 1, 64–66.

Heun, C. 2000. "No Web Bargains for Kmart," *InformationWeek*, August 21, 18.

Heun, C. 2001. "World Wrestling Federation Turns to Escalate to Wrestle with Web Sales," *InformationWeek*, November 19, 69.

Hudgins-Bonafield, C. 1998. "The Electronic Crane: E-Commerce Infrastructure Builds Upward," *Network Computing*, 9(23), December 15, 44–48.

Kalakota, R. and M. Robinson. 1999. *E-Business: Roadmap for Success*. Reading, MA: Addison-Wesley.

Kambil, A., E. Eselius, and K. Monteiro. 2000. "Fast Venturing: The Quick Way to Start a Web Business," *Sloan Management Review*, 41(4), Summer, 55–67.

Kara, D. 1999. "Sourcing Solutions for Wired World Emerging," *Software Magazine*, 19(1), June, 60–71.

Karpinski, R. 1999. "Cost Conscious Configurator," *InternetWeek*, April 12, 9.

Karpinski, R. 2001. "Vanished into Thin Air," *InternetWeek.com*, June 13. Available online at: (http://www.internetweek.com/transtoday01/ttoday061301.htm).

Karpinski, R. 2001. "Vanishing Vendors Are Common Concern," *InternetWeek*, June 25, 15.

Keen, P. 2000. "Six Months—or Else," *Computerworld*, 34(15), April 10, 48.

Keen, P. 2000. "Back to Process," *Computerworld*, 34(19), May 8, 50.

Keil, M., P. Cule, K. Lyytinen, and R. Schmidt. 1998. "A Framework for Identifying Software Project Risks," *Communications of the ACM*, 41(11), November, 76–83.

Keil, M. and D. Robey, 1999. "Turning Around Troubled Software Projects: An Exploratory Study of the De-Escalation of Commitment to Failing Courses of Action," *Journal of Management Information Systems*, 15(4), 63–87.

Kerzner, H. 2000. *Advanced Project Management: Best Practices*. New York: John Wiley & Sons.

Kling, A. 2000. "Penny Wise, Project Foolish," *PlanetIT*, August 18. Available online at: (http://www.PlanetIT.com/docs/PIT20000818S0013).

Lewis, D. 2000. "Some Retailers De-Emphasize Web Payback," *InternetWeek*, October 23, 94.

Lewis, D. 2001. "Servers As Utility? One User Believes," *InternetWeek*, August 20, 1–2.

Maitra, A. 1999. *Project Management for Internet Businesses*. New York: John Wiley & Sons.

McConnell, S. 1996. *Rapid Development: Taming Wild Software Schedules*. Redmond, WA: Microsoft Press.

Melymuka, K. 2000. "Born to Lead Projects," *Computerworld*, 34(13), March 27, 62–63.

Mollison, C. 2002. "To Outsource or Not to Outsource: That Is the Question," *Internet World*, January 1, 23–42.

Netto, M. 2001. "When Does an ASP Pay Off?" *Computerworld*, 35(34), August 20, 48.

Neuwirth, R. 1998. "Race into Cyberspace Gushes $80M Red Ink," *Editor & Publisher*, 131(51), December 19, 12–13.

Radcliff, D. 2000. "The Web's Master Builders," *Computerworld*, 34(11), March 13, 81.

Ramsey, C. 2000. "Managing Web Sites as Dynamic Business Applications," *Intranet Design Magazine*, June. Available online at: (http://idm.internet.com/articles/200006/wm_index.html).

Randall, L. 1999. "Average E-Commerce Web Site Costs US $1 Million," *Computing Canada*, 25(24), June 18, 11.

Rogers, A. 1999. "Up-Front Web Costs Are Half the Story," *Computer Reseller News*, June 7, 3.

Ruud, M. and J. Deutz. 1999. "Moving your Company Online," *Management Accounting*, 80(8), February, 28–32.

Schactman, N. 1999. "E-Business Demands A New Outlook on ROI," *Information Week*, October 18, 154–156.

507

Schultz, K. 1998. "Winners Craft Creative Solutions for Business," *InfoWorld*, 20(39), September 28, 73–74.

Schwalbe, K. 2000. *Information Technology Project Management*. Cambridge, MA: Course Technology.

Siebel, T. and P. House. 1999. *Cyber Rules: Strategies for Excelling at E-Business*. New York: Currency-Doubleday.

Sisco, M. 2001. "Improve Your IT Department's Performance by Calculating Key Equations," *TechRepublic*, December 6. Available online at: (http://www.techrepublic.com/article.jhtml?id=r00620011206MS01.htm).

Stavropoulos, A. and S. Jurvetson. 2000. "Does Your Idea Make Sense? Seven Questions to Ask Yourself Before Chasing Your Dot-Com Dreams," *Business 2.0*, March 1, 37–39.

Stewart, T. 1997. *Intellectual Capital: The New Wealth of Nations*. New York: Doubleday Dell.

Stewart, T. 1999. "Larry Bossidy's New Role Model: Michael Dell," *Fortune*, 139(7), April 12, 166–167.

Stoiber, J. 1999. "Maximizing IT Investments," *CIO Enterprise Magazine*, July 15. Available online at (http://www.cio.com/archive/enterprise/071599_checks.html).

Tattum, L. 2000. "Chemical Industry E-Strategies and Implementations," *Chemical Week*, July 26, S11–S15.

Violino, B. 2000. "Payback Time for E-Business—Net Projects No Longer Too 'Strategic' for ROI," *InternetWeek*, May 1, 1.

The Wall Street Journal. 1999. "Spending Campaign Is Set for Newspaper's Web Site," June 22, B16.

Walsh, B. 1998. "Building a Business Plan for an E-Commerce Project," *Network Computing*, 9(17), September 15, 69–73.

Wexler, J. 2000. "Lands of Opportunity," *Computerworld*, 34(26), June 26, 72–73.

Wilder, C. 1999. "ROI: E-Business Strategic Investment," *InformationWeek*, May 24, 48–56.

Willard, C. 2000. "Taking Techies to their Limits," *Computerworld*, 34(26), June 26, 74–75.

Wysocki, B. 2000. "U.S. Incubators Help Japan Hatch Ideas," *The Wall Street Journal*, June12, A1.

Yourdon, E. 2000. "Success in E-Projects," *Computerworld*, 34(34), August 21, 36.

Yourdon, E. and P. Becker. 1997. *Death March: The Complete Software Developer's Guide to Surviving "Mission Impossible" Projects*. Upper Saddle River, NJ: Prentice Hall.

Chapter 13

APPENDIX: A COMPARISON OF HTML AND XML FEATURES

MARKUP LANGUAGES ON THE WEB

In Chapter 2, you learned about two markup languages—HTML and XML—that are in common use on the Web today. In this Appendix you will learn more about these languages and when to use them. This Appendix does not provide a detailed description of either language and it is not a tutorial in markup languages. The Online Companion includes links to tutorials and other resources that you can use if you want to learn more about HTML and XML.

Tim Berners-Lee developed HTML, a simplified subset of SGML, in the early 1990s. Publishers had been using SGML for many years to mark up text so that it could be easily reformatted for printing in various types of documents. Although HTML is easy to learn and provides a useful tool for creating Web pages, it does not provide an easy way to express the semantics, or meaning, of information included on a Web page.

XML was developed to overcome this limitation of HTML. XML is also derived from SGML and therefore resembles HTML somewhat. For example, both HTML and XML use angle brackets to enclose the text markup tags that are the main components of both languages. Despite the similarities between the two languages, XML is quite different from HTML in both its purpose and its use.

HTML

HTML is a markup language that was designed to make the presentation of text on Web pages easy to accomplish. Another goal of HTML was to create a set of standards for text presentation so that Web browsers could render text files from any Web server as a readable Web page, no matter what type of Web server software the server was running.

Because HTML is designed to create and enforce standards, the original intent of Berners-Lee was that the tags would have specific meanings on which the Web community would agree. In general, that has occurred. The World Wide Web Consortium (W3C) reviews proposed changes to existing tags and specifications for new tags, and periodically publishes new versions of the HTML tag definitions on its Web site. Since the review process takes considerable time, however, the companies that create Web browser software have not always been willing to wait for approval of proposed tags.

At various times during the history of HTML, both Microsoft and Netscape (now a part of AOL Time Warner) enabled their Web browsers to use new tags before those tags were approved by the W3C. In some cases, these tags were enabled in one browser and not the other. In other cases, the tags used were never approved by the W3C, or were approved in a different form than had been implemented in the Web browser software. Web page designers who wanted to use the latest available tags were often frustrated by this state of affairs. Many of these Web designers had to create separate sets of Web pages for the different types of browsers, which was inefficient and expensive. Most of these tag difference issues were resolved when the W3C issued the specification for HTML version 4.0 in 1997, although enough of them remain to cause occasional problems for Web designers.

The purpose of most HTML tags is to format text and control its appearance on the Web page that appears in the browser window. Figure A-1 shows some sample text marked up with HTML tags, and Figure A-2 shows this text as it appears in a Web browser. The tags in these two figures are among the most common HTML tags in use today on the Web.

```
<html>

  <head>

    <title>HTML Tag Examples</title>

  </head>

  <body>

    <h1>This text is set in Heading One tags</h1>
    <h2>This text is set in Heading Two tags</h2>
    <h3>This text is set in Heading Three tags</h3>


    <p>
    This text is set within Paragraph tags. It will appear
    as one paragraph; the text will wrap at the end of each
    line that is rendered in the web browser no matter where
    the typed text ends. The text inside Paragraph tags is
    rendered without regard to extra spaces typed in the text,
    such as these:          Character formatting can also
    be applied within Paragraph tags. For example, <b>this
    text is set in bold</b>, <font face="Arial">this text is
    set in Arial</font>, and <i>this text is set in italics</i>.
    </p>

    <pre>
The Preformatted tag instructs the web browser to render the text
    exactly    the    way    it    is    typed,
as in this example.
    </pre>

    <p>
    HTML includes tags that instruct the web browser to place
    text in bulleted or numbered lists:
    </p>

    <ul>
      <li>Bulleted list item one</li>
      <li>Bulleted list item two</li>
      <li>Bulleted list item three</li>
    </ul>

    <ol>
      <li>Numbered list item one</li>
      <li>Numbered list item two</li>
      <li>Numbered list item three</li>
    </ol>

    <p>
    HTML also includes tags for left, center, and right justification:
    </p>

    <div align="left">Text aligned left</div>
    <div align="center">Text aligned center</div>
    <div align="right">Text aligned right</div>

    <p>
    The most important tag in HTML is the Anchor Hypertext
    Reference tag, which is the tag that provides a link to
    another web page (or another location in the same web page).
    For example, the underlined text
    <a href="http://www.course.com/">Course Technology</a>
    is a hyperlink to the web page of the publisher of this book.
    </p>

  </body>

</html>
```

Figure A-1   *Text marked up with HTML tags*

This text is set in Heading One tags

This text is set in Heading Two tags

This text is set in Heading Three tags

This text is set within Paragraph tags. It will appear as one paragraph; the text will wrap at the end of each line that is rendered in the Web browser no matter where the typed text ends. The text inside Paragraph tags is rendered without regard to extra spaces typed in the text, such as these: Character formatting can also be applied within Paragraph tags. For example, **this text is set in bold**, this text is set in Arial, and *this text is set in italics.*

```
The Preformatted tag instructs the Web browser to render the text
      exactly    the    way    it    is    typed,
as in this example.
```

HTML includes tags that instruct the Web browser to place text in bulleted or numbered lists:

- Bulleted list item one
- Bulleted list item two
- Bulleted list item three

1. Numbered list item one
2. Numbered list item two
3. Numbered list item three

HTML also includes tags for left, center, and right justification:

Text aligned left

Text aligned center

Text aligned right

The most important tag in HTML is the Anchor Hypertext Reference tag, which is the tag that provides a link to another Web page (or another location in the same Web page). For example, the underlined text Course Technology is a hyperlink to the Web page of the publisher of this book.

Figure A-2 *Text marked up with HTML tags as it appears in a Web browser*

Other frequently used HTML tags (not shown in the figures) let Web designers include graphics on Web pages and format text in the form of tables. The text and HTML tags that form a Web page can be viewed when the page is open in a Web browser by using the menu commands View, Source (in Internet Explorer) or View, Page Source (in Netscape Navigator).

HTML was designed to help people display information on the Web, and it is well-suited to that task. You can learn more about the latest HTML specification by following the Online Companion link to the **W3C HTML Pages**. As the Web grew, HTML continued to provide a useful tool for Web designers who wanted to create attractive layouts of text and graphics on their pages. However, as companies began to conduct electronic commerce on the Web, the need to present large amounts of data on Web pages also became important. Companies created Web sites that contained lists of inventory items, sales invoices, purchase orders, and other business data. The need to keep these lists updated was also important and posed a new challenge for many Web designers.

The tool that had helped these Web designers create useful Web pages—HTML—was not such a good tool for presenting or maintaining information lists. Consider the simple example of a Web page that includes a list of countries and some basic information about each country. A Web designer might decide to use HTML tags to show each information item the same way for each country. Each information item would use a different tag. Assume that the Web designer in this case decided to use the HTML Heading tags to present the data. Figure A-3 shows the data and the HTML Heading tags for four countries (this is only an example; the actual list would include more than 150 countries). The first item in the list provides the definitions for each tag. Figure A-4 shows this HTML document as it appears in a Web browser.

```html
<html>
    <head>
      <title>Countries</title>
    </head>
    <body>
      <h1>Countries</h1>

      <h2>CountryName</h2>
      <h3>CapitalCity</h3>
      <h4>AreaInSquareKilometers</h4>
      <h5>OfficialLanguage</h5>
      <h6>VotingAge</h6>

      <h2>Argentina</h2>
      <h3>Buenos Aires</h3>
      <h4>2,766,890</h4>
      <h5>Spanish</h5>
      <h6>18</h6>

      <h2>Austria</h2>
      <h3>Vienna</h3>
      <h4>83,858</h4>
      <h5>German</h5>
      <h6>19</h6>

      <h2>Barbados</h2>
      <h3>Bridgetown</h3>
      <h4>430</h4>
      <h5>English</h5>
      <h6>18</h6>

      <h2>Belarus</h2>
      <h3>Minsk</h3>
      <h4>207,600</h4>
      <h5>Byelorussian</h5>
      <h6>18</h6>
    </body>
</html>
```

Figure A-3   *Country list data with HTML tags*

Countries

CountryName

CapitalCity

AreaInSquareKilometers

OfficialLanguage

VotingAge

Argentina

Buenos Aires

2,766,890

Spanish

18

Austria

Vienna

83,858

German

19

Barbados

Bridgetown

430

English

18

Belarus

Minsk

207,600

Byelorussian

18

Figure A-4 *Country list data as it appears in a Web browser*

These figures reveal some of the shortcomings of using HTML to present a list of items when the meaning of each item in the list is important. The Web designer in this case used HTML Heading tags. HTML only has six levels of heading tags; thus, if the individual items had any more information elements than shown in this example, this approach would not work. The Web designer could have used various combinations of text attributes such as size, font, color, bold, or italics to distinguish among items, but none of these tags would convey the meaning of the individual data elements. The only information about the meaning of each country's listing appears in the first list item, which includes the definitions for each element. In the mid-1990s, Web professionals began to consider XML as a list formatting alternative to HTML that would communicate the meaning of data more effectively.

XML

As you learned in Chapter 2, XML was introduced in 1996 and released as a version recommendation by the W3C in 1998. XML differs from HTML in two important respects. First, XML is not a markup language with defined tags. It is a framework within which individuals, companies, and other organizations can create their own sets of tags. Second, XML tags do not specify how text will appear on a Web page; instead, the tags convey the meaning (the semantics) of the information included within them.

To understand this distinction between appearance and semantics, consider the list of countries example from the previous section. In XML, tags that define the meaning of a fact can be created for each fact. Figure A-5 shows the country data marked up with XML tags. XML files are not typically displayed directly in a Web browser; they are usually translated using another file that contains formatting instructions or are read by a program. Formatting instructions are often written in the extensible stylesheet language (XSL) and the programs that read or transform XML files are usually written in Java.

Some browsers, such as Internet Explorer, can render XML files directly without additional instructions. Figure A-6 shows the Country List XML file displayed in Internet Explorer.

```
declaration ─── <?xml version="1.0"?>

              ┌─ <CountriesList>
root          │
element ──────┘  <Country Name = "Argentina">
                  <CapitalCity>Buenos Aires</CapitalCity>
                  <AreaInSquareKilometers>2,766,890</AreaInSquareKilometers>
                  <OfficialLanguage>Spanish</OfficialLanguage>
                  <VotingAge>18</VotingAge>
                 </Country>

                 <Country Name = "Austria">
                  <CapitalCity>Vienna</CapitalCity>
                  <AreaInSquareKilometers>83,858</AreaInSquareKilometers>
                  <OfficialLanguage>German</OfficialLanguage>
                  <VotingAge>19</VotingAge>
                 </Country>

                 <Country Name = "Barbados">
                  <CapitalCity>Bridgetown</CapitalCity>
                  <AreaInSquareKilometers>430</AreaInSquareKilometers>
                  <OfficialLanguage>English</OfficialLanguage>
                  <VotingAge>18</VotingAge>
                 </Country>

                 <Country Name = "Belarus">
                  <CapitalCity>Minsk</CapitalCity>
                  <AreaInSquareKilometers>207,600</AreaInSquareKilometers>
                  <OfficialLanguage>Byelorussian</OfficialLanguage>
                  <VotingAge>18</VotingAge>
                 </Country>

                </CountriesList>
```

Figure A-5 *Country list data with XML tags*

```
    <?xml version="1.0" ?>
  - <CountriesList>
    - <Country Name="Argentina">
        <CapitalCity>Buenos Aires</CapitalCity>
        <AreaInSquareKilometers>2,766,890</AreaInSquareKilometers>
        <OfficialLanguage>Spanish</OfficialLanguage>
        <VotingAge>18</VotingAge>
      </Country>
    - <Country Name="Austria">
        <CapitalCity>Vienna</CapitalCity>
        <AreaInSquareKilometers>83,858</AreaInSquareKilometers>
        <OfficialLanguage>German</OfficialLanguage>
        <VotingAge>19</VotingAge>
      </Country>
    - <Country Name="Barbados">
        <CapitalCity>Bridgetown</CapitalCity>
        <AreaInSquareKilometers>430</AreaInSquareKilometers>
        <OfficialLanguage>English</OfficialLanguage>
        <VotingAge>18</VotingAge>
      </Country>
    - <Country Name="Belarus">
        <CapitalCity>Minsk</CapitalCity>
        <AreaInSquareKilometers>207,600</AreaInSquareKilometers>
        <OfficialLanguage>Byelorussian</OfficialLanguage>
        <VotingAge>18</VotingAge>
      </Country>
    </CountriesList>
```

Figure A-6 *Country list XML file displayed in Internet Explorer*

The first line in the file is the declaration, which indicates that the file uses version 1.0 of XML. XML markup tags are similar to SGML markup tags; thus, the declaration can help avoid confusion in organizations that use both. The second line and the last line are the root element tags. The root element of an XML file contains all of the other elements in that file and is usually assigned a name that describes the purpose or meaning of the file.

The other elements are called child elements; for example, Country is a child element of CountriesList. Each of the other attributes is, in turn, a child element of the Country element. The names of these child elements were created for use in this file. If programmers in another organization were to create a file with country information, they would likely use different names for these elements, which would make it difficult for the two organizations to share information. Thus, the greatest strength of XML—that it allows users to define their own tags—is also its greatest weakness.

To overcome that weakness, many companies have agreed to follow common standards for XML tags. These standards, in the form of data type definitions (DTDs) or XML schemas, are available for a number of industries, including the **ebXML** initiative for electronic commerce standards, the **eXtensible Business Reporting Language (XBRL)** for accounting and financial information standards, **LegalXML** for information in the legal profession, and **MathML** for mathematical and scientific information.

24/7 operation The operation of a site or service 24 hours a day, seven days a week.

A

Acceptance An expression of willingness to take an offer, including all of its stated terms.

Access control list (ACL) A list of resources and the usernames of people who are permitted access to those resources within a computer system.

Account aggregation A feature offered by some online banks that allows a customer to obtain bank, investment, loan, and other financial account information from multiple Web sites and display it all in one location at the bank's Web site.

Account manager A person who keeps track of multiple Web sites in use by a project or keeps track of the projects that will combine to create a larger Web site.

Accredited Standards Committee X12 (ASC X12) A committee that develops and maintains uniform EDI standards in the United States.

Acquiring bank Synonymous with merchant bank, which is a bank that does business with merchants who want to accept credit cards.

Active ad A Web ad that generates graphical activity that appears to float over the Web page itself instead of opening in a separate window.

Active content Programs that are embedded transparently in Web pages that cause an action to occur.

Active Server Pages (ASP) Applications that generate dynamic content within Web pages using either Jscript code or Visual Basic.

Active wiretapping An integrity threat that exists when an unauthorized party can alter a message stream of information.

ActiveX An object, or control, that contains programs and properties that are put in Web pages to perform particular tasks.

Activity A task performed by a worker in the course of doing his or her job.

Ad-blocking software A program that prevents banner ads and pop-up ads from loading.

Ad view The loading of a Web page that contains an ad into a Web browser.

Addressable media Advertising efforts sent to a known addressee; these include direct mail, telephone calls, and e-mail.

Advertising-subscription revenue model A business model in which subscribers pay a fee and accept some level of advertising.

Affiliate marketing An advertising technique in which one Web site (called an "affiliate") includes descriptions, reviews, ratings, or other information about products that are sold on another Web site. The affiliate site includes links to the selling site, which pays the affiliate site a commission on sales made to visitors who arrived from a link on the affiliate site.

Affiliate program broker A company that serves as a clearinghouse or marketplace for sites that run affiliate marketing programs and sites that want to become affiliates.

American National Standards Institute (ANSI) The coordinating body for electrical, mechanical, and other technical standards in the United States.

Anchor tag The HTML tag used to specify hyperlinks.

Animated GIF Moving graphic images that are often used in Web ads to attract users' attention.

Anonymous electronic cash Electronic cash that cannot be traced back to the person who spent it.

Anonymous FTP A protocol that allows users to access limited parts of a remote computer using FTP without having an account on the remote computer.

Applet A program that executes within another program; it cannot execute directly on a computer.

Application Program Interface (API) A set of routines, protocols, and tools for building software applications.

Application server A middle-tier software and hardware combination that lies between the Internet and a corporate backend server.

Application service provider (ASP) A Web-based site that provides management of applications such as

spreadsheets, human resources management, or e-mail to companies for a fee.

Application services Internet-provided services to users, such as Web page delivery, network management tools, remote login, file transmission, electronic mail, and directory services.

Applications specialist The member of an electronic commerce team who is responsible for maintenance of software that performs a specific function, such as catalog, payment processing, accounting, human resources, and logistics software.

Archiving Saving a log on a storage device.

Ascending-price auction A type of auction in which bidders publicly announce their successively higher bids until no higher bid is forthcoming, also called an English auction.

ASCII text file A file that contains only characters available through the keyboard but no formatting.

Asymmetric connection An Internet connection that provides different bandwidths for each direction.

Asymmetric Digital Subscriber Line (ADSL) Internet connections using the DSL protocol with bandwidths from 16 to 640 Kbps upstream and 1.5 to 9 Mbps downstream.

Asymmetric encryption Synonymous with public-key encryption, which is the encoding of messages using two mathematically related but distinct numeric keys.

Asynchronous transfer mode (ATM) Internet connections with bandwidths of up to 622 Gbps.

Attachment A data file (document, spreadsheet, or other) that is appended to an e-mail message.

Auctioneer The person who manages an auction.

Authority to bind The power of an individual to commit his or her company to a contract.

Automated Clearing House (ACH) One of several systems set up by banks or government agencies, such as the U.S. Federal Reserve Board, that process high volumes of low dollar amount electronic fund transfers.

B

Backbone The main network of connections that carry most of the traffic on the Internet.

Backbone router Computers that handle packet traffic along the Internet's main connecting points and that can each handle more than 50 million packets per second.

Backdoor An electronic hole in electronic commerce software left open by accident or intentionally.

Bandwidth The amount of data that can be transmitted in a fixed amount of time. Also, the number of simultaneous site visitors that a Web site can accommodate without degrading service.

Banner ad A small rectangular object on a Web page that displays a stationary or moving graphic and includes a hyperlink to the advertiser's Web site.

Banner advertising network An organization that acts as a broker between advertisers and Web sites that carry ads.

Banner exchange network An organization that arranges for Web sites to run reciprocal ads for each other.

Base 2 A number system in which each digit is either a 0 or a 1, corresponding to a condition of either "off" or "on." Also known as a binary system.

Behavioral segmentation The creation of a separate Web site experience for customers based on their behavior.

Benchmarking Testing that compares hardware and software performances.

Bid An offer of a certain price made on an item that is up for auction.

Bidder A potential buyer at an auction; one who places bids.

Bill presentment A Web site feature that allows customers to view and pay bills online.

Binary Synonymous with Base 2.

Binary data Types of data including files containing word-processed documents, worksheets, graphics, and other data.

Bluetooth A wireless transmission protocol that is used for short distances and lower bandwidth connections.

Brand Customers' perceptions of a product.

Brand leveraging A strategy in which a well-established Web site extends its dominant position to other products and services.

Broadband Connections that operate at speeds of greater than about 200 Kbps.

Browser Synonymous with Web browser, which is software that lets users read HTML documents and move from one HTML document to another using hyperlinks.

Buffer An area of a computer's memory that is set aside to hold data read from a file or database.

Bulk mail Synonymous with spam, which is electronic junk mail.

Business logic Rules of a particular business.

Business manager The member of an electronic commerce team who is responsible for implementing the elements of the business plan and reaching the objectives set by the internal team. The business manager should have experience in and knowledge of the business activity being implemented in the site.

Business process patent A patent that protects a specific set of procedures for conducting a particular business activity.

Business processes The activities in which businesses engage as they conduct commerce.

Business-to-business (B2B) Transactions conducted between businesses on the Web.

Business-to-consumer (B2C) Transactions conducted between shoppers and businesses on the Web.

Byte an 8-bit number (in most computer applications).

C

Call center A company that handles customer telephone calls and e-mails for other companies.

Cannibalization The loss of traditional sales of a product to its electronic counterpart.

Capital investment A major outlay of funds made by a company to purchase fixed assets such as property, a factory, or equipment.

Capital project Synonymous with capital investment.

Card not present A payment card transaction in which the seller does not see the card; for example, a telephone or Web transaction.

Cascading style sheets Technology that allows designers to apply many predefined page display styles to Web pages.

Catalog On electronic commerce sites, a listing of goods or services that may include photographs and descriptions, often stored in a database.

Catalog model A business model in which the seller establishes a brand image, then uses the strength of that image to sell through printed catalogs mailed to prospective buyers. Buyers place orders by mail or by calling the seller's toll-free telephone number.

Cause marketing An affiliate marketing program that benefits a charitable organization.

Centralized architecture A server structure that uses few very large and fast computers.

Certification authority (CA) An agency that issues digital certificates to organizations or individuals.

Channel conflict Sales activities on a company's Web site that interfere with its existing sales outlets.

Charge card A payment card with no preset spending limit. The entire amount charged to the card must be paid in full each month.

Chargeback The process in which a merchant bank retrieves the money it placed in a merchant account as a result of a cardholder successfully contesting a charge.

Cipher text Text that comprises a seemingly random assemblage of bits. Cipher text is what messages become after they are encrypted.

Circuit A specific route between source and destination along which data travels.

Circuit switching A way of connecting computers or other devices that uses a centrally controlled single connection. In this method, which is used by telephone companies to provide voice telephone service, the connection is made, data is transferred, and the connection is terminated.

Clear text Text that has not been encrypted or encoded in any way.

Click Synonymous with click-through.

Clickstream Data about the pages viewed by Web site visitors, including the time viewed and the sequence of viewing.

Click-through The loading of an advertiser's Web page that results from a visitor clicking a banner advertisement on another Web page.

Client A networked PC or workstation on which users run applications.

Client-server model A network in which each computer on the network is either a client or a server.

Client-side electronic wallet An electronic wallet that stores a consumer's information on the consumer's own computer.

Client-side scripting The embedding in a Web page of a small program that executes on the client computer when it loads the Web page.

Clipping service A company or service that collects and groups copies of articles and summaries of news coverage about particular companies.

Closed loop system A payment card arrangement involving a consumer, a merchant, and a payment card company (such as American Express or Discover) that processes transactions between the consumer and merchant without involving banks.

Closing tag The second half of a two-sided HTML tag; it is identified by a slash (/) that precedes the tag's name.

Collision The occurrence of two messages resulting in the same hash value; the probability of this happening is extremely small.

Co-located hosting Self-hosting wherein the server is owned by the online store but is located at the Web host's site, and the Web host provides maintenance based on the level of service the online business requires.

Co-location (collocation, colocation) An Internet service arrangement in which the service provider rents a physical space and an Internet connection to the client to install its own server hardware.

Commerce A negotiated exchange of valuable objects or services between two or more parties. Commerce includes all activities that each party undertakes to complete the transaction.

Commerce service provider (CSP) A Web host service that also provides commerce hosting services on its computer.

Commodity A product or service that has become so standardized and well-known that buyers cannot detect a difference in the offerings of various sellers, and they decide to buy based on price.

Common gateway interface (CGI) A protocol that allows Web servers to interact dynamically with other software packages to create custom Web pages.

Common law The part of English and U.S. law that is established by previous court decisions and historical practices.

Communication modes Ways of identifying and reaching customers.

Company A business engaged in commerce; synonymous with firm.

Component-based application server A business logic approach that separates presentation logic from business logic.

Component outsourcing Synonymous with partial outsourcing, which is the outsourcing of the design, development, implementation, or operation of specific portions of an electronic commerce system.

Computer forensics The field responsible for the collection, preservation, and analysis of computer-related evidence.

Computer forensics expert An individual hired to access client computers to locate information that can be used in legal proceedings.

Computer security The protection of computer resources from various types of threats.

Computer virus (virus) Synonymous with virus, which is software that attaches itself to another program and can cause damage when the host program is activated.

Configuration table Information about connections that lead to particular groups of routers, specifications on which connections to use first, and rules for handling instances of heavy packet traffic and network congestion.

Consideration The bargained-for exchange of something valuable, such as money, property, or future services.

Constructive notice The idea that citizens should know that when they leave one area and enter another, they become subject to the laws of the new area.

Consumer-to-business Electronic commerce that can occur in general consumer auctions when the seller is a consumer and the bidders are businesses.

Consumer-to-consumer (C2C) A category of electronic commerce that includes individuals who buy and sell items among themselves.

Content creator A person who writes original content for a Web site.

Content editor A person who purchases and adapts existing material for use on a Web site.

Content manager Synonymous with content editor.

Contract An agreement between two or more legal entities that provides for an exchange of value between or among them.

Conversion rate Used in advertising to calculate the percentage of recipients that respond to an ad or promotion.

Cookie Bits of information about Web site visitors created by Web sites and stored on client computers.

Cookie blocker A third-party program that selectively prevents cookie storage.

Copy control An electronic mechanism for providing a fixed upper limit to the number of copies that one can make of a digital work.

Copyright A legal protection of intellectual property.

Cost per thousand (CPM) An advertising pricing metric that equals the dollar amount paid to reach 1000 people in an estimated audience.

Countermeasure A physical or logical procedure that recognizes, reduces, or eliminates a threat.

Cracker A technologically skilled person who uses those skills to obtain unauthorized entry into computers or network systems, usually with the intent of stealing information or damaging the information, the system's software, or the system's hardware.

Crawler Synonymous with spider, which is the first part of a search engine.

Credit card A card with a preset spending limit based on the card holder's credit limit. A minimum monthly payment must be made against the balance on the card, and interest is charged on the unpaid balance.

Cryptography The science that studies encryption, which is the hiding of messages so that only the sender and receiver can read them.

Culture The combination of language and customs that are unique to a particular population.

Customer-centric The Web site development approach of putting the customer at the center of all site designs.

Customer life cycle The five stages of customer loyalty.

Customer portal A corporate Web site designed to meet the needs of customers by offering additional services such as private stores, part number cross-referencing, product use guidelines, and safety information.

Customer relationship management Synonymous with technology-enabled relationship management, which is the obtaining and use of detailed customer information.

Customer service The persons within an electronic commerce team who are responsible for managing customer relationships in the electronic commerce operation.

Customer value The cost that a customer pays for a product, minus the benefits the customer gains from the product.

Cyber vandalism The electronic defacing of an existing Web site page.

Cybersquatting The practice of registering a domain name that is the trademark of another person or company with the hope that the trademark owner will pay huge amounts of money for the domain rights.

Cycling Replacing the oldest log with the newest log.

D

Data Encryption Standard (DES) An encryption standard adopted by the U.S. government for encrypting sensitive information.

Data mining Looking for hidden patterns in data.

Database The storage element of a search engine.

Database administration The function within an electronic commerce team that is responsible for defining the data elements in the database design and the operation of the database management software.

Database manager Software that stores information in a highly structured way.

Data-grade line The quality of telephone wiring in most urban and suburban areas; made more carefully of higher grade copper than voice-grade lines.

Dead link A Web link that when clicked displays an error message instead of a Web page.

Debit card A payment card that removes the amount of the charge from the cardholder's bank account and transfers it to the seller's bank account.

Decentralized architecture A server structure that uses a large number of less-powerful computers and divides the workload among them.

Decrypted Information that has been decoded. The opposite of encrypted.

Decryption program A procedure to reverse the encryption process, resulting in the decoding of an encrypted message.

Dedicated hosting Web hosting wherein a firm has an individually dedicated server that is owned and administered by a service provider who dictates terms of usage.

Defamatory statement A statement that is false and injures the reputation of a person or company.

Delay threat The disruption of normal computer processing.

Demographic information Characteristics that marketers use to group visitors, including address, age, gender, income level, type of job held, hobbies, and religion.

Demographic segmentation The grouping of customers by characteristics such as age, gender, family size, income, education, religion, or ethnicity.

Denial of Service threat (DoS) Synonymous with necessity threat, which is the disruption of normal computer processing.

Denial threat The prohibition of normal computer processing.

Descending-price auction Synonymous with Dutch auction, which is an open auction in which bidding starts at a high price and drops until a bidder accepts the price.

Dictionary attack program A program that cycles through an electronic dictionary, trying every word in the book as a password.

Digital certificate An attachment to an e-mail message or data embedded in a Web page that verifies the identity of a sender or Web site.

Digital signature An encryption message digest.

Digital subscriber line A telephone-based Internet connection method that uses networking equipment for connectivity and is higher in quality than POTS.

Digital subscriber loop (DSL) Synonymous with digital subscriber line.

Digital watermark A digital code or stream embedded undetectably in a digital image or audio file.

Direct connection EDI The form of EDI in which EDI translator computers at each company are linked directly to each other through modems and dial-up telephone lines or leased lines.

Direct materials Materials that become part of the finished product in a manufacturing process.

Disintermediation The removal of an intermediary from a value chain.

Distributed architecture Synonymous with decentralized architecture.

Distributed database system A database that stores the same data in many different physical locations.

Distributed information system A large information system that stores the same data in many different physical locations.

Distribution Synonymous with place, which is the need to have products or services available in many different locations.

Doing business Synonymous with commerce, which is a negotiated exchange of valuable objects or services between two or more parties.

Domain name The address of a Web page; it can contain two or more word groups separated by periods. Components of domain names become more general from right to left.

525

Domain name ownership change
The changing of owner information maintained by a public domain registrar in the registrar's database to reflect the new owner's name and business address. This usually only happens when safeguards are not in place.

Dotted decimal The IP address notation in which addresses appear as a four separate numbers separated by periods.

Double auction A type of auction in which buyers and sellers each submit combined price-quantity bids to an auctioneer. The auctioneer matches the sellers' offers (starting with the lowest price, then going up) to the buyers' offers (starting with the highest price, then going down) until all of the quantities are sold.

Double spending The spending of the same unit of electronic cash twice by submitting the same electronic currency to two different vendors.

Download To receive a file from another computer.

Downstream The connection that occurs when information is sent to a user's computer from an ISP.

Downstream bandwidth (downlink bandwidth) The connection that occurs when information travels to your computer from your ISP.

Downstream strategies Strategies that improve the value that a business provides to its customers.

Due diligence Research procedures to check an entity's background.

Dynamic catalog An area of a Web site that stores information about products in a database.

Dynamic content Nonstatic information constructed in response to a Web client's request.

Dynamic page A Web page whose content is shaped by a program in response to a user request.

E

Early outsourcing The hiring of an external company to do initial electronic commerce site design and development. The external team then trains the original company's information systems professionals in the new technology, eventually handing over complete responsibility of the site to the internal team.

Eavesdropper A person or device who is able to listen in on and copy Internet transmissions.

E-business (electronic business) software Commerce software that provides tools for both business-to-consumer and business-to-business commerce, it is often designed to interface with existing back office systems.

EDI for Administration, Commerce, and Transport (EDIFACT) The 1987 publication that summarizes the United Nations' standard transaction sets for international EDI.

EDI-capable banks Banks that can exchange payment and remittance through value-added networks.

EDI-compatible Firms that can exchange data in specific standard electronic formats with other firms.

E-government The use of electronic commerce by governments and government agencies to perform business-like activities.

Electronic business (e-business) Another term for electronic commerce; sometimes used to mean business-to-business electronic commerce.

Electronic cash A form of electronic payment that is anonymous and can be spent only once.

Electronic commerce (e-commerce) Business activities conducted using electronic data transmission over the Internet and the World Wide Web.

Electronic customer relationship management (eCRM) Synonymous with technology-enabled relationship management, which is the practice of obtaining details about stakeholders and using the information to provide customized interactions.

Electronic data interchange (EDI) Exchange between businesses of computer-readable data in a standard format.

Electronic funds transfer (EFT) Electronic transfer of account exchange information over secure private communications networks.

Electronic mail (e-mail) Messages that are sent from one user to another (or multiple recipients) using particular mail programs and protocols.

Electronic wallet A software utility that holds electronic cash, credit card information, owner identification and address information, and provides this data automatically at electronic commerce sites.

Element The HTML components that specify how a document or a part of a document should be formatted and arranged onscreen.

EMV standard A standard for the handling of payment card transactions; agreed upon by Europay, MasterCard, and Visa.

Encapsulation The process that occurs when VPN software encrypts a packet content, then places the encrypted packets inside an IP wrapper in another packet.

Encryption The coding of information using a mathematical-based program and secret key; it makes a message illegible to casual observers or those without the decoding key.

Encryption algorithm The logic that implements an encryption program.

Encryption program A program that transforms clear text into cipher text.

English auction A type of auction in which bidders publicly announce their successively higher bids until no higher bid is forthcoming.

Enterprise resource planning (ERP) Business software that integrates all facets of a business, including planning, manufacturing, sales, and marketing.

Entity body An optional, though almost always present, part of a message from a client that contains the HTML page requested by the client and passes bulk information to the server.

E-procurement software Software that allows a company to manage its purchasing function through a Web interface.

Escrow service An independent third party who holds an auction buyer's payment until the buyer receives the purchased item and is satisfied that it is what the seller represented it to be.

Ethical hacker A computer specialist hired to probe computer systems for security weaknesses or to locate information that can be used in legal proceedings.

Extensible The property of XML that allows users to extend the language by creating their own tags.

Extensible Hypertext Markup Language (XHTML) A new markup language proposed by the WC3 that is a reformulation of HTML version 4.0 as an XML application.

527

Extensible Markup Language (XML) A language that describes the semantics of a page's contents and defines data records on a page.

Extensible system Any system that can easily be enhanced without voiding earlier work done on the system.

Extranet A network system that extends a company's intranet and allows it to connect with the networks of business partners or other designated associates.

E-zine An electronic magazine.

F

Fair use The approved limited use of copyright material when certain conditions are met.

Fast venturing The joining of an existing company that wants to launch an electronic commerce initiative with external equity partners and operational partners who provide the experience and skills needed to develop and scale up the project very rapidly.

Fee-for-service revenue model A business model in which payment is based on the value of the service provided.

Fee-for-transaction revenue model A business model in which businesses charge a fee for services based on the number or size of the transactions they process.

File Transfer Protocol (FTP) A protocol that enables users to transfer files over the Internet.

Financial VANS (FVANS) Value-added networks that are not banks but can translate financial transaction sets into ACH formats and transmit them to banks that are not EDI capable.

Finger An Internet utility program that runs on UNIX computers and allows a user to obtain limited information about other network users.

Firewall A computer that provides a defense between one network (inside the firewall) and another network (outside the firewall, such as the Internet) that could pose a threat to the inside network. All traffic to and from the network must pass through the firewall. Only authorized traffic, as defined by the local security policy, is allowed to pass through the firewall, which itself is immune to penetration.

Firm A business engaged in commerce.

First-price sealed-bid auction A type of auction in which bidders submit their bids independently and privately, with the highest bidder winning the auction.

Flat-rate access The monthly fee paid by a consumer or business for unlimited telephone line usage.

Float Money deposited in a customer's account that earns interest for the merchant.

Forum selection clause A statement within a contract that dictates that the contract will be enforced according to the laws of a particular state or other jurisdiction; signing a contract with a forum selection clause constitutes voluntary submission to the jurisdiction named in the forum selection clause.

Four Ps of marketing The essential issues of marketing: product, price, promotion, and place.

Frame relay A routing technology.

Full banner A Web banner ad that measures 468 × 60 pixels.

Full-privilege FTP A protocol that allows users to upload files to and download files from a remote computer using FTP.

G

Gateway computers Synonymous with routers, which are computers that determine the best way for data packets to travel.

Gateway server A firewall that filters traffic based on applications requested by clients on the trusted network.

Geographic segmentation The grouping of customers by location of home or workplace.

Graphical user interface (GUI) Computer program control functions that are displayed using pictures, icons, and other easy-to-use graphical elements.

Group purchasing site A type of auction Web site that negotiates with a seller to obtain lower prices on an item as individual buyers enter bids on that item.

H

Hacker A dedicated programmer who writes complex code that tests the limits of technology; usually meant as a compliment. Sometimes, however, used as a synonym for cracker.

Half banner A Web banner ad that measures 234 × 60 pixels.

Hash algorithm A security utility that mathematically combines every character in a message to create a fixed-length number (usually 128 bits in length) that is a condensation, or fingerprint, of the original message.

Hash coding The process used to calculate a number from a message.

Hash value The number that results when a message is hash coded.

Hexadecimal A number system that uses 16 digits.

Hierarchical business organizations Firms that include a number of levels with cumulative responsibility. These organizations are typically headed by a top-level president or officer. A number of vice presidents report to the president. A larger number of middle managers report to the vice presidents.

Hierarchical hyperlink structure A hyperlink structure in which the user starts from a home page and follows links to other pages in whatever order they wish.

High availability servers Spare server computers for handling high traffic volumes that occur periodically.

High reliability servers Servers that are available 24 hours a day, seven days a week.

High-speed DSL (HDSL) An Internet connection service that provides 768 Kbps of symmetric bandwidth.

Hops The number of hosts between two Internet-connected computers.

HTML extensions Developer-created Web page features that only work in certain browsers.

Hyperlink A type of tag that points to another location in the same or another HTML document. Also called a hypertext link.

Hypertext A system of navigating between HTML pages using links.

Hypertext elements HTML text elements that are related to each other within one document or among several documents.

Hypertext link (hyperlink) A pointer in an HTML document to another location within the same document or to a different HTML document.

Hypertext Markup Language (HTML)
The language of the Internet; it contains codes attached to text that describe text elements and their relation to one another.

Hypertext server Synonymous with Web server, which is a computer that is connected to the Internet and that stores files written in HTML that are publicly available through an Internet connection.

Hypertext Transfer Protocol (HTTP)
The Internet protocol responsible for transferring and displaying Web pages.

I

Implied contract The agreement between two parties that a contract exists, even if no contract has been written and signed.

Impression The loading of a banner ad on a Web page.

Incubator A company that offers start-up businesses a physical location with offices, accounting and legal assistance, computers, and Internet connections at a very low monthly cost in exchange for an ownership interest in the startup business.

Independent exchange A vertical portal that is not controlled by a company that was already an established buyer or seller in the industry.

Index A list containing every Web page found by a spider, crawler, or bot.

Indirect connection EDI The form of EDI in which each company transmits and receives EDI messages through a value-added network.

Indirect materials Materials and supplies that are used in or to support manufacturing activities but that do not become part of the manufactured product.

Industry Multiple firms selling similar products to similar customers.

Industry consortia-sponsored marketplace A marketplace formed by several large buyers in a particular industry.

Industry value chain The larger stream of activities in which a particular business unit's value chain is embedded.

Integrated Services Digital Network (ISDN) High-grade telephone service that uses the DSL protocol and offers bandwidths of up to 128 Kbps.

Integrity The category of computer security that addresses the validity of data; confirmation that data has not been modified.

Integrity violation A security violation that occurs whenever a message is altered while in transit between sender and receiver.

Intelligent agent (agent) A program that performs information gathering, information filtering, and/or mediation on behalf of a person or entity.

Interactive marketing unit (IMU) ad format The standard banner sizes that most Web sites have voluntarily agreed to use.

Interactive Message Access Protocol (IMAP) A newer e-mail protocol with improvements over POP.

Internet A global system of interconnected computer networks.

Internet2 A successor to the Internet with bandwidths in excess of 1 Gbps.

Internet access provider (IAP) Synonymous with Internet service provider.

Internet backbone Routers that handle packet traffic along the Internet's main connecting points.

Internet commerce Another term for electronic commerce; sometimes used to distinguish electronic commerce conducted on the Internet or World Wide Web instead of on private networks.

Internet host A computer that is directly connected to the Internet.

Internet protocol See TCP/IP.

Internet service provider (ISP) A company that sells Internet access rights directly to Internet users.

Interoperability The coordination of a company's information systems so that they all work together.

Interoperable software Software that runs transparently on a variety of hardware and software configurations.

Interstitial ad An intrusive Web ad that opens in its own browser window, instead of the page that the user intended to load.

Intranet An interconnected network of computers operated within a single company or organization.

IP address The number that represents the address of a particular location (computer) on the Internet.

IP tunneling The creation of a private passageway through the public Internet that provides secure transmission from one extranet partner to another.

IP version 4 (IPv4) The version of IP that has been in use for the past 20 years on the Internet; it uses a 32-bit number to identify the computers connected to the Internet.

IP version 6 (IPv6) The protocol that will replace IPv4.

J

Java sandbox A security model that confines Java applet actions to a security model-defined set of rules.

Java server pages A server-side scripting program developed by Sun Microsystems.

Java servlet An application that runs on a Web server and generates dynamic content.

JavaScript A scripting language developed by Netscape to enable Web page designers to build active content.

Jurisdiction A government's ability to exert control over a person or corporation.

K

Key A number used to encode or decode messages.

Knowledge management The intentional collection, classification, and dissemination of information about a company, its products, and its processes.

L

Late outsourcing The hiring of an external company to maintain an electronic commerce site that has been designed and developed by an internal information systems team.

Leased line A permanent telephone connection between two points; it is always active.

Legitimacy The idea that those subject to laws should have some role in formulating them.

Life-cycle segmentation The use of customer life cycle stages to create groups of customers that are in each stage.

Linear hyperlink structure A hyperlink structure that resembles conventional paper documents whereby the user reads pages in serial order.

Link checker A site management tool that examines each page on the site and reports on any URLs that are broken, that seem to be broken, or that are in some way incorrect.

Liquidation broker An agent that finds buyers for unusable and excess inventory.

Load-balancing switch A piece of network hardware that monitors the workloads of servers attached to it and assigns incoming Web traffic to the server that has the most available capacity at that instant in time.

Local area network (LAN) A network that connects workstations and PCs within a single physical location.

Localization A type of language translation that considers multiple elements of the local environment, such as business and cultural practices, in addition to local dialect variations in the language.

Lock-in effect The inherent greater value to customers of Web sites or technologies that they already use.

Log file A collection of data that shows information about Web site visitors' access habits.

Logical security The protection of assets using nonphysical means.

Long-arm statute A state law that creates personal jurisdiction for courts.

M

Machine translation Software translation of one language to another.

Macro virus A virus that is written in the macro language of a particular program. It is transmitted in a file that is downloaded or attached to an e-mail message and executes when that file is opened.

Mail bomb A security attack wherein many people (hundreds or thousands) each send a message to a particular address, exceeding the recipient's allowable mail limit and causing mail systems to malfunction.

Mail order model Synonymous with catalog model.

Mailing list An e-mail address that forwards messages to certain users who are subscribers.

Maintenance, Repair, and Operating (MRO) Commodity supplies, including general industrial merchandise and standard machine tools that are used in a variety of industries.

Managed service provider (MSP) A Web site hosting service firm; synonymous with ASP and CSP.

Many-to-many communications A model of communications in which a number of entities communicate with a number of other entities.

Many-to-one communications A model of communications in which a number of entities communicate with a single other entity.

Market A real or virtual space in which potential buyers and sellers come into contact with each other and agree on a medium of exchange (such as currency or barter).

Market research Activities undertaken by firms—such as conducting surveys, engaging in conversations with customers, and holding focus groups—to help identify customer needs.

Market segmentation The identification by advertisers of specific subsets of their markets that have common characteristics.

Marketing mix The combination of elements that companies use to achieve their goals for selling and promoting their products and services.

Marketing strategy A particular marketing mix that is used to promote a company or product.

Marketspace A market that occurs in the virtual world instead of in the physical world.

Masquerading Pretending to be someone you are not (for example, by sending an e-mail that shows someone else as the sender) or representing a Web site as that of another person or organization.

Mass media The method of contacting potential customers through the distribution of broadcast, printed, billboard, or mailed advertising materials.

Merchandising The combination of store design, layout, and product display intended to create an environment that encourages customers to buy.

Merchant account An account that a merchant must hold with a bank that allows the merchant to process credit or debit card transactions.

Merchant bank A bank that does business with merchants who want to accept credit or debit cards.

Message digest Synonymous with hash value, which is the number that results when a message is hash coded.

Meta language A language that comprises a set of language elements and can be used to define other languages.

Micromarketing The practice of targeting very small and well-defined market segments.

Micropayments Internet payments for items costing very little—usually $1 or less.

Middleware Software that handles connections between electronic commerce software and accounting systems.

Minimum bid In an English auction, the price for an item at which the auctioning begins.

Minimum bid increment The amount by which one bid must exceed the previous bid.

Mobile commerce (m-commerce) Resources accessed using devices that have wireless connections, such as stock quotes, directions, weather forecasts, and airline flight schedules.

Monetizing The conversion of existing site visitors seeking free information or services into fee-paying subscribers or purchasers of services.

Multipurpose Internet Mail Extension (MIME) An e-mail protocol that allows users to attach binary files to e-mail messages.

Multi-vector virus A virus that can enter a computer system in several different ways.

N

Name changing A problem that occurs when someone registers purposely misspelled variations of well-known domain names. These variants sometimes lure consumers who make typographical errors when entering a URL.

Name stealing Theft of a Web site's name that occurs when someone, posing as a site's administrator, changes the ownership of the domain name assigned to the site to another site and owner.

National Center for Supercomputing Applications (NCSA) Housed at the University of Illinois, Urbana-Champaign, the NCSA is one of the five original centers in the National Science Foundation's Supercomputer Centers Program. Mosaic, the first graphic Web browser program and predecessor to the Netscape browser, was invented at NCSA.

Necessity The category of computer security that includes data delay or data denial threats.

Necessity threat The disruption of normal computer processing or denial of processing.

Net bandwidth The actual speed information travels, taking into account traffic on the communication channel at any given time.

Network access points (NAPs) The four primary connection points for access to the Internet backbone in the United States.

Network access providers The four companies (Pacific Bell, Sprint, Ameritech, MFS Corporation) that are primary providers of Internet access rights; they sell these rights to smaller Internet service providers.

Network address translation (NAT) device A computer that converts private IP addresses into normal IP addresses when it forwards packets to the Internet.

Network Control Protocol (NCP) Used by ARPANET in the early 1970s to route messages in its experimental wide area network.

Network economic structure A business structure in which firms coordinate their strategies, resources, and skill sets by forming a long-term, stable relationship based on a shared purpose.

Network operations Web site staff whose responsibilities include load estimation and monitoring, resolving network problems as they arise, designing and implementing fault-resistance technologies, and managing any network operations that are outsourced to ISPs, CSPs, or telephone companies.

Network specification The set of rules that equipment connected to a network must follow.

Newsgroup A topic area in Usenet where people read and post articles.

Nexus The association between a tax-paying entity and a governmental tax-ing authority.

Nonrepudiation Verification that a particular transaction actually occurred; this prevents parties from denying a transaction's validity or its existence.

N-tier architecture Higher-order client-server architectures that have more than three tiers.

O

Occasion segmentation Behavioral segmentation that is based on things that happen at a specific time or on a specific occasion.

Octet An 8-bit number.

Offer A declaration of willingness to buy or sell a product or service; it includes sufficient details to be firm, precise, and unambiguous.

One-sided tag HTML tags that require only an opening tag.

One-to-many communication model A model of communications in which one entity communicates with a number of other entities.

One-to-one communication model A model of communications in which one entity communicates with one other entity.

One-to-one marketing A highly customized approach to offering products and services that match the needs of a particular customer.

One-way function An algorithm that cannot be converted back to its original value.

Online community Synonymous with virtual community, which is an electronic gathering place for people with common interests.

Open architecture The philosophy behind the Internet that dictates that independent networks should not require any internal changes to be connected to the network, packets that do not arrive at their destinations must be retransmitted from their source network, routers do not retain information about the packets they handle, and no global control exists over the network.

Open auction (open outcry auction) An auction in which bids are publicly announced.

Open EDI EDI conducted on the Internet instead of over private leased lines.

Open loop system A payment card arrangement involving a consumer and his or her bank, a merchant and its bank, and a third party (such as Visa or MasterCard) that processes transactions between the consumer and merchant.

Open source Software that can be downloaded and used at no cost.

Opening tag An HTML tag that precedes the text that a tag affects.

Opportunity cost Lost benefits from an action not taken.

Optical fiber An Internet connectivity material with bandwidths up to 10 Gbps.

Opt-in e-mail The practice of sending e-mail messages to people who have requested information on a particular topic or about a specific product.

Orphan file A file on a Web site that is not linked to any page.

Outsourcing The hiring of another company to perform design, implementation, or operational tasks for an information systems project.

P

Packet filter firewall A firewall that examines all data flowing back and forth between a trusted network and the Internet.

Packets The small pieces of files and e-mail messages that travel over the Internet.

Packet-switched A network in which packets are labeled electronically with their origin, sequence, and destination addresses. Packets travel from computer to computer along the interconnected networks until they reach their destination. Each packet can take a different path through the interconnected networks and the packets may arrive out of order. The destination computer collects the packets and reassembles the original file or e-mail message from the pieces in each packet.

Packet switching A network system of data transmission in which files and messages are broken down into small units called "packets." These packets are labeled electronically with codes indicating origin and destination. Packets travel from computer to computer along the network and upon reaching their destination are reassembled.

Page-based application server Application server software that returns pages generated by scripts that include the rules for presenting data on the Web page with the business logic.

Page view A page request made by a Web site visitor.

Paid placement (sponsorship) The purchased right to a listing at or near the top of a search engine's results page for a particular set of search terms.

Partial outsourcing The outsourcing of the design, development, implementation, or operation of specific portions of an electronic commerce system.

Patent An exclusive right to make, use, and sell an invention granted by a government to the inventor.

Pay-per-click model A business model in which an affiliate earns payment each time a site visitor clicks a link to load the seller's page.

Pay-per-conversion model A business model in which an affiliate earns payment each time a site visitor is converted from a visitor into either a qualified prospect or a customer.

Peer-to-peer payment system Payments from one type of entity to another of the same type.

Permission marketing A marketing strategy that only sends specific information to persons who have indicated an interest in receiving information about the product or service being promoted.

Per se defamation A legal cause of action in which a court deems some types of statements to be so negative that injury is assumed.

Persistent cookie A cookie that exists indefinitely.

Personal contact A method of identifying and reaching customers that involves searching for, qualifying, and contacting potential customers.

Personal identification number (PIN) A random assemblage of digits, chosen by the customer, that serves as a password for monetary transactions.

Personal jurisdiction A court's authority to hear a case based on the residency of the defendant; a court has personal jurisdiction over a case if the defendant is a resident of the state in which the court is located.

PHP: Hypertext Preprocessor (PHP) A server-side scripting technology developed by the Apache Software Foundation.

Physical security Tangible protection devices such as alarms, guards, fireproof doors, fences, and vaults.

Ping (Packet Internet Groper) A program that tests the connectivity between two computers connected to the Internet.

Place (distribution) The need to have products or services available in many different locations.

Plain old telephone service (POTS) The network connecting telephones; it provides a reliable data transmission bandwidth of about 56 Kbps.

Plug-in An application that helps a browser to display information (such as video or animation) but that is not part of the browser.

Pop-behind ad A pop-up ad that is followed very quickly by a command that returns the focus to the original browser window, resulting in an ad that is parked behind the user's browser waiting to appear when the browser is closed.

Pop-up ad An ad that appears in its own window when the user opens or closes a Web page.

Portal A Web site that serves as a customizable home base from which users do their searching, navigating, and other Web-based activity.

Post-implementation audit A formal review of a project after it is up and running.

Post Office Protocol (POP) The protocol responsible for retrieving e-mail from a mail server.

Presence The public image conveyed by an organization to its stakeholders.

Price The amount a customer pays for a product.

Primary activities Activities that are required to do business: design, production, promotion, marketing, delivery, and support of products or services.

Privacy The protection of individual rights to nondisclosure of information.

Private company marketplace A marketplace that provides auctions, requests for quotes postings, and other features to companies that want to operate their own marketplace.

Private IP addresses A series of IP numbers that have been set aside for subnet use and are not permitted on packets that travel on the Internet.

Private key A single key that is used to encrypt and decrypt messages. Synonymous with symmetric key.

Private-key encryption The encoding of a message using a single numeric key to encode and decode data; it requires both the sender and receiver of the message to know the key, which must be guarded from public disclosure.

Private network A private, leased-line connection between two companies that physically links their individual computers or intranets.

Private store A password-protected area of a Web site that offers individual customers negotiated price reductions on a limited selection of products and other customized features.

Private valuation The amount a bidder is willing to pay for an item that is up for auction.

Procurement The business activity that includes all purchasing activities plus the monitoring of all elements of purchase transactions.

Product The physical item or service that a company is selling.

Product disparagement A statement that is false and injures the reputation of a product or service.

Project management Formal techniques for planning and controlling activities undertaken to achieve a specific goal.

Project management software Application software that provides built-in tools for managing people, resources, and schedules.

Project portfolio management A technique in which each project is monitored as if it were an investment in a financial portfolio.

Promotion Any means of spreading the word about a product.

Prospecting The part of personal contact selling in which the salesperson identifies potential customers.

Protocol A collection of rules for formatting, ordering, and error-checking data sent across a network.

Proxy bid In an electronic auction, a predetermined maximum bid submitted by a bidder.

Proxy server A firewall that communicates with the Internet on behalf of the trusted network.

Psychographic segmentation The grouping of customers by variables such as social class, personality, or approach to life.

Public key One of a pair of mathematically related numeric keys, it is used to encrypt messages and is freely distributed to the public.

Public-key encryption The encoding of messages using two mathematically related but distinct numeric keys.

Public marketplace A vertical portal that is open to new buyers and sellers just entering an industry.

Public network An extranet that allows the public to access its intranet, or when two or more companies link their intranets.

R

Rational branding An advertising strategy that substitutes an offer to help Web users in some way in exchange for their viewing an ad.

Reintermediation The introduction of a new intermediary into a value chain.

Remote server administration Control of a Web site by an administrator from any Internet-connected computer.

Repeat visits Subsequent visits a Web site visitor makes to a particular page.

Request header The part of a message from a client to a server that contains additional information about the client and more information about the request.

Request line The part of a message from a client to a server that contains a command, the name of the target resource (without the protocol or domain name), and the protocol name and version.

Request message The message that a Web client sends to request a file or files from a Web server.

Reserve Synonymous with reserve price.

Reserve price The minimum price a seller will accept for an item sold at auction.

Response header field In a client/server transmission, it follows the response header line and returns information describing the server's attributes.

Response header line The part of a message from a server to a client that indicates the HTTP version used by the server, status of the response, and an explanation of the status information.

Response message The reply that a Web server sends in response to a client request.

Response time The amount of time a server requires to process one request.

Return on investment A method for evaluating the potential costs and benefits of a proposed capital investment.

Revenue model The combination of strategies and techniques that a company uses to generate cash flow into the business from customers.

Reverse auction A type of auction in which sellers bid prices for which they are willing to sell items or services.

Reverse bid The process in which an auction customer seeks products by describing an item or service in which he or she is interested, and then entertains responses from merchants who offer to supply the item at a particular price.

Reverse link checker A Web site management program that checks on sites with which a company has entered a link exchange program and ensures that link exchange partners are fulfilling their obligation to include a link back to the company's Web site.

Rich media objects Programming components of attention-grabbing Web banner ads.

Robot (bot, spider, crawler) A program that automatically searches the Web to find Web pages that might be interesting to people.

Routers (router computers, routing computers) Computers that determine the best way for data packets to move forward to their destination. Synonymous with gateway computers.

Routing algorithm The program used by a router to determine the best path for data packets to travel.

Routing table Synonymous with configuration table.

S

Save area The location of a computer where programs store critical information before control of that information is passed to another program.

Scalable A system's ability to be adapted to meet changing requirements.

Scaling problem The exponential increase in cost that results from the expansion of a private network.

Scrip Digital cash minted by a small number of third-party organizations.

Sealed-bid auction An auction in which bidders submit their bids independently and are usually prohibited from sharing information with each other.

Search engine Web software that finds other pages based on key word matching.

Search engine optimization (search engine positioning, search engine placement) The combined art and science of having a particular URL listed near the top of search engine results.

Search utility The part of a search engine that finds matching Web pages for search terms.

Second-price sealed-bid auction A type of auction in which bidders submit their bids independently and privately; the highest bidder wins the auction but pays only the amount bid by the second-highest bidder.

Secrecy The category of computer security that addresses the protection of data from unauthorized disclosure and confirmation of data source authenticity.

Secure Electronic Transaction (SET) A protocol that provides security for card payments as they traverse the Internet between merchant sites and processing banks.

Secure envelope A security utility that encapsulates a message and provides secrecy, integrity, and client/server authentication.

Secure Sockets Layer (SSL) A protocol for transmitting private information securely over the Internet.

Security policy A written statement describing assets to be protected, the reasons for protecting the assets, the parties responsible for protection, and acceptable and unacceptable behaviors.

Segment Also called a market segment; a subset of a company's potential customer pool that has common demographic characteristics.

Self hosting A system of Web hosting in which the online business owns and maintains the server and all its software.

Semantics The meaning of text (rather than its display characteristics) in XML pages, as described by XML tags.

Server A powerful computer dedicated to managing disk drives, printers, or network traffic.

Server architecture The different ways that servers can be connected to each other and to related hardware such as routers and switches.

Server farm A large collection of electronic commerce Web site servers.

Server-side electronic wallet An electronic wallet that stores a customer's information on a remote server that belongs to a particular merchant or to the wallet's publisher.

Server Side Includes (SSI) A type of HTML comment that directs the Web server to dynamically generate data for a Web page when it is requested.

Server-side scripting (server-side includes [SSI] or server-side technologies) A Web page response approach in which programs running on the Web server create Web pages before sending them back to the requesting Web clients as parts of response messages.

Server software The software that a server computer uses to make files and programs available to other computers on the same network.

Service mark A distinctive mark, device, motto, or implement used to identify services provided by a company.

Session cookie A cookie that exists only until the user shuts down his or her browser.

Session key A key used by an encryption algorithm to create cipher text from plain text during a single secure session.

Shared hosting A Web hosting arrangement in which a corporate Web site is on a server that hosts other Web sites simultaneously and is controlled by a third-party service provider.

Shill bidder An individual employed by a seller or auctioneer who makes bids on behalf of the seller, sometimes artificially inflating an item's price. Shill bidders may be prohibited by the rules of a particular auction.

Shipping profile The collection of attributes that affect how easily a product can be packaged and delivered.

Shopping cart An electronic commerce utility that keeps track of selected items for purchase and automates the purchasing process.

Signature Any symbol executed or adopted for the purpose of authenticating a writing.

Signed Java applet A Java applet that contains an embedded digital signature from a trusted third party; it is proof of the identity of the applet's source.

Signed message or code The status of a message or Web page when it contains an attached digital certificate.

Simple Mail Transfer Protocol (SMTP) A standardized protocol used by a mail server to format and administer e-mail.

Skyscraper ad A large banner ad on the side of a Web page that remains visible as the user scrolls down through the page.

Small payment Any payment of less than $10.

Smart card A plastic card with an embedded microchip that contains information about the card owner.

Sniffer program A program that taps into the Internet and records information that passes through a router from the data's source to its destination.

Sniping software Software that observes auction progress until the last second or two of the auction clock, then places a bid high enough to win the auction.

Sourcing The part of procurement devoted to identifying suppliers and determining the qualifications of those suppliers.

Spam Electronic junk mail.

Spider The first part of a search engine; it automatically and frequently searches the Web to find pages and updates its database of information about old Web sites.

Sponsorship Synonymous with paid placement.

Square button A Web banner ad measuring 125×125 pixels.

Stakeholders The various entities involved in a business; these include customers, suppliers, employees, stockholders, neighbors, and the general public.

Static catalog A simple list of products written in HTML and displayed on a Web page or a series of Web pages.

Static page A Web page that displays unchanging information retrieved from disk.

Statute of Frauds State laws that specify that contracts for the sale of goods worth more than a specified amount (usually $500) and contracts that require actions that cannot be completed within one year must be created by a signed writing.

Statutory law That part of British and U.S. law that comprises laws passed by elected legislative bodies.

Steganography The hiding of information within another piece of information, such as a graphic.

Stickiness The ability of a Web site to keep visitors at its site and to attract repeat visitors.

Sticky The condition of having stickiness.

Stored value card Either an elaborate smart card or a simple plastic card with a magnetic strip that records the currency balance, such as a prepaid phone, copy, subway, or bus card.

Strategic alliance The coordination of strategies, resources, and skill sets by companies into long-term, stable relationships with other companies and individuals based on shared purposes.

Strategic business unit (business unit) A unit within a company that is organized around a specific combination of product, distribution channel, and customer type.

Strategic partners The entities taking part in a strategic alliance.

Strategic partnership Synonymous with strategic alliance.

Subject matter jurisdiction A court's authority to decide a dispute between entities based on the issue of dispute.

Subnetting The use of reserved private IP addresses within LANs and WANs to provide additional address space.

Supply alliances Long-term relationships among participants in the supply chain.

Supply chain The part of an industry value chain that precedes a particular strategic business unit. It includes the network of suppliers, transportation firms, and brokers that combine to provide a material or service to the strategic business unit.

Supply chain management The process of taking an active role in working with suppliers and other participants in the supply chain to improve products and processes.

Supply management Synonymous with procurement, which is the business activity that includes all purchasing activities plus the monitoring of all elements of purchase transactions.

Supporting activities Secondary activities that back up primary business activities. These include human resource management, purchasing, and technology development.

SWOT (strengths, weaknesses, opportunities, threats) analysis Evaluation of the strengths and weaknesses of a business unit, and identification of the opportunities presented by the markets of the business unit and threats posed by competitors of the business unit.

Symmetric connection An Internet connection that provides the same bandwidth in both directions.

Systems administrator A member of an electronic commerce team who understands the server hardware and software and is responsible for the system's reliable and secure operation.

T

T1 High-bandwidth telephone company connections that operate at 1.544 Mbps.

T3 High-bandwidth telephone company connections that operate at 44.736 Mbps.

Tags HTML codes inserted into documents that specify formatting and arrangement of page elements.

TCP/IP The set of protocols that provide the basis for the operation of the Internet. The TCP protocol includes rules that computers on a network use to establish and break connections. The IP protocol determines routing of data packets.

Technology-enabled customer management The business practice of obtaining detailed information about a customer's behavior, preferences, needs, and buying patterns, and using that information to set prices, negotiate terms, tailor promotions, add product features, and provide other customized interactions.

Technology-enabled relationship management The business practice of obtaining detailed information about the organization's stakeholders (customers, suppliers, employees, and others) and using that information to customize the organization's interactions with those stakeholders.

Telnet A protocol that allows users to log on to a computer and access its contents from a remote location.

Terms of service Rules and regulations intended to limit the Web site owner's liability for what a visitor might do with information obtained from the site.

Third-party assurance provider An independent organization that attests to the quality of business and privacy policies of Web sites.

Threat An act or object that poses a danger to assets.

Three-tier architecture A client/server architecture that builds on the two-tier architecture by adding applications and their associated databases that supply non-HTML information to the Web server on request.

Throughput The number of HTTP requests that a particular hardware and software combination can process in a unit of time.

Tier-one suppliers The capable suppliers that work directly with and have long-term relationships with businesses.

Tier-three suppliers Suppliers that provide components and raw materials to tier-two suppliers.

Tier-two suppliers Suppliers that provide components and raw materials to tier-one suppliers.

Top-level domain (TLD) The last part of a domain name; the most general identifier in the name.

Tort An action taken by a legal entity that causes harm to another legal entity.

Trademark A distinctive mark, device, motto, or implement that a company affixes to the goods it produces for identification purposes.

Trademark dilution The reduction of the distinctive quality of a trademark by alternative uses.

Trade name The name (or a part of that name) that a business uses to identify itself.

Trading partners Businesses that engage in EDI with one another.

Transaction An exchange of value.

Transaction costs The total of all costs incurred by a buyer and seller as they gather information and negotiate a transaction.

Transaction processing Processes that occur as part of completing a sale; these include calculation of any discounts, taxes, or shipping costs and transmission of payment data (such as a credit card number).

Transaction sets Formats for specific business data interchanges using EDI.

Transmission Control Protocol See TCP/IP.

Trial visit The first visit a Web site visitor makes to a particular page.

Trigger word Keywords used to jog the memory of visitors and remind them of something they want to buy on the site.

Triple Data Encryption Standard (3DES) A version of the Data Encryption Standard that cannot be cracked even with today's supercomputers.

Trojan horse A program hidden inside another program or Web page that masks its true purpose (usually destructive).

Trusted (network) A network that is within a firewall.

Trusted applet A Java applet that has full access to system resources on a client computer.

Two-sided tags HTML tags that require both an opening and a closing tag.

Two-tier architecture A client/server architecture in which only a client and server are involved in the requests and responses that flow between them over the Internet.

U

Ultimate consumer orientation A focus on the needs of the consumer who is at the end of a supply chain.

Uniform resource locator (URL) Names and abbreviations representing the IP address of a particular Web page. Contains the protocol used to access the page and the page's location. Used in place of dotted quad notations.

Untrusted (network) A network that is outside a firewall.

Untrusted applet A Java applet that is not known to be secure.

Upload bandwidth Synonymous with upstream bandwidth.

Upstream bandwidth The connection that occurs when you send information from your connection to your ISP.

Upstream strategies Strategies that focus on reducing costs or generating value by working with suppliers or inbound logistics.

URL broker A business that sells or auctions domain names that it believes others will find valuable.

Usability testing The testing and evaluation of a company's Web site for ease of use by visitors.

Usage-based market segmentation Customizing visitor experiences to match the site usage behavior patterns of each visitor or type of visitor.

Usenet (User's News Network) One of the first mailing lists; it allows subscribers to read and post articles within topic areas.

V

Value-added bank A bank that offers value-added network services for nonfinancial transactions.

Value-added network (VAN) An independent company that provides connection and EDI transaction forwarding services to businesses engaged in EDI.

Value chain A way of organizing the activities that each strategic business unit undertakes to design, produce, promote, market, deliver, and support the products or services it sells.

Value system Synonymous with industry value chain.

VBScript A programming language that can create dynamic pages within HTML documents.

Vertical integration The practice of an existing firm replacing one of its suppliers with its own strategic business unit that creates the supplied product.

Vertical portal (vortal) A vertically integrated Web information hub focusing on an individual industry.

Vickrey auction Synonymous with second-price sealed-bid auction. Named for William Vickrey, who won the 1996 Nobel Prize in Economics for his studies of the properties of this auction type.

Viral marketing Tactics that rely on existing customers to tell other persons—the company's prospective customers—about the products or services they have enjoyed using.

Virtual community An electronic gathering place for people with common interests.

Virtual company A strategic alliance occurring among companies that operate on the Internet.

Virtual host Multiple servers that exist on a single computer.

Virtual private network (VPN) A network that uses public networks and their protocols to transmit sensitive data using a system called "tunneling" or "encapsulation."

Virtual server Synonymous with virtual host.

Virus Software that attaches itself to another program and can cause damage when the host program is activated.

Visit The request of a Web site visitor for a page from a Web site.

Voice-grade line Telephone wiring that costs less than lines designed to carry data, is made of lower-grade copper, and was never intended to carry data. These lines can only carry limited bandwidth—usually less than 14 Kbps.

W

Warranty disclaimer A statement indicating that the seller will not honor some or all implied warranties.

Web browser Software that lets users read HTML documents and move from one HTML document to another using hyperlinks.

Web bug A tiny, invisible Web page graphic that provides a way for a Web site to place cookies.

Web catalog model A business model of selling goods and services on the Web wherein the seller establishes a brand image that conveys quality and uses the strength of that image to sell through catalogs mailed to prospective buyers. Buyers place orders by mail or by calling the seller's toll-free telephone number.

Web client A computer that is connected to the Internet and used to download Web pages.

Web community Synonymous with virtual community.

Web directory A listing of hyperlinks to Web pages that is organized into hierarchical categories.

Web graphics designer A person trained in art, layout, and composition who also understands how Web pages are constructed and who ensures that the Web pages are visually appealing, easy-to-use, and make consistent use of graphics elements from page to page.

Web server A computer that is connected to the Internet and that stores files written in HTML that are publicly available through an Internet connection.

Web server software Software that makes files available to other computers on the Internet.

Wide area network (WAN) A network of computers that are connected over great distances.

Winner's curse A psychological phenomenon that causes bidders to become caught up in the excitement of competitive bidding and bid more than their private valuations.

Wireless Ethernet (802.11b Wi-Fi) A wireless standard that can connect devices over several hundred feet through many kinds of walls and other barriers.

Wire transfer Synonymous with electronic funds transfer, which is the electronic transfer of account exchange information over secure private communications networks.

World Wide Web (Web) The subset of Internet computers that connects computers and their contents in a specific way, and that allows for easy sharing of data using a standard interface.

World Wide Web Consortium A not-for-profit group that maintains standards for the Web.

Worm A virus that replicates itself on other machines.

Writing A tangible representation of the terms of a contract.

Y

Yankee auction A type of English auction that offers multiple units of an item for sale and that allows bidders to specify the quantity of items they want to buy.

Z

Zombie A program that secretly takes over another computer for the purpose of launching attacks on other computers. Zombie attacks cannot be traced to their creators.

cookies, 285–286
   digital certificates, 407
   Sothby's and, 228
   terms of service agreement, 275–276
American Express, 445, 447, 475
American Red Cross, 112
American Society of Mechanical Engineers, 170
Ameritech, 30
Ameritrade, 102
Ameritrade Holding, 167, 234
Amphire Solutions, 207
Analog Web server log file analyzer, 307
anchor tag, 60–61
Andale, 237
Andreessen, Marc, 32
animated GIFs, 153
Anonymizer, 382
anonymous electronic cash, 454–455
anonymous FTP, 52
ANSI (American National Standards Institute), 189
Answer Financial, 102
Anticypersquatting Consumer Protection Act, 368
antivirus software, 416
AOL, 96, 168, 246
AOL (Time Warner), 32
AOL Policy on Unsolicited Bulk E-mail, 50
Apache, 313
Apache Cocoon Project, 311
Apache HTTP Server, 313
Apache Web server, 298
API (application program interface), 306
APNIC (Asia-Pacific Network Information Center), 43
apparel retailers and catalog model, 86–87
applets, 370, 372–375
application servers, 339–341

application services, 45
applications specialists, 502
archiving log files, 313
Ariba, 211
ARIN (American Registry for Internet Numbers), 43
ARIS MusiCode system, 403
ARPANET, 42, 49
Art.com, 170
Artuframe, 170
ASC X12 (Accredited Standards Committee X12), 189–190
ascending-price auctions, 220
ASCII text files, 52
ASP (Active Server Pages), 311, 339
ASPs (application service providers), 300, 341, 495
Association of American Publishers, 54
asymmetric connections, 72
asymmetric encryption, 419
ATM (Asynchronous Transfer Mode), 75
Atrieva, 185
AT&T IP Services, 199
attachments, 49
Auction Block site, 232–233
Auction Universe, 226
AuctionBeagle, 237
auctionet, 220
Auctionguide.com, 235
AuctionHawk, 237
auction-related services, 235–237
auctions
   ascending-price, 220
   auctioneer, 220
   basics, 219–222
   bidders, 110
   bids, 220
   consumer-to-business, 223
   consumer-to-consumer, 223
   descending-price, 221
   double, 221–222

   Dutch, 221
   English, 220–221
   first-price sealed-bid, 221
   general consumer, 223–229
   open, 220
   open-outcry, 220
   origins, 220
   private valuations, 220
   reverse, 237–239
   sealed-bid, 221
   second price sealed-bid, 221
   seller-bid (reverse), 237–239
   shill bidders, 220
   specialty consumer, 229–232
   Vickrey, 221
   Yankee, 220–221
AuctionWatch site, 235
AuctionWeb, 218
audio watermarks, 403
authentication, 432–434
Authenticode technology and digital certificates, 410–411
authority to bind, 275
Authorize.Net, 449
Autobytel, 1–2, 13, 101
automated e-mail, 158
automobile sales, 101
Autoweb.com, 101
AvantGo, 240
Avendra marketplace, 211
Aventail, 71
awareness stage, 150

B

B2B (business-to-business), 4
B2C (business-to-consumer), 4
B2C (business-to-consumer) commerce site software, 330
Baan, 340
backbone routers, 41
backdoors, 381
bandwidth, 72

business process patent, 278

business processes, 10–13

business rules, 339

business units, 22

  evaluating opportunities, 25–26

  primary activities, 23

  value chain, 24

Business Week online Web site, 98

Business.com, 170

business-to-business Web auctions, 232–235

Buy.com, 88

buyers, 6–7, 147

bytes, 43

C

C2B (consumer-to-business) auctions, 223

C2C commerce (consumer-to-consumer), 4

c2it, 465

CA (certification authority), 409–410, 453

cable modems, 73

Caldera, 303

California Voter Foundation, 112

call centers, 502–503

Calomiris, Michael, 263

Cambridge Technology Partners, 119

cannabalization, 90

capital investments, 491

capital projects, 491

card not present transactions, 445

CareerSite, 96

Cars.com, 170

CarsDirect.com, 101

Cart, 32, 381

CartIt!, 334

cash, 443

Castells, Manuel, 21

catalog display, 331–332

catalog model, 84

  apparel retailers, 86–87

  cannibalization, 90

  channel conflict, 90

  computer manufacturers, 85

  flowers and gifts, 87

  general discounters, 88

  luxury goods, 85

  strategic alliances, 91

catalogs, 331–332

Catatonic, 362

Catchings, Bill, 285

cause marketing, 165

CBOT (Chicago Board of Trade), 234

CDenergy, 231

CDnow, 159–161, 164

cellular-satellite communications technology, 240

Center for Supercomputing, 441

centralized architecture, 319

Cerf, Vincent, 42

CERN: European Laboratory for Particle Physics, 31

CERT (Computer Emergency Response Team), 391

CGI (Common Gateway Interface)

  programs, 384

  threats, 388–389

CGI (common gateway interface), 306

Chambers, John, 482

channel conflict, 90

charge cards, 445

chargeback, 448

Charles Schwab, 102

CheckFree, 456

checks, 443

CheMatch.com, 207, 234

Chemdex, 207

Chicago Board of Trade, 222

Chicago NAP, 30

Children's Online Privacy Protection Act of 1998, 284

Christie's, 220

Cigarbid.com, 231

CIOs (chief information officers), 500, 501

cipher text, 418

circuit, 40

circuit switching, 40

circuit-switched networks, 40

Cirque du Soleil, 346

Cisco Global Networked Business Model, 13

Cisco Systems, 13, 210, 482

Citibank Online, 105

CityAuction, 232

Claritas, 151

Clark, James, 32

Classified Ventures, 226

clear text, 418

ClearCross, 265

click, 155

clickshare, 457

Clickshare Example Scenario, 457

clickstream, 354

click-through, 155

client, 45

client computers

  protecting, 406–416

  threats, 369–380

client-server architecture, 64

  n-tier, 67–68

  three-tier, 67–68

  two-tier, 65–67

client-server communication, 65–68

client/server model, 45

client-side electronic wallets, 466

client-side scripting, 62

Clinton, Bill, 430

closed loop systems, 447

closing tag, 58

Giovanni, 403

Global Computer Supplies, 183

Global Reach, 256

GML (Generalized Markup Language), 54

Godin, Seth, 158

Golf Club Exchange, 231

Gollent, Manfred, 496

Gomez.com, 118

Google, 168

Gordon Brothers, 233–234

government, 186–187

Grainger, 210

graphics and security, 376–380

GreatDomains.com, 171

Green party Web site, 113

group purchasing sites, 237–239

GTE, 477

guaranteeing transaction delivery, 430

GUI (graphical user interface), 32

Gump's, 166

H

hackers, 364–365

Haggle Online, 230

half banner, 153

hardware, 299–302

   benchmarking, 305

   determining type, 317

   scalable, 303

   technologies, 39

Harry and David, 87

hash algorithm, 419

hash coding, 419

hash functions, 428

HDSL (high-speed DSL), 73

Headspace, Inc., 414

Hewlett-Packard, 54, 186

hexadecimal numbering system, 44

hierarchical business organizations, 17

hierarchical hyperlink structure, 61

hierarchies, 18–20, 187

high availability servers, 300

high reliability servers, 300

HomeSite, 62

Honeywell, 482

Hong Kong Citybus Web page, 474

hops, 48

host services, 344

HostCompare.com, 497

HostIndex site, 301

hosting services, 341–343

HotDog Professional, 62

Hotel Reservations Network, 100

HotMail, 163, 468

HoTMetaL Pro, 62

HotOffice, 185

HTML (Hypertext Markup Language), 31–32, 54–55, 509, 510–515

   anchor tag, 60–61

   links, 60–61

   scripting language and style sheet capabilities, 62–63

   tags, 57–60, 510–512

HTML documents, 32, 57

HTML editors, 62–63

HTML extensions, 54

HTTP (Hypertext Transfer Protocol), 45

human resources, 24, 184

Human Rights Watch, 262

Hyatt, 211

hybrid EDI solutions, 200–201

hyperlinks, 32, 60–61

   e-mail, 158

   hierarchical structure, 61

   linear structure, 61

HyperMart, 344

hypertext, 31–33, 54

hypertext server, 31

I

i2 Technologies, 356

IAB (Interactive Advertising Bureau), 153

IAB Web site, 153

IANA (Internet Assigned Numbers Authority), 43

IAPs (Internet access providers), 71–72

iAuthorizer, 449

IBM, 10, 26, 298, 303, 450, 477

IBM Global Services, 201

IBM Home Page Reader, 115

IBM Tivoli Systems, 337

ICANN (Internet Corporation for Assigned Names and Numbers), 44, 171

ICANNWatch Web site, 44

icons, 259

ICVERIFY, 448

identifying customers, 23

Idiom Technologies, 258

IETF (Internet Engineering Task Force), 43

IIS (Internet Information Server), 314–315, 339, 379

"ILOVEYOU" virus, 378

IMAP (Interactive Mail Access Protocol), 46

IMAP Connection Web site, 46

implied contracts, 272

impostors, 364

impressions, 155

IMU (interactive marketing unit) ad formats, 153

income taxes, 287–289

incubators, 498

independent exchanges, 207

index, 168, 307

indexing programs, 307

indirect connection EDI, 196

indirect materials procurement, 182–183

industry consortia-sponsored marketplaces, 211

Index

price, 134

product, 134

product-based, 135–137

promotion, 134

marketplaces, 207–208, 210–211

markets, 17, 18–20, 187

creation of, 22

improvement in, 21–22

marketspace, 159

markup languages, 53–63

Marriott, 211

masquerading, 383–384

mass media

approach, 122

obtaining information, 123–124

trust, 140

MasterCard, 164, 445, 446, 447, 448, 467, 477

MasterCard International, 475

MathML, 517

Maytag, 90, 119, 120

McAfee, 378

McCool, Rob, 313

MCI Mail, 29

McKinsey & Company, 488

m-commerce, 76

media choice, 139–140

medium of exchange, 17

Memex, 31

Mercata, 239

merchandising, 11

merchant account, 447–448

merchant bank, 447

Merriam-Webster's Collegiate Dictionary, 94

Merrick, Phillip, 328

Merrill Lynch, 102

message digests, 426

meta language, 53

META tags, 168

MetalSite, 209

Metropolitan Transportation Commission, 474

MFS Corporation, 30

Michelin North America, 206

micromarketing, 140

MicronPC, 252–253

micropayments, 450, 457

Microsoft, 32, 477

Microsoft Corporation Web site, 47

Microsoft FrontPage, 63, 299, 308

Microsoft Internet Explorer, 32, 410–413

Microsoft Project, 500

Microsoft Security Pages, 379

Microsoft server products, 303

Microsoft Windows 2000 Server Edition, 303

Microsoft Windows NT Server, 303

Microsoft .NET Passport, 468

middleware, 337

midrange packages, 349–353

Milacron, 183, 210, 242

Milpro Web site, 183, 242–243

MIME (Multipurpose Internet Mail Extensions), 46

Mindcraft, 303

mini-mills, 209

minimum bids, 220, 225

MIT Media Lab Software Agents Group, 240

MIT Software Agents Group, 322

mixed-model Web portals, 246

MMC (Microsoft Management Console), 315

mobile commerce, 76

MoMA (Museum of Modern Art), 113

Mondex, 475–476

monitoring active content, 406–415

Monster.com, 97

Monteizing, 246

Montgomery Ward & Company, 84

Monty Python, 50

Morris, Robert, Jr., 362

mortgage loan brokers, 104

mortgagebot.com, 104

MRO (maintenance, repair, and operations) supplies, 182, 242

MSN, 96, 168

MSN Carpoint, 101

MSN MoneyCentral, 105

MSPs (managed service providers), 300

multi-vector virus, 378

My Virtual Model, 298

my.ca.gov, 186–187

MySAP CRM, 356

mySimon, 322

mySimon Web site, 240

mySQL, 340

N

NACHA - The Electronic Payments Association link, 448

name changing, 367, 368

name stealing, 367, 369

NAPs (network access points), 30

NAPs (network access providers), 30, 75

NAT device (network address translation), 43

National Association for Purchasing Management, 203

National Bureau of Economic Research, 2

National Computer Security Center, 391

National Infrastructure Protection Center, 379

National Security Agency, 418

NCP (Network Control Protocol), 42

NCSA (National Center for Supercomputing Applications), 311–312, 313

necessity, 365

necessity threats, 384

price, 117

    service element, 117

professional services, 106

profit-driven organizations, 109–111

Proflowers.com, 165, 166

programming threats, 389–390

project management, 499–500

Project Management Institute, 500

project portfolio management, 500–501

Promedix, 207

promotion, 134

ProQuest, 93

prospecting, 122

protecting

    assets, 365, 398–400

    communications channels, 417–430

    intellectual property, 401–403

    privacy, 404–406

protocols, 42

proxy bid, 225

proxy servers, 436

Prudential, 104

PSINet, 430

psychographic segmentation, 141

Public Broadcasting System, 112

public key, 419

public marketplaces, 207

public networks, 70

public profiles, 468

public Web servers, 311

public-key encryption, 419, 421, 424

purchase transaction and traditional commerce, 6–7, 9

Purchasing, 239

purchasing

    activities, 180–183

    direct vs indirect materials procurement, 182–183

EDI process, 193–195

    paper-based process, 192–193

Q

Quaker Oats Web site, 109–111

Question.com, 170

Quicken.com My Finances service, 105

QuickTime, 376

Quotesmith.com, 102

R

rational branding, 163–164

real estate, 104

RealAge, 107

RealArcade Central, 106

RealNetworks, 374–375

RealPlayer GoldPass, 106

Red Hat, 303

Red Hat Linux Buffer Overflow Attacks Page link, 390

RedEnvelope, 133, 151

refdesk.com, 96

Reform party Web site, 113

Regulation M, 252, 253

reintermediation, 101

relevance, 162–163

remote server administration, 309–310

repeat visits, 155

Republican party Web site, 113

request headers, 66

request line, 66

request message, 66

reserve, 220

reserve price, 220

resources, allocating, 17

Respond.com, 239

response header line, 66–67

response message, 66

response time, 305

retail stores, 143

RetailExhcange.com, 233

Reuters wire service, 98

revenue models for selling

    advertising-subscription mixed model, 98–100

    advertising-supported model, 95–97

    fee-for-services revenue models, 105–107

    fee-for-transaction models, 100–105

    selling information or other digital content, 92–94

    Web catalog model, 84–92

reverse auctions, 237–239

reverse link checker, 309

rich media objects, 153

Richter, R. Gene, 238

RIPE (Reseaux IP Europeens), 43

Rivest, Ronald, 419

robot (bot), 167

Roebuck, Alvah, 84

ROI (return on investment), 491

Rolling Stone magazine, 346

RosettaNet, 57

Ross-Dove Company, 233

router computers, 40

routers, 40, 41, 48

routiners, 149

routing algorithms, 40

routing computers, 40

routing packets, 40–41

routing tables, 41

RSA Public Key Cryptosystem, 419

Rubric, Ltd., 257

Ryder System, 184

S

Sabre system, 100

SafeBuyer.com, 235

SAIC, 477

sales taxes, 289

SalesCart, 334

Salomon Smith Barney, 102

563

virtual host, 307

virtual server, 307

Virtual Vineyards, 258

Virus Information, 378

viruses, 50, 378

Visa, 445, 446, 447, 448, 477

visits, 155

VisualRoute, 48

voice-grade lines, 73

voice-grade telephone
connections, 72

vortals, 207

VPNs (virtual private
networks), 70–71

W

W3C (World Wide Web
Consortium), 54, 55, 510

W3C Getting Started with
HTML page, 59

W3C HTML Page, 54, 512

W3C Security FAQ, 387

W3C Web Accessibility
Initiative site, 115

W3C XML Pages, 57

Wall Street Journal, The, 21,
98, 300

Wall Street Journal, The Web
site, 434

wallets, 443

Wal-Mart, 5, 89

Walmart.com, 89

Walton, Sam, 89

WANs (wide area networks),
40–42

Ward, Aaron Montgomery, 84

warfare, 282–283

warranties, 274

warranty disclaimer, 274

Washington, D.C. NAP, 30

Washington Post, The, 98, 431

Watchfile, 309

Web

See also World Wide Web

customer-contact
medium, 123

market segmentation,
142–143

marketing strategies,
134–135

nature of communication,
122–125

obtaining information,
124–125

offering customers choice,
143–145

personal contact, 140

stateless system, 336

virtual communities, 13

written contracts, 273–274

Web auctions

business-to-business,
232–235

strategies, 222–239

Web browsers, 32

active content, 371

HTML extensions, 54

Web bug, 405

Web clients, 63–68

Web community, 241

Web directories, 95, 167–168

Web graphics designer, 502

Web hosting choices, 300–302

Web pages

active content, 369–372

delivery, 45

dynamic, 305, 310–311

scripting languages, 62

static, 305

translation services and
translation software, 257

Web portals, 95–96, 244–246

Web presence

achieving goals, 108–113

anonymity, 254

business objectives and, 108

Coca Cola, 108

connecting with customers,
122–125

effective, 107–122

failure to understand,
113–114

identifying goals, 108

interactivity, 114

MoMA (Museum of Modern
Art), 113

not-for-profit organizations,
111–113

Pepsi, 108

political parties, 112–113

profit-driven organizations,
109–111

trust and loyalty, 117

Web servers, 31, 63–68

access control, 432–434

architectures, 318–321

authentication, 432–434

benchmarking, 304–306

callback systems, 433

core capabilities, 306–307

data analysis, 307

default language setting of
browser, 256

desirable features, 306–311

determining hardware and
software information, 317

displayed folder names,
385–386

dynamic content, 310–311

eBay, 321

hardware, 299–303

indexing, 307

load-balancing systems,
319–320

operating systems, 302–303

performance evaluation,
304–306

privilege levels, 385

protecting, 430–436

remote server
administration, 309–310

response time, 305

scalable, 302

searching, 307

secrecy violation, 385

self-hosting, 300

software, 64, 311–317

submitting username and password, 386–387

superuser, 385

threats, 385–388

throughput, 305

two-tier client-server architecture, 65–67

username and password pairs, 387

Web hosting choices, 300–302

Web Site Garage, 310

Web site visitors

making Web sites accessible, 115–117

meeting needs, 114–117

motivations, 114–115

sending e-mail messages to, 157–158

Web sites

accessibility, 115–117

affiliate marketing, 164–166

animated graphics, 115–116

builders, 63

business infrastructure adjustments, 113

buyers, 147

content issues, 277–282

content-delivery, 300

custom-centric design, 121–122

development, 299, 492–493

ECML-compliant, 471

flexible interface, 115

goals for constructing, 116–117

growth of, 33

internal vs. outsourcing, 493–499

intranets, 299–300

key words, 167–168

link checking, 308–309

links to multiple language versions, 256

local language versions, 255

management tools, 308

multiple information formats, 116

naming issues, 169–170

poor performance of, 39

rating, 118

redesign of, 119–120

search engines, 322

selectable level of detail, 115

shoppers, 147

shopping cart page, 147

standard information set, 114

stickiness, 95, 245

surfing or browsing, 147

text box control, 147

transaction-processing, 300

translating pages, 256

trigger words, 147

types, 299–300

usability testing, 119

visitors, 114–117

visits, 155

Web storefronts, 341

Web stores, 338

WebAuthorize, 448

Web-based crime, 282–283

Webcams, 243

Web-enabled automated warehousing operations, 184

WebGenie Software, 334

WebMD, 107

WebMethods, 328–329, 357

WebSideStory, 405

Website Recycling Company site, 234

WebSphere Application Server, 339

WebSphere Commerce Suite, 351, 354

WebStone, 305–306

WebTrends Web server log file analyzer, 307

WebWasher, 406

WeddingChannel.com, 166

Weinberger, David, 114

WELL (whole earth 'lectronic link), 243

Wells Fargo, 464

Whole Earth Review, 243

Williamson, Oliver, 19

Williams-Sonoma, 166

Winebid, 231

Wine.com, 258

winner's curse, 221

WIPO (World Intellectual Property Association), 368

wire transfers, 4

wireless connections, 75–76

wireless Ethernet, 75

Wisdom.com, 170

WML (Wireless Markup Language), 311

Women in Music Store, 331–332

World Wide Web, 3, 27, 32

See also Web

emergence of, 31–33

markup languages, 53–63

platform neutrality, 63

revenue models for selling, 84–107

Worldpoint Interactive, 257

WorldServer software, 258

worms, 362–363, 378, 397

writing, 273

WSJ.com, 492

W.W. Grainger, 182–183

X

Xandau, 31

XBRL (Extensible Business Reporting Language), 517

XHTML (extensible hypertext markup language), 55

XML (Extensible Markup Language), 55–57, 199, 311, 509, 515–517

XSL (extensible stylesheet language), 515